Ecological Transition and Circular Economy

Ecological Transition and Circular Economy

Editors

Nicola Raimo
Filippo Vitolla
Ornella Malandrino
Benedetta Esposito

MDPI • Basel • Beijing • Wuhan • Barcelona • Belgrade • Manchester • Tokyo • Cluj • Tianjin

Editors
Nicola Raimo
LUM University
Italy

Filippo Vitolla
LUM University
Italy

Ornella Malandrino
University of Salerno
Italy

Benedetta Esposito
University of Salerno
Italy

Editorial Office
MDPI
St. Alban-Anlage 66
4052 Basel, Switzerland

This is a reprint of articles from the Special Issue published online in the open access journal *Sustainability* (ISSN 2071-1050) (available at: https://www.mdpi.com/journal/sustainability/special_issues/ecological_transition).

For citation purposes, cite each article independently as indicated on the article page online and as indicated below:

LastName, A.A.; LastName, B.B.; LastName, C.C. Article Title. *Journal Name* **Year**, *Volume Number*, Page Range.

ISBN 978-3-0365-7962-7 (Hbk)
ISBN 978-3-0365-7963-4 (PDF)

Contents

About the Editors

Nicola Raimo

Nicola Raimo (Prof.) is a Senior Researcher in Business Administration in the Department of Management, Finance and Technology at LUM University, Casamassima, Italy. His main research interests are corporate social responsibility, non-financial disclosure, integrated reporting, intellectual capital and digitalisation. He is an author of several published articles and serves as reviewer for a number of international journals. He is also a member of the Italian Society of Accounting and Business Administrations Scholars (SIDREA).

Filippo Vitolla

Filippo Vitolla (Prof.) is a Full Professor of Business Administration at the LUM University, Casamassima, Italy. He is a member of the Faculty of LUM University International PhD in Economics and Management of Sustainability and Innovation. He is also a member of the Italian Academic Association of Business Administration (AIDEA and SIDREA). He is an author of many books, books chapters and articles published in national and international academic journals.

Ornella Malandrino

Ornella Malandrino (Prof.) is a Full professor of Commodity Science at the University of Salerno, where she teaches and conducts research on quality, sustainability and corporate social responsibility. She is the inter-departmental observatory director for Gender Studies and Equal Opportunities at the University of Salerno.

Benedetta Esposito

Benedetta Esposito (Dr) is a PhD student in big data management in the department of management and innovation systems at the University of Salerno. Her research is mainly focused on corporate social responsibility, sustainability and the circular economy in the agri-food sector.

Preface to "Ecological Transition and Circular Economy"

This Special Issue has addressed the need to go in-depth into the Ecological Transition and Circular Economy literature using a multidimensional approach. Articles from multiple scientific fields have been presented, providing a comprehensive picture of the complexity of the circular transition. This collection highlights how a circular economy requires combining different knowledge, skills, perspectives and approaches to shift towards sustainable consumption and production patterns successfully.

Special attention is given to the multiple dimensions of circular economy applications, with articles focused on different circular economy application levels, such as: processes, companies, supply chains and cities. Furthermore, the book explores in-depth the concept of the circular economy when applied to the energy, construction and agri-food sectors that are particularly committed to this circular transition.

More specifically, the book comprises 15 papers developed using different research methods.

The majority of the articles have been focused on the circular transition in the energy sector and greenhouse gases emissions (Khatami and Goharian, 2022; Kumar Jha et al., 2022; Ma et al., 2022; Marszowki and Iwaszenko, 2022; Møller et al., 2022; Rej et al., 2022). Another stream of published papers investigated the spatial dimension of the circular economy (Stričík and Čonková, 2021; Zhang et al., 2022; Paoli et al., 2022). Two articles have focused on circular economy application at a supply-chain level (Fassio et al., 202; Marchione Saes et al., 2023). Furthermore, one critical literature review was developed for the construction industry (Charef et al., 2021). Lastly, three articles have investigated the circular economy from a company (Garcia-Sanchez et al., 2022; Vitolla et al., 2023) and a consumption perspective (Rafiq et al., 2022).

Nicola Raimo, Filippo Vitolla, Ornella Malandrino, and Benedetta Esposito
Editors

sustainability

MDPI

Article

Circular Economy Projects and Firm Disclosures in an Encouraging Institutional Environment

Isabel-María García-Sánchez [1,*], Francisco-Manuel Somohano-Rodríguez [2], Víctor Amor-Esteban [3] and Benito Gonzalez-Valdueza [4]

1 Instituto Multidisciplinar de Empresa—IME, Universidad de Salamanca, Campus Miguel de Unamuno, Paseo Francisco Tomás y Valiente, s/n, Edificio FES, 37007 Salamanca, Spain

2 Departamento de Administración de Empresas, Universidad de Cantabria, Avda. de los Castros 56, 39005 Santander, Spain; fm.somohano@unican.es

3 Departamento de Estadística, Universidad de Salamanca, Campus Miguel de Unamuno, c/Alfonso X El Sabio, s/n, 37007 Salamanca, Spain; vamor@usal.es

4 Facultad de Economía y Empresa, Universidad de Salamanca, Campus Miguel de Unamuno, Paseo Francisco Tomás y Valiente, s/n, 37007 Salamanca, Spain; benitogv@usal.es

* Correspondence: lajefa@usal.es

Abstract: This paper analyses the strategies implemented by listed Spanish companies that are leaders in their industrial environments to inform shareholders and the public about their circular economy projects. It uses content and textual approaches through the factorial correspondence analysis of all the information about the circular economy presented on corporate websites. The analysis of the 17,510 resulting terms suggests that companies prioritise discourses about a sustainable future, their commitment to the proper use of resources and the reduction or elimination of greenhouse gases. The sectors most sensitive to institutional pressures, such as oil and energy companies, are more active in disclosing the problems and desired solutions of their projects.

Keywords: circular economy; resource use; greenhouse; industries; textual analysis

Citation: García-Sánchez, I.-M.; Somohano-Rodríguez, F.-M.; Amor-Esteban, V.; Gonzalez-Valdueza, B. Circular Economy Projects and Firm Disclosures in an Encouraging Institutional Environment. *Sustainability* **2022**, *14*, 1149. https://doi.org/10.3390/su14031149

Academic Editors: Nicola Raimo, Filippo Vitolla, Benedetta Esposito and Ornella Malandrino

Received: 7 December 2021
Accepted: 13 January 2022
Published: 20 January 2022

1. Introduction

Our world is technologically and scientifically connected. We know that the industrial revolution has had very positive effects overall in the reduction of poverty and the improvement of well-being, but it has also had negative effects, such as environmental degradation. Transition to the circular economy (CE) is a technical evolution from the linear production–consumption model. The circular economy interconnects business cycles to maintain the value of products and services as long as possible, diminishes production costs by reducing the flow of materials (raw materials are replaced with recycled ones), saves energy and is based on the idea that natural and social capital must be constantly renewed through multiple phases [1,2]. CE is one of the pillars of sustainable policies in many countries, especially those in the European Union, and it is an important driver of the transition towards global sustainability. The NextGenerationEU Plan establishes circular transformation as one of the engines of a post-COVID-19 economic recovery through a package of measures (structural funds, research and innovation financing programmes and others). It is part of the Green Deal or the European Green Pact to help European businesses and consumers transition to a more sustainable economy and the Eurozone's roadmap for designing strategic recovery and resilience plans for a green, digital and climate-resilient future [3].

European institutions understand that the current system of production and consumption does not environmentally bring into equilibrium raw materials, consumption, goods produced and consumed, and the waste generated. Hence, an ecological transition is an evolution from a polluting production model to a more environmentally friendly one, driving the sustainable development objectives in economic, social and environmental dimensions. It requires integrating "sustainable circularity" into the socioeconomic system to ensure that the regenerative management of the resource–waste cycle improves competitiveness, creates new jobs, reduces dependence on raw materials and minimises environmental impacts.

The CE model increases the efficient use of resources, minimising waste and emissions (reduce, reuse, recycle) in favour of an extended "R-Typology" (reject, rethink, repair renew, remanufacture, retrofit, recover). In Spain, the Ministry for Ecological Transition and Demographic Challenge (MITECO) has developed the Recovery, Transformation and Resilience Plan to reactivate an economy affected by the COVID-19 pandemic. This plan has four fundamental pillars: ecological transition, digital transformation, social and territorial cohesion and equality. The Spanish Circular Economy Strategy (EEEC), Spain Circular 2030, lays the foundations for a new production and consumption model where the value of products, materials and resources is maintained in the economy for as long as possible waste generation is reduced to a minimum and unavoidable waste is used to the greatest extent possible. This strategy thus contributes to Spain's efforts to achieve a sustainable decarbonised, resource-efficient and competitive economy.

The CE has received little attention from the academic community until recently. Currently, however, quite a lot of research has appeared considering different aspects of the CE, such as the efficient use of materials and their productivity [4–10], its adoption by countries and sectors [2,11–16], business networks [17] and transitions towards the circular model [2]. Once the need for the transition was accepted, authors explored the factors that drive or hinder this process, opening a wide range of opportunities to advance the research. One of these new areas is the analysis of company initiatives in the CE and their real impacts and legitimacy [18–20]. Ref. [7] explain that although studies show that social institutions and legitimacy are relevant aspects of the transition to the CE, our understanding of how these factors affect initiatives is still limited.

Economic neo-institutionalism explains that the economic reality of the market is where companies, consumers and governments act. They are all determinants of business decisions and economic results. Therefore, these economic agents are not individualistic and their decisions are not only determined by rationality since membership in social organisations is conditioned by restrictions inherent to the institutional structure in which they operate [21]. This institutional framework, according to [22], comprises formal and informal rules that guide individual behaviour and reduce uncertainty. Decision-making is aimed at guaranteeing compliance with the coercive, normative and cognitive elements in place Institutions are, therefore, enforcing the rules of the game, or the constraints conventionally constructed to frame human and organisational interaction in a given society.

Although the CE has a theoretical–conceptual basis, there are still many aspects to research [2]. Among these aspects are why companies select particular CE projects, the effects these projects have on the companies themselves and how they can counteract the negative effect of their corporate activity. The objective of our research is to explain the relationship between institutional pressures and the amount and type of information companies provide about the CE projects they carry out in a proactive and favourable institutional environment. To do this, we have categorised the companies in our sample as symbolic or active according to the real value of the CE information they make publicly available [23], taking the companies' activity sector into account [24]. Faced with substantive projects, managers may prefer symbolic actions that appear to comply with the rules and thus send signals to stakeholders using a "green discourse" [25,26]. According to our hypothesis, these symbolic messages are stronger in the industries most affected by institutional pressures.

The results show that industrial strategies can be identified in CE disclosures, as they are broader and more detailed in companies that belong to the sectors most sensitive to institutional, especially coercive, pressures, such as the oil and energy industries. Our work makes a significant contribution to the area of corporate information and the CE. It complements previous studies such as those by [16,27–29], who found a high correlation between sensitive industries and the sustainable initiatives voluntarily carried out by companies. In fact, our work broadens the application of semantic metrics, i.e., the measurement of distances on the ontology level, focussing on compliance with sustainability indicators in an organisational context [19,20,30,31].

2. Theoretical Framework

The CE implies an evolution of the current economic model characterised by linearity in the extraction, manufacturing, commercialisation, consumption and disposal processes typical of an "open" planet towards a new paradigm of extraction, manufacturing, commercialisation, consumption and reuse (recovery and recycling) of a "limited" planet. On this limited planet, human beings must find their place in a cyclical ecological system capable of continually reconverting materials and conserving their values [32]. The McKinsey Center for Business and Environment and the Ellen MacArthur Foundation explain that the CE has its raison d'être in the best preservation of the usefulness of products, components and materials. It is based on three basic principles [33]: (1) the preservation and appreciation of natural resources through controlling finite reserves and balancing the flows of renewable resources; (2) the optimisation of resource productivity, considering the technical and biological cycles of products, components and materials; (3) improving the efficiency of the system by protecting human well-being with regard to food, mobility, housing, education, health and entertainment and managing land, air, water and noise pollution, the release of toxic substances and climate change.

Why the CE influences companies and how they face the challenge of converting their businesses can be analysed from the perspective of the institutional theory [34], incorporating the social and legitimation aspects of sustainability [35] to explain what factors help and what factors hinder transformation to the CE [7]. According to the resource-based approach (RBV), companies must configure their resources to develop the capacities they need to sustain a competitive advantage, i.e., a competitive advantage depends on the match between distinctive internal capabilities and changing external environmental circumstances [36], such as climate change. Separating institutions into three pillars, based on Scott's institutional theory [37], we identify three types of institutional pressures (coercive, normative and mimetic) that are exerted on companies by their external environment. These pressures force companies to reconfigure their key resources, and thus, to standardise their functioning [38].

However, in this process, organisations must maintain their competitive advantages and obtain new ones, theoretically through cost reduction, sustainable manufacturing and material circularity [39]. These two approaches, institutionalism and the competitive advantage of RBV, can be addressed in the characterisation of the CE research scenario. Both approaches have been used simultaneously in several studies to consider the environmental and social pressures that can affect the heterogeneity of companies' environmental strategies [40,41] and their effects on business performance [42,43].

Assuming the positive intention of environmental and social pressures, institutional changes are slow to come about due to, for example, the circumstances noted by [44] that prevent institutional pressures from working well. These circumstances are (1) little regulation regarding reuse and its inconsistent application in different countries; (2) the lack of legal indications beyond recycling, such as for reuse; (3) the lack of a reuse culture and people's preference for new products. Factors such as the capacities of managers and staff and resource availability also affect the integration of climate change in company strategies, determining the advantages or disadvantages in the transition [45–47].

Business initiatives often do not find followers among consumers due to customers' concerns for their own benefit, health and safety [48] or because these initiatives are not perceived as being beneficial for the environment [49]. From a performance perspective, the changes required of established companies to transform to the CE are significant and costly [50], especially in environmentally sensitive sectors [51]. These affect all organisational levels and include production systems, anticipating impacts, innovating, collaborating both internally and externally, optimising and reporting openly and completely, with clarity, precision, honesty and promptness [52].

Firms often disseminate information to increase the value of their brands [53] and offset the costs of their transition to the CE, at least partially. In doing so, companies can collaborate with external agents and disseminate information on sustainability and CE aspects, among other measures. These actions are part of mimetic institutionalism, and they help legitimise an organisation [54]. For example, although recycling is one of the central processes of a reverse logistics business integrated into the value chain, for a manufacturer, it represents an increase in costs that can become a competitive advantage if this practice is recognised as responsible behaviour by the markets [55]. Therefore, this information is used to capture value [56]. Companies have different ways of increasing the information available to stakeholders to reduce uncertainty and favour company interests by aligning their brands with the social behaviours that are the heart of the transition to a "new" socioeconomic system [57].

The use of information to improve reputations and enhance the value of brands is explained by the signal theory [53], while the legitimacy theory [28,58] posits that the extent of a firm's disclosures is also a product of the firm's exposure to public pressure from stakeholder groups in the social, political and regulatory environment. For instance, reputation improvement and brand value enhancement can be carried out internally, motivating workers to create a culture of innovation internally [59], or externally, favouring a culture of sustainability in industry and institutions [60]. Our research is focused on corporations from industries whose activities have a negative influence on the environment (environmentally sensitive industries according to the National Pollutant Release Inventory [61]). In general, corporations from oil and energy and basic material industries (steel, cement and concrete product manufacturing, foundries, etc.) emphasise information on environmental, health and safety issues [62,63]. In a proactive and favourable institutional environment, institutional pressures contribute to the amount and type of information leading companies provide about their CE projects. Thus, our hypothesis is formulated as follows:

Hypothesis 1 (H1). *Corporations from environmentally sensitive industries disclose and report more CE information than firms from other sectors do.*

Organisations search their environment for clues about who to follow, looking for emerging leaders who have the prestige and respect to inspire imitation, that is, they engage in institutional mimicry. These emerging leaders align with other leaders, opinion makers and governments, who make up a tacit or explicit network to drive a shift towards the CE, highlighting its environmental benefits to increase consumer acceptance [49]. According to [64], these networks are important to establish subjective norms that shape a social system while spreading new ideas, in this case, about the CE [65]. Rogers identifies structure, opinion leadership and types of decision as the most important areas of interest in social systems, resulting in certain processes being internally created and maintained, linking the elements of the social system to the adoption process. Ultimately, social systems are not action systems derived from an approach based on resources and capabilities and directed by the thoughts and behaviours of people in the organisation. Social systems are communication systems where communication itself determines the process of change [66].

In the following sections, we analyse CE business initiatives and the relationship between institutional pressures and the volume and type of information provided about implemented CE projects.

3. Research Design

3.1. Population and Sample

We selected the companies listed on the IBEX-35 of the Madrid Stock Exchange as the target population with which to analyse the initiatives and projects developed in the area of CE and the information disclosed. The reasons for this selection are that these companies operate in the institutional environment established in the previous section, and they are immersed in a system that fosters a sustainable business model by promoting different public initiatives. This framework determines the existence of institutional pressures in favour of circular transformation and, therefore, the decisions that these companies make regarding changes in their economic activity to legitimise themselves in their environment. These companies are also the most liquid and the most active in sustainability issues [29].

Based on the analysis of the information available on these companies' corporate websites—specific sites and sustainability or similar reports—the analysis sample corresponds to 17 listed Spanish companies operating in all sectors, although the majority are involved in the oil and energy industries (see Table 1, the 13th company today comprises two merged companies). Compared to the population, the companies in the sample represent 52% of the listed companies and are the only ones that offer public information about their current and future circular transformation projects and processes (Appendix A Table A1).

Table 1. Sample description.

Industry	Freq.	%
Consumer goods	2	11.8%
Basic materials, industry and construction	3	17.6%
Oil and energy	7	41.2%
Consumer services	1	5.8%
Financial services	2	11.8%
Technology and telecommunications	2	11.8%
Total	17	100%

3.2. Methodology

Information for the analysis and identification of the key issues was obtained through content and textual analysis [67,68]. Specifically, the information on the CE was all first disclosed by the companies on their corporate websites and is verbatim. The information was collected in March 2021 and has been subsequently processed using the analysis methodologies indicated below.

Projects, initiatives and other aspects have been identified through content analysis. They allowed us to identify which CE initiatives have been promoted by companies, the CE phase involved and the benefits derived from them. The information downloaded from the websites was also processed through a textual analysis programme based on our own R code. It allowed us to determine the technical and quantitative content of the information reported to stakeholders.

For the textual analysis of information, we applied a descending hierarchical classification according to the method described by [69], where lexical classes are defined, and each one represents a subject that can be described by the vocabulary that defines the subject. By using the proximity of text segments to cluster the terms and identifying the central points of company communiques, we can understand firms' priorities in terms of circular transformation.

In this analysis, we also constructed a matrix of lexical data, where the rows include the different words used in the reports and the columns include the companies issuing the reports. Their confluence corresponds to absolute frequency. We did not work with this matrix, however, since pre-processing, which consists of eliminating empty words like conjunctions, prepositions, etc. and keeping only semantically loaded elements, was applied to the initial data. Lemmatisation was carried out on this group of elements, reducing several words to a single term based on semantic relationships (e.g., I said, I will say, let us say). Once all the words were obtained, and the data cleaning and debugging processes were complete, we obtained the final lexical data matrix with which we worked.

Given the multidimensional structure of this matrix, we used the correspondence factor analysis (CFA), which is a classic multivariate technique in text statistics to detect associations and oppositions between individuals and observations. These associations and oppositions can be visualised separately or simultaneously on two-dimensional factor maps [70]. Since the CFA works on profiles (the distribution of relative frequencies of a line in a table, row or column in relation to the marginal total) and captures structures we carried out a statistical analysis of textual data to identify the most relevant terms in the analysed reports. This analysis resulted in a geometric representation that facilitated the interpretation of the numerical information of the lexical table, looking for axes of the maximum dispersion of point-profiles around the centroid, with a minimum loss of information. The chi-square distance metric was used, a weighted Euclidean distance which allows infrequent words to be weighted more and the most frequent words, less. The chi-square distance neutralises all distortions in the graphical representation.

The words and companies—or activity sectors by groups of companies—are shown in their representation on the factorial plane. We then looked for similarities according to the proximity of the words to the companies on the plane. In this way, we characterised each company based on the terms they used the most, which indicates the differences between companies and their main focuses. We did the same for the sectors of activity, discovering what really concerns these sectors and where they use most of their resources. The overall methodological process is outlined in Figure 1.

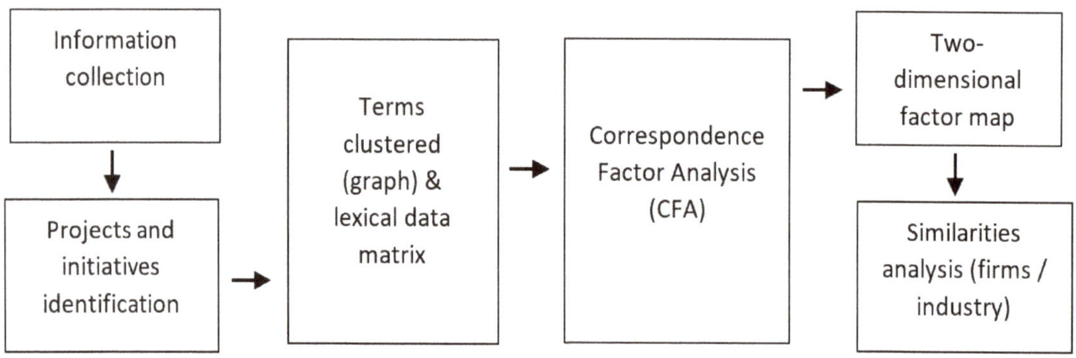

Figure 1. Methodological process outline.

4. Results

4.1. Descriptive Textual Analysis

The analysis of the texts resulted in a total of 17,510 terms, represented in 2341 different ways. We first represented the most frequent terms used by companies in a similarity graph, similar to classification trees, representing the links of the selected type with the other class types (see Figure 2). Before commenting on the results, note that certain words assumed to be common to the topic have been deleted to obtain only the most relevant information. These deleted words include *circular economy, sustainability, project, environmental, management, waste, commitment* and *responsible*. This type of discourse focuses on the term *use*, as it is linked with the rest of the terms; *the use of matter, product, water*, in general, the importance of how to treat all resources. The CE represents a new model of society that uses and optimises the stock and flow of materials, energy and waste, and its objective is the efficient use of resources.

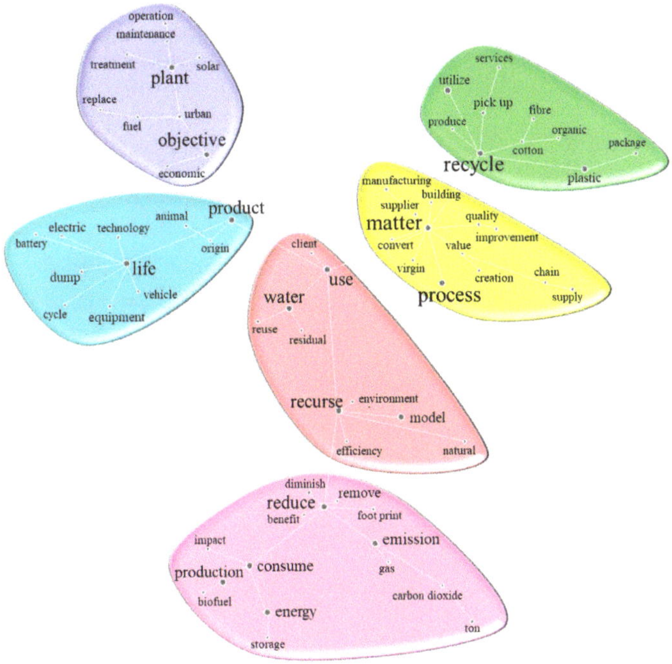

Figure 2. Representation of the most frequently used terms.

The word *matter* is close to the central term *use.* It is connected to the terms *quality, processes in its creation, manufacture* or *supply.* Following its trail, it connects with the verb *recycle.* A little further away, we find the word *plastic* linked to the term *package,* highlighting the importance of plastic in this movement. On the left, we have the term *product* and the importance of *animal* and *origin*; close to these, we find the discourse on the life cycle of certain *equipment, vehicles* and the relevance of *electricity* and new *technology.* In the upper area, we find the paragraph dedicated to objectives, naturally including the economy. It talks about the importance of replacing urban fuel or the creation of solar plants and their maintenance. Finally, in the lower part, we observe how the CE is a strategy that aims to reduce both the use of virgin materials and the production of waste, closing the loops or economic and ecological flows of resources. Here, we can find terms such as *reduce* or *remove,* linked to *emissions, gas, carbon dioxide,* or the impact or *footprint* left by tons of waste. On the other hand, we have the importance of reducing production and consumption, especially energy, its storage and the increasingly frequent alternative of *biofuel.*

Secondly, and in view of what has already been noted, we grouped all the terms into four clearly differentiated clusters, discovering the four topics that companies prioritise in their CE reports (see Figure 3):

- Red cluster: represents 22.2% of companies' discourses, focussed on extending the useful life of products, with a notable emphasis on recycling, especially plastic.
- Blue cluster: companies focus 26.1% of their reports on pollution, with the aim of avoiding or reducing gas emissions (carbon dioxide, fuel, refinery, etc.) and favouring processes for prudent water consumption.
- Purple cluster: occupying 20.7% of the reports, this cluster focuses on business actions. There are references to electrical installations, community relations and many references to the management (cleaning) of generated waste.
- Green cluster: companies use 31% of their discourse explaining the importance of the CE for a sustainable future. They focus on the search for a commitment to and

the development of sustainable activities, where they prioritise and promote models, make plans to change the environmental impact of their current activities and promote the adequate use of resources over time.

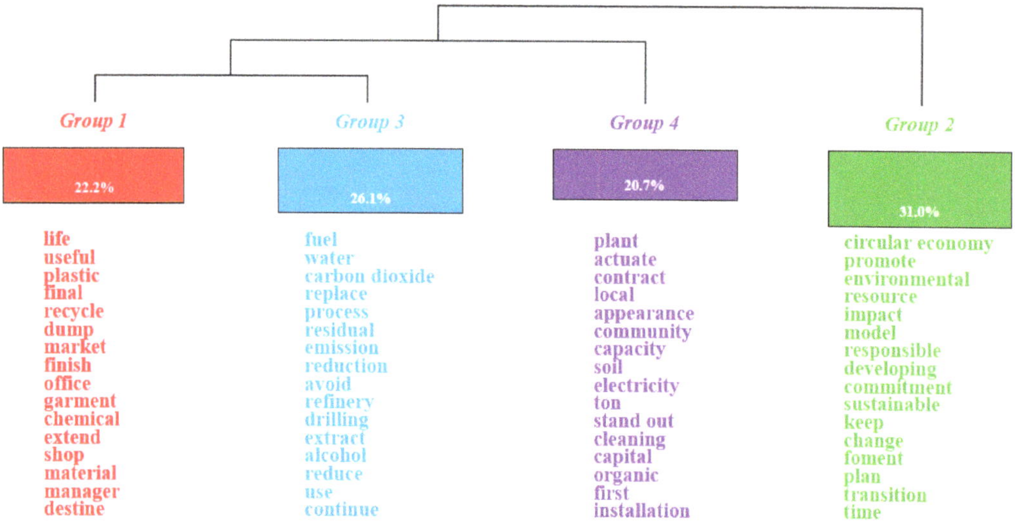

Figure 3. Grouping of clearly differentiated clustered terms.

4.2. Analysis of CE Projects

The number of projects promoted, the CE phase to which they refer and a description of the projects and their benefits were identified through content analysis for each of the 17 companies reporting on their commitment to the CE. This information is detailed in Annex 1. These companies have been classified into three levels according to the specificity of their CE initiatives. Level 0 includes those companies that do not present a description of their CE phases and do not determine circular benefits; that is, they make symbolic speeches [23]. Level 1 includes active companies that present a description of their projects but not the benefits. Level 2 includes very active companies that report all the information, a description and the benefits of their CE projects. Twenty-two percent of the companies in the data sample used for this research belong to Level 0, with no description of benefits. Forty-four percent correspond to Level 1, with projects seeking to reduce landfill waste, recycle vehicles, give a second life to obsolete machines, revalue technology equipment, eliminate single-use plastics, etc.

The remaining thirty-four percent are Level 2 companies, where the projects promoted are mainly linked to the production of biogas, second-life electric vehicle batteries, the production of green hydrogen, the creation of new textile fibres from recycled garments and digital solution projects to reduce the consumption of energy, water and CO_2. All these projects create benefits by removing tons of urban waste, generating large amounts of biogas, lengthening the life of batteries and avoiding large amounts of CO_2 emissions. They result in a 10% reduction in fuel consumption, a 20% reduction in water consumption in agriculture, and a 30% reduction in public lighting, among many other advantages. They also help to decrease 85% of the black spots where there is poor waste separation.

4.3. Business Strategies in CE Reporting

We grouped the information disclosed according to levels 0, 1 and 2, into which we have classified the companies. As expected, there is a greater volume of information at the higher levels (Level 0: 843 words, Level 1: 3146 words and Level 2: 13,521 words), with very concise reports from firms in Level 0, at an average of 211 words; reports that are nearly twice the length from companies in Level 1, at 393 words, and reports in Level 2 that clearly explain the phases and benefits of their CE projects, at 2254 words.

Studying the most commonly used terms (see Table 2) allows us to see that companies in Level 0 use a discourse aimed at actions that should not be carried out, talking about reducing water and energy consumption and emphasising the footprint of gas emissions, carbon dioxide, etc. The terms used by companies in Level 1 show a different discourse, focussing on new objectives, such as the use of electrical material, raw materials versus products and their manufacturing processes and the importance of recycling. Finally, the information included by companies classified in Level 2 involves the most important aspects of the CE. This production and consumption model is mainly concerned with reducing the amount of waste generated, which involves improving raw material use or consumption, reusing and giving a second or third life to certain products, and recycling as the last phase a product goes through: the three Rs on which the CE is based in search of a sustainable economy.

Table 2. Word frequency distribution by levels.

	Level 0	Freq.		Level 1	Freq.		Level 2	Freq.
1	Water	9	1	Residual	51	1	Residual	147
2	Consume	9	2	Objective	22	2	Raw material	108
3	Residual	8	3	Recurse	17	3	Life	48
4	Emission	8	4	Use	15	4	New	48
5	Reduce	6	5	Material	15	5	Use	43
6	Energy	6	6	Electronic	13	6	Water	43
7	Carbon dioxide	6	7	Equipment	13	7	Reduce	42
8	Efficiency	5	8	Recycle	12	8	Product	42
9	Recurse	4	9	Consume	11	9	Recycle	41
10	Model	4	10	Model	11	10	Process	40
11	Impact	4	11	Generation	11	11	Recurse	37
12	Gas	4	12	Net	11	12	Material	37
13	Environment	4	13	Product	11	13	Objective	36
14	Use	3	14	Raw Material	10	14	Plant	36
15	Generate	3	15	Process	9	15	Plastic	33
16	Natural	3	16	Life	9	16	Energy	31
17	Foot Print	3	17	Natural	8	17	Reuse	28
18	Biodiversity	3	18	Building	8	18	Pick up	25
19	Material	2	19	Value	8	19	Production	24
20	Plant	2	20	Reduce	7	20	Ton	24
21	Building	2	21	Impact	7	21	Consume	23
22	Production	2	22	Pick Up	7	22	Model	22
23	Ton	2	23	Dump	7	23	Building	22
24	Ecoeffiency	2	24	New	6	24	Emission	22
25	Hydric	2	25	Plastic	6	25	Service	22
26	Raw material	2	26	Economic	6	26	Produce	22
27	Objective	1	27	Ecodesign	6	27	Urban	22
28	Recycle	1	28	Gas	5	28	Technology	22
29	Process	1	29	Reuse	5	29	Fuel	22
30	Production	1	30	Chain	5	30	Dump	21
31	Urban	1	31	Use	5	31	Animal	21
32	Technology	1	32	Maintenance	5	32	Obtain	20
33	Reuse	1	33	Water	4	33	Quality	20
34	Origen	1	34	Efficiency	4	34	Cotton	19
35	Treatment	1	35	Services	4	35	Electric	17
36	Organic	1	36	Energy	3	36	Carbon dioxide	17
37	Solar	1	37	Treatment	3	37	Origin	17
38	Client	1	38	Oil	3	38	Manufacturing	17
39	Chain	1	39	Operation	3	39	Generate	16
40	Creation	1	40	Convert	3	40	Equipment	16

Table 2. *Cont.*

	Level 0	Freq.		Level 1	Freq.		Level 2	Freq.
41	Net	1	41	Dangerous	3	41	Value	16
42	Collaboration	1	42	Plant	2	42	Use	16
43	Remove	1	43	Production	2	43	Treatment	16
44	Residual	1	44	Produce	2	44	Organic	16
45	Storage	1	45	Origin	2	45	Vehicle	16
46	Supply	1	46	Client	2	46	Cycle	15
47	Species	1	47	Creation	2	47	Benefit	15
48	Symbiosis	1	48	Collaboration	2	48	Solar	15
49	Photovoltaic	1	49	Remove	2	49	Package	15
50	Biomethane	1	50	Residual	2	50	Efficiency	14
51	Ecosystem	1	51	Supply	2	51	Economic	14
52	Life	0	52	Biomethane	2	52	Supplier	14
53	New	0	53	Obtain	2	53	Biofuel	14
54	Product	0	54	Cycle	2	54	Foot print	13
55	Plastic	0	55	Diminish	2	55	Fibre	13
56	Use	0	56	Improvement	2	56	Battery	13
57	Pick up	0	57	Decarbonisation	2	57	Garments	13
58	Electric	0	58	Underground	2	58	Oil	12
59	Equipment	0	59	Regasification	2	59	Client	12
60	Dump	0	60	Emissions	1	60	Operation	12

If we delve a little deeper into the information obtained by levels (see Table 3), we find no difference in the use of numbers in the reports, with values around 1.4–1.7%. However we do find differences in the use of technical words, and the proportion of these words is much higher for firms in Level 2.

Table 3. Proportion of numbers and technical words by levels, CEO and work team training.

Level	Level 0	Level 1	Level 2
Average words	211	393	2254
% Numbers	1.7%	1.4%	1.7%
% Techniques	4.9%	3.5%	6.2%
CEO training			
% Engineers or chemists (CEO)	50.0%	37.5%	50.0%
% Lawyers (CEO)	0.0%	25.0%	16.7%
% Economists and managers (CEO)	50.0%	33.3%	33.3%
Board member training			
% Male	72.4%	68.0%	70.1%
Total with training information (M)	7	9	8
% Engineers or chemists (M)	35.9%	19.4%	21.9%
% Lawyers (M)	25.1%	28.4%	30.9%
% Economists and managers (M)	39.0%	52.2%	47.1%
% Female	27.6%	32.0%	29.9%
Total with training information (F)	3	4	3
% Engineers or chemists (F)	25.0%	23.8%	15.0%
% Lawyers (F)	6.3%	23.8%	18.6%
% Economists and managers (F)	68.8%	52.5%	66.4%
Total Board Members	10	14	12

We analysed the possible effect of CEOs' and directors' training and gender. The CEOs were all male, and we found a notable difference in the training of those in Level 2, where there are high rates of engineers or chemists. The teams in the companies have higher proportions of men at all levels, at around 70%, and higher rates of economists in training, especially in Level 2 companies, with greater differences over other degrees among women.

4.4. Industrial Strategies in CE Reporting

Finally, we grouped the information by activity sectors: consumer goods, basic materials, industry and construction, oil and energy, consumer services, financial services, and technology and telecommunications. We first observe a very large difference in the length of the reports, where companies in the oil and energy sector use an average of 1561 terms, followed by consumer goods with 1195, both well distanced from the rest. Companies producing basic materials use 785 words, technology and telecommunications use 545 and consumer services and financial services companies provide the most concise reports on the CE, with around 185 words (see Table 4).

Table 4. Proportion of numbers and technical words by sectors, CEO and work team training.

Sectors	Consumer Goods	Basic Materials	Oil and Energy	Consumer Services	Financial Services	Technology and Telecom
Average words	1195	785	1561	185	187	545
% Numbers	1.1%	1.5%	1.8%	0.0%	1.7%	2.1%
% Techniques	5.3%	6.4%	4.9%	7.0%	2.7%	2.8%
CEO training						
% Engineers or chemists (CEO)	0.0%	66.7%	85.7%	0.0%	0.0%	0.0%
% Lawyers (CEO)	50.0%	0.0%	0.0%	0.0%	33.3%	50.0%
% Economists and managers (CEO)	50.0%	33.3%	14.3%	100.0%	66.7%	50.0%
Board member training						
% Male	61.8%	76.8%	72.7%	70.0%	63.9%	65.1%
Total with training information (M)	7	8	7	7	10	10
% Engineers or chemists (M)	12.5%	20.4%	33.5%	28.6%	16.7%	15.6%
% Lawyers (M)	32.5%	16.2%	29.2%	42.9%	36.7%	21.1%
% Economists and managers (M)	55.0%	63.4%	37.3%	28.6%	46.7%	63.3%
% Female	38.2%	23.2%	27.3%	30.0%	36.1%	34.9%
Total with training information (F)	4	3	3	3	4	5
% Engineers or chemists (F)	25.0%	16.7%	20.7%	0.0%	25.0%	30.0%
% Lawyers (F)	29.2%	0.0%	18.3%	0.0%	33.3%	20.0%
% Economists and managers (F)	45.8%	83.3%	61.0%	100.0%	41.7%	50.0%
Board Members	11	11	11	10	16	15
Average Level	1.0	1.3	1.1	0.0	1.0	1.5
Level 0	50.0%	0.0%	28.6%	100.0%	0.0%	0.0%
Level 1	0.0%	66.7%	28.6%	0.0%	100.0%	50.0%
Level 2	50.0%	33.3%	42.9%	0.0%	0.0%	50.0%

The majority of the percentages of numbers in companies' information are between 1.5% and 2%, with technology and telecommunications in the lead. Consumer goods (1.1%) and the consumer services (0%) sectors are far below this. The companies with the most concise reports are those that use technical terms to a greater extent, with 7% in consumer services. Close to these values are basic materials (6.4%), consumer goods (5.3%) and oil and energy (4.9%). Financial services and technology and telecommunications use lower percentages, at around 2.7%.

The information about the heads of companies shows us that training in engineering or chemicals corresponds to the CEOs of sectors such as basic materials and oil and energy. In the rest of the sectors, most of the CEOs have training in economics and business. We find a greater number of company team members in the financial services and technology and telecommunications sectors, with around 15–16 members, whereas the rest of the sectors have 10–11 members. These are mostly men (62–77%). The greatest difference is found in basic materials and oil and energy, whose company team members are 23% and 27% women, respectively; 30% in consumer services and around 35–38% in the rest of the sectors, with consumer goods being the sector with the highest percentages of women (38%). The highest percentages of women are trained in economics and business in all the sectors. The same is true for men, with smaller differences, and there are higher proportions of law graduates in consumer services.

Correspondence factorial analysis allowed the terms used most often by the companies to be simultaneously represented with the technical words discussed in the previous table. This study focuses on the representation of Figure 4, where, according to the location of the companies close to the sectors of activity, we can characterise the CE reports they each make and observe the most important differences.

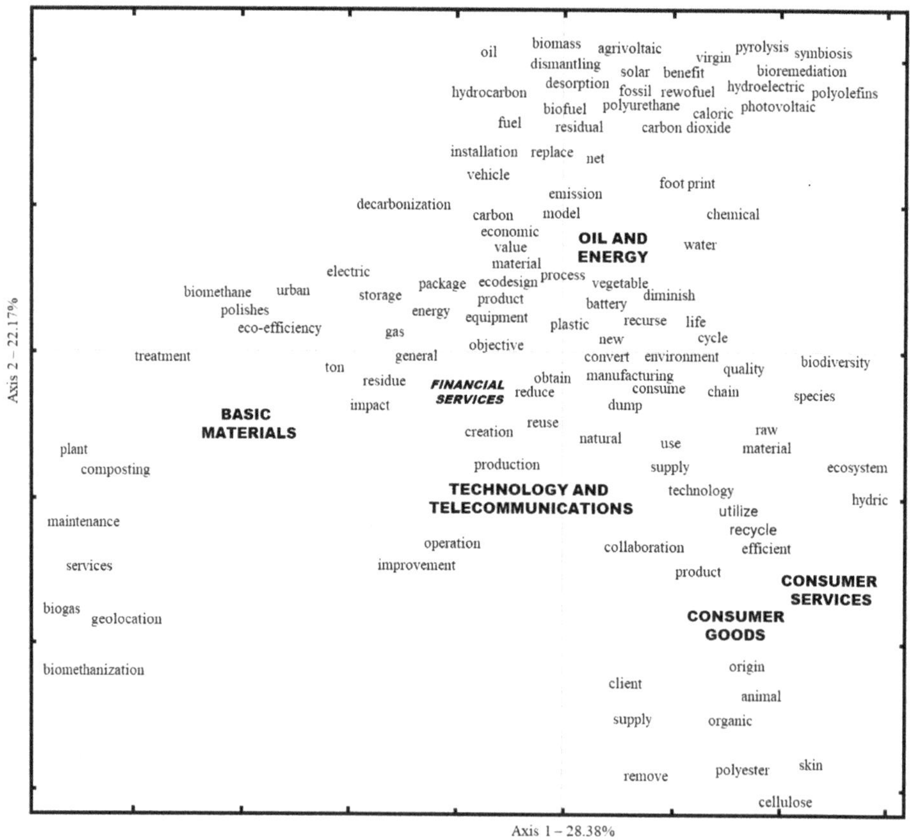

Figure 4. Representation of frequent and technical words according to sector activity.

- Oil and energy: these companies are environmentally sensitive industries [51]. They focus on the negative effects of emissions and pollution, the footprint left by their actions, carbon dioxide, plastics, chemical agents, vehicles, and so on, but they also comment on their new sustainable processes and models based on the substitution of fuels or methods of generating energy, placing emphasis on eco-design, the use of biofuels, the construction of solar plants, hydroelectric plants, electric vehicles, and others, with the idea of planning tomorrow's energy transitions today.
- Basic materials: the companies belonging to this sector are environmentally sensitive industries [51] and focus on the impact of the tons of waste produced. Their objective is to treat this waste and reduce and reuse materials before recycling. They speak about the importance of treating plastic containers or the use of alternative energies. An important part of this discourse is directed towards reducing environmental impacts, the treatment and maintenance of gas, switching from biogas to biomethane, the importance of composting and eco-efficiency.
- Financial services: these companies make concise and vague speeches about the CE.
- Technology and telecommunications: these companies talk about improvements in production, creation and transport operations and the search for innovative technologies to help improve global waste management, reduce environmental impact and optimise costs.
- Consumer goods: these companies focus their reports on products. They explain that we live in an age of consumption, anchored to an economic model based on "produce, buy, use, discard" and, in turn, they urge us to move towards a new model where products and raw materials can have a second life without the need to produce new consumer goods. In this way, waste and waste reduction, reuse and recycling result in increased resource efficiency, respecting ecosystems and biodiversity.
- Consumer services: these companies make similar speeches to those of companies in the consumer goods sector, proclaiming the importance of product life. Although they are true, the reports analysed were very concise, with little specific information.

According to these results, we accept the null hypothesis since the environmentally sensitive industries (oil and energy industry and basic materials) disclose and report more CE information (average words, % numbers, % techniques and firms in level 2) than firms from other sectors do. However, consumer goods is in the top three, in our opinion, because they are closer to demand and can directly increase the value of the brands.

As a last point, and to explain how it is done, we present the words most commonly used by companies in their speeches (around 50), making up 11.6%, together with the most repeated technical words (around 40), although these make up only 1.5% of the sample (see Figure 5).

This representation collects about 50% of the information. The most frequent words (in blue) are less weighted and are located in the centre of the figure. The technical words (in orange), with a higher weighting, are located closer to the edges of the figure.

- On the right side of Figure 5, we can see a discourse on the *waste* and *footprint* that different *emissions,* such as *carbon dioxide,* leave on the road, compared to *alternative fuel* for the sake of *environmental benefit*, such as the use of *biomass* and *biofuel*. Company 1 stands out in this regard.
- Other discourses, such as that of Company 7, are more closely related to the *product*, establishing its *origin*, the *quality* standards in production, responsible use of *material*, and everything related to *recycling*.
- Company 2, Company 3 and Company 9 are the main energy companies, and their speeches are related to *biogas* or *biomethane* and *regasification*.
- Company 4 and Company 6, two large electricity companies in the Spanish energy market, focus on *social responsibility* and seek to contribute to *change*, so they promote *photovoltaic* solar energy and *wind* energy, refer to *decarbonisation* and talk about the use of *hydroelectric* energy and harming any *ecosystem* (*bryozoans*, *biotope* etc.) as little as possible.

- The rest of the companies produce a less specific discourse, where they prioritise the term *circular economy*, the *efficient* use of *resources* and *water*, and speak of *eco-efficiency*, *biodiversity*, *ecodesign*, etc.

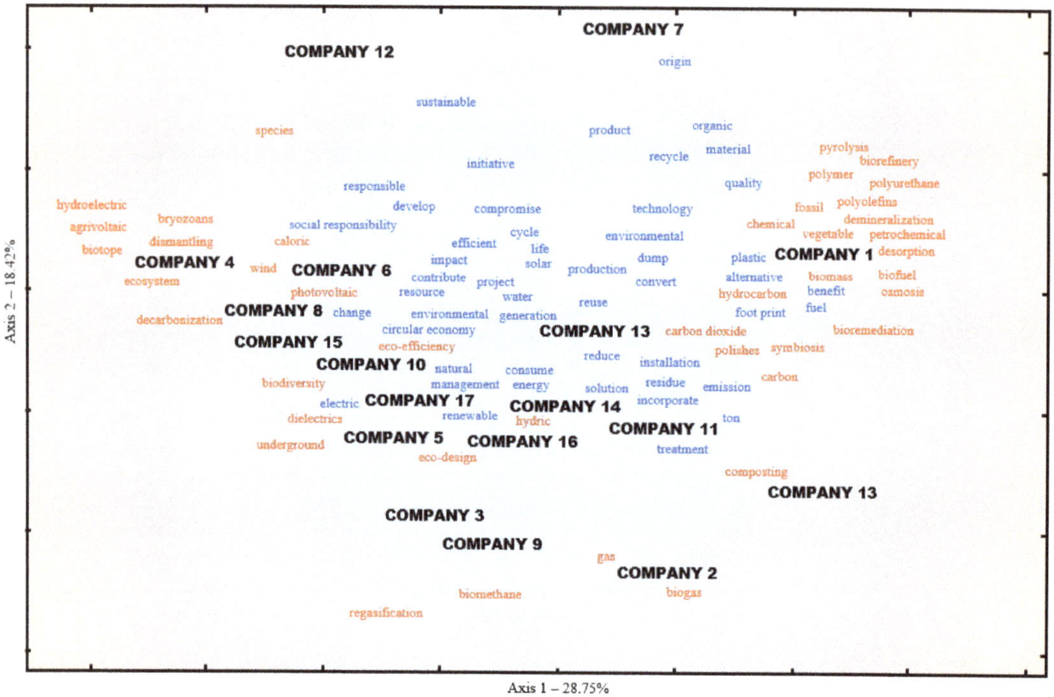

Figure 5. Factorial analysis of correspondences, frequent words and technical words by companies (Note: Company 13 resulted from the acquisition of one financial company, purchased outright from another).

5. Discussion and Conclusions

Company initiatives and impacts are important factors in the speed of the transition to the CE [7,18]. In our research, we categorised the information published in the corporate web pages of 17 companies listed on the Spanish stock market according to their content value. These companies are considered leaders in their sectors.

The analysis of the texts show that the companies, especially environmentally sensitive industries, prioritise their commitment to a sustainable future based on the proper use of resources in their speeches. The next most important topic is the reduction or elimination of greenhouse gases and the prudent consumption of water, followed by lengthening the useful life of products with a special emphasis on recycling. Finally, companies mention specific technical and social initiatives that affect the environment. Thirty-four percent of the companies describe their CE projects in detail and report the benefits of these projects (Level 2). Forty-four percent describe only their initiatives, but not the benefits, focusing on new objectives in terms of the use of raw materials and the importance of recycling (Level 1). The remaining twenty-two percent do not do so in either case (Level 0), so they are considered symbolic projects. The latter refer mainly to the regulatory aspects of unacceptable actions and the objectives of reducing water and energy consumption and the carbon footprint.

We found differences in CEO training, with a notable prevalence of engineers or chemists, as well as higher proportions of male and female economists in management teams in the companies that provide more and more detailed information (Level 2).

The result of the industry analysis is that the oil and energy companies provide more information on the negative aspects of their activities but offer information on their solutions, with an emphasis on eco-design, biofuels, alternative energies and energy transition planning. Basic materials companies highlight the importance of waste and its treatment, reduction and reuse, as do consumer goods and consumer services companies, which promote the use of goods and services until the end of their useful life to reduce waste. Finally, technology and telecommunications companies include some information on improvements in waste management and the reduction of environmental impacts with innovative technologies, and financial services hardly include any specific information on CE initiatives or projects.

Our results show that institutional mimicry is feasible in the oil and energy and consumer goods sectors because the listed companies disclose and report more CE information, and they are profitable. However, much remains to be done to move towards the "new" socioeconomic system suggested by [57]. The companies in our study act as opinion leaders due to their size, so the information provided on their corporate websites benefits the transformation process towards the CE because they explain the benefits produced by corporate initiatives using a significant percentage of easily understandable terms [48]. The results are in the line with those in the work of [24], as most companies focus on reducing materials as well as air and water emissions. They used a data sample consisting of 220 large manufacturing companies in the EU that provided CE reporting information in 2016. Their results were that the keywords Reduce and Recycle were more frequent than Reuse (156, 61 and 183 companies mentioned them more than ten times, respectively), while the majority of the companies included almost one KPI about Emission, Energy and Water (200, 163 and 146, respectively). Considering the same keywords, our results in Table 2 show that from 17 listed companies, Reduce (55 times in Levels 0, 1 and 2) and Recycle (54 times) were also more frequent than Reuse (34 times). However, Water was more frequent than Emission or Energy (56, 49 and 40, respectively). Furthermore, our paper makes an important contribution since it incorporates the sector of activity into the work done by [24]. The analysis suggests that institutional pressures contribute positively to the amount and type of information leading companies provide about their CE projects. Hence, industrial strategies can be identified in CE disclosures as they are broader and more detailed in companies that belong to the sectors most sensitive to institutional, especially coercive, pressures, such as the oil and energy industries. Our work makes a significant contribution to the area of corporate information and the CE because these industrial strategies do not signal a direct embodiment of CE principles. However, they are key to accelerating the transition to a CE from our current linear economic model [19].

The findings of this study lead to some useful implications for research, practitioners and policymakers. For research, the empirical results are consistent with the literature since large, listed firms are more active in implementing CE practices to mitigate negative environmental effects [20], especially in the sectors most sensitive to institutional pressures with worse environmental performance [28]. For policymakers, the offer of tax incentives and the NextGenerationEU funds should have a guiding effect on enterprises [71]. Hence, companies' CE disclosures reveal their eagerness to comply with CE principles, and their willingness to establish guidelines for the industry. For practitioners, the study confirms that corporate managers consider the adoption of CE practices a win–win strategy to achieve environmental and economic goals simultaneously.

There are two limitations, however, which will be future developments of the work. First of all, the sample is limited since the cost of data processing means that we have reviewed a small number of companies. Second, in connection with this, we are looking for a way to digitise the information collection process and expand the sample. This would make it possible to complete the information since sometimes companies use channels other than their corporate websites to update data about their initiatives.

Author Contributions: Conceptualization, I.-M.G.-S. and F.-M.S.-R.; software, V.A.-E.; validation, I.-M.G.-S., F.-M.S.-R., B.G.-V. and V.A.-E.; formal analysis, B.G.-V. and V.A.-E.; investigation, I.-M.G.-S., F.-M.S.-R., B.G.-V. and V.A.-E.; writing—original draft preparation, I.-M.G.-S., F.-M.S.-R. and V.A.-E.; writing—review and editing, F.-M.S.-R.; project administration, I.-M.G.-S.; funding acquisition, I.-M.G.-S. All authors have read and agreed to the published version of the manuscript.

Funding: Ministerio de Ciencia, Innovación y Universidades [Grant/Award No. RTI2018-093423-B-I00]; Consejería de Educación, Junta de Castilla y León [Grant/Award No. SA069G18]; Junta de Castilla y León y Fondo Europeo de Desarrollo Regional [Grant/Award No. CLU-2019-03 Unidad de Excelencia "Gestión Económica para la Sostenibilidad" (GECOS)].

Institutional Review Board Statement: Not applicable.

Informed Consent Statement: Not applicable.

Data Availability Statement: Table A1 includes the original data used in this research.

Conflicts of Interest: The authors declare no conflict of interest.

Appendix A

Table A1. Information about Ce Projects by Company.

COMPANY	INDUSTRY	Nº PROJECTS	CE STAGES	DESCRIPTION	CIRCULAR BENEFITS
COMPANY 1	Oil and energy	More than 230 CE projects in the different business units.	Efficiency and process innovation, renewable energy and alternative raw materials	FUEL GENERATION FROM URBAN WASTE. Through the use of technology, different waste is heated to high temperatures and in the absence of oxygen, produces gas that can be used to replace traditional fuels.	Remove 10,000 tons per year of urban waste from landfills in the first phase. Remove 100,000 tons of waste per year in the following phases.
COMPANY 2	Basic materials, industry and construction	13 projects	Reuse	ECOPARK. It is a plant designed and built by Company 2 for the mechanical-biological treatment of solid urban waste to produce biogas and compost that are later converted into energy and fertiliser, respectively.	10,000 tons per year of garbage from selective collection 35,000 tons per year of dehydrated sludge Biogas generation: 8,900,000 Nm^3/year6 composting tunnels with a capacity of 35,000 tons per year
COMPANY 3	Oil and gas	Unknown	Reuse	It is launching a project to take advantage of the cold from liquefied natural gas that allows it to be channelled to refrigeration facilities.	Unknown
COMPANY 4	Oil and gas	17 Projects in Spain	Reuse	ELECTRIC VEHICLE BATTERIES TO LIGHT UP A CITY. This project is based on an energy storage system using electric vehicle batteries at the City X thermal power plant to guarantee the city's electricity supply in the event of a power outage. In this way, a second life is being given to the batteries of electric vehicles.	It is a more economical and sustainable alternative to stationary power storage batteries Contributes to improving the quality of supply Extends the life of already spent batteries in electric vehicles

Table A1. *Cont.*

COMPANY	INDUSTRY	N° PROJECTS	CE STAGES	DESCRIPTION	CIRCULAR BENEFITS
COMPANY 5	Basic materials, industry and construction	Unknown	Reuse	Zero Waste to Landfill project: a project in collaboration with Ford that seeks to reduce the waste that is taken to the landfill to zero by looking for a new location for the waste that is generated.	Unknown
COMPANY 6	Oil and gas	Not available	Ecodesign	Green Hydrogen Plant This project will consist of a photovoltaic solar plant, a lithium-ion battery system and a hydrogen production system through electrolysis. The green hydrogen produced at this plant will be used in Brand's ammonia factory to produce green fertilisers.	It will avoid emissions of 39,000 tons per year of CO_2. It will reduce the plant's natural gas needs by more than 10%.
COMPANY 7	Consumer goods	Unknown	Reuse and recycle	CLOSING THE LOOP. Installing collection points, mitigating the generation of waste and promoting the reuse of clothing, marketing it for social purposes and recycling for industrial materials. This initiative is carried out through cooperation with nonprofit organisations and research programs in technologies to create new textile fibres from recycled garments.	100% of the company's waste is recycled or treated with another environmentally sustainable management method. 1201 million security alarms recycled. 120 million hangers recycled.
COMPANY 8	Financial services	Unknown	Recycle	Company 8 has a Road Safety and Experimentation Centre. This is a global technological centre for the design, insurance, use, maintenance, repair and recycling of vehicles and other solutions for the mobility of goods and people.	Unknown
COMPANY 9	Oil and gas	Unknown	Unknown	Unknown	Unknown
COMPANY 10	Oil and gas	Unknown	Reuse	Obsolete Machines Marketing Project. This project promotes the relocation of power machines that are no longer used in the value chain as resources or raw materials, giving them a second life and preventing them from ending up in a landfill.	Unknown

Table A1. *Cont.*

COMPANY	INDUSTRY	N° PROJECTS	CE STAGES	DESCRIPTION	CIRCULAR BENEFITS
COMPANY 11	Technology and telecommunications	Unknown	Ecodesign	ECO SMART It is a project of digital solutions for Company 11 client companies that helps them to see their contribution to the environment when they are installed. Based on a 100% renewable and low-emission network to reduce energy, water and CO_2 consumption.	15% reduction in fuel consumption with fleet management services 20% reduction in water consumption in agriculture with Smart Agro 30% reduction in public lighting consumption with Smart Lighting 43% reduction in the average time spent searching for parking with Smart Parking, 26.6% reduction of energy consumption with the Energy Efficiency service and LUCA Energy, 10% reduction in fuel and 85% development of black spots where there is poor waste separation, thanks to Smart Waste
COMPANY 12	Basic materials, industry and construction	Unknown	Unknown	Company 12's Positive Impact 360° project is a plan approved by the Board of Directors that summarises a series of actions that add value in five main areas: ethical, responsible and transparent governance, eco-efficiency and the fight against climate change, the CE and sustainable products, committed teams, culture, diversity and safety, supply chain and community impact.	Unknown
COMPANY 13 *(Resulted of the acquisition of one financial company that purchases another outright)*	Financial services	Unknown	Reuse and recycle	It carries out a selective collection of waste that allows it to be recovered and recycled both in corporate buildings and in the office network. The comprehensive plan for the revaluation of technology equipment also promotes the transfer of said equipment to nonprofit organisations, and in 2020 this plan was extended to office furniture.	Unknown
COMPANY 14	Technology and telecommunications	Unknown	Consume	Zero Plastic InitiativeThis consists of gradually eliminating single-use plastics in all the entity's offices.	Unknown
COMPANY 15	Consumer services	Unknown	Unknown	Unknown	Unknown
COMPANY 16	Oil and Gas	Unknown	Unknown	Unknown	Unknown
COMPANY 17	Consumer goods	Unknown	Unknown	Unknown	Unknown

References

1. Yuan, Z.; Bi, J.; Moriguichi, Y. The Circular Economy: A New Development Strategy in China. *J. Ind. Ecol.* **2013**, *2*, 23–44. [CrossRef]
2. Aranda-Usón, A.; Portillo-Tarragona, P.; Scarpellini, S.; Llena-Macarulla, F. The progressive adoption of a circular economy by businesses for cleaner production: An approach from a regional study in Spain. *J. Clean. Prod.* **2020**, *247*, 119648. [CrossRef]
3. García-Sánchez, I.-M.; Raimo, N.; Vitolla, F. Are Environmentally Innovative Companies Inclined towards Integrated Environmental Disclosure Policies? *Adm. Sci.* **2021**, *11*, 29. [CrossRef]
4. Weisz, H.; Krausmann, F.; Amann, C.; Eisenmenger, N.; Erb, K.-H.; Hubacek, K.; Fischer-Kowalski, M. The physical economy of the European Union: Cross-country comparison and determinants of material consumption. *Ecol. Econ.* **2006**, *58*, 676–698. [CrossRef]
5. Preston, F. A global redesign? Shaping the circular economy. *Energy Environ. Resour. Gov.* **2012**, *2*, 1–20.
6. West, J.; Schandl, H.; Krausmann, F.; Kovanda, J.; Hak, T. Patterns of change in material use and material efficiency in the successor states of the former Soviet Union. *Ecol. Econ.* **2014**, *105*, 211–219. [CrossRef]
7. Ranta, V.; Aarikka-Stenroos, L.; Ritala, P.; Mäkinen, S.J. Exploring institutional drivers and barriers of the circular economy: A cross-regional comparison of China, the US, and Europe. *Resour. Conserv. Recycl.* **2018**, *135*, 70–82. [CrossRef]
8. Dong, L.; Dai, M.; Liang, H.; Zhang, N.; Mancheri, N.; Ren, J.; Dou, Y.; Mingming, H. Material flows and resource productivity in China, South Korea and Japan from 1970 to 2008: A transitional perspective. *J. Clean. Prod.* **2017**, *141*, 1164–1177. [CrossRef]
9. Awasthi, A.K.; Cucchiella, F.; D'Adamo, I.; Li, J.; Rosa, P.; Terzi, S.; Wei, G.; Zeng, X. Modelling the correlations of e-waste quantity with economic increase. *Sci. Total Environ.* **2018**, *613* (Suppl. C), 46–53. [CrossRef]
10. Robaina, M.; Villar, J.; Pereira, E.T. The determinants for a circular economy in Europe. *Environ. Sci. Pollut. Res.* **2020**, *27*, 12566–12578. [CrossRef]
11. Di Maio, F.; Rem, P.C. A robust indicator for promoting circular economy through recycling. *J. Environ. Prot.* **2015**, *6*, 1095. [CrossRef]
12. Smol, M.; Kulczycka, J.; Avdiushchenko, A. Circular economy indicators in Relation to eco-innovation in European regions. *Clean Technol. Environ. Policy* **2017**, *19*, 669–678. [CrossRef]
13. Elia, V.; Gnoni, M.G.; Tornese, F. Measuring circular economy strategies through index methods: A critical analysis. *J. Clean. Prod.* **2017**, *142*, 2741–2751. [CrossRef]
14. Linder, M.; Sarasini, S.; van Loon, P. A Metric for Quantifying Product-Level Circularity. *J. Ind. Ecol.* **2017**, *21*, 545–558. [CrossRef]
15. Katz-Gerro, T.; Sintas, J.L. Mapping circular economy activities in the European Union: Patterns of implementation and their correlates in small and medium-sized enterprises. *Bus. Strat. Environ.* **2018**, *28*, 485–496. [CrossRef]
16. García-Sánchez, I.-M.; Somohano-Rodríguez, F.-M.; Amor-Esteban, V.; Frías-Aceituno, J.-V. Which region and which sector leads the circular economy? CEBIX, a multivariant index based on business actions. *J. Environ. Manag.* **2021**, *297*, 113299. [CrossRef]
17. Moraga, G.; Huysveld, S.; Mathieux, F.; Blengini, G.A.; Alaerts, L.; Van Acker, K.; DeMeester, S.R.; Dewulf, J. Circular economy indicators: What do they measure? *Resour. Conserv. Recycl.* **2019**, *146*, 452–461. [CrossRef] [PubMed]
18. Geissdoerfer, M.; Savaget, P.; Bocken, N.M.; Hultink, E.J. The Circular Economy—A new sustainability paradigm? *J. Clean. Prod.* **2017**, *143*, 757–768. [CrossRef]
19. Gunarathne, N.; Wijayasundara, M.; Senaratne, S.; Kanchana, P.K.; Cooray, T. Uncovering corporate disclosure for a circular economy: An analysis of sustainability and integrated reporting by Sri Lankan companies. *Sustain. Prod. Consum.* **2021**, *27*, 787–801. [CrossRef]
20. Kuo, L.; Chang, B.G. The affecting factors of circular economy information and its impact on corporate economic sustainability-Evidence from China. *Sustain. Prod. Consum.* **2021**, *27*, 986–997. [CrossRef]
21. García-Sánchez, I.-M. Corporate social reporting and assurance: The state of the art. *Rev. Contab.* **2021**, *24*, 241–269. [CrossRef]
22. North, D. *Transaction Costs, Institutions, and Economic Performance*; ICS Press: San Francisco, CA, USA, 1992.
23. Walker, K.; Wan, F. The Harm of Symbolic Actions and Green-Washing: Corporate Actions and Communications on Environmental Performance and Their Financial Implications. *J. Bus. Ethics* **2011**, *109*, 227–242. [CrossRef]
24. Dagiliene, L.; Frendzel, M.; Sutiene, K.; Wnuk-Pel, T. Wise managers think about circular economy, wiser report and analyze it. Research of environmental reporting practices in EU manufacturing companies. *J. Clean. Prod.* **2020**, *274*, 121968. [CrossRef]
25. Ramus, C.A.; Montiel, I. When are corporate environmental policies a form of greenwashing? *Bus. Soc.* **2005**, *44*, 377–414. [CrossRef]
26. Connelly, B.L.; Certo, S.T.; Ireland, R.D.; Reutzel, C.R. Signaling Theory: A Review and Assessment. *J. Manag.* **2010**, *37*, 39–67. [CrossRef]
27. Delmas, M.A. The diffusion of environmental management standards in Europe and in the United States: An institutional perspective. *Policy Sci.* **2002**, *35*, 91–119. [CrossRef]
28. Cho, C.H.; Patten, D.M. The role of environmental disclosures as tools of legitimacy: A research note. *Account. Organ. Soc.* **2007**, *32*, 639–647. [CrossRef]
29. García-Sánchez, I.-M.; Amor-Esteban, V.; Galindo-Álvarez, D. Communication Strategies for the 2030 Agenda Commitments: A Multivariate Approach. *Sustainability* **2020**, *12*, 10554. [CrossRef]
30. Tate, W.L.; Dooley, K.J.; Ellram, L.M. Transaction cost and institutional drivers of supplier adoption of environmental practices. *J. Bus. Logist.* **2011**, *32*, 6–16. [CrossRef]

31. Parra-Domínguez, J.; David, F.; Azevedo, T. Family Firms and Coupling among CSR Disclosures and Performance. *Adm. Sci.* **2021**, *11*, 30. [CrossRef]
32. Boulding, K.E. *The Economics of the Coming Spaceship Earth*; Taylor & Francis: New York, NY, USA, 1966.
33. Ellen MacArthur Foundation. Towards a Circular Economy: Business Rationale for an Accelerated Transition. 2015. Available online: www.ellenmacarthurfoundation.org (accessed on 1 July 2021).
34. Zeng, H.; Chen, X.; Xiao, X.; Zhou, Z. Institutional pressures, sustainable supply chain management, and circular economy capability: Empirical evidence from Chinese eco-industrial park firms. *J. Clean. Prod.* **2017**, *155*, 54–65. [CrossRef]
35. Oliver, C. Strategic responses to institutional processes. *Acad. Manag. Rev.* **1991**, *16*, 145–179. [CrossRef]
36. Hart, S.L. A natural-resource-based view of the firm *Acad. Manag. Rev.* **1995**, *20*, 986–1014.
37. Scott, W.R. The adolescence of institutional theory. *Adm. Sci. Q.* **1987**, *32*, 493–511. [CrossRef]
38. DiMaggio, P.J.; Powell, W.W. The iron cage revisited: Institutional isomorphism and collective rationality in organizational fields. *Am. Sociol. Rev.* **1983**, *48*, 147–160. [CrossRef]
39. Molina-Azorín, J.F. Microfoundations of strategic management: Toward micro-macro research in the resource-based theory. *BRQ Bus. Res. Q.* **2014**, *17*, 102–114. [CrossRef]
40. Daddi, T.; Todaro, N.M.; De Giacomo, M.R.; Frey, M. A Systematic Review of the Use of Organization and Management Theories in Climate Change Studies. *Bus. Strat. Environ.* **2018**, *27*, 456–474. [CrossRef]
41. Aibar-Guzmán, B.; Frías-Aceituno, J.-V. Is It Necessary to Centralize Power in the CEO to Ensure Environmental Innovation? *Adm. Sci.* **2021**, *11*, 27. [CrossRef]
42. Aibar-Guzmán, C.; Somohano-Rodríguez, F.M. Do Consumers Value Environmental Innovation in Product? *Adm. Sci.* **2021**, *11*, 33. [CrossRef]
43. Pérez-Calderón, E.; Pache-Durán, M.; Milanés-Montero, P. Inversión ecoeficiente: Efectos sobre el desempeño económico y financiero de los grupos empresariales del Dow Jones Sustainability World Index. *Rev. Contab.* **2021**, *24*, 220–230. [CrossRef]
44. Widmer, T.; Prior, D. Institutional Isomorphism, Institutional Logics and Organisational Fields: An Institutionalist Perspective on Circular Economy. In Proceedings of the Spring Servitization, Linköping, Sweden, 13–15 May 2019.
45. Escobar, L.F.; Vredenburg, H. Multinational oil companies and the adoption of sustainable development: A resource-based and institutional theory interpretation of adoption heterogeneity. *J. Bus. Ethics* **2011**, *98*, 39–65. [CrossRef]
46. Amran, A.; Ooi, S.K.; Wong, C.Y.; Hashim, F. Business strategy for climate change: An ASEAN perspective. *Corp. Soc. Responsib. Environ. Manag.* **2016**, *23*, 213–227. [CrossRef]
47. Wahyuni, D.; Ratnatunga, J. Carbon strategies and management practices in an uncertain carbonomic environment—Lessons learned from the coal-face. *J. Clean. Prod.* **2015**, *96*, 397–406. [CrossRef]
48. Liu, Q.; Li, H.M.; Zuo, X.L.; Zhang, F.F.; Wang, L. A survey and analysis on public awareness and performance for promoting circular economy in China: A case study from Tianjin. *J. Clean. Prod.* **2009**, *17*, 265–270. [CrossRef]
49. Elzinga, R.; Reike, D.; Negro, S.O.; Boon, W.P. Consumer acceptance of circular business models. *J. Clean. Prod.* **2020**, *254*, 119988. [CrossRef]
50. Bag, S.; Pretorius, J.H.C.; Gupta, S.; Dwivedi, Y.K. Role of institutional pressures and resources in the adoption of big data analytics powered artificial intelligence, sustainable manufacturing practices and circular economy capabilities. *Technol. Forecast. Soc. Chang.* **2020**, *163*, 120420. [CrossRef]
51. Radu, C.; Smaili, N. Corporate performance patterns of Canadian listed firms: Balancing financial and corporate social responsibility outcomes. *Bus. Strat. Environ.* **2021**, *30*, 3344–3359. [CrossRef]
52. BS 8001:2017—Framework for Implementing the Principles of the Circular Economy in Organizations. Guide. Available online: www.aenor.com (accessed on 23 September 2021).
53. Etzion, D.; Pe'Er, A. Mixed signals: A dynamic analysis of warranty provision in the automotive industry, 1960–2008. *Strat. Manag. J.* **2013**, *35*, 1605–1625. [CrossRef]
54. Scott, W.R. *Institutions and Organizations: Ideas, Interests, and Identities*; Sage Publications: Thousand Oaks, CA, USA, 2013.
55. Bocken, N.M.P.; Short, S.W.; Rana, P.; Evans, S. A literature and practice review to develop sustainable business model archetypes. *J. Clean. Prod.* **2014**, *65*, 42–56. [CrossRef]
56. Teece, D.J. Technological innovation and the theory of the firm: The role of enterprise-level knowledge, complementarities, and (dynamic) capabilities. In *Handbook of the Economics of Innovation*, 1st ed.; Elsevier B.V.: Amsterdam, The Netherlands, 2010; Volume 1.
57. Milios, L. Advancing to a circular economy: Three essential ingredients for a comprehensive policy mix. *Sustain. Sci.* **2018**, *13*, 861–878. [CrossRef] [PubMed]
58. Patten, D.M. The relation between environmental performance and environmental disclosure: A research note. *Account. Organ. Soc.* **2002**, *27*, 763–773. [CrossRef]
59. Benner, M.J.; Tushman, M.L. Exploitation, exploration, and process management: The productivity dilemma revisited. *Acad. Manag. Rev.* **2003**, *28*, 238–256. [CrossRef]
60. Jalonen, H. The uncertainty of innovation: A systematic review of the literature. *J. Manag. Res.* **2011**, *4*, 1. [CrossRef]
61. Berthiaume, A. Use of the National Pollutant Release Inventory in environmental research: A scoping review. *Environ. Rev.* **2021**, *29*, 329–339. [CrossRef]

2. Reverte, C. Corporate social responsibility disclosure and market valuation: Evidence from Spanish listed firms. *Rev. Manag. Sci.* **2014**, *10*, 411–435. [CrossRef]
3. Li, Y.; Richardson, G.D.; Thornton, D.B. Corporate disclosure of environmental liability information: Theory and evidence. *Contemp. Acc. Res.* **1997**, *14*, 435–474. [CrossRef]
4. Karahanna, E.; Straub, D.W.; Chervany, N.L. Information Technology Adoption Across Time: A Cross-Sectional Comparison of Pre-Adoption and Post-Adoption Beliefs. *MIS Q.* **1999**, *23*, 183. [CrossRef]
5. Rogers, E.M. Evolution: Diffusion of Innovations. In *International Encyclopedia of the Social & Behavioral Sciences*, 2nd ed.; Wright, J.D., Ed.; Elsevier: Oxford, UK, 2015; pp. 378–381.
6. Luhmann, N. *Social Systems*; Stanford University Press: Stanford, CA, USA, 1995.
7. Melón-Izco, Á.; Ruiz-Cabestre, F.J.; Ruiz-Olalla, C. Readabilty in management reports: Extension and good governance practices. *Rev. Contab.* **2021**, *24*, 19–30. [CrossRef]
8. Hadro, D.; Klimczak, K.M.; Pauka, M. Management's choice of tone in letters to shareholders: Sincerity, bias and incentives. *Rev. Contab.* **2021**, *24*, 202–219. [CrossRef]
9. Reinert, M. Système Alceste: Une méthodologie d'analyse des données textuelles. *Actes Jornades Int. D'analisi Dades Textuals* **1991**, 44–161.
0. Benzécri, J.P. *L'analyse des Données*; Dunod: Paris, France, 1973; Volume 2, p. 1.
1. Zarbà, C.; Chinnici, G.; La Via, G.; Bracco, S.; Pecorino, B.; D'Amico, M. Regulatory Elements on the Circular Economy: Driving into the Agri-Food System. *Sustainability* **2021**, *13*, 8350. [CrossRef]

sustainability

MDPI

Article

Circular Economy Disclosure in Sustainability Reporting: The Effect of Firm Characteristics

Filippo Vitolla [1], Vitiana L'Abate [1], Felice Petruzzella [2], Nicola Raimo [1,*] and Antonio Salvi [3]

[1] Department of Management, Finance and Technology, LUM University, 70010 Casamassima, Italy
[2] Department of Corporate Finance, Giustino Fortunato Telematic University, 82100 Benevento, Italy
[3] Department of Management, University of Turin, 10124 Turin, Italy
* Correspondence: raimo@lum.it

Abstract: The circular economy is increasingly establishing itself as a model capable of overcoming the current linear economy of production and consumption recognized as unsustainable by society. Its relevance has also attracted the attention of academics, interested not only in the implementation methods of the circular economy, but also in the ways in which companies communicate information about them. However, although in recent years some scholars have begun to investigate the circular economy disclosure (CED), research on this topic is still in an embryonic state. In fact, in the academic literature there are only a few studies related to the CED and its drivers. This study aims to fill this gap by investigating, under the lens of stakeholder theory, the effect of firm characteristics on the level of CED. To this end, it firstly involves the use of a manual content analysis of the sustainability reports drawn up by 88 international companies to measure the level of CED and, secondly, a regression model to test the impact of the firm characteristics. Empirical results demonstrate a positive effect of firm size, financial leverage and firm profitability on the level of CED. The results have important practical implications for firms and policymakers.

Keywords: circular economy; sustainability reporting; disclosure; sustainability; stakeholder theory

Citation: Vitolla, F.; L'Abate, V.; Petruzzella, F.; Raimo, N.; Salvi, A. Circular Economy Disclosure in Sustainability Reporting: The Effect of Firm Characteristics. *Sustainability* **2023**, *15*, 2200. https://doi.org/10.3390/su15032200

Academic Editor: Luigi Aldieri

Received: 16 December 2022
Revised: 12 January 2023
Accepted: 20 January 2023
Published: 24 January 2023

1. Introduction

Ecosystems are essential to sustain life on the planet, but these precious resources and habitats are not always treated with respect. Decades of overuse, overconsumption and neglect have led to the destruction or serious degradation of many vital ecosystems. The spread of the COVID-19 pandemic and the Russia–Ukraine war have aggravated the state of emergency and have shown even more the need to rationalize resources and avoid waste.

In this context, the circular economy is increasingly establishing itself as a response to environmental problems, especially in terms of recycling and resource recovery. To date, in many countries of the world, and especially in those of the European Union, it represents one of the pillars of sustainable policies and one of the main drivers of the ecological transition towards global sustainability [1]. The NextGenerationEU Plan identified the circular economy as one of the elements capable of guaranteeing economic recovery following the COVID-19 pandemic [1]. The circular economy deals with designing and directing economic activity towards the regeneration of resources, keeping components, products and materials always at their maximum utility and value [2]. In fact, the circular economy, unlike traditional linear systems [3], "represents a development strategy that enables economic growth while aiming to optimize the chain of consumption of biological and technical materials" [4]. It requires a profound transformation for firms and consumers who are called to move from a linear production and consumption system to a new approach based on reduction, reuse, recycling and recovery, and therefore to a strong reduction of waste [5–7]. The transition to a circular economy must involve changes in various parts of the value chain. This creates the need to develop the knowledge base to support the transition in order to derive real benefits from it [8]. The role of digital technologies in this process is crucial to overcome barriers to the circular economy transition and support companies' decision-making process [9,10], fostering the implementation of circular business models [11]. Therefore, the circular economy involves, on the one hand, the correct management of the scarce resources available to a company for its production processes and, on the other hand, the creation of value from the waste produced [7,12]. Nevertheless, there is a lack of consensus among researchers on the actual benefits of the circular economy, which appears to be far from being as promising as its supporters claim [13].

Despite the importance that the circular economy assumes for stakeholders [14], companies still do not communicate data and information relating to this production and consumption model in a completely effective and structured way [7]. In fact, even if circular economy reporting practices are starting to take hold among companies [7], there are still problems and delays linked, on the one hand, to the absence of tailored measurement tools and specific reporting standards [15] and, on the other hand, to the absence of a widely adopted and recognized underlying language [16]. These circumstances have led to the proliferation of heterogeneous forms of processes for measuring and representing information related to the circular economy [17–19], which take the form of quantitative and qualitative documents based on different reporting techniques and tools [7,20]. In light of this, there is a need to conduct in-depth analysis and studies capable of identifying the best reporting solutions for companies interested in disseminating information relating to the circular economy.

However, from an academic point of view, while some studies have focused on different research lines, such as the efficient use of materials and their productivity, e.g., [21–23], the transition to circular models, e.g., [24,25] and the implementation of circular economy strategies in different countries and sectors, e.g., [26–29], the communication of such strategies and its drivers represent issues still poorly examined by the literature. This study aims to fill the knowledge gap related to the circular economy disclosure (CED). Specifically, this study intends to answer the following research questions:

- RQ1: What is the level of circular economy information disseminated within the sustainability reports?
- RQ2: What are the main firm characteristics capable of affecting the level of CED?

The choice of sustainability reports as containers of the CED derives from the thought of several authors [30,31], according to whom the information relating to the circular economy included by companies within the sustainability reports represent a clear expression of the corporate thinking around the concept and allow us to fully understand the trends and insights related to the adoption of such production and consumption systems [30,31]. Among the possible factors capable of affecting the level of CED, under the lens of stakeholder theory, this study focuses attention on the main financial firm characteristics. This choice is connected to the huge costs associated with both the implementation of circular economy strategies and the collection and representation of information relating to these production and consumption systems.

The remainder of this work is organized as follows: Section 2 offers the literature review, while Section 3 offers the theoretical background and hypotheses development. Section 4 introduces the research methodology, while Section 5 presents and discusses the results. Finally, Section 6 draws conclusions.

2. Literature Review

In the last decades, the concept of circular economy has gained the attention of academics, institutions and organizations, which have provided several definitions [10,32]. The concept of circular economy was introduced by Pearce et al. [33] to express the need to evaluate the interlinkage between economic and environmental activities. They have introduced an economic system based on closed-loop material flow, according to which everything is an input to everything else [34]. Several definitions of circular economy have been provided over time following different emerging trends, such as the green economy and bio-economy, e.g., [35], industrial ecology and industrial symbiosis, e.g., [36,37] and cradle-to-cradle design, e.g., [38]. Accordingly, the circular economy can be considered an umbrella concept resulting from the ongoing scholarly debate [39]. Although many studies have analyzed the circular economy from different perspectives, such as waste management, supply chain or digital technology, further critical analysis of the factors driving organizations to adopt the circular economy and the related benefits is needed [40].

The relevance of the circular economy has gained the attention of scholars towards the way companies communicate information relating to this production and consumption system. Scholars have examined the CED within different corporate documents. Among these, the most examined are the integrated reports and the sustainability reports, with some scholars who have carried out a comparison between the different corporate documents.

Among the contributions that have examined the integrated reports, a relevant study is the one conducted by Barnabè and Nazir [7]. This study investigated, through a content analysis, the ability of integrated reporting to provide the principles, key elements and concepts to ensure a good representation of the aspects related to the circular economy and then examined, through a case study, the existing interactions between integrated reporting and the circular economy. The authors found important differences in the choices related to the dissemination of information by companies with an important role played by integrated reporting whose principles support the CED. The adequacy of this reporting tool for the dissemination of circular economy information is also confirmed by Kunc et al. [41] through a multiple case study analysis. Barnabè and Nazir [7] instead investigated, through a case study analysis, the way in which companies can enable and conceptualize the opportunities and principles of the circular economy through the dissemination of an integrated report. The study conducted by the authors demonstrated that the adoption of this reporting tool allows for a better understanding of the opportunities and activities related to the circular economy, also in relation to the development of future strategies.

Among the contributions examining the sustainability reports, Opferkuch et al. [42] highlighted a clear disconnect between the circular economy and sustainability disclosure by virtue of the absence of precise guidelines and the consequent subjectivity in the selection of the information to be disclosed and the reporting approaches to be adopted. The scarce attention paid to circular economy information in sustainability reports has been confirmed by different studies [31,43–45]. In particular, Opferkuch et al. [43] found that only a limited number of companies integrate the circular economy information into their sustainability reports through the use of indicators and the clarification of the objectives. In addition, Janik et al. [44], examining the energy sector, found the presence of numerous information relating to greenhouse gases and limited attention to circular economy information. Furthermore, Tiscini et al. [31] found that the circular economy is still under-examined in relation to strategy, governance, performance and management. Finally, Marco–Fondevila et al. [45] found that around half of the largest Spanish companies do not include any elements related to the circular economy in their sustainability reports.

Another part of the studies investigated the type of circular economy information disseminated within the sustainability reports. In this regard, Stewart and Niero [30], examining 46 reports of companies operating in the fast-moving consumer goods sector, mainly found the presence of information regarding the main product and packaging, procurement strategies and end-of-life management, while less attention was paid to business model strategies and circular product design. Istudor and Suciu [46], examining the food retail sector, found a propensity of companies to disclose information related to waste and emissions reduction and recycling. Finally, García–Sánchez et al. [1] found that firms give priority to information relating to a sustainable future, the reduction or elimination of greenhouse gases and the commitment to the proper use of resources.

Another part of the studies examined the determinants of CED in sustainability reports. In this regard, Dagiliene et al. [47] found that regulations and mimetic pressures have a strong influence on environmental reporting from the circular economy perspective, while the effect of coercive factors is not significant. Still in relation to external pressures, Wang et al. [48] found a positive impact of environmental institutional protecting pressures on the amount of circular economy information disseminated within sustainability reports. In addition, the authors found a positive influence of ownership concentration and institutional ownership.

Finally, some authors have conducted studies analyzing different types of corporate documents. Among these, Gunarathne et al. [2], examining the integrated and sustainability reports, found a low level of CED among companies located in Sri Lanka. In addition, Kuo and Chang [49] investigated the determinants and effects of the CED by examining the annual and sustainability reports in the Chinese context. The authors demonstrated a greater propensity towards the CED by firms operating in environmentally-sensitive industries, larger firms and state-owned enterprises. As regards the effects, the authors instead found a positive impact of the CED on the growth rate and profitability. Still examining the annual and sustainability reports, Roberts et al. [50] found higher CED levels among automotive companies, while less circular economy information is disseminated by companies operating in the transportation, aerospace and defense sectors.

3. Theoretical Background and Hypotheses Development

This study uses the stakeholder theory to explain the choice of companies to disseminate circular economy information. It represents one of the most used socio-political theories in the accounting area and is based on the key assumption, according to which the economic dimension cannot be investigated without considering the social, environmental, institutional and political domain in which organizations operate [51,52]. In particular, the stakeholder theory focuses on the analysis of the existing relationships between firms and the various actors who have interests in the business activity, defined as stakeholders [51,53,54]. They, according to Freeman [53] can be considered as "any group or individual who can affect or is affected by the achievement of the firm's objectives". According to Donaldson and Preston [55], there are three theoretical approaches to considering stakeholder claims: descriptive, normative and instrumental.

The descriptive perspective examines whether and how companies actually consider the needs of stakeholders [56,57]. The normative perspective, on the other hand, is based on the ethical principles of the stakeholder theory, and therefore considers all stakeholders worthy of attention as moral subjects with their own rights, regardless of their strength and their ability to influence firm performance [57,58]. Finally, according to the instrumental perspective of the stakeholder theory, firms consider and satisfy the needs and requirements of stakeholders for mere reasons of strategic expediency [57,58]. Therefore, the management of relations with stakeholders is based on accurate assessments of their influence and relevance [59] and has the main objective of obtaining legitimacy and consensus among stakeholders, especially the most relevant ones [57,60]. Building a climate of trust with stakeholders is, in fact, fundamental for the survival of the firm [52,58] and for obtaining advantages, such as better access to capital [61], higher profitability [62] and better corporate reputation [63]. This study uses the instrumental perspective to explain the choice of companies to disseminate circular economy information within the sustainability reports. CED can, in fact, be understood as a tool available to companies capable of allowing dialogue with stakeholders, satisfying their information needs and obtaining consensus and legitimacy. Some companies may have a greater need to disseminate information on the circular economy, due to the greater number of stakeholders and the greater pressures they are exposed to. This study focuses on three firm characteristics that could affect the level of CED: firm size, financial leverage and firm profitability. The individual hypotheses are formulated below.

The academic literature agrees on the positive influence of firm size on the level and quality of corporate disclosure, e.g., [54,57,64,65]. The main principle on which this positive influence is based, connected to the stakeholder theory, is attributable to the greater pressures to which larger companies are exposed [66,67]. Such firms are usually more visible, due to the strong impacts they have on the community, and have a wider stakeholder base [58,66]. Furthermore, they are exposed to greater risks, not only of a reputational nature, but also connected to a possible interference by governments in corporate management in the event of non-compliance with the social contract [54,67]. These reasons, in line with the stakeholder theory, could push larger companies to disclose a greater amount of information in order to mitigate the greater pressures to which they are exposed and reduce the risk level. In particular, the dissemination of circular economy information can represent a solution for informing stakeholders about the production and consumption models that are increasingly at the center of the information needs of stakeholders. Therefore, in the light of the above, we introduce the following hypothesis:

H1: *Firm size positively affects the level of CED.*

The literature also agrees on the positive influence of financial leverage on the level and quality of corporate disclosure, e.g., [57,66,68,69]. The main principle on which this positive influence is based, connected to the stakeholder theory, is linked to the greater pressures to which firms with greater financial leverage are exposed [70]. These companies are, in fact, more exposed by virtue of the broader base of stakeholders and, in particular, of debt-holders to whom they need to demonstrate the ability to fulfil the financial obligations [57,65]. Furthermore, firms characterized by higher financial leverage are more exposed to the risk of bankruptcy [57]. These reasons, in line with the stakeholder theory, could push firms with higher financial leverage to disclose a greater amount of information in order to mitigate the greater pressures to which they are exposed and reduce the associated risk level. In particular, the dissemination of circular economy information could represent a solution for informing stakeholders and, in particular, debt-holders about the production and consumption models. Therefore, in light of the above, we introduce the following hypothesis:

H2: *Financial leverage positively affects the level of CED.*

Finally, the literature has also highlighted a positive influence of firm profitability on the level and quality of corporate disclosure, e.g., [65,69,71]. The main principle on which this positive influence is based, connected to the stakeholder theory, is attributable to the desire of the most profitable companies to promote their positive image among stakeholders [69] and to distinguish their financial results from those of competitors [64]. Furthermore, the most profitable companies, by virtue of the greater financial resources, can be more advanced in the implementation of circular models and, therefore, could have greater incentives to adopt transparent behaviors in relation to the circular economy. Another motivation behind the positive influence is represented by the availability of greater monetary resources that could also be used for the disclosure processes [72,73]. In this regard, the most profitable companies could have better internal information collection systems and include more qualified personnel able to better manage the disclosure processes [57,73]. Finally, a further reason is represented by the greater public scrutiny to which the most profitable firms are exposed [74]. These reasons, in line with the stakeholder theory, could push the most profitable companies to disclose a greater amount of information. Therefore, in light of the above, we introduce the following hypothesis:

H3: *Firm profitability positively affects the level of CED.*

4. Research Methodology

4.1. Sample

The sample of this study consists of 88 international companies that published a sustainability report in 2021. The Global Reporting Initiative (GRI) website was chosen as the source for identifying the companies to be included in the sample. The GRI is an independent international organization whose purpose is to support companies in communicating their impact on sustainability. The choice of GRI website guarantees that the sustainability reports have been prepared in compliance with the GRI Standards. In this regard, the GRI Standards represent the best practices at a global level for the preparation of the sustainability report. In fact, sustainability reporting based on the GRI Standards allows organizations to provide information about the positive or negative contribution to sustainable development, also facilitating the reporting of their environmental, social and economic impacts. In particular, the "Community Members" section of the GRI website was chosen. This section includes a list of 500 organizations that are part of the GRI Community From this list, only the companies of the GRI Community that drafted the sustainability report in 2021 were selected. This process led to the construction of the final sample that comprises 88 international companies.

The sample is heterogeneous in terms of sector and geographical location. In fact, the selected companies operate in 15 different sectors and are located in 28 different countries and 5 regions.

4.2. Dependent Variable

The dependent variable of this study is the CED score (CEDS). It was measured through a manual content analysis of the sustainability reports. According to Krippendorff [75], content analysis is a "research technique for making replicable and valid inferences from data according to their context". Content analysis is widely used in studies related to disclosure and reporting, as it is considered reliable and objective [75,76] and allows data and information to be collected quickly and cheaply [77]. Furthermore, it allows the subsequent implementation of econometric models aimed at analyzing the variables capable of influencing the level or quality of the disclosure examined [78]. For these reasons, content analysis has been used both in studies relating to environmental disclosure, e.g., [79,80] and in those relating to the CED, e.g., [7,49]. The construction of a disclosure index represents one of the most used methodologies to quantify the information collected through content analysis techniques [81–83]. In order to ensure consistency and comparability of the results obtained, the use of a disclosure index related to the CED already present in the literature was preferred. In particular, the disclosure index developed by Wang et al. [48] was chosen. This disclosure index includes 10 items divided into 4 categories: (1) honors and performances; (2) investment and expenditure; (3) policies and implementation; and (4) resource reuse.

The first category includes information relating to the honors that companies have obtained in implementing circular economy strategies and policies and information relating to the performance achieved in terms of emissions reduction, energy saving and cost savings. The second category includes information relating to relevant investments and recurrent expenditure in circular economy strategies and policies. The third category includes information relating to national policies relating to the circular economy and the tax advantages and financial aid received by the company for the implementation of circular economy strategies and policies. Finally, the fourth category includes information related to the reuse of resources in the context of circular economy strategies and policies. The items identified were evaluated using a scale ranging from 0 to 3. For each item, a score of 0 is assigned in the absence of information, a score of 1 is assigned when a qualitative description is present, a score of 2 is assigned when there is only a quantitative description and, finally, a score of 3 is assigned when there is both a qualitative and quantitative description. In light of this, the dependent variable of this study can assume a score between 0 and 30. Table 1 presents an overview of the items classified in the different categories.

Table 1. Circular economy disclosure index.

Categories	Items	Scores
1. Honors and performances of circular economy	Definition of circular economy and honors	0–3
	Performances of energy conservation and emission reduction	0–3
	Performances of resource reuse	0–3
	Other performances on emission reduction	0–3
	Cost saving due to the implementation of circular economy	0–3
2. Investments in the circular economy	Circular economy significant investment	0–3
	Other circular economy recurrent expenditure	0–3
3. Circular economy policies and implementation	Circular economy financial aids and tax concessions policies	0–3
	Implementation of circular economy related provisions	0–3
4. Resource reuse of circular economy	Circular economy utilization ratio information	0–3

4.3. Independent and Control Variables

The independent variables used in this study are: firm size (LNFS), financial leverage (FINLEV) and firm profitability (PROF).

LNFS is a proxy of firm size and is computed as the natural logarithm of a firm's total assets. FINLEV is a proxy of firms' leverage and is computed as total assets divided by shareholders' equity. PROF is a proxy of firm profitability. To this purpose we use firms' return on equity (ROE), calculated as net income divided by shareholders' equity.

Following the literature in the field, in order to improve the reliability of our econometric analysis we included a set of control variables. To this purpose, a first set of governance variables have been included. Specifically, we included: board size (BS), board gender diversity (BGD), and board age (BA). BS is a proxy of the board of directors' size and is calculated as the number of executive and non-executive members within the board of directors [80,84]. Following the literature in the field, we expect a positive relationship between BS and CEDS, because a larger board of directors should be able to better carry out its control and monitoring functions [85,86]. Additionally, larger boards tend to have more experienced members and diversified skills that enhance the voluntary dissemination of information. BGD represents the percentage of female members within the board of directors [5]. We expect a positive relationship between BGD and CEDS, because a more diversified board of directors should be able to better carry out their controlling and monitoring functions, due to their ability to better represent the interests of all corporate stakeholders [87]. BA, instead, refers to the average age of the members of the board of directors [88,89]. In this regard, we expect a negative relationship between BA and CEDS, as younger directors could be smarter, dynamic and oriented towards values such as sustainability and corporate transparency [90].

A second set of control variables complete the econometric model: industry environmental sensitivity (IES), firm age (FA) and civil law system (CIV). IES represents a dummy variable that expresses the environmental sensitivity of the industry in which the company operates. This variable assumes a score equal to 1 for the firms operating in highly polluting sectors, and 0 otherwise. Following Branco and Rodrigues [66] and Tagesson et al. [91], we consider the following sectors as environmentally sensitive: automotive, agriculture, chemical, construction and construction materials, aviation, energy and energy utilities, logistics, forest and paper products, waste management, mining, metal products, railroad, and water utilities. Firms operating in an industry with a relevant impact on the environment should disclose a greater amount of information, in order to mitigate the strong pressures to which they are exposed [58]. FA is a proxy of the stability of the company and is calculated as the number of years that have passed since the company was established. There is a lack of unanimous consensus among past researchers on the impact of FA on the disclosure quality [29]. Despite this, we expect a positive relationship between FA and CEDS, because long-term companies should have more consolidated information collection and representation systems and, therefore, could provide a higher level of CED [58,65].

Finally, CIV is a dummy variable that assumes a value equal to 1, if the firm is headquartere in a country with a civil law system, and 0 otherwise [29]. In this regard, according to Frías Aceituno et al. [84], companies located in civil law countries tend to disclose more informatio

All independent and control variables were manually collected by the authors usin the Refinitiv database and by analyzing the companies' sustainability reports.

4.4. Model Specification

We perform a regression analysis in order to investigate the determinants of the leve of CED. The regression model is reflected in the following equation:

$$CEDS = \beta_0 + \beta_1\ LNFS + \beta_2\ FINLEV + \beta_3\ PROF + \beta_4\ BS + \beta_5\ BGD + \beta_6\ BA + \beta_7\ IES + \beta_8\ FA + \beta_9\ CIV + \varepsilon$$

5. Results and Discussion

5.1. Descriptive Statistics and Correlation Analysis

Table 2 reports the descriptive statistics, the matrix correlation and the results of th variance inflation factor (VIF) analysis. The mean value of the dependent variable is 16.1· This value demonstrates an adequate dissemination of circular economy information b companies within the sustainability reports. However, the need for further efforts b companies to achieve full transparency in relation to the circular economy is evident. Th independent variables LNFS, FINLEV and PROF, instead, have a mean value equal t 8.96, 2.75 and 0.194, respectively. The control variables relating to the board of director characteristics present the following values: the board of directors is, on average, made u of 15.63 members (BS), with a female percentage of 29.80 and an average directors' age o 56.72 years. Finally, IES has a mean value of 0.51, FA of 63.17, and the dummy CIV of 0.6

Table 2. Descriptive statistics, VIF and correlation analysis.

	Variable	Mean	S.D.	VIF	1	2	3	4	5	6	7	8	9	1
1	CEDS	16.13933	4.52297	-	1									
2	LNFS	8.961515	3.236341	1.29	0.3882 ***	1								
3	FINLEV	2.751584	1.950219	1.19	0.1465	−0.1120	1							
4	PROF	0.1944845	0.3763681	1.17	0.3399 ***	−0.1030	0.2189 **	1						
5	BS	15.62791	8.56813	1.51	0.3765 ***	0.3906 ***	−0.0076	−0.0082	1					
6	BGD	0.2980493	0.3747293	1.08	−0.0765	−0.1526	−0.0843	0.0423	−0.0681	1				
7	BA	56.72049	3.51547	1.34	0.3134 ***	0.2595 **	−0.0168	0.0341	0.4714 ***	0.0505	1			
8	IEA	0.5113636	0.5027355	1.16	0.2986 ***	0.1118	0.0443	0.1380	0.2031 *	−0.0274	0.1177	1		
9	FA	63.17241	47.71115	1.16	0.2168 **	0.0870	0.0283	0.2327 **	0.2123 *	0.0608	0.2089 *	0.1126	1	
10	CIV	0.6136364	0.4897059	1.29	−0.0005	0.1880 *	0.2637 **	0.0366	0.1087	−0.2047 *	−0.0004	0.2515 **	0.0729	1

Notes: *** Significant at the 1% level; ** Significant at the 5% level; * Significant at the 10% level.

The correlation coefficients are quite low, in fact, the higher coefficient is equal tc 0.47 between BA and BS, a value below the threshold recognized by the literature [92] Furthermore, the absence of concerns about multicollinearity issues is confirmed by the VIF analysis results, being the higher coefficient equal to 1.51 (BS), a value lower than 10 the maximum threshold beyond which issues of multicollinearity arise [93].

5.2. Regression Analysis

Table 3 displays the results of the regression analysis. The impact of the three depen dent variables is positive and statistically significant, confirming our initial expectations The adjusted R^2 is equal to 0.362, which demonstrates the ability of the econometric model to explain about 36.2% of the variance in the dependent variable.

Hypothesis 1 (H1) is supported by the regression results. In fact, the coefficient of LNFS is positive (0.4368258) and statistically significant (p = 0.003), indicating a positive impact of firm size on the level of CED. This result demonstrates that larger companies disclose a greater amount of information related to the circular economy within their sustainability reports. This result, from a stakeholder theory perspective, can be explained by the greater pressures from stakeholders and the higher exposure to risks that characterizes these companies. In this regard, the dissemination of circular economy information represents a solution to inform stakeholders about production and consumption models and, therefore, mitigate pressures and reduce the risk level. The positive impact of the firm size on the level of circular economy information confirms what was found by Kuo and Chang [49] in relation to the CED in the Chinese context, and by numerous studies in relation to environmental disclosure, e.g., [94–96] and sustainability disclosure, e.g., [66,68,69,97,98].

Table 3. Regression model results.

Variables	Coefficient	Standard Error	*p*-Value	Sign.
Cons	1.783727	6.914648	0.797	
LNFS	0.4368258	0.1397557	0.003	***
FINLEV	0.3754967	0.2172297	0.088	*
PROF	3.931364	1.132437	0.001	***
BS	0.0959489	0.0559512	0.090	*
BGD	−0.7212378	1.081539	0.507	
BA	0.1313353	0.1288267	0.311	
IES	1.818563	0.8423015	0.034	**
FA	0.0034553	0.0087815	0.695	
CIV	−1.896228	0. 9101816	0.041	**
N	88			
Adj. R^2	0.362			

Notes: *** Significant at the 1% level; ** Significant at the 5% level; * Significant at the 10% level.

Hypothesis 2 (H2) is also supported by the regression results. In fact, the coefficient of FINLEV is positive (0.3754967) and statistically significant (p = 0.088), indicating a positive impact of financial leverage on the level of CED. This result demonstrates that firms with higher financial leverage tend to disseminate a higher level of circular economy information within their sustainability reports. This result, from a stakeholder theory perspective, can be explained by the greater pressures that these companies receive from stakeholders and capital providers and by the greater risk of bankruptcy to which they are exposed. In this regard, the dissemination of circular economy information represents a solution to inform stakeholders and, in particular, capital providers about production and consumption models, and therefore mitigate pressures and reduce the associated risks. The positive impact of financial leverage on the level of circular economy information is in line with previous studies in the field of environmental disclosure [94,95] and sustainability disclosure [68,69].

Finally, Hypothesis 3 (H3) is also supported by the regression results. In fact, the coefficient of PROF is positive (3.931364) and statistically significant (p = 0.001), indicating a positive impact of firm profitability on the level of CED. This result demonstrates that the most profitable companies disclose a greater amount of information related to the circular economy within their sustainability reports. This result, from a stakeholder theory perspective, can be explained by the willingness of the most profitable companies to promote their positive image among stakeholders and by the possible greater involvement in circular economy practices. Furthermore, other possible explanations can be traced back to the better processes for collecting and representing circular economy information, deriving from the greater monetary resources enjoyed by these companies and the greater public scrutiny to which the most profitable companies are exposed. In this regard, the dissemination of circular economy information represents a solution to inform stakeholders about production and consumption models, and therefore promote a positive image and satisfy the information needs of the public at large. The positive impact of firm profitability on the level of circular economy information confirms the results obtained by Kuo and Chang [49] in relation to the CED in the Chinese context, and by several studies concerning social and environmental disclosure [95–97] and sustainability disclosure [69,88,99].

6. Conclusions

6.1. Objectives and Summary of Results

The circular economy represents an important opportunity to reduce costs, improve efficiency and reduce the environmental impact of business activities [100]. By adopting circular economy practices, companies can reduce the costs linked to the purchase of raw materials, increase efficiency in the use of resources and improve their reputation and their attractiveness in the eyes of consumers and investors, thus aiming for economic growth [101]. In this context, information relating to the circular economy is becoming increasingly important [102,103].

The aim of this study was to investigate the impact of firm characteristics on the level of CED. In particular, this study examined the impact of firm size, financial leverage and firm profitability on the level of circular economy information disseminated by the companies within the sustainability reports. The results highlighted that all the independent variables have a positive impact on the level of CED. In other words, this study demonstrated that larger companies, those with a higher degree of financial leverage, and those with a higher level of profitability, provide more circular economy information in their sustainability reports.

6.2. Theoretical and Practical Implications

This study enriches the academic literature in several ways. First, this study contributes to the debate about the CED, which is increasingly attracting the interest of academics. In this regard, this study, by investigating an international sample, considerably extends the academic literature that is still scarce and often focused only on national samples. Secondly, this study extends the analysis of the CED to a tool, such as the sustainability report, that is still little explored by the academic literature as a potential container of circular economy information. This contribution is particularly relevant by virtue of the proximity of the topics of circular economy and sustainability, which makes sustainability reports ideal containers for the dissemination of circular economy information, and therefore increases the value of the analysis conducted in this study. Thirdly, this study contributes to extending knowledge about the factors capable of influencing firms' CED policies by showing the role played by firm characteristics. In this regard, therefore, this study enriches the scarce literature on the determinants of the CED. Finally, this study extends the field of application of the stakeholder theory, still little used to frame the choices of firms to provide circular economy information. In fact, this theory, although it has been widely used to frame sustainability disclosure policies, was still poorly anchored to the firms' CED policies.

The findings of this study also have important practical implications for firms. The relevance of the CED for the stakeholders should, in fact, push firms to make important efforts to achieve full transparency in relation to this production and consumption model. In this regard, firms should provide information regarding different aspects of the circular economy, such as performance, investments, policies, implementation, certifications and standards. Furthermore, they should also provide specific indicators relating to the circular economy within the corporate reports to allow stakeholders to immediately assess the companies' commitment to implementing circular models and to make comparisons with other companies in the sector.

The findings also have important implications for policymakers. In light of the importance of the circular economy, they should, first of all, encourage the implementation of this model of production and consumption by companies with specific regulations and incentives. Secondly, policymakers, through specific interventions, should encourage the transparency of companies in relation to circular economy strategies and policies, due to the relevance that such information has for stakeholders.

6.3. Limitations and Future Research Directions

However, this study is not without limitations. The main limitations are related to the sample size, the time horizon of the econometric analysis and the limited number of determinants examined. In fact, the sample includes a limited number of companies and the econometric analysis covers only a single year. Furthermore, due to the lack of data, this study only examines the impact of firm size, leverage and profitability on the level of CED. However, these limitations do not reduce the overall quality of the study and offer important insights for future research. In fact, on the one hand, they will be able to extend the size of the sample using different criteria for the identification of companies and, on the other hand, they will be able to extend the time horizon of the econometric analysis in order to also investigate the evolution of the CED over the years. In addition, using different databases, future research will be able to collect other data and analyze the impact of further firm-level variables, such as sales growth or market-to-book value on the level of circular economy information disseminated. They will also be able to analyze the impact of other country- and sector-level variables. Future research will also be able to extend the analysis of the CED to other corporate documents and websites to understand which methods and tools are most used by companies to disseminate circular economy information.

Author Contributions: Conceptualization, V.L., F.P. and N.R.; methodology, V.L., F.P.; validation, V.L., F.P. and N.R.; investigation, V.L.; data curation, V.L.; writing—original draft preparation, V.L., F.P. and N.R.; writing—review and editing, F.V. and A.S.; visualization, F.V. and A.S.; supervision, F.V. and A.S.; project administration, F.V. and A.S. All authors have read and agreed to the published version of the manuscript.

Funding: This research received no external funding.

Institutional Review Board Statement: Not applicable.

Informed Consent Statement: Not applicable.

Data Availability Statement: Not applicable.

Conflicts of Interest: The authors declare no conflict of interest.

References

1. García-Sánchez, I.M.; Somohano-Rodríguez, F.M.; Amor-Esteban, V.; Gonzalez-Valdueza, B. Circular Economy Projects and Firm Disclosures in an Encouraging Institutional Environment. *Sustainability* **2022**, *14*, 1149. [CrossRef]
2. Gunarathne, N.; Wijayasundara, M.; Senaratne, S.; Kanchana, P.K.; Cooray, T. Uncovering corporate disclosure for a circular economy: An analysis of sustainability and integrated reporting by Sri Lankan companies. *Sustain. Prod. Consum.* **2021**, *27*, 787–801. [CrossRef]
3. Millar, N.; McLaughlin, E.; Börger, T. The circular economy: Swings and roundabouts? *Ecol. Econ.* **2019**, *158*, 11–19. [CrossRef]
4. European Commission. Closing the Loop—An EU Action Plan for the Circular Economy. 2014. Available online: https://www.eea.europa.eu/policy-documents/com-2015-0614-final (accessed on 13 December 2022).
5. Andersen, F.M.; Larsen, H.; Skovgaard, M.; Moll, S.; Isoard, S. A European model for waste and material flows. *Resour. Conserv. Recycl.* **2007**, *49*, 421–435. [CrossRef]
6. Geissdoerfer, M.; Savaget, P.; Bocken, N.M.; Hultink, E.J. The Circular Economy–A new sustainability paradigm? *J. Clean. Prod.* **2017**, *143*, 757–768. [CrossRef]
7. Barnabè, F.; Nazir, S. Investigating the interplays between integrated reporting practices and circular economy disclosure. *Int. J. Product. Perform. Manag.* **2021**, *70*, 2001–2031. [CrossRef]
8. Johansen, M.R.; Christensen, T.B.; Ramos, T.M.; Syberg, K. A review of the plastic value chain from a circular economy perspective. *J. Environ. Manag.* **2022**, *302*, 113975. [CrossRef] [PubMed]
9. Liu, Q.; Trevisan, A.H.; Yang, M.; Mascarenhas, J. A framework of digital technologies for the circular economy: Digital functions and mechanisms. *Bus. Strategy Environ.* **2022**, *31*, 2171–2192. [CrossRef]
10. Lobo, A.; Trevisan, A.H.; Liu, Q.; Yang, M.; Mascarenhas, J. Barriers to transitioning towards smart circular economy: A systematic literature review. In *Sustainable Design and Manufacturing, Proceedings of the 8th International Conference on Sustainable Design and Manufacturing (KES-SDM 2021), Split, Croatia, 15–17 September 2021*; Springer: Singapore, 2022; pp. 245–256.
11. Neligan, A.; Baumgartner, R.J.; Geissdoerfer, M.; Schöggl, J.P. Circular disruption: Digitalisation as a driver of circular economy business models. *Bus. Strategy Environ.* **2022**, *online ahead of print*. [CrossRef]
12. Corvellec, H.; Stowell, A.F.; Johansson, N. Critiques of the circular economy. *J. Ind. Ecol.* **2022**, *26*, 421–432. [CrossRef]

13. Abuabara, L.; Paucar-Caceres, A.; Burrowes-Cromwell, T. Consumers' values and behaviour in the Brazilian coffee-in-capsules market: Promoting circular economy. *Int. J. Prod. Res.* **2019**, *57*, 7269–7288. [CrossRef]
14. Beck, R.; Avital, M.; Rossi, M.; Thatcher, J.B. Blockchain technology in business and information systems research. *Bus. Inf. Syst. Eng.* **2017**, *59*, 381–384. [CrossRef]
15. MacArthur, E.; Zumwinkel, K.; Stuchtey, M.R. *Growth within: A Circular Economy Vision for a Competitive Europe*; Ellen MacArthur Foundation: Cowes, UK, 2015.
16. Korhonen, J.; Nuur, C.; Feldmann, A.; Birkie, S.E. Circular economy as an essentially contested concept. *J. Clean. Prod.* **2018**, *175*, 544–552. [CrossRef]
17. Elia, V.; Gnoni, M.G.; Tornese, F. Measuring circular economy strategies through index methods: A critical analysis. *J. Clean. Prod.* **2017**, *142*, 2741–2751. [CrossRef]
18. Svensson, N.; Funck, E.K. Management control in circular economy. Exploring and theorizing the adaptation of management control to circular business models. *J. Clean. Prod.* **2019**, *233*, 390–398. [CrossRef]
19. Ünal, E.; Urbinati, A.; Chiaroni, D. Managerial practices for designing circular economy business models: The case of an Italian SME in the office supply industry. *J. Manuf. Technol. Manag.* **2018**, *30*, 561–589. [CrossRef]
20. Garza-Reyes, J.A.; Salomé Valls, A.; Peter Nadeem, S.; Anosike, A.; Kumar, V. A circularity measurement toolkit for manufacturing SMEs. *Int. J. Prod. Res.* **2019**, *57*, 7319–7343. [CrossRef]
21. Preston, F. A Global Redesign? Shaping the Circular Economy. 2012. Available online: http://www.chathamhouse.org/sites/default/files/public/Research/Energy%20Environment%20and%20Development/bp0312_preston.pdf (accessed on 12 December 2022).
22. West, J.; Schandl, H.; Krausmann, F.; Kovanda, J.; Hak, T. Patterns of change in material use and material efficiency in the successor states of the former Soviet Union. *Ecol. Econ.* **2014**, *105*, 211–219. [CrossRef]
23. Robaina, M.; Murillo, K.; Rocha, E.; Villar, J. Circular economy in plastic waste-Efficiency analysis of European countries. *Sci. Total Environ.* **2020**, *730*, 139038. [CrossRef] [PubMed]
24. Ranta, V.; Aarikka-Stenroos, L.; Mäkinen, S.J. Creating value in the circular economy: A structured multiple-case analysis of business models. *J. Clean. Prod.* **2018**, *201*, 988–1000. [CrossRef]
25. Aranda-Usón, A.; Portillo-Tarragona, P.; Scarpellini, S.; Llena-Macarulla, F. The progressive adoption of a circular economy by businesses for cleaner production: An approach from a regional study in Spain. *J. Clean. Prod.* **2020**, *247*, 119648. [CrossRef]
26. Dong, S.; Wang, Z.; Li, Y.; Li, F.; Li, Z.; Chen, F.; Cheng, H. Assessment of comprehensive effects and optimization of a circular economy system of coal power and cement in Kongtong District, Pingliang City, Gansu Province, China. *Sustainability* **2017**, *9*, 787. [CrossRef]
27. Smol, M.; Kulczycka, J.; Avdiushchenko, A. Circular economy indicators in relation to eco-innovation in European regions. *Clean Technol. Environ. Policy* **2017**, *19*, 669–678. [CrossRef]
28. Katz-Gerro, T.; Lopez Sintas, J. Mapping circular economy activities in the European Union: Patterns of implementation and their correlates in small and medium-sized enterprises. *Bus. Strategy Environ.* **2019**, *28*, 485–496. [CrossRef]
29. Garcia-Sanchez, I.M.; Raimo, N.; Vitolla, F. CEO power and integrated reporting. *Meditari Account. Res.* **2021**, *29*, 908–942. [CrossRef]
30. Stewart, R.; Niero, M. Circular economy in corporate sustainability strategies: A review of corporate sustainability reports in the fast-moving consumer goods sector. *Bus. Strategy Environ.* **2018**, *27*, 1005–1022. [CrossRef]
31. Tiscini, R.; Martiniello, L.; Lombardi, R. Circular economy and environmental disclosure in sustainability reports: Empirical evidence in cosmetic companies. *Bus. Strategy Environ.* **2022**, *31*, 892–907. [CrossRef]
32. Merli, R.; Preziosi, M.; Acampora, A. How do scholars approach the circular economy? A systematic literature review. *J. Clean. Prod.* **2018**, *178*, 703–722. [CrossRef]
33. Pearce, D.W.; Turner, R.K.; Turner, R.K. *Economics of Natural Resources and the Environment*; Johns Hopkins University Press: Baltimore, MD, USA, 1990.
34. Su, B.; Heshmati, A.; Geng, Y.; Yu, X. A review of the circular economy in China: Moving from rhetoric to implementation. *J. Clean. Prod.* **2013**, *42*, 215–227. [CrossRef]
35. Hina, M.; Chauhan, C.; Kaur, P.; Kraus, S.; Dhir, A. Drivers and barriers of circular economy business models: Where we are now, and where we are heading. *J. Clean. Prod.* **2022**, *333*, 130049. [CrossRef]
36. Mohan, S.V.; Nikhil, G.N.; Chiranjeevi, P.; Reddy, C.N.; Rohit, M.V.; Kumar, A.N.; Sarkar, O. Waste biorefinery models towards sustainable circular bioeconomy: Critical review and future perspectives. *Bioresour. Technol.* **2016**, *215*, 2–12. [CrossRef] [PubMed]
37. Alvarez, R.; Ruiz-Puente, C. Development of the tool symbiosys to support the transition towards a circular economy based on industrial symbiosis strategies. *Waste Biomass Valorization* **2017**, *8*, 1521–1530. [CrossRef]
38. Saavedra, Y.M.; Iritani, D.R.; Pavan, A.L.; Ometto, A.R. Theoretical contribution of industrial ecology to circular economy. *J. Clean. Prod.* **2018**, *170*, 1514–1522. [CrossRef]
39. Braungart, M.; McDonough, W.; Bollinger, A. Cradle-to-cradle design: Creating healthy emissions–a strategy for eco-effective product and system design. *J. Clean. Prod.* **2007**, *15*, 1337–1348. [CrossRef]
40. Blomsma, F.; Brennan, G. The emergence of circular economy: A new framing around prolonging resource productivity. *J. Ind. Ecol.* **2017**, *21*, 603–614. [CrossRef]
41. Kunc, M.H.; Giorgino, M.C.; Barnabè, F. Developing forward-looking orientation in integrated reporting. *Meditari Account. Res.* **2020**, *29*, 823–850. [CrossRef]

42. Opferkuch, K.; Caeiro, S.; Salomone, R.; Ramos, T.B. Circular economy in corporate sustainability reporting: A review of organisational approaches. *Bus. Strategy Environ.* **2021**, *30*, 4015–4036. [CrossRef]
43. Opferkuch, K.; Caeiro, S.; Salomone, R.; Ramos, T.B. Circular economy disclosure in corporate sustainability reports: The case of European companies in sustainability rankings. *Sustain. Prod. Consum.* **2022**, *32*, 436–456. [CrossRef]
44. Janik, A.; Ryszko, A.; Szafraniec, M. Greenhouse gases and circular economy issues in sustainability reports from the energy sector in the European Union. *Energies* **2020**, *13*, 5993. [CrossRef]
45. Marco-Fondevila, M.; Llena-Macarulla, F.; Callao-Gastón, S.; Jarne-Jarne, J.I. Are circular economy policies actually reaching organizations? Evidence from the largest Spanish companies. *J. Clean. Prod.* **2021**, *285*, 124858. [CrossRef]
46. Istudor, L.G.; Suciu, M.C. Bioeconomy and Circular Economy in the European Food Retail Sector. *Eur. J. Sustain. Dev.* **2020**, *9*, 501. [CrossRef]
47. Dagiliene, L.; Frendzel, M.; Sutiene, K.; Wnuk-Pel, T. Wise managers think about circular economy, wiser report and analyze it. Research of environmental reporting practices in EU manufacturing companies. *J. Clean. Prod.* **2020**, *274*, 121968. [CrossRef]
48. Wang, P.; Che, F.; Fan, S.; Gu, C. Ownership governance, institutional pressures and circular economy accounting information disclosure: An institutional theory and corporate governance theory perspective. *Chin. Manag. Stud.* **2014**, *8*, 487–501. [CrossRef]
49. Kuo, L.; Chang, B.G. The affecting factors of circular economy information and its impact on corporate economic sustainability-Evidence from China. *Sustain. Prod. Consum.* **2021**, *27*, 986–997. [CrossRef]
50. Roberts, L.; Georgiou, N.; Hassan, A.M. Investigating biodiversity and circular economy disclosure practices: Insights from global firms. *Corp. Soc. Responsib. Environ. Manag.* **2022**, *online ahead of print*. [CrossRef]
51. Gray, R.; Kouhy, R.; Lavers, S. Corporate social and environmental reporting: A review of the literature and a longitudinal study of UK disclosure. *Account. Audit. Account. J.* **1995**, *8*, 47–77. [CrossRef]
52. Deegan, C.; Blomquist, C. Stakeholder influence on corporate reporting: An exploration of the interaction between WWF-Australia and the Australian minerals industry. *Account. Organ. Soc.* **2006**, *31*, 343–372. [CrossRef]
53. Freeman, R.E. *Strategic Management: A Stakeholder Approach*; Pitman: Boston, MA, USA, 1984.
54. Nicolò, G.; Ricciardelli, A.; Raimo, N.; Vitolla, F. Visual disclosure through integrated reporting. *Manag. Decis.* **2021**, *60*, 976–994. [CrossRef]
55. Donaldson, T.; Preston, L.E. The stakeholder theory of the corporation: Concepts, evidence, and implications. *Acad. Manag. Rev.* **1995**, *20*, 65–91. [CrossRef]
56. Brenner, S.N.; Cochran, P. The stakeholder theory of the firm: Implications for business and society theory and research. *In Proc. Int. Assoc. Bus. Soc.* **1991**, *2*, 897–933. [CrossRef]
57. Raimo, N.; Vitolla, F.; Minutiello, V.; Marrone, A.; Tettamanzi, P. Readability of integrated reports: Evidence from worldwide adopters. *Corp. Soc. Responsib. Environ. Manag.* **2022**, *29*, 524–534. [CrossRef]
58. Vitolla, F.; Raimo, N.; Rubino, M.; Garzoni, A. How pressure from stakeholders affects integrated reporting quality. *Corp. Soc. Responsib. Environ. Manag.* **2019**, *26*, 1591–1606. [CrossRef]
59. Mitchell, R.K.; Agle, B.R.; Wood, D.J. Toward a theory of stakeholder identification and salience: Defining the principle of who and what really counts. *Acad. Manag. Rev.* **1997**, *22*, 853–886. [CrossRef]
60. Ogden, S.; Watson, R. Corporate performance and stakeholder management: Balancing shareholder and customer interests in the UK privatized water industry. *Acad. Manag. J.* **1999**, *42*, 526–538. [CrossRef]
61. Torugsa, N.A.; O'Donohue, W.; Hecker, R. Proactive CSR: An empirical analysis of the role of its economic, social and environmental dimensions on the association between capabilities and performance. *J. Bus. Ethics* **2013**, *115*, 383–402. [CrossRef]
62. Kumar, V.; Lahiri, A.; Dogan, O.B. A strategic framework for a profitable business model in the sharing economy. *Ind. Mark. Manag.* **2016**, *69*, 147–160. [CrossRef]
63. Aksak, E.O.; Ferguson, M.A.; Duman, S.A. Corporate social responsibility and CSR fit as predictors of corporate reputation: A global perspective. *Public Relat. Rev.* **2016**, *42*, 79–81. [CrossRef]
64. Frias-Aceituno, J.V.; Rodríguez-Ariza, L.; Garcia-Sánchez, I.M. Explanatory factors of integrated sustainability and financial reporting. *Bus. Strategy Environ.* **2014**, *23*, 56–72. [CrossRef]
65. Vitolla, F.; Raimo, N.; Marrone, A.; Rubino, M. The role of board of directors in intellectual capital disclosure after the advent of integrated reporting. *Corp. Soc. Responsib. Environ. Manag.* **2020**, *27*, 2188–2200. [CrossRef]
66. Branco, M.C.; Rodrigues, L.L. Factors influencing social responsibility disclosure by Portuguese companies. *J. Bus. Ethics* **2008**, *83*, 685–701. [CrossRef]
67. Reverte, C. Determinants of corporate social responsibility disclosure ratings by Spanish listed firms. *J. Bus. Ethics* **2009**, *88*, 351–366. [CrossRef]
68. Andrikopoulos, A.; Samitas, A.; Bekiaris, M. Corporate social responsibility reporting in financial institutions: Evidence from Euronext. *Res. Int. Bus. Financ.* **2014**, *32*, 27–35. [CrossRef]
69. Sharif, M.; Rashid, K. Corporate governance and corporate social responsibility (CSR) reporting: An empirical evidence from commercial banks (CB) of Pakistan. *Qual. Quant.* **2014**, *48*, 2501–2521. [CrossRef]
70. Liu, X.; Anbumozhi, V. Determinant factors of corporate environmental information disclosure: An empirical study of Chinese listed companies. *J. Clean. Prod.* **2009**, *17*, 593–600. [CrossRef]
71. Wallace, R.O.; Naser, K. Firm-specific determinants of the comprehensiveness of mandatory disclosure in the corporate annual reports of firms listed on the stock exchange of Hong Kong. *J. Account. Public Policy* **1995**, *14*, 311–368. [CrossRef]

72. Brammer, S.; Millington, A. Firm size, organizational visibility and corporate philanthropy: An empirical analysis. *Bus. Ethics: A Eur. Rev.* **2006**, *15*, 6–18. [CrossRef]
73. Schreck, P.; Raithel, S. Corporate social performance, firm size, and organizational visibility: Distinct and joint effects on voluntary sustainability reporting. *Bus. Soc.* **2018**, *57*, 742–778. [CrossRef]
74. Ng, E.J.; Koh, H.C. An agency theory and probit analytic approach to corporate non-mandatory disclosure compliance. *Asia Pac. J. Account.* **1994**, *1*, 29–44. [CrossRef]
75. Krippendorff, K. *Content Analysis: An Introduction to Its Methodology*; Sage: Beverly Hills, CA, USA, 1980.
76. Guthrie, J.; Petty, R.; Yongvanich, K.; Ricceri, F. Using content analysis as a research method to inquire into intellectual capital reporting. *J. Intellect. Cap.* **2004**, *5*, 282–293. [CrossRef]
77. Salvi, A.; Petruzzella, F.; Raimo, N.; Vitolla, F. Transparency in the digitalization choices and the cost of equity capital. *Qual. Res. Financ. Mark.* **2022**, *online ahead of print*. [CrossRef]
78. Raimo, N.; Vitolla, F.; Marrone, A.; Rubino, M. The role of ownership structure in integrated reporting policies. *Bus. Strategy Environ.* **2020**, *29*, 2238–2250. [CrossRef]
79. Giacomini, D.; Zola, P.; Paredi, D.; Mazzoleni, M. Environmental disclosure and stakeholder engagement via social media: State of the art and potential in public utilities. *Corp. Soc. Responsib. Environ. Manag.* **2020**, *27*, 1552–1564. [CrossRef]
80. Raimo, N.; de Nuccio, E.; Vitolla, F. Corporate governance and environmental disclosure through integrated reporting. *Meas. Bus. Excell.* **2022**, *26*, 451–470. [CrossRef]
81. Vitolla, F.; Raimo, N.; Rubino, M.; Garzoni, A. Broadening the horizons of intellectual capital disclosure to the sports industry: Evidence from top UEFA clubs. *Meditari Account. Res.* **2022**, *30*, 142–162. [CrossRef]
82. Jakhar, S.K.; Mangla, S.K.; Luthra, S.; Kusi-Sarpong, S. When stakeholder pressure drives the circular economy: Measuring the mediating role of innovation capabilities. *Manag. Decis.* **2019**, *4*, 904–920. [CrossRef]
83. Kristensen, H.S.; Mosgaard, M.A. A review of micro level indicators for a circular economy–moving away from the three dimensions of sustainability? *J. Clean. Prod.* **2020**, *243*, 118531. [CrossRef]
84. Frias-Aceituno, J.V.; Rodriguez-Ariza, L.; Garcia-Sanchez, I.M. The role of the board in the dissemination of integrated corporate social reporting. *Corp. Soc. Responsib. Environ. Manag.* **2013**, *20*, 219–233. [CrossRef]
85. Hill, C.W.; Jones, T.M. Stakeholder-agency theory. *J. Manag. Stud.* **1992**, *29*, 131–154. [CrossRef]
86. Vitolla, F.; Raimo, N.; Rubino, M. Board characteristics and integrated reporting quality: An agency theory perspective. *Corp. Soc. Responsib. Environ. Manag.* **2020**, *27*, 1152–1163. [CrossRef]
87. Rodrigues, L.L.; Tejedo-Romero, F.; Craig, R. Corporate governance and intellectual capital reporting in a period of financial crisis: Evidence from Portugal. *Int. J. Discl. Gov.* **2017**, *14*, 1–29. [CrossRef]
88. Giannarakis, G. The determinants influencing the extent of CSR disclosure. *Int. J. Law Manag.* **2014**, *56*, 393–416. [CrossRef]
89. Cucari, N.; Esposito de Falco, S.; Orlando, B. Diversity of board of directors and environmental social governance: Evidence from Italian listed companies. *Corp. Soc. Responsib. Environ. Manag.* **2018**, *25*, 250–266. [CrossRef]
90. Handajani, L.; Subroto, B.; Sutrisno, T.; Saraswati, E. Does board diversity matter on corporate social disclosure? An Indonesian evidence. *J. Econ. Sustain. Dev.* **2014**, *5*, 8–16.
91. Tagesson, T.; Blank, V.; Broberg, P.; Collin, S.O. What explains the extent and content of social and environmental disclosures on corporate websites: A study of social and environmental reporting in Swedish listed corporations. *Corp. Soc. Responsib. Environ. Manag.* **2009**, *16*, 352–364. [CrossRef]
92. Kennedy, P. *A Guide to Econometrics*, 4th ed.; Blackwell Publishers: Oxford, UK, 1999.
93. Myers, R.H. *Classical and Modern Regression with Applications*, 2nd ed.; Duxbury: Belmont, CA, USA, 1990.
94. Huang, C.L.; Kung, F.H. Drivers of environmental disclosure and stakeholder expectation: Evidence from Taiwan. *J. Bus. Ethics* **2010**, *96*, 435–451. [CrossRef]
95. Andrikopoulos, A.; Kriklani, N. Environmental disclosure and financial characteristics of the firm: The case of Denmark. *Corp. Soc. Responsib. Environ. Manag.* **2013**, *20*, 55–64. [CrossRef]
96. Lu, Y.; Abeysekera, I. Stakeholders' power, corporate characteristics, and social and environmental disclosure: Evidence from China. *J. Clean. Prod.* **2014**, *64*, 426–436. [CrossRef]
97. Gamerschlag, R.; Möller, K.; Verbeeten, F. Determinants of voluntary CSR disclosure: Empirical evidence from Germany. *Rev. Manag. Sci.* **2011**, *5*, 233–262. [CrossRef]
98. Tamimi, N.; Sebastianelli, R. Transparency among S&P 500 companies: An analysis of ESG disclosure scores. *Manag. Decis.* **2017**, *55*, 1660–1680.
99. Sharma, P.; Panday, P.; Dangwal, R.C. Determinants of environmental, social and corporate governance (ESG) disclosure: A study of Indian companies. *Int. J. Discl. Gov.* **2020**, *17*, 208–217. [CrossRef]
100. Özbay, R.D.; Athari, S.A.; Saliba, C.; Kirikkaleli, D. Towards Environmental Sustainability in China: Role of Globalization and Hydroelectricity Consumption. *Sustainability* **2022**, *14*, 4182. [CrossRef]
101. Saliba, C.B.; Hassanein, F.R.; Athari, S.A.; Dördüncü, H.; Agyekum, E.B.; Adadi, P. The Dynamic Impact of Renewable Energy and Economic Growth on CO2 Emissions in China: Do Remittances and Technological Innovations Matter? *Sustainability* **2022**, *14*, 14629. [CrossRef]

102. García-Sánchez, I.-M.; Raimo, N.; Vitolla, F. Are Environmentally Innovative Companies Inclined towards Integrated Environmental Disclosure Policies? *Adm. Sci.* **2021**, *11*, 29. [CrossRef]
103. Raimo, N.; Vitolla, F.; Marrone, A.; Rubino, M. Do audit committee attributes influence integrated reporting quality? An agency theory viewpoint. *Bus. Strategy Environ.* **2021**, *30*, 522–534. [CrossRef]

sustainability

MDPI

Article

Indicators for the Circular City: A Review and a Proposal

Federica Paoli [1,2,*], Francesca Pirlone [1] and Ilenia Spadaro [1]

1 DICCA-Department of Civil, Chemical and Environmental Engineering, University of Genoa, 16145 Genoa, Italy
2 Higher University School IUSS, 27100 Pavia, Italy
* Correspondence: federica.paoli@edu.unige.it

Abstract: The theme of the circular city is currently much debated in the literature as a possible strategy for achieving sustainability in urban areas. However, as a recent development it still has many features in the making, one of the most important being the issue concerning monitoring and the tool through which to achieve it. In the paper, therefore, the "indicator" tool is explored in depth. Metrics represent a fundamental and complex aspect that is foundational to measuring and quantifying the progress of results achieved with respect to the goals set. Currently, most existing indicators are associated with specific aspects of the circular economy; there have been few examples of indicators designed to assess the circularity of an entire city. The paper aims to identify priority themes and describe a set of indicators to be used at the urban level. In the absence of an established reference frame, themes and indicators were identified through a methodology starting with an extensive literature search and careful analysis, including statistical analysis, of the scientific literature as well as international and European strategies on the subject. A particular result of this research is the definition ofa minimum set of indicators common to all cities, which can be applied for comparative purposes.

Keywords: urban sustainability; circular cities; indicators

Citation: Paoli, F.; Pirlone, F.; Spadaro, I. Indicators for the Circular City: A Review and a Proposal. *Sustainability* **2022**, *14*, 11848. https://doi.org/10.3390/su141911848

Academic Editors: Nicola Raimo, Filippo Vitolla, Ornella Malandrino and Benedetta Esposito

Received: 18 July 2022
Accepted: 15 September 2022
Published: 20 September 2022

1. Introduction

The topic of the circular city is currently much debated in the literature and is seen as one of the possible solutions for achieving sustainability in urban areas. The transition to circular cities is at the center of this debate. Specifically, this transition involves the creation of an integrated city in which the principles of circular economy are applied to all local government divisions, in a process facilitated by political initiative and support that through good example promotes change among residents and various stakeholders. Thus, the basis of this vision is the employment of circular economy ideals, which include in their foundational design the concepts of second use, remanufacturing, efficient use of resources, elimination of waste, avoidance of toxic materials, and improving and making sustainable waste management through the utilization of the 9Rs strategy (reduce, reuse, recycle, recover, reject, repair, refurbish, remanufacture, and reuse) [1]. It has been reported that by applying the circular city model, Europe's gross domestic product (GDP) can be increased by 7%, with yearly savings of 600 billion euros, benefits of 1.8 trillion euros each year, and the creation of 170,000 jobs by 2035 [2]. Furthermore, carbon dioxide discharges could be diminished by 48% by 2030 and 83% by 2050 [3]. Over a similar period, raw material utilization can be decreased by 32% and 53% [4]. This study framework applies analysis at the city level, discretizing it into key sectors for the transition which were identified in the Circular Economy Action Plan (electronics and ICT, batteries and vehicles, packaging, plastics, textiles, construction and buildings, food, water, and nutrients). These sectors are key because they have the greatest potential for circular innovation, the greatest environmental impact, and the greatest demand on resources. For each of these sectors, there is a need to be able to close the loop, as well as to create circular activity in general that connects the different sectors. However, effective circular planning and decision-making requires an understanding of the flows of materials and energy that leave and enter cities, and are consumed, processed, or stored there. The ability to collect and analyze this data helps to identify where and how to intervene and which circuits to close, provides cities with information about their economic activities, and allows them to link current initiatives and their potential to make the city more circular. In this context, identifying operational indicators is a priority for cities to plan their transition to circularity, to understand what parameters can be measured and where the population and other actors in the cities can have the greatest impact [5]. Today there are several tools that have been adapted for assessing circularity in urban settings, but few that have been designed specifically for the purpose. However, this is an important area for development as the fielding of a circular model needs a tool that can take into account the multidimensional impacts involved. Above all, such a tool should include all the actors and the various sectors that are involved in the process. This requirement is fundamental [6] to oversee and weigh the positive and adverse consequences [7–9] and balance them against the commitment of every actor involved (public administration, research, merchants, companies, and population). This appraisal instrument should be able to overcome the constraints of the ongoing financial methodology [10], by "catching" social perspectives. Moreover, renewal and change of the regulative and administrative structure is expected to help the shift to circularity [11]. Cities currently play a dominant role in the world economy; they are not only the centers of present-day living and the centers of gross domestic product production, but are also greatly responsible for the consumption of natural resources, greenhouse gas production, and waste. It is natural, then, that research continues to focus on attempts to improve existing urban conditions. As previously mentioned, to try to limit such harmful effects, it would be desirable for cities to become ecosystems in which closed cycles predominate, so that no waste is produced. This would lead to improvements in several areas, including ecological footprint and greenhouse gas emissions, urban safety, and public health [12]. This goal is facilitated by the fact that cities are centers where progress, sharing, experimentation, and encounters are favored, therefore where solutions to the social, economic, and environmental problems of our time are more likely to be found [13]. However, for such progress to be called sustainable, which it should be, there is a need for cities to change and to do so quickly. Such change to date has been brought about by the enforcement of the circular economy in urban settings to achieve what is called a circular city. Underlying this concept is the need to create closed and regenerative cycles that can optimize use of available resources while avoiding production of waste: "A circular city is one that promotes the transition from a linear to a circular economy in an integrated way across all its functions in collaboration with citizens, businesses and the research community. [. . .] to improve human well-being, reduce emissions, protect and enhance biodiversity, and promote social justice, in line with the Sustainable Development Goals" [5]. At present, the concept of the circular city encompasses within it several others, most relevant of which are those of the smart city and the resilient city. The concept of resilience has been expanded from its original meaning, that is related solely to the ability to adapt to a natural event, including man-made events within the definition. This concept is taken up in the Urban Agenda and is fundamental to risk reduction. The smart city concept, on the other hand, is currently widely used and defines cities based on their ability to resolve problems such as crime, traffic congestion, inefficient services, and economic stagnation. According to the European Union's definition: "A smart city is a place where traditional networks and services are made more efficient with the use of digital solutions for the benefit of its inhabitants and businesses".

Therefore, this study paper approaches the topic of circular cities by identifying indicators common for all cities, so that they can be compared with each other.

2. Materials and Methods

2.1. The Importance of Indicators in a Circular City

The European Commission, with 75% of its citizens living in urban areas [14], has understood the potential that the circular economy can have for its cities. So, in accordance with the commitments made with the adoption of the "Action Plan for the Circular Economy" in April 2017, it launched a working group to create a set of indicators to measure the levels of "circularity" in 27 European countries. In January 2018, it adopted the "Circular Economy Observational Framework," which turned out to be well thought out for estimating progress towards a circular economy. This tool addresses different aspects at all stages of the lifecycles of renewable and nonrenewable resources (materials, water, and energy) used in products and services [15]. This model of the circular economy consists of ten indicators, some of which are themselves divided into a number of sub-indicators, divided overall into four thematic areas: Production and Consumption, Waste Management, Secondary Raw Materials, and Competitiveness and Innovation. The ten indicators have been designed to provide a general overview of some of the key elements required to succeed in increasing circularity in the EU economy [15]. The Green New Deal of January 2020 emphasized the central role of these issues by putting the circular economy at the center of its policies aiming to achieve the goals agreed in Paris in 2015. This Circular Economy Action Plan defines a future-oriented agenda to achieve a cleaner and more competitive Europe, implemented through co-creation conducted by economic actors, consumers, citizens, and civil society organizations [16].

It is therefore clear that to be successfully complete urban ecological transition through the concept of a circular city, one must consider technical and also social elements. The circular city is not only an economic model for the efficient management of resources, but offers a holistic model for the promotion of environmental sustainability and the improvement of social cooperation between the actors involved. Regarding elements of a more technical nature, as well as aspects of a purely economic nature, indicators are currently attracting particular attention. How to measure circularity in the urban context is one of the questions frequently asked in the recent literature, and the answer is by no means obvious. Indicators are invaluable both for analyzing the impact of new legislative or regulatory proposals and also for assessing the posthumous effectiveness of the measures adopted, as city populations must understand how their decisions impact both themselves and the environment around them, to assess the expected, achieved and missed targets for cycle closures [5]. Thus, circularity indicators should be considered, which although not designed for direct application to the city, can nonetheless provide a good picture of some of its areas. In the context of circular cities, we need to think about defining what we consider relevant as well as thinking about what can be measured and is worth measuring [17]. Some indicators have been made available for circular cities [18]. However, it is often not possible to apply these indicators, because of a lack of data or because similar analyses have been carried out using different indicators that do not allow for comparison and monitoring [19]. Therefore, it is essential to capture the important aspects of a city in its various areas of activity, and for each of them to identify indicators that are universal and allow comparison between different sectors and case studies [20]. The indicators described in the literature and those proposed as part of circularity initiatives cover at least environmental, social, economic and cultural aspects. However, features related to a city's circularity are not fully measurable and therefore require specific indicators which may be subjective and qualitative. Although each member state is subject to binding EU targets, there is currently no set of indicators shared on either a national or European scale. The lack of a widespread participatory implementation strategy makes these targets difficult to achieve. There is a need to create a shared base of data and knowledge, that can measure the effectiveness of actions implemented in urban areas, by measuring flows, to identify where and how action can be taken towards the closing of cycles, provide the city with a clear vision of the policies in place, and secure the involvement of relevant actors. Aiming to identify levels of depth, steps to evaluate the level of the circular economy in a city can be developed from various analyses:

1. Quantifying the circularity of the individual proposed project with respect to the issue in which it fits;
2. then, assessing the impact of the project with respect to priority issues such as mobility, waste, energy, and reduction of inputs (land, water, and energy consumption) and outputs (waste and pollutant production);
3. finally, considering the urban neighborhood, it is essential to assess impacts on different related priority issues (waste, mobility, energy, etc.) to finally quantify the different projects implemented in the area.

Therefore, this study aims firstly to implement a reconnaissance of the existing indicators used in the literature, and secondly to propose a useful set of indicators to evaluate circularity at the urban level.

2.2. Proposed Methodology

The objective of this research paper is to identify indicators useful for monitoring actions aimed at planning a circular city. To do this we initially investigated the literature, including strategy documents, to identify indicators that are currently in use. These indicators were separated within a framework according to the different priority areas at the urban level. In particular, the review was carried out on scientific publications that dealt with the topic of circularity and referred to indicators (research carried out in 2019 by the University of Naples Federico II [21] was developed and implemented using data provided by 46 publications). Specifically, the search was conducted on scientific articles published in *Sustainability* and *Science Direct* from 2015 to 4 April 2022 by running an analysis for the keywords "circular city" and "indicators."

The approach began with investigation of the literature and the main documents and strategies, to identify the indicators currently used in relation to circularity. These indicators were categorized within a framework according to the different priority areas at the urban level. A structure was then applied through which information was gathered about these indicators, the key sector they refer to, their unit of measurement, the country of publication of the article, and number of relevant articles found. The following were identified as key sectors: Built environment, organic material and biowaste, energy, mobility, electronics and ICT, packaging and plastic, textiles, and water. Three other sectors were also identified that cut across that list; namely, the environmental dimension, economic and financial dimension, and social and cultural dimension. This database was designed to be implemented in Office or Excel as an updatable and implementable tool. An extrapolation is given below (Figure 1) and, for the purpose of providing a practical example, in the form of a table in Appendix A (Tables A1 and A2).

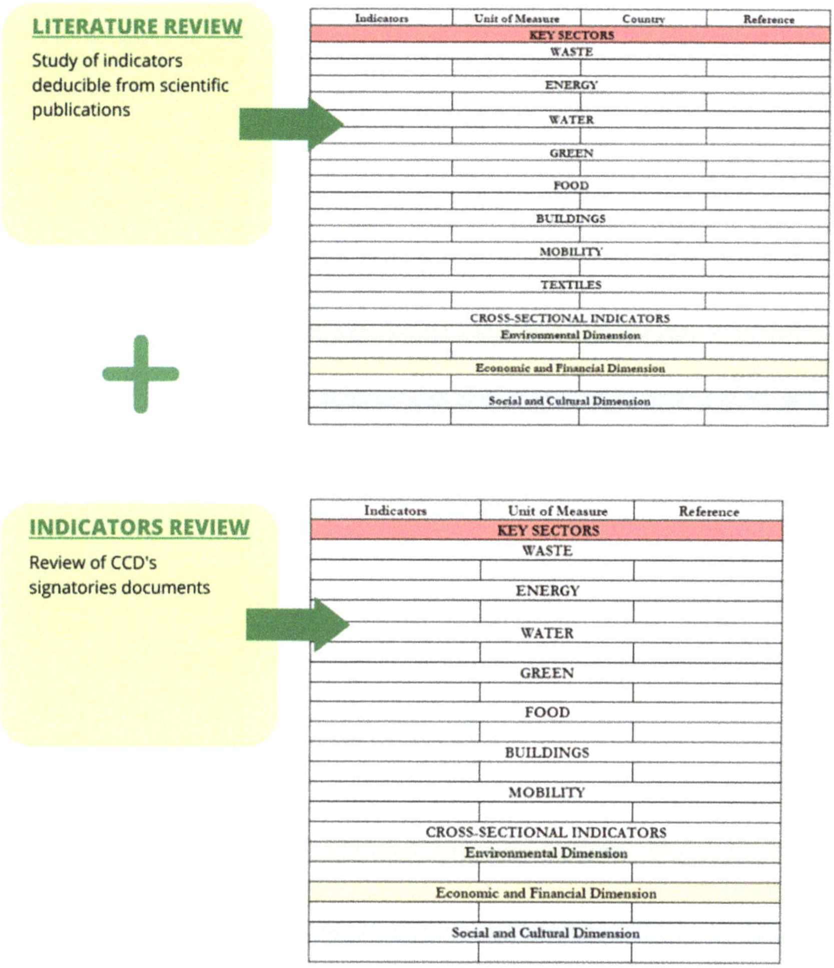

Figure 1. Proposed methodology.

Subsequently, the process continued by considering the indicators present in the international landscape, focusing on official documents issued by some of the signatory cities of the Circular City Declaration. This choice was made as it was deemed important to focus on the analysis of those cities that have already declared their commitment to circularity. The Declaration of European Circular Cities is a crucial tool in the changeover to the circular city, and was relaunched on 1 October 2020 at the European Sustainable Cities and Towns Conference in Mannheim. It was created to ensure that the transition from a linear to a circular economy can be accelerated, by having signatory cities commit to and, importantly, pioneer this new vision. This declaration stems from the idea that the methodical passage to a regenerative economic model is crucial if climate neutrality is to be achieved, a sustainable society promoted, and resource consumption contained within planetary limits. Among the difficulties encountered during the literature search was that indicators capable of capturing relevant aspects of the circular city have often been often used with extremely specific applications related to individual projects, and have not been applied to the urban context as a whole. Moreover, in some cases these indicators appear similar but are not identical, and different units of measurement make their comparison and identification more difficult, while for others the unit of measurement is obscure. When it came to the analysis of official documents, the difficulty encountered was different. In cases where such documents exist, they do not always provide a clear view of the studies behind them, often they contain no reference to the indicators used, if indeed they were used. In the end, the two lists that were produced contained some 295 indicators for analysis.

However, despite difficulties related to the non-homogeneity of the material present, through the analysis of the indicators we are able in this paper to propose a possible selection of indicators that takes into account the aspects mentioned above, and is directly applicable to different urban realities. As already mentioned, the need to focus analysis on the key sectors of the city makes it necessary to break down the indicators in the different sectors under consideration. In this perspective it is therefore also necessary to have a tool that can assess the multidimensional impacts created in the different sectors for all the actors involved [6]. The indicators that are proposed have therefore been selected based primarily on the number of sources found for these indicators, therefore including those that are already most widely used in the literature and official documents. Secondly, we avoided selecting indicators whose unit of measurement was unclear or even unknown. Finally, indicators were chosen with their applicability in the urban context in mind while maintaining their division into the key priority areas (waste, mobility, energy, etc.), for application at the neighborhood or urban area scale. Again, a table is presented within which are listed indicators, the key sectors they belong to, and their units of measurement. In addition, these indicators were chosen by reasoning about the impacts of the key sectors, cross referencing waste versus energy, water, mobility, etc., thus allowing database users to search for indicators by themes of interest. This database was also designed to be accessible in Office or Excel, for easily updating and implementation.

3. Results

Based on the creation of existing indicator databases and their in-depth analysis, despite the various difficulties encountered and mentioned above, 33 indicators were selected that can be applied to urban realities with different conditions. Criteria for the selection of these indicators first included the number of sources found that propose the use of the specific indicator; those that were already most widely used in the literature and official documents were adopted. Second, indicators were chosen whose unit of measurement was clear and easily measurable (and thus monitorable). Finally, it was essential to consider their applicability in the urban context by their breakdown into key areas (this necessity emerged because of specific analysis).

3.1. Databases Existing Indicators

In terms of the presentation of the research results, as illustrated at approach level, the studies of existing indicators were divided into two different databases, the first created from scientific publications, and the second from official documents produced by various cities. All the implementations made are reported in the Appendix A, dedicated to this purpose to allow greater readability of the text. Seven different tables are provided; the first table is related to the circular cities indicator database deduced from the scientific publications mentioned above.

Additional indicators cross cutting those previously mentioned were then identified as encapsulating the whole selection, i.e., environmental, economic and financial, and social and cultural dimensions. Table A2 of Appendix A shows the indicators found in the literature, separated for these three cross-cutting indicators.

To identify the priority indicators in order to focus on measuring circularity at the urban level, the criteria identified by the proposed approach (see Section 3) were applied. To do this, the indicators shown in Tables A1 and A2 in the Appendix A were analyzed and a series of statistical analyses were conducted, described below.

First, analysis was conducted to assess the key sectors concerned (Figure 2). The most frequently used were related to energy (23.9%), waste (21.1%), and water (12.7%). Continuing to investigate the cross-cutting sectors, in first place with as much as 50% were those concerning the social and cultural dimension (Figure 3). This may be significant in that it underscores how research is not currently focused on technical or economic aspects compared with the social side, thus paying special attention to the wellbeing of the population and the quality of services offered by the city.

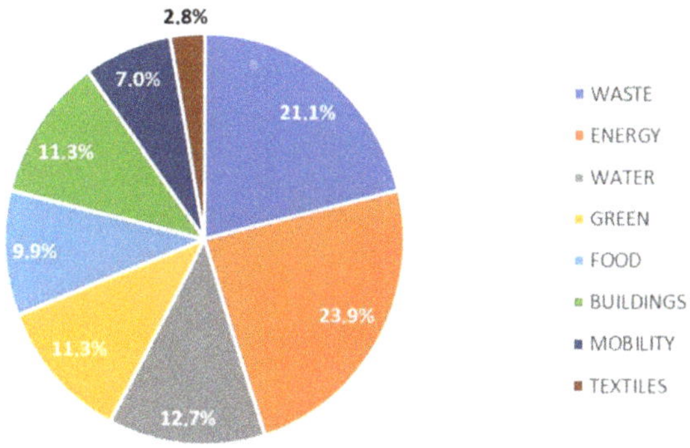

Figure 2. Analysis of indicators and key sectors.

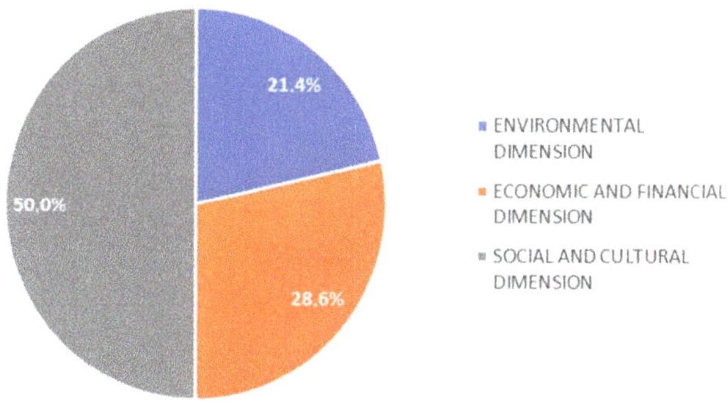

Figure 3. Analysis of indicators and cross-sectional sectors.

For the creation of the second database, indicators found in official documents published by 10 signatory cities to the Circular City Declaration were studied in depth. Currently, 60 cities have already signed the Declaration of European Circular Cities, thus demonstrating their leadership in picking an asset-efficient, low-carbon, and socially capable method of improvement, but for only 10 of these could relevant official documents be found online, shown in Table 1.

The main difficulty encountered during the analysis of official documents, after first finding them, was that those that do exist do not always provide a clear view of their objectives and the indicators they intend to use to achieve them. Tables A3 and A4 of the Appendix A provide the data showing the indicators that were found, separated into key sectors (see Appendix A, Table A3) and cross-cutting sectors (see Appendix A, Table A4).

For the indicators presented in Tables A2 and A3, analysis was again conducted by separating them according to key sectors (Figure 4) and cross-cutting key sectors (Figure 5). As can be seen in the first pie chart (Figure 4), the breakdown obtained was found to be comparable to that shown in Figure 2, including the percentages of each sector. In general, among the indicators deduced from scientific articles and those from official documents, in the first five positions were the categories of energy, waste, mobility, water, and buildings.

On the other hand, regarding the comparison of cross-sectoral sectors, shown in the graphs in Figures 3 and 5, the greater attention given to the indicators deduced from scientific articles for the economic-financial sector could be a further indication that, as already mentioned, these indicators were very often found in publications specifically related to individual projects in the business field, and not generally those related to the urban context as a whole.

Table 1. Circular City Declaration: city and official documents on circularity analyzed.

City (Country)	Document Name
Glasgow (United Kingdom)	Circular Glasgow [22]
Ljubljana (Slovenia)	Roadmap towards the circular economy in Slovenia [23]
Maribor (Slovenia)	Roadmap towards the circular economy in Slovenia [23]
Oulu (Finland)	Making City [24]
Paris (France)	Circular Paris [25]
Prague (Czech Republic)	Circular Prague [26]
Rotterdam (Netherlands)	Circular Rotterdam [27]
Tampere (Finland)	Carbon Neutral Tampere 2030 Roadmap [28]
Tirana (Albania)	Green City Action Plan of Tirana [29]
Umeå (Sweden)	The Circular Economy in Umeå, Sweden [30]

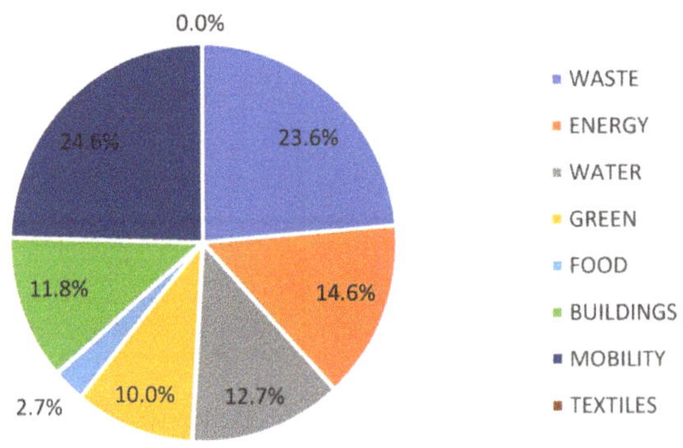

Figure 4. Breakdown by key sectors of indicators deduced from official documents.

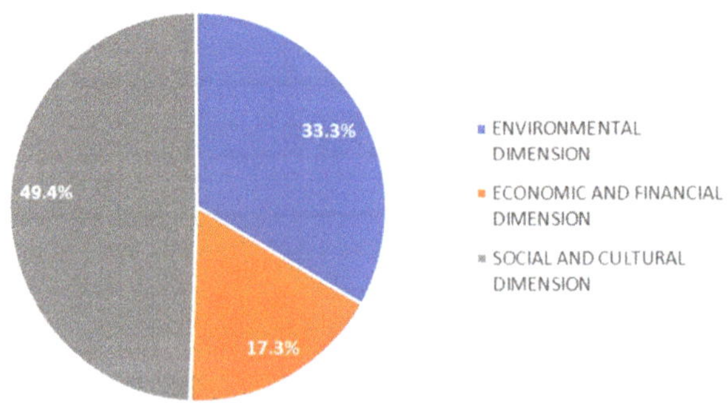

Figure 5. Breakdown by cross-cutting sectors of indicators deduced from official documents.

For these indicators, moreover, it was decided to assess whether any relationship existed between their number and the sizes of the cities from which they were taken, in terms of number of inhabitants (Figure 6). From this analysis it can be seen, for example, that the number of indicators was the lowest for the largest city examined, i.e., Paris, and reached its maximum for the medium-sized city of Tirana. This supports the idea of proposing a minimum set of indicators applicable to all cities regardless of their characteristics, which can thus allow comparison of their progress and planned actions.

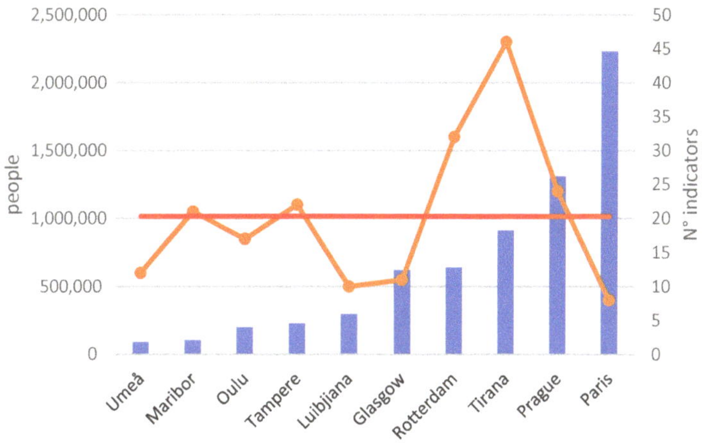

Figure 6. City size and number of indicators deduced from official documents.

3.2. *Proposed Indicators*

Following the obvious need for indicators quantifying the circularity of a city, it was decided to propose a selection of indicators (see Appendix A, Tables A5 and A6). Aiming to enable in-depth analysis on a neighborhood/urban area scale, the database of indicators that was designed was divided into the key priority areas identified in the urban sphere (waste, mobility, energy, etc.), in order to assess their impacts on each priority issue, and to arrive in conclusion at a unified vision of the different issues in the urban sphere. As mentioned above, it is indeed particularly important to consider the interaction of the various sectors with each other, so that we can respond to an organic issue with tools that are designed for organic use. It was also decided to select some of the transversal indicators (environmental, economic and financial, and social and cultural dimensions) while maintaining their characteristics of transversality, with a view to being applied at lower thematic or project levels. This route was taken because, as mentioned earlier, the circular economy also incorporates effects that refer to the social and environmental dimension, in addition to the economic dimension. Therefore, research in the urban context cannot be exempt from monitoring these sectors since they represent additional concerns and influence various aspects of the decisional procedures.

Through the analysis of the indicators examined, it was possible to propose a selection of indicators useful for assessing the circularity of actions planned and subsequently implemented in a city. From the analyses reported in Section 3.1, specific key areas emerged to be considered for assessing the circularity of urban areas, i.e., waste, energy, water, built environment, and mobility. These indicators were identified also for consideration of their applicability to different urban realities.

In addition, to make it easier to use the database and thus to improve the analysis and collection of data for measuring indicators, it was decided to group the data according to these five areas. The data can thus also be analyzed cross-sectionally, to evaluate the impact of, for example, waste compared with energy or mobility. In addition, indicators were considered for the so-called cross sectors (environmental, economic–financial and social–environmental) and a sector was added representing the crossover of those mentioned above, for further analysis of results from the territory under consideration. This choice was motivated by the fact that, in this researcher's opinion, to promote sustainable innovation it is necessary to ensure that each city incentivizes and fosters the development of patents for advanced and sustainable technical innovations. In this sense, therefore, these areas represent very important sectors for evaluation when analyzing the circularity of an urban area.

All indicators were selected keeping in mind the definition of circular city described above. Based on the discussion in the literature and in case studies, these 33 indicators (20 for the identified key sectors, 13 for cross-cutting key sectors, and a single indicator for certifications) thus attempt to describe some of the different aspects by which circularity in a city is defined and their interactions with one another. Moreover, these indicators are designed to be quantified initially at the smallest scale (neighborhood) and then later to be re-proposed at larger scales (urban, municipal).

The motivation for this work and the creation of the dataset was to identify indicators as generic as possible, so that through the systematic use of this tool the circularity of any urban reality can be determined. Moreover, the design provides adequate differentiation between the areas under consideration that involve various problems and particular needs. The differentiated and simultaneously general character of these indicators have makes it possible for them to be compared among different case studies. Precisely with this in mind, this tool has been designed to be applied on a national or, even better, a European scale, where it can create a common basis that allows the evaluation of how an urban reality is positioned in the international panorama in relation to the other realities present. As far as specific cases are concerned, we refer instead to the possibility of using, in parallel with those proposed, indicators that are instead more pertinent to the context under consideration and therefore designed specifically for it. Moreover, it would be useful in general to be able to identify a circularity index from these indicators, to qualitatively assess the trends of urban transition in different cities examined. A particular problem that has not been addressed in the course of this work, but which must be kept in mind during implementation, is that of the necessary data collection. Data should be collected from the different sectors, requiring contact with the various bodies involved in their collection, and finally arriving at the creation of a common database for public access. This vision is, unfortunately, still far from being achieved, since the bodies responsible for these measurements at the sub-national level differ in their natures, leading to confusion caused by different definitions of indicators, lack of shared protocols in the construction of "elementary indicators", and consequent loss of relationship with the local context [31].

Precisely for this reason, indicators at the local level must remain as close as possible to those designed at the national and European levels, and should only later be specified at the detailed urban scale.

4. Discussion

This report shows that the changeover to the circular city is ongoing in its development, and that several key issues are still being researched, for example, the question of indicators. In this regard, as mentioned earlier, the assessment of circularity in urban areas is a multidimensional task that must necessarily consider all the dimensions and sectors involved. There is a need for an integrated assessment tool that can represent all the sectors involved, incorporate the relational aspects between the sectors, and capture their multidimensional impacts.

As seen in the previous chapters, effective tools to monitor and evaluate circularity in the urban context have been lacking. Within the scientific and international literature, examples can often be found relating to very narrow contexts (business applications, individual projects, etc.) but rarely has the topic been addressed by suggesting robust and integrated indicators.

In addition, aiming to comparatively assess propensities toward these issues, analysis was carried out into the geographical distribution of the examined papers, from which it emerged that the countries where the debate seems to be most heated are Italy (22.2%) and the United Kingdom (12.7%), as can be seen in Figure 7. In addition, European countries in general have shown strong sensitivity to the topic, providing 77.8% of the papers (Figure 8).

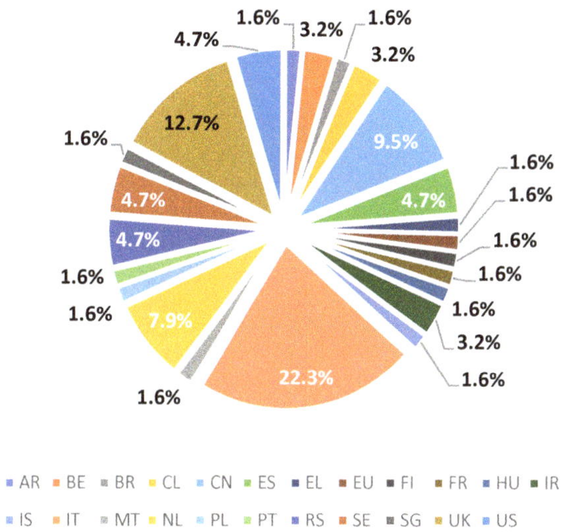

Figure 7. Distinct papers based on their country of origin.

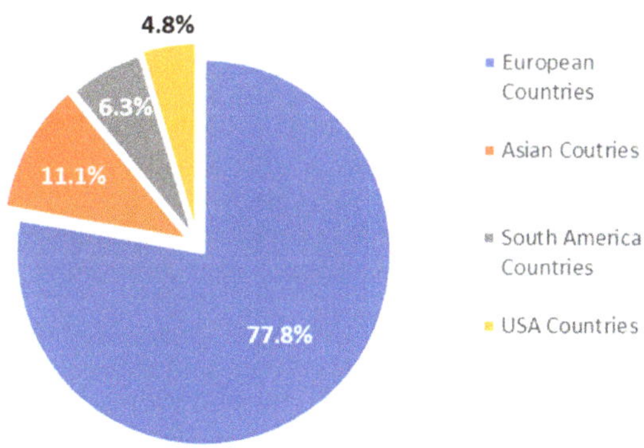

Figure 8. Distinct indicators based on their continent of origin.

Even when considering some of the more promising existing cases, the situation hardly changes. In this context we are often faced with the total nonexistence of documents that address the issue of urban circularity, or the published studies that do exist are too often lacking in detail and/or have been subject to censorship of data and topics that should be in the public domain, where they can foster the knowledge and interest of citizens and businesses.

The aim of this work, to identify indicators that are as generic as possible and therefore applicable to any urban reality, is thus confirmed as a possible solution to the problem highlighted, in that the general character of these indicators makes it possible for them to be compared between different case studies, with the intention that they be applied on a national or, even better, European scale. From these indicators it would be necessary to create a circularity index to allow the qualitative and comparative evaluation of strategies applied in different cities.

Aiming finally to assess an initial analysis of the applicability of the proposed tool, taking advantage of the integrated assessment document on the quality of the urban environment "Cities in transition: Italian capitals towards environmental sustainability" produced by the National System for Environmental Protection [32], we undertook an assessment of how many of the proposed indicators would be immediately usable and implementable through existing numerical data. The result obtained showed that by using the database provided, 45.5% of the proposed indicators would already be implementable.

5. Conclusions

In conclusion, this study aims to bring to attention the highlighted problem and to suggest, if not a solution, at least a contribution to consider and a basis on which to build active and participatory discussion. The creation of a minimum set of indicators representative of the main sectors within the urban sphere is, in the authors' opinion, useful for comparing cities with each other, regardless of their particular characteristics, and avoiding the risk of circular cities becoming self-referential through the creation of their own indicators designed ad hoc for actions implemented by the administration. In addition, another point to keep in mind when approaching these issues is that of participation. It is difficult to imagine how these strategies can be implemented, or even discussed, without placing the citizen at their center. However, too often we are confronted with a lack of data and difficulty in accessing relevant information. Development of user-friendly tools for all would greatly shorten the time needed to reach completion of the urban ecological transition, as it would enabledirect interaction of the three pillars on which the circular city is based, i.e., public, private, and social. Within this framework, monitoring tools should also necessarily be designed to be as inclusive as possible, both with regard to the various issues under consideration and to the dissemination of information. The tool proposed here, due to its features of implementability and updatability, could be a first step in the right direction. The dissemination and use of such tools could help by providing continuous and diverse feedback, thus creating insights for new implementable and beneficial solutions with multiple scales of application.

In general, the current institutional system is often an obstacle, and even if some cities are trying to move toward the circular model, it remains a concept surrounded by some ambiguity. It is necessary to develop a strategy that can put these guidelines into practice, creating a plan at the regulatory level that brings together and communicates the various circular themes, which must integrate and coordinate with each other. Indeed, regarding sustainability, one cannot act only on individual sectors, but must develop a plan that represents common action. This plan of action must set out a forward-looking agenda co-created with economic and research personnel, citizens, and public administrators. It must present a series of interconnected initiatives aimed at establishing a strategic and coherent framework in which sustainable products, services, and business models will be the new normal, and transform old consumption patterns to make sure that we succeed in avoiding generation of waste.

Consequently, for all the reasons mentioned above, the discussion and exploration of circular city realization, and specifically means for its implementation and subsequent monitoring, is a fertile area of activity. The growing number of circular cities and increased attention to the topic will fuel further research and enable implementation and improvement of the indicators presented in this study.

Author Contributions: Conceptualization, F.P. (Federica Paoli) and F.P. (Francesca Pirlone); methodology, F.P. (Federica Paoli), F.P. (Francesca Pirlone) and I.S.; validation, F.P. (Francesca Pirlone); formal analysis I.S.; investigation, F.P. (Federica Paoli); data curation, F.P. (Federica Paoli); writing original draft preparation, F.P. (Federica Paoli) and I.S.; writing review and editing, F.P. (Federica Paoli) and F.P. (Francesca Pirlone); supervision, F.P. (Francesca Pirlone). All authors have read and agreed to the published version of the manuscript.

Funding: This research received no external funding.

Institutional Review Board Statement: Not applicable.

Informed Consent Statement: Not applicable.

Data Availability Statement: Not applicable.

Conflicts of Interest: The authors declare no conflict of interest.

Appendix A

In this section, the six implemented tables analyzed in Sections 3.1 and 3.2 are shown.

Table A1. Circular cities indicator database deduced from scientific publications (for the years 2015–2022).

Indicator	Unit of Measure	Country	Reference
	Key Sectors		
	Waste		
Waste Quality Index	-	BR, AR	[33]
Waste generation	tons	BR, AR, CL, RS, IR, IT, UK	[33–37]
Recycling rate of municipal waste	%/year	US, CL, IR, SE, RS, NL, IT, FR, IS, CN, ES	[3,20,34,35,38–41]
Recycling rate of packaging waste	%/year	US, IT	[3,42]
Amount of landfilled waste Percentage of material solid waste landfilled Percentage of household waste landfilled	%/year or tons/year	US, SE, CL, RS, IR, NL, IT, FR, IS, CN, UK, EU, HU, EL	[3,20,34,38–40,43–49]
Percentage of material solid waste incinerated	%/year	UK, IT, SE	[43,47]
Percentage of material solid waste composted	%/year	CL, RS, IR, UK, EU, CN	[34,43,44,46]
Use of recycled goods in municipal administration	%/year	EU	[44]
Use of recycled goods in industrial production	%/year	EU	[44]
Percentage of material solid waste reused or recycled	%/year	EU, CN, IT, SE	[44,46,47]
Percentage of household waste reused or recycled	%/year	EU, SE, IR	[44,50]
Amount of recycled goods sold	N°/month (or year)	EU	[44]
Separated waste (recovery and treatment of waste generated in city)	kg/year	SE, US, IT, CN, NL	[20,36,46,48,51,52]
	Energy		
Energy saved due to the use of recycled goods in industrial production	%/year or kWh/year	CL, RS, IR, EU, EL, IT, NL	[34,38,44,49,52]
Energy consumption	kWh inhabitant^{-1} year^{-1}	BR, AR, CN	[33,53]
Non-renewable energy use	%/year or kWh/year	UK	[43]
Renewable energy use	%/year or kWh/year	NL, IT, ES	[38,41,48,52,54]
Input (energy, materials) in production processes using renewable sources	-	SE, IR, IT	[50,51]

Table A1. *Cont.*

Indicator	Unit of Measure	Country	Reference
Input in production processes involving reused materials	-	IT	[55]
Input in production processes using recycled materials	-	IT	[55]
Output from production processes using renewable sources	-	IT	[55]
Output from production processes involving reused materials	-	IT	[55]
Output from production processes using recycled materials	-	IT	[55]
Volume (amount) of resource flow	-	UK	[56]
Amount of recycled resources	-	UK	[56]
Amount of reused resources	-	UK	[56]
Amount of resources saved	-	BE	[57]
Amount of waste heat from industry used for heating the city and for horticulture	kWh/year	NL	[58]
Amount of groundwater warmed in the earth and used to heat homes and offices	m^3/year	NL	[58]
Number of homes receiving their energy (heat and electricity) from biogas (i.e., fermenting the manure of cows)	N°/total	NL	[58]
	Water		
Water use	Mm^3	UK, IT, SE	[37,47]
Dispersion from municipal water supply	-	IT, SE	[47,48]
Water consumption productivity	water consumption (m^3)/revenues (V)	ES	[41]
Water consumption for habitation (for example, reduction due to harvesting rainwater on the roofs)	%/year or l/year	UK, IT, NL, BE, CN, US, SG	[43,52,57,59,60]
Safe water accessibility (water issues regarding treatment and distribution)	-	SE, US, UK, SG	[20,60]
Water efficiency (water issues regarding treatment and distribution)	-	SE, US	[20]
Saving water due to the use of recycled goods in industrial production	%/year mc/year	NL, EU, UK, SG, US, CN	[38,44,60,61]
Amount of phosphate recovered from sewage water	kg/day	NL	[58]
Percentage of water consumption for habitation (for example, reduction due to harvesting rainwater on the roofs)	%/year	UK, BE	[43,57]
	Green		
Utilized agricultural area—SAU	km^2	IT, SE, UK,	[47,51,62]
Number of farms		IT, UK	[51,62]
Ecological and sustainable land-use regeneration	% (m^2 of regenerated land/m^2 of abandoned land)	IT	[63]

Table A1. *Cont.*

Indicator	Unit of Measure	Country	Reference
Mixed functionality	-	IT	[64]
Permeable surface area	m^2	IT	[64]
Green space area per capita	m^2/person	IT, SE, CN, PT	[47,48,51,59,65,66]
Density of the urban fabric (sqm of built environment on the total)	-	IT	[51]
Percentage of green roofs	%/total city surface	SE, US, PT	[20,66]
	Food		
- Amount of food waste treated - Food waste treated in small and medium-sized enterprises (SMEs)	%/total food waste	SE, US, IR, FI, PL	[20,50,67]
Recycling surplus food/food waste		MT, UK	[68]
Unsold products recovered every day for redistribution through the market itself or nearby community facilities	kg/day	EU	[44]
Percentage of local nutrient recovery	%	NL	[54]
Food waste	-	BE	[69]
Circular markets	-	IT, FR, IS	[39]
	Buildings		
Number of new buildings	N°	IT, SE	[47]
Buildings designed for complete disassembly	N°	NL	[54]
Reuse of building components at their end of life	%	NL	[54]
Design for flexibility by using modular systems	%	NL	[54]
Recycling rate of recyclable materials and constructions	%	NL	[54]
Percentage of retrofitting interventions on buildings	%/total building	SE, US, IT	[20,63]
Percentage of degraded buildings	%/total building	SE, US, IT	[20,51,64]
Percentage of reuse or recycling of recyclable demolition materials	%	IT, NL	[52,58]
	Mobility		
Public transport usage	% of inhabitants using public transport	SE, US	[20]
Electrical energy consumed in the transport sector	% of transport sector using electrical energy	SE, US, IT	[20,47]
Integration of new transport systems	-	IT	[70]
Proximity to public transport	-	IT	[64]
Pedestrian connections	-	IT	[64]
	Textiles		
Low-impact and non-toxic materials used in production processes	%	NL	[54]
Sustainable materials sourced from certified or eco-verified sources	%	NL	[54]

Table A2. Database of cross-sectional indicators for circular cities deduced from scientific publications (for the years 2015–2022).

Indicators	Unit of Measure	Country	Reference
Environmental Dimension			
Carbon footprint	$MtCO_{2eq}$	UK	[37]
- CO_2 emissions - CO_2 consumption footprint - GHG emission per capita	-	BR, AR, NL, SE, IR, CN, IT, ES,	[33,38,50,53,71–73]
- Annual amount of greenhouse gas emissions, annual amount of CO_2 emissions - Percentage reduction of greenhouse gas emissions	%/year or tons/year	IT, UK, EU, NL	[42–44,74]
Air quality	mg/Nm^3	IT, NL	[48,52,65]
Economic and Financial Dimension			
Disposable income of households (improvement through reduced costs of products and services)	€/year	IT	[42]
Revenue from recycled goods sold	€/month €/year	EU	[44]
Potential value of material after recovery and re-use	€	UK	[56]
Circular economy innovation budget (in relation to platforms and businesses leading to innovation in areas of the circular economy)	%/year	SE, US	[20]
Investment costs	m^2/€	IT	[64]
Payback period (PBP)	year	IT	[64]
Green investment	-	NL	[38]
Attractiveness	-	IT	[64]
Synergies among industries	N°	SE, US	[20]
Social and Cultural Dimension			
Livability (e.g., improvement through reduction of time lost from congestion, reduction of air pollution, improved waste, wastewater treatment)	-	IT, UK, SG, US	[42,60]
Walkability (length of pedestrian path)	km	IT	[65]
Percentage of CE patents	[CE patents/total patents] × 100	ES	[41]
Percentage of CE investment	CE investment in tangible goods (V)/total investment in tangible goods	ES	[41]
Percentage of CE jobs	%	ES	[41]
- Job creation - Employment opportunities	N° of jobs	IT, EU, BE, UK	[42,44,57,64,75]
Number of events and dissemination activities about circular economy	N° of events/year	EU, IT	[44,64]
Participants in events about circular economy (including public bodies, companies, universities, research centers, professional associations, etc.)	N° of participants/year	EU, IT, UK	[44,75]
Cultural and Recreational Services	N°	IT	[64]
Socio-cultural Associations	N°	IT	[64]
Potential for cultural initiatives	-	IT	[64]
Integration of compact adaptive space design in urban strategies	Qualitative (yes/no)	IT, UK	[75]
Adoption of nature-based solutions	N° of practices in the city	IT, NL, UK	[52,75]
Attractiveness	-	IT	[64]

Table A3. Circular city indicator database deduced from official documents.

Indicators	Unit of Measure	Reference
	Key Sectors	
	Waste	
Amount or percentage of recycled material	Tons/year or %/year	Circular Rotterdam, Roadmap towards the circular economy in Slovenia (Maribor), Roadmap towards the circular economy in Slovenia (Ljubljana)
Amount or percentage of products reused	Tons/year or %/year	Circular Rotterdam, Roadmap towards the circular economy in Slovenia (Maribor)
Amount or percentage of products recovered	Tons/year, or %/year, or T/inhabitant/year, or %	Circular Prague, Roadmap towards the circular economy in Slovenia (Maribor), Roadmap towards the circular economy in Slovenia (Ljubljana), The Circular Economy in Umeå, Sweden
Waste diverted from landfill	Tons/inhabitant/year or %	The Circular Economy in Umeå, Sweden
Mixed waste composition	-	Carbon Neutral Tampere 2030 Roadmap
Percentage of incoming/outgoing flows	%/year	Circular Paris
Average amount of products going to landfill or incineration	Tons/year	Circular Prague
Percentage of MSW landfilled disposed of in EU-compliant sanitary landfills	%	Green City Action Plan of Tirana
Percentage of collected MSW composted	%	Green City Action Plan of Tirana
Waste reduction in production of goods–raw material efficiency	kg of waste per €1000 output	Circular Prague
Amount or percentage of waste separation	%/year or tons/year	Circular Rotterdam, Circular Prague
Increase in clean plastics and drink packaging streams from residual waste	%/year	Circular Rotterdam
- Percentage of recycling of solid waste generated in the city - Percentage of packaging waste recycled - Percentage of municipal waste recycled	%/year or %	Circular Rotterdam, Roadmap towards the circular economy in Slovenia (Maribor), Roadmap towards the circular economy in Slovenia (Ljubljana), Green City Action Plan of Tirana, Carbon Neutral Tampere 2030 Roadmap, Making City (Oulu)
Tonnage of waste diverted via repair, reuse, recovery, and upcycling activities (recycling centers, artisans, second-hand goods stores, fab labs, etc.)	tons/year	Circular Paris
Traceability of hazardous waste	-	Roadmap towards the circular economy in Slovenia (Maribor)
- Amount of waste produced in the city - Amount of waste generated per capita	Tons/year, or tons/per capita/year, or kg/year/capita	Circular Rotterdam, Green City Action Plan of Tirana
Amount of waste produced in the city and treated within the city itself	tons/year or %/year	Circular Prague
Amount of solid waste reused	Tons/year or %/year	Circular Prague; Roadmap towards the circular economy in Slovenia (Maribor), Roadmap towards the circular economy in Slovenia (Ljubljana), Circular Glasgow
- Amount or percentage of waste avoided - Amount of household waste reduced by preventing waste and encouraging reuse	Tons/year or %/year	Circular Prague; Maribor (Slovenia), Circular Glasgow, Circular Rotterdam, Circular Paris

Table A3. *Cont.*

Indicators	Unit of Measure	Reference
Amount of biowaste processed in biogas facilities	% or tons/year	Circular Prague
Share of the population with weekly municipal solid waste (MSW) collection	%	Green City Action Plan of Tirana
Difference between quantity of waste and quantity of products consumed	Tons of waste/tons of products consumed	Circular Rotterdam
	Energy	
Energy consumption of city properties	total consumption and consumption per m^2	Carbon Neutral Tampere 2030 Roadmap, Making City (Oulu)
Energy savings per year	%/year, or kg/inhabitant/year, or %	Circular Glasgow, Circular Paris, The Circular Economy in Umeå, Sweden, Making City (Oulu)
Energy requirement per capita	GJ/person/year	Circular Rotterdam
GDP per energy requirement	€/GJ	Circular Rotterdam
Supply of renewable energy	%	Circular Rotterdam, Carbon Neutral Tampere 2030 Roadmap
Embedded energy use	tons/capita	Circular Rotterdam
Percentage of renewable or recycled energy use	%/year	Green City Action Plan of Tirana
Percentage of renewable electricity and heat supply for all municipal operations	%	Carbon Neutral Tampere 2030 Roadmap
Electricity consumption per capita	MWh per Capita/year	Making City (Oulu)
Primary energy consumption per capita	MWh per Capita/year	Making City (Oulu)
Primary energy sources (share)	% or MWh/cap	Making City (Oulu)
Percentage of buildings heated mainly by natural gas	%	Circular Prague
Percentage of buildings heated mainly by energy from incineration	%	Circular Prague
Electricity consumption in industry, per unit of industrial GDP	kWh/2010 USD	Green City Action Plan of Tirana
Emissions from centralized energy production	t CO_{2e}	Carbon Neutral Tampere 2030 Roadmap
Emissions from oil heating	t CO_{2e}	Carbon Neutral Tampere 2030 Roadmap
	Water	
Water consumption per capita	l/day/capita	Green City Action Plan of Tirana
Water consumption per unit of city GDP	l/day/USD	Green City Action Plan of Tirana
Water savings	Ml/inhabitant/year or %	The Circular Economy in Umeå, Sweden
-Biochemical oxygen demand (BOD) in rivers and lakes -Ammonium (NH4) concentration in rivers and lakes	µg/l	Green City Action Plan of Tirana
Percentage of water samples in a year that comply with national potable water quality standards	%	Green City Action Plan of Tirana
Water Exploitation Index	%	Green City Action Plan of Tirana
Unit of water consumed in power plants, per unit of primary energy generated	l/MW/h	Green City Action Plan of Tirana

Table A3. *Cont.*

Indicators	Unit of Measure	Reference
Industrial water consumption as percent of total urban water consumption	%	Green City Action Plan of Tirana
Non-revenue water	%	Green City Action Plan of Tirana
Annual average of daily number of hours of continuous water supply per household	h/day	Green City Action Plan of Tirana
Percentage of residential and commercial wastewater that is treated according to applicable national standards	%	Green City Action Plan of Tirana
Percentage of buildings (non-industrial) equipped to reuse grey water	%	Green City Action Plan of Tirana
Percentage of wastewater from energy generation activities that is treated according to applicable national standards	%	Green City Action Plan of Tirana
	Green	
Proportion of green and recreational areas per capita	%	Roadmap towards the circular economy in Slovenia (Maribor)
Number of contaminated sites	CSs/1000 inch (or km^2)	Green City Action Plan of Tirana
- Concentration of mercury in soil - Concentration of cadmium in soil - Concentration of zinc in soil - Concentration of mineral oil in soil (using infrared spectroscopy)	mg/kg	Green City Action Plan of Tirana
Open green space area ratio per 100,000 inhabitants	Hectares or m^2 per resident	Green City Action Plan of Tirana, Carbon Neutral Tampere 2030 Roadmap
Share of green space areas within urban limits	%	Green City Action Plan of Tirana
- Abundance of bird species (all species) - Abundance of other species	Annual % of change	Green City Action Plan of Tirana
Ecosystem services provided by green spaces	-	Carbon Neutral Tampere 2030 Roadmap
	Food	
Percentage of sustainable food	%	Circular Rotterdam
Amount of food waste	%	Carbon Neutral Tampere 2030 Roadmap
Share of units offering vegetarian options	%	Carbon Neutral Tampere 2030 Roadmap
	Buildings	
Construction materials from secondary sources	%	Circular Prague
Tons of residual materials not utilized (construction sector)	Tons/total	Circular Prague
Percentage of reduction of emissions due to smart and clean building logistics (construction sector)	%	Circular Rotterdam
Amount of construction waste saved by implementing interventions related to circular economy	tons/year	Circular Rotterdam
- Electricity consumption in residential buildings - Electricity consumption in non-residential buildings	kWh/m^2 or kWh per resident	Green City Action Plan of Tirana, Carbon Neutral Tampere 2030 Roadmap

Table A3. *Cont.*

Indicators	Unit of Measure	Reference
- Heating and cooling consumption in buildings, fossil fuel use in residential buildings, fossil fuels - Heating and cooling consumption in residential buildings, fossil fuels - Heating and cooling consumption in non-residential buildings, fossil fuels	kWh/m^2	Green City Action Plan of Tirana
Share of energy class A in new residential buildings	%	Carbon Neutral Tampere 2030 Roadmap
Share of recovered materials in construction	%	Carbon Neutral Tampere 2030 Roadmap
Building connected to the DH network or renewable energy grid	%	Oulu
Low-emission new materials, verification with CO_2 calculations	%	Carbon Neutral Tampere 2030 Roadmap
	Mobility	
Access to public transport	%	Making City (Oulu)
Energy consumption in transport sector	kWh/year or MWh/cap	Roadmap towards the circular economy in Slovenia (Maribor), Making City (Oulu)
Average age of car fleet (total and by type)	Year	Green City Action Plan of Tirana
Percentage of diesel cars in total vehicle fleet	%	Green City Action Plan of Tirana
Fuel standards for light passenger and commercial vehicles	€	Green City Action Plan of Tirana
Share of total passenger car fleet run by electric, hybrid fuel cell, liquefied petroleum gas (LPG) and compressed natural gas (CNG) energy	%	Green City Action Plan of Tirana, Carbon Neutral Tampere 2030 Roadmap
Percentage of low-emission buses in bus fleet	%	Green City Action Plan of Tirana
Transport modal share of commuting (cars, motorcycles, taxi, bus, metro, tram, bicycle, pedestrian)	Private transport %	Green City Action Plan of Tirana
Transport modal share of total trips	%	Green City Action Plan of Tirana
Motorization rate	Number of vehicles per capita	Green City Action Plan of Tirana
Average number of vehicles (cars and motorbikes) per household	Number of vehicles per household	Green City Action Plan of Tirana
- Kilometers of road dedicated exclusively to public transit per 100,000 population - Kilometers of bicycle path per 100,000 population	Km	Green City Action Plan of Tirana
Share of population having access to public transport within 15 min by foot	%	Green City Action Plan of Tirana
Frequency of bus service	Average number of passengers at station per hour in bus network	Green City Action Plan of Tirana
- Average travel speed on primary thoroughfares during peak hour - Travel speed of bus service on major thoroughfares (daily average)	km/h	Green City Action Plan of Tirana
Share of households within 300 m or 700 m of the main public services	%	Carbon Neutral Tampere 2030 Roadmap

Table A3. *Cont.*

Indicators	Unit of Measure	Reference
- Modal share of public transport on an autumn weekday - Modal share of walking on an autumn weekday - Modal share of cycling on an autumn weekday - Modal share of travel by car on an autumn weekday	%	Carbon Neutral Tampere 2030 Roadmap
Amount of outsourced transport services using low emission fuel sources	line km	Carbon Neutral Tampere 2030 Roadmap
Car travel output	km/person	Carbon Neutral Tampere 2030 Roadmap
Modal split	%	Making City (Oulu)
Fuel mix in mobility	%	Making City (Oulu)
Public infrastructure promoting low-carbon mobility	km/100,000 people	Making City (Oulu)

Table A4. Cross-sectional indicators database of circular cities deduced from official documents.

Indicators	Unit of Measure	Reference
Environmental Dimension		
- Amount of CO_2 emissions - Amount of greenhouses gases emissions - Annual CO_2 equivalent emissions per capita	kg of CO_2/year or tons/year/capita	Circular Glasgow, Circular Prague, Green City Action Plan of Tirana, Making City (Oulu)
Annual CO_2 emissions per unit of GDP	Tons/m. USD of GDP	Green City Action Plan of Tirana
- CO2 (or CO_2 equivalent) emissions saved (also through industrial and urban symbiosis) - GHG emissions saved (for example, by an increase in circularity)	Tons/year, or T CO_2 equivalent/year, or %/year, or tons CO_2/capita, or %	Circular Glasgow, Circular Prague, The Circular Economy in Umeå, Sweden, Making City (Oulu)
Amount of NOx emissions	Tons/year	Circular Prague
- Amount of fine dust emissions - Annual average air quality particulate matter	Tons/year or PM2.5 $\mu g/m^3$	Circular Prague Circular Rotterdam
CO_2 intensity	tons/capita	Circular Rotterdam
Embedded CO_2 emissions	tons/capita	Circular Rotterdam
- Average annual concentration of PM2.5 - Average annual concentration of PM10 - Average daily concentration of SO2 - Average daily concentration of NOx	$\mu g/m^3$	Green City Action Plan of Tirana
Climate change adaptation	-	Circular Prague
- Primary resources used - Raw material avoided - Virgin resources used - Amount of primary resource use avoided	Tons/year, or %/year, or T/inhabitant/year, or %	Circular Rotterdam, Circular Prague, Circular Glasgow, The Circular Economy in Umeå, Sweden
Use of renewable resources	%/year	Circular Rotterdam
Primary raw material demand per capita	ton/capita	Circular Rotterdam
Raw material consumption	%/year	Circular Prague
Raw materials with high risk for impact on biodiversity	%	Circular Rotterdam
Percentage of dwellings damaged by the most intense flooding in the last 10 years	%	Green City Action Plan of Tirana

Table A4. *Cont.*

Indicators	Unit of Measure	Reference
Awareness and preparedness for natural disasters	-	Green City Action Plan of Tirana
Annual number of storm water/sewerage overflows per 100 km of network length	Number of events per year	Green City Action Plan of Tirana
Economic and Financial Dimension		
Gross added value	€/year	Circular Prague
Total investments	-	Making City (Oulu)
Return on investment	€	Circular Prague, Making City (Oulu)
Sustainability of investments from the municipality	-	Green City Action Plan of Tirana
Value creation from growth of circular economy models	€/year	Circular Paris
Volume of sales from growth of circular economy models	Amount/year or €/year	Circular Glasgow
Sales of locally produced goods	Amount/year or €/year	Circular Glasgow
Revenues through sales from growth of circular economy models	€/year	Circular Glasgow
Change in GDP through circular activities	%	Circular Rotterdam
Resources productivity	-	Roadmap towards the circular economy in Slovenia (Maribor)
Creating added value and economic growth	€/year	Roadmap towards the circular economy in Slovenia (Maribor), Roadmap towards the circular economy in Slovenia (Ljubljana)
Turnover of organizations working in the circular economy (including all sectors and types)	€/year	Circular Paris
Existence of funding programs and economic incentives for circular economy projects with specific objectives, prioritized sectors, and a monitoring framework for the outcomes	qualitative	The Circular Economy in Umeå, Sweden
Estimated economic damage from natural disasters (floods, droughts, earthquakes etc.) as a share of GDP	%	Green City Action Plan of Tirana
Social and Cultural Dimension		
- Number of new jobs - Share of circular jobs (full- or part-time jobs related to one of the seven basic principles of circular employment) - Percentage of new jobs related to the circular economy - Number of new jobs from recycling of packaging - Number of new jobs from industrial ecology - Number of new green jobs	N°/year or %/year	Circular Rotterdam, Circular Paris, Circular Glasgow, Roadmap towards the circular economy in Slovenia (Maribor), Circular Prague, Roadmap towards the circular economy in Slovenia (Ljubljana)
- New business opportunities - New businesses that have integrated circularity into their development process	N°/year or %/year	Circular Rotterdam, Circular Paris, Circular Glasgow, Roadmap towards the circular economy in Slovenia (Maribor), Circular Prague, Roadmap towards the circular economy in Slovenia (Ljubljana)

Table A4. *Cont.*

Indicators	Unit of Measure	Reference
Unemployment rate	%/year	Circular Rotterdam, Roadmap towards the circular economy in Slovenia (Maribor), Malmö (Sweden), Roadmap towards the circular economy in Slovenia (Ljubljana)
Change in circular jobs	%	Circular Rotterdam
Percentage of population that shows an increase in circular behavior	%	Circular Rotterdam
Social cohesion (objective participation)	-	Circular Rotterdam
Percentage of population that describes their own health as good or very good	%/year	Circular Rotterdam
Percentage of population dying from diseases of the respiratory system (diseases of the respiratory system can be an air quality indicator, but also of habits such as smoking)	%/year	Circular Rotterdam
Number of new circular initiatives	%/year	Circular Rotterdam
Percentage of residents participated in dialogue and/or design related to circular economy	N°/year	Malmö (Sweden), Making City (Oulu)
Development of cooperative economy	-	Roadmap towards the circular economy in Slovenia (Maribor)
Number of new forms of enterprises (SMEs, start-ups, incubators, etc.)	N°/year	Roadmap towards the circular economy in Slovenia (Maribor), Roadmap towards the circular economy in Slovenia (Ljubljana)
Level of citizens' satisfaction with the administration services	qualitative	Roadmap towards the circular economy in Slovenia (Maribor)
Transformation of neighborhoods and local community	-	Roadmap towards the circular economy in Slovenia (Maribor)
Competitiveness of the economy	-	Roadmap towards the circular economy in Slovenia (Maribor)
Professional and managerial transformation of the city administration	-	Roadmap towards the circular economy in Slovenia (Maribor)
Attractiveness in terms of tourist visits	N° of visitors/year	Roadmap towards the circular economy in Slovenia (Maribor), Roadmap towards the circular economy in Slovenia (Ljubljana)
Annual number of visitors (with active engagement) to the reuse hubs	N°/year	Circular Prague
Number of public administrations/departments involved in the design of the circular economy imitative.	N°	The Circular Economy in Umeå, Sweden
- Number of actions identified to achieve the objectives. - Number of circular economy projects to implement the actions.	N°	The Circular Economy in Umeå, Sweden
- Number of staff employed for the circular economy initiative's design within the city, region, or administration. - Number of stakeholders involved to co-create the circular economy imitative.	N°	The Circular Economy in Umeå, Sweden
- Number of projects financed by the city or regional government/total number of projects. - Number of projects financed by the private sectors/total number of projects.	N°	The Circular Economy in Umeå, Sweden

Table A4. *Cont.*

Indicators	Unit of Measure	Reference
- Existence of a circular economy strategy with specific goals and priorities, actions, sectors, and a monitoring framework. - Co-ordination mechanisms across levels of governments to set and implement a circular economy strategy or initiative are well established and functioning. - Existence of overall policy coherence between circular economy initiatives and related policy areas (e.g., climate change, sustainable development, and air quality). - Regular capacity-building programs for activities associated with designing, setting, implementing, and monitoring the circular economy strategy. - Existence of a circular public procurement framework (e.g., waste diversion from procurement activities, raw materials avoided, and percentage of recycled content). - Existence of an information system on the circular economy. Data are publicly available and citizens and business informed of the opportunities related to circular business models and behaviors. - Existence of a monitoring and evaluation framework that includes environmental, economic and social aspects.	qualitative	The Circular Economy in Umeå, Sweden
Residents' satisfaction with the attractiveness and functionality of the urban environment in the continuous resident survey	-	Carbon Neutral Tampere 2030 Roadmap
Share of recreational areas in the total detailed planning area of the inner city	%	Carbon Neutral Tampere 2030 Roadmap

Table A5. Database of proposed circularity indicators.

	Indicators and [Unit of Measure] for Each Key Sectors				
Thematics	**Waste**	**Energy**	**Water**	**Built Environment**	**Mobility**
Waste	- Waste generation per capita [tons/Ab *year] - Recycling percentage (recycling, repair, reuse, recovery, and upcycling activities) [%]	- Saving energy due to the use of recycled goods [kWh/year] - Renewable energy use [%/year or kWh/year]	- Water consumption per capita [l/year/capita]	- Percentage of reuse or recycling of recyclable demolition materials [%]	- Amount of waste produced in the city and treated within the city itself [tons/year or %/year]
Energy		- Total energy consumption [kWh inhabitant^{-1} year^{-1}] - Energy requirement per capita [GJ/person/year]		- Electricity consumption in residential buildings [kWh/m^2 or kWh per resident] - Electricity consumption in non-residential buildings [kWh/m^2 or kWh per resident] - Percentage of building heating mainly through renewable sources [%/total buildings]	- Share of total passenger car fleet run by electric, hybrid fuel cell, liquefied petroleum gas (LPG) and compressed natural gas (CNG) energy [%]
Water			- Annual average of daily number of hours of continuous water supply per household [h/day]	- Dispersion from municipal water supply [%]	
Built Environment				- Percentage of degraded buildings [%/total buildings] - Percentage of retrofitting interventions on buildings [%/total buildings]	- Public space density: Pedestrian areas, squares, and green spaces [% of municipal area/neighborhood] - Kilometers of road dedicated exclusively to public transit per 100,000 population [km]
Mobility					- Public transport usage [% of inhabitants using public transport]

Table A6. Database of proposed cross-cutting circularity indicators.

Environmental Dimension	Economic and Financial Dimension	Social and Cultural Dimension
Indicators	**Indicators**	**Indicators**
- Annual amount of greenhouse gas emissions [%/year or tons/year]	- Green investment	- Job creation [N°]
- Annual amount of CO_2 emissions [%/year or tons/year]	- Investment costs [m^2/€]	- Employment opportunities [N°]
- Percentage of reduction of greenhouse gas emissions [%/year or tons/year]	- Payback period (PBP) [year]	- Number of events and dissemination activities about circular economy [N° of events/year]
	- Return on investment [€]	- Participants in events about circular economy (including public bodies, companies, universities, research centers, professional associations, etc.) [N° of participants/year]
		- Nature-based solutions adoption [N°]
	Certifications	
	- Density of certifications produced in the territory [N°/Ab]	

References

1. Kalmykova, Y.; Sadagopan, M.; Rpsado, L. Circular economy—From review of theories and practices to development of implementation tools. *Res. Conserv. Recycl.* **2018**, *135*, 190–201. [CrossRef]
2. Bourguignon, D. *Closing the Loop: NEW Circular Economy Package*; European Parliamentary Research Sevice: Brussels, Belgium, 2016.
3. Ellen MacArthur Foundation. *Growth Within: A Circular Economy Vision for a Competitive Europe*; Ellen MacArthur Foundation: Cowes, UK, 2015.
4. Ellen MacArthur Foundation. *Towards a Circular Economy: Business Rationale for an Accelerated Transition*; Ellen MacArthur Foundation: Cowes, UK, 2015.
5. What Is a Circular City, A Circular City Definition. Circular City Declaration. 2020. Available online: https://circularcitiesdeclaration.eu/ (accessed on 20 March 2022).
6. Fusco Girard, L.; De Rosa, F.; Nocca, F. *Verso il Piano Strategico di una Città Storica: Viterbo*; BDC Bollettino del Centro Calza Bini: Napoli, Italy, 2014; pp. 11–33.
7. Lichfield, N. *Community Impact Evaluation: Principles and Practice*; Routledge: London, UK, 2005.
8. Coccossis, H.; Nijkamp, P. *Sustainable Tourism Development*; Avebury Press: Aldershot, Avebury, UK, 1995.
9. Fusco Girard, L.; Nijkamp, P. *Cultural Tourism and Sustainable Local Development*; Routledge: London, UK, 2016.
10. Porter, M.E.; Kramer, M.R. Creating Shared Value. *Harv. Bus. Rev.* **2011**, *89*, 62–77.
11. Ezzat, A.M. Sustainable Development of Seaport Cities through Circular Economy: A Comparative Study with Implications to Suez Canal Corridor Project. *Eur. J. Sustain. Dev.* **2016**, *5*, 509–522. [CrossRef]
12. Williams, J. Circular Cities: What Are the Benefits of Circular Development? *Sustainability* **2021**, *13*, 5725. [CrossRef]
13. Brescia, R.; Marshall, J.T. *How Cities Will Save the World*; Routledge: London, UK, 2018.
14. Ellen MacArthur Foundation. *Circular Economy in Cities Project Guide*; Ellen MacArthur Foundation: Cowes, UK, 2019.
15. Ministero dell'Ambiente e della Tutela del Territorio e del Mare; Ministero dello Sviluppo Economico; ENEA. *Economia Circolare ed uso Efficiente Delle Risorse—Indicatori per la Misurazione Dell'economia Circolare*; Ministero dell'Ambiente e della Tutela del Territorio e del Mare, Ministero dello Sviluppo Economico: Rome, Italy; ENEA: Poznań, Poland, 2018.
16. European Commission, Directorate-General for Communication 2020. *Circular Economy Action Plan: For a Cleaner and More Competitive Europe*; Publications Office of the European Union: Luxembourg, 2020.
17. European Academies' Science Advisory (EASAC). *Indicator for Circular Economy*; EASAC: Halle, Germany, 2016.
18. Saidani, M.; Yannou, B.; Leroy, Y.; Cluzel, F.; Kendall, A. A taxonomy of circular economy indicators. *J. Clean. Prod.* **2019**, *207*, 542–559. [CrossRef]
19. Douma, A.; de Winter, J.; Ramkumar, S.; Raspail, N.; Dufourmont, J. *Circular Jobs*; Glodshmeding Foundation Amsterdam: Amsterdam, The Netherlands, 2015.
20. Cavaleiro de Ferreira, A.; Fuso-Nerini, F. A Framework for Implementing and Tracking Circular Economy in Cities: The Case of Porto. *Sustainability* **2019**, *11*, 1813. [CrossRef]
21. Fusco Girard, L.; Nocca, F. Moving towards the circular Economy/city model: Which tools for operationalizing this model? *Sustainability* **2019**, *11*, 6253. [CrossRef]

22. Circle Economy. *Circular Glasgow a Vision and Action Plan for the City of Glasgow*; Circle Economy: Amsterdam, The Netherlands, 2016.
23. Circular Change. *Roadmap towards the Circular Economy in Slovenia*; Circular Change: Ljubljana, Slovenia, 2018.
24. Making City. *H2020 Programme*; Making City: Oulu, Finland, 2018.
25. Mairie de Paris. *White Paper on the Circular Economy of Greater Paris*; Circular Economy Club Paris: Paris, France, 2017.
26. Circle Economy. *Circular Prague*; Circle Economy: Amsterdam, The Netherland, 2019.
27. Gemeente Rotterdam. *Roadmap Circular Economy Rotterdam*; Rotterdam Circulair: Rotterdam, The Netherlands, 2016.
28. Tampere City Board. *Carbon Neutral Tampere 2030*; Tampere City Board: Tampere, Finland, 2020.
29. Bashkia Tiran. *Green City Action Plan of Tirana*; Bashkia Tirane: Tirana, Albania, 2018.
30. OECD Urban Studies. *The Circular Economy in Umeå, Sweden*; OECD Urban Studies: Paris, France, 2020.
31. Richiedei, A.; Pezzagno, M. Territorializing and Monitoring of Sustainable Development Goals in Italy: An Overview. *Sustainability* **2022**, *14*, 3056. [CrossRef]
32. Sistema Nazionale per la Protezione dell'Ambiente. *Città in Transizione: I Capoluoghi Italiani Verso la Sostenibilità Ambientale*; Sistema Nazionale per la Protezione dell'Ambiente: Rome, Italy, 2022.
33. Deus, R.M.; Mele, F.D.; Bezerra, B.S.; Battistelle, R.A.G. A municipal solid waste indicator for environmental impact: Assessment and identification of best management practices. *J. Clean. Prod.* **2020**, *242*, 118433. [CrossRef]
34. Behzad, M.; Hashemkhani Zolfani, S.; Pamucar, D.; Behzad, M. A comparative assessment of solid waste management performance in the Nordic countries based on BWM-EDAS. *J. Clean. Prod.* **2020**, *266*, 122008. [CrossRef]
35. Stankovic, J.J.; Jankovic-Milic, V.; Marjanovic, I.; Janjic, J. An integrated approach of PCA and PROMETHEE in spatial assessment of circular economy indicators. *Waste Manag.* **2021**, *128*, 154–166. [CrossRef]
36. Lombardi, G.V.; Gastaldi, M.; Rapposelli, A.; Romano, G. Assessing efficiency of urban waste services and the role of tariff in a circular economy perspective: An empirical application for Italian municipalities. *J. Clean. Prod.* **2021**, *323*, 129097. [CrossRef]
37. Salemdeeb, R.; Saint, R.; Pomponi, F.; Pratt, K.; Lenaghan, M. Beyond recycling: An LCA-based decision-support tool to accelerate Scotland's transition to a circular economy. *Res. Conserv. Recycl. Adv.* **2022**, *13*, 200069. [CrossRef]
38. Nurdiana, J.; Franco-Garcia, M.L.; Heldeweg, M.A. How Shall We Start? The Importance of General Indices for circular cities in Indonesia. *Sustainability* **2021**, *13*, 11168. [CrossRef]
39. Taruffi, A.; Sciullo, A.; Diemer, A.; Nedelciu, C.E. Integrating Circular Bioeconomy and Urban Dynamics to Define an Innovative Management of Bio-Waste: The Study Case of Turin. *Sustainability* **2021**, *13*, 6221. [CrossRef]
40. Meng, M.; Wen, Z. Approaches and Policies to Promote Zero-Waste City Approaches and Policies to Promote Zero-Waste City. *Sustainability* **2021**, *13*, 13537. [CrossRef]
41. Rincon-Moreno, J.; Ormazabal, M.; Alvarez, M.J.; Jaca, C. Advancing circular economy performance indicators and their application in Spanish companies. *J. Clean. Prod.* **2021**, *279*, 123605. [CrossRef]
42. Croci, E. Circular Cities: Setting the Agenda. In Proceedings of the International Conference Circular Cities: Promoting Sustainable Innovation in Urban Systems and Service Within the Energy-Food-Water-Climate Nexus, Milan, Italy, 12 March 2018.
43. Williams, J. Circular Cities: Challenges to Implementing Looping Actions. *Sustainability* **2019**, *11*, 423. [CrossRef]
44. Eurocities. *Full Circle. Cities and the Circular Economy*; Eurocities: Brussels, Belgium, 2017.
45. Hoang, N.H.; Fogarassy, C. Sustainability Evaluation of Municipal Solid Waste Management System for Hanoi (Vietnam)—Why to Choose the 'Waste-to-Energy' Concept. *Sustainability* **2020**, *12*, 1085. [CrossRef]
46. Wang, D.; He, J.; Ting Tang, Y. Assessing the transition of municipal solid waste management by combining material flow analysis and life cycle assessment. *Res. Conserv. Recycl.* **2022**, *177*, 105966. [CrossRef]
47. Maranghi, S.; Parisi, M.L.; Facchini, A.; Rubino, A.; Kordas, O.; Basosi, R. Integrating urban metabolism and life cycle assessment to analyse urban sustainability. *Ecol. Indic.* **2020**, *112*, 106074. [CrossRef]
48. Perchinunno, P.; Mongelli, L.; Rotondo, F. A multidimensional approach for the identification of environmental and health wellness territories. *Ecol. Indic.* **2020**, *110*, 105911. [CrossRef]
49. Banias, G.; Batsioula, M.; Achillas, C.; Patsios, S.I.; Kontogiannopoulos, K.N.; Bochtis, D.; Moussiopoulos, N. A Life Cycle Analysis Approach for the Evaluation of Municipal Solid Waste Management Practices: The Case Study of the Region of Central Macedonia, Greece. *Sustainability* **2020**, *12*, 8221. [CrossRef]
50. Heshmati, A.; Rashidghalam, M. Assessment of the urban circular economy in Sweden. *J. Clean. Prod.* **2021**, *310*, 127475. [CrossRef]
51. Fabbricatti, K.; Biancamano, P.F. Circular Economy and Resilience Thinking for Historic Urban Landscape Regeneration: The Case of Torre Annunziata, Naples. *Sustainability* **2019**, *11*, 3391. [CrossRef]
52. Gravagnuolo, A.; Girard, L.F.; Kourtit, K.; Nijkamp, P. Adaptive re-use of urban cultural resources: Contours of circular city planning. *City Culture Soc.* **2021**, *26*, 100416. [CrossRef]
53. Liu, C.; Sun, W.; Li, P. Characteristics of spatiotemporal variations in coupling coordination between integrated carbon emission and sequestration index: A case study of the Yangtze River Delta, China. *Ecol. Indic.* **2022**, *135*, 108520. [CrossRef]
54. Metabolic, Circular Cities: Hello Jobs, Goodbye Waste. 2018. Available online: https://envisioncharlotte.com/wp-content/uploads/2018/03/20180201_S_Charlotte_EcoNetworkSeries_EN_Eva.pdf (accessed on 23 March 2022).

55. Enel (Ente nazionale per l'energia elettrica), Cities of Tomorrow. Circular Cities. 2018. Available online: https://www.enelamericas.com/content/dam/enel-americas/about_us/econimia-circular/en/cities-of-tomorrow-en-2018.pdf (accessed on 20 March 2022).
56. Owen, A.; Liddell, J. Implementing a Circular Economy at City Scale—A Challenge for Data and Decision Making, not Technology. In Proceedings of the International Sustainable Ecological Engineering Design for Society (SEEDS) Conference, Leeds Beckett University, Leeds, UK, 14–15 September 2016.
57. Marin, J.; De Meulder, B. Interpreting Circularity. Circular City Representations Concealing Transition Drivers. *Sustainability* **2018**, *10*, 1310. [CrossRef]
58. Agenda Stad. *The Perspective of the Circular City*; Agenda Stad: Amsterdam, The Netherlands, 2019.
59. Yang, Z. Sustainability of Urban Development with Population Decline in Different Policy Scenarios: A Case Study of Northeast China. *Sustainability* **2019**, *11*, 6442. [CrossRef]
60. Arora, M.; Yeow, L.W.; Cheah, L.; Derrible, S. Assessing water circularity in cities: Methodological framework with a case study. *Res. Conserv. Recycl.* **2022**, *178*, 106042. [CrossRef]
61. Lu, C.; Pam, X.; Mao, J.; Pang, J.; Xue, B. Modeling of Waste Flow in Industrial Symbiosis System at City-Region Level: A Case Study of Jinchang, China. *Sustainability* **2021**, *13*, 466. [CrossRef]
62. Williams, J. The Circular Regeneration of a Seaport. *Sustainability* **2019**, *11*, 3424. [CrossRef]
63. Bosone, M.; Ciampa, F. Human-Centred Indicators (HCI) to Regenerate Vulnerable Cultural Heritage and Landscape towards a Circular City: From the Bronx (NY) to Ercolano (IT). *Sustainability* **2021**, *13*, 5505. [CrossRef]
64. Della Spina, L. Multidimensional Assessment for "Culture-Led" and "Community-Driven" Urban Regeneration as Driver for Trigger Economic Vitality in Urban Historic Centers. *Sustainability* **2019**, *11*, 7237. [CrossRef]
65. Cerreta, M.; Giovene di Girasole, E.; Poli, G.; Regalbuto, S. Operationalizing the Circular City Model for Naples' City-Port: A Hybrid Development Strategy. *Sustainability* **2020**, *12*, s2927. [CrossRef]
66. Rodrigues, M.; Antunes, C. Best Management Practices for the Transition to a Water-Sensitive City in the South of Portugal. *Sustainability* **2021**, *13*, 2983. [CrossRef]
67. Erälinna, L.; Szymoniuk, B. Managing a Circular Food System in Sustainable Urban Farming. Experimental Research at the Turku University Campus (Finland). *Sustainability* **2021**, *13*, 6231. [CrossRef]
68. Camilleri, M.A. Sustainable Production and Consumption of Food. Mise-en-Place Circular Economy Policies and Waste Management Practices in Tourism Cities. *Sustainability* **2021**, *13*, 9986. [CrossRef]
69. Zeller, V.; Lavigne, C.; D'Ans, P.; Towa, E.; Achten, W.M.J. Assessing the environmental performance for more local and more circular biowaste management options at city-region level. *Sci. Total Environ.* **2020**, *25*, 140690. [CrossRef]
70. Dell'Ovo, M.; Dell'Anna, F.; Simonelli, R.; Sdino, L. Enhancing the Cultural Heritage through Adaptive Reuse. A Multicriteria Approach to Evaluate the Castello Visconteo in Cusago (Italy). *Sustainability* **2021**, *13*, 4440. [CrossRef]
71. Zhu, H.; Pan, K.; Liu, Y.; Chang, Z.; Jiang, P.; Li, Y. Analyzing Temporal and Spatial Characteristics and Determinant Factors of Energy-Related CO_2 Determinant Factors of Energy-Related CO_2 High-Resolution Gridded Data. *Sustainability* **2019**, *11*, 4766. [CrossRef]
72. Longato, D.; Lucertini, G.; Dalla Fontana, M.; Musco, F. Including Urban Metabolism Principles in Decision-Making: A Methodology for Planning Waste and Resource Management. *Sustainability* **2019**, *11*, 2101. [CrossRef]
73. Medina-Tapia, M.; Robusté, F. Implementation of Connected and Autonomous Vehicles in Cities Could Have Neutral Effects on the Total Travel Time Costs: Modeling and Analysis for a Circular City. *Sustainability* **2019**, *11*, 482. [CrossRef]
74. Prendeville, S.; Bocken, N. Circular Cities: Mapping Six Cities in Transition. *Environ. Innov. Soc. Trans.* **2018**, *26*, 171–194. [CrossRef]
75. Foglia, A.; Bruni, C.; Cipolletta, G.; Eusebi, A.; Frison, N.; Katsou, E.; Akyol, C.; Fatone, F. Assessing socio-economic value of innovative materials recovery solutions validated in existing wastewater treatment plants. *J. Clean. Prod.* **2021**, *322*, 129048. [CrossRef]

sustainability

Article

When Do Supply Chains Strengthen Biological and Cultural Diversity? Methods and Indicators for the Socio-Biodiversity Bioeconomy

Maria Sylvia Macchione Saes [1,*], Beatriz Macchione Saes [2,*], Elis Regina Monte Feitosa [1], Peter Poschen [3], Adalberto Luis Val [4] and Jacques Marcovitch [1]

1 School of Economics, Management, Accounting and Actuarial Sciences, University of São Paulo (USP), São Paulo 05508-011, SP, Brazil
2 School of Politics, Economics and Business, Federal University of São Paulo (UNIFESP), Osasco 06120-042, SP, Brazil
3 Faculty Environment and Natural Resources, University of Freiburg, 79098 Freiburg, Germany
4 Laboratory of Ecophysiology and Molecular Evolution, National Institute for Research in the Amazon (INPA), Manaus 69067-375, AM, Brazil
* Correspondence: ssaes@usp.br (M.S.M.S.); beatriz.saes@unifesp.br (B.M.S.)

Abstract: The bioeconomy has gained traction among the broader discourses on sustainable development, ecological transition, and the circular economy. Governments in the Global North and international institutions maintain that the bioeconomy can gradually replace fossil-based raw materials and nonrenewable resources with biomass and biological renewables. The Global South has increasingly adopted the approach, but with important variations across mega-biodiverse regions. In these regions, the bioeconomy must encourage economic activities that preserve biodiversity and strengthen local communities, promoting their well-being and cultural diversity. This paper argues that conventional research methods and indicators are not fit for this purpose. We therefore propose an alternative method and indicators and present an initial validation of the approach with an application to the pirarucu (*Arapaima gigas*) value chain in the Brazilian Amazon. By applying a bottom-up approach to evaluation that considers the perspective of the individuals and communities involved, the proposed methodology captures relevant dimensions of the value chain—including trade-offs—while identifying bottlenecks and the role of institutions. It also allows for verification of the achievement of the objectives of the socio-biodiversity bioeconomy in this model. The application to the case study finds that the managed pirarucu fisheries are a viable value chain associated with improved fish stocks and lower than average forest loss. Socio-economic benefits include the generation of reasonable income and greater participation by women. Income remains a complement to other sources of livelihood, however, and attractiveness to local communities is an issue. Positive outcomes are owed largely to local knowledge, collective action, and the role played by meta-organizations, while negative ones such as overfishing have resulted from institutional failures. Conventional analysis would likely not have considered these factors and missed these policy lessons. This corroborates the view that alternative methods and indicators are needed for the socio-biodiversity bioeconomy. While the application to the case study suggests the method and the indicators are conceptually suitable, we identify a number of shortcomings regarding the identification of interventions, attribution, and monitoring of the sustainability of the model.

Keywords: methods; evaluating metrics; supply chains; socio-biodiversity; bioeconomy

Citation: Saes, M.S.M.; Saes, B.M.; Feitosa, E.R.M.; Poschen, P.; Val, A.L.; Marcovitch, J. When Do Supply Chains Strengthen Biological and Cultural Diversity? Methods and Indicators for the Socio-Biodiversity Bioeconomy. *Sustainability* **2023**, *15*, 8053. https://doi.org/10.3390/su15108053

Academic Editors: Nicola Raimo, Filippo Vitolla, Ornella Malandrino and Benedetta Esposito

Received: 9 March 2023
Revised: 5 May 2023
Accepted: 10 May 2023
Published: 15 May 2023

1. Introduction

The bioeconomy has been attracting growing attention as an alternative path to face the environmental challenges of the twenty-first century. Recent literature presents pathways for a sustainable and decarbonized economy founded on the use of biomass and other renewable natural resources [1–3]. The term "bioeconomy" is controversial, however, and used with several different meanings. Specific challenges in the Global South—especially in mega-biodiverse regions—have been fostering reflection on social, cultural, and local biodiversity issues [4–8]. This emerging bioeconomy focused on socio-biodiversity still lacks a systematic and consistent analytical framework.

One major challenge for adequately mapping and assessing socio-biodiversity bioeconomy value chains lies in the diverse, sometimes conflicting objectives and, thus, multiple criteria that emerge from value chain actors. Testimonials by local actors have shown that socio-biodiverse value chains cannot aim only at short-term efficiency and enhancing job creation, income generation, or equitable market access; it is paramount that these go hand-in-hand with ecosystem resilience and the conservation of biological and cultural diversity. For instance, value chains in the Amazon, such as palm oil, açaí, coffee, or cocoa, can either promote biological and cultural diversity or undermine it, especially by encouraging monoculture and compromising the autonomy of communities [9]. A qualitative assessment of socio-biodiversity bioeconomy chains is important to bring up inconsistencies and potential incompatibilities.

Traditional and mainstream value chain studies, which were developed based on economic theories such as industrial organizations, transaction costs, and resource-based theory [10], deal with the operational efficiency of the production system as a whole: the chain's function is to serve consumers in a way that offers, simultaneously, lower cost and higher quality products. Nonetheless, inefficiency reduction is often not related to income inequality reduction or improved environmental sustainability. On the contrary, income distribution and environmental preservation are generally seen as restrictions imposed, for instance, by legislative bodies. Only exceptionally, the development of a strategic subsystem is seen as an opportunity for capturing value, such as a brand that builds on social or environmental attributes [11–13]. Even then, a focus on competitiveness imposes a vision of maximizing output in the short term, disregarding long-run socio-environmental limits and risks.

In this paper, we propose a research method using suitable indicators for the analysis and diagnosis of biodiversity value chains, aiming to assess and evaluate paths and obstacles to achieving the objectives of the socio-biodiversity bioeconomy and to advise public policy. Bioeconomy objectives are discussed and defined in Section 2 by contrasting different bioeconomy perspectives and analytical methods that are found in the literature. The need for a distinct socio-biodiversity bioeconomy and its features are emphasized in Section 2.3. Section 3 presents the choice and development of the method and indicators for analyzing socio-biodiversity bioeconomy value chains, encompassing indicators that consider the perspectives of local actors and qualitative criteria to evaluate bioeconomy value chains. The merits of the methodology proposed are discussed in Section 4, based on its application to the pirarucu (*Arapaima gigas*) value chain in the Amazon Region (Brazil). The analysis reveals that sustainable management of pirarucu is the basis of a viable bioeconomy chain, despite relevant trade-offs, bottlenecks, and institutional challenges. Section 5 discusses how possible pathways towards the socio-biodiversity bioeconomy should respond to these ambiguous results, along with our main contributions and limitations, both methodological and concerning the case study.

2. The Need for Socio-Biodiversity Bioeconomy

2.1. Economic-Ecological Bioeconomy: A Biophysical View of the Economy

In the 1970s, Nicholas Georgescu-Roegen [14] advocated for a revolution in economic theory that would consider biophysical aspects of the economic process. In biophysical terms, an economy does not create energy or matter but rather transforms resources extracted from nature, dissipating energy and generating polluting waste. Initially, the term "bioeconomics" was employed to designate a new scientific paradigm to replace neoclassical economics, in which the economy was considered part of nature. Subsequently, "bioeconomy" was used to refer to political and technological recommendations related to Georgescu-Roegen's theoretical contributions and to his "minimum bioeconomic program" [2,3].

Those recommendations are based on the entropic vision of bioeconomics and, more recently, ecological economics. Like a living being, the economic process depends on the input of low entropy matter and the output of degraded matter and heat to keep itself organized. It is a metabolic view of the economy. Technology may improve the economy's environmental efficiency, but it does not eliminate the dependence on new natural resource inputs. At least until solar energy use becomes viable and widespread, the alternative is to meet humanity's needs with minimum natural resource depletion and energy consumption [15].

Analytical methods based on this bioeconomy perspective seek to generate indicators that represent socio-economic metabolism, accounting for material flows and stocks in economic systems (their physical quantity in tons), as well as the energy associated with economic transformations [16–18]. Technological change and development throughout history are characterized by transformations of countries' socio-metabolic profiles and, occasionally, by transitions to different socio-ecological regimes [19,20].

2.2. Mainstream Bioeconomy: Economic Use of Biological Resources

Mainstream bioeconomics originated in the wake of the biotechnological revolution in the 1990s. Advances in genetic engineering were supposed to revolutionize fields such as pharmaceutics, medicine, agronomy, and chemistry, generating wealth and jobs. The concern with environmental benefits was initially not at the core of policy discussions and strategies [1,2,21,22]. We define this strand as mainstream because it is currently the most widespread view in reports from international institutions and governments around the world. However, in contrast to the ecological economics perspective, this strand aligns more closely with the concepts of circular bioeconomy or circular economy, precisely because it also emphasizes the notion of a circular flow of resource use.

In the last decade, this view has become associated with the energy transition and decarbonization of economies and has been incorporated into mainstream policy [23–25]. In this framework, bioeconomy may be defined as the set of economic activities connected to the invention, production, and use of renewable biological resources [23], leading to the progressive substitution of fossil-based raw materials and nonrenewable resources and to circular production methods [1,24,26]. It may encompass a broad range of economic activities, including agriculture, forestry, fisheries, commerce, waste management, and several industries [25,27].

Analytical methods associated with this perspective aim to quantify and qualify the environmental sustainability of production chains as well as their ability to create wealth and jobs. Life cycle analysis (LCA) is one of the most commonly used methods to assess environmental performance [28,29]. LCA quantifies environmental impacts from resource extraction to the end of product life as well as the possible results from more systemic changes, such as the transition to a circular bioeconomy [30]. Regarding economic potential, several methods and models are used to measure the size of the bioeconomy of countries or regions. The most common ones are gross value added, input–output analysis, and computable general equilibrium models [31]. There are estimates for several countries, especially in the Global North [32,33]. In the US, for instance, the bioeconomy accounted for around 5% of the country's gross domestic product (GDP) in 2016 [31]. In Germany, it reached 7.6% of GDP in 2007, a share close to that in the Netherlands (6.6–7.2%) [34].

2.3. Socio-Biodiversity Bioeconomy: The Cultural and Natural Richness of "Poor" Regions

A third, more recent, perspective on bioeconomy is emerging in mega-biodiverse countries of the Global South [4–8]. In addition to its emphasis on biodiversity, an important distinctive trait of this socio-biodiversity bioeconomy is the recognition of local populations whose livelihoods depend on nature and biodiversity conservation. Indigenous populations in mega-diverse regions are often vulnerable to the expansion of economic activities such as mineral and agricultural exploitation. Here, bioeconomy is seen as a way to preserve the forest and protect biodiversity while empowering local communities and securing their well-being [7,8,35–37]. The knowledge and culture of indigenous, fishermen and fisherwomen, riverside, and peasant populations are also often described as part of this bioeconomy, with local populations and social movements holding local knowledge for scientific and technological advancements connected to biodiversity [38,39].

The Amazon region, with Its important biological diversity and relevance to climate regulation, stands out in the debate on a socio-biodiverse bioeconomy [7,8,37]. Among the principles for a bioeconomy in the region, the following loom: zero deforestation; biodiversity conservation; strengthening of ancient practices of the region; science and technology (S&T) for the sustainable use of socio-biodiversity; and reduction in social and territorial inequality [36–40].

Socio-biodiversity bioeconomy contrasts with the previous mainstream perspective for both its emphasis on people and biodiversity. A bioeconomy based on the use of renewable biological resources may contribute to the energy transition but does not necessarily ensure biodiversity conservation. For instance, a bioeconomy based on biofuels or forest monocultures is generally harmful to biodiversity [1,41,42] and therefore counterproductive in mega-biodiverse regions. In those regions, socio-biodiverse economies should be based on value chains that respect ecosystem resilience.

In contrast to both previous perspectives, the socio-biodiversity bioeconomy is based on concrete experiences and activities conducted by populations living in mega-biodiverse regions. There is a consolidated literature on such activities, particularly non-timber forest products (NTFPs). Several case studies analyze whether NTFPs bring about development or improvements for local communities without overloading forest resources or ecosystems, with mixed results [43–45]. These empirical results are often ignored by proponents of the socio-biodiversity economy, which has a more normative character when proposing a new economic model, despite not yet having delineated clear strategies to overcome the social and environmental challenges identified in the literature.

3. Materials and Methods

The first step towards methodological approaches consistent with a socio-biodiversity bioeconomy was to consider the social and cultural features of the various territorial contexts in the Amazon region. To focus on the socio-environmental and economic criteria compatible with bioeconomy targets, the method had to capture local values and indicators. Indicators used to analyze value chains in general are not based on the perspective of producers or the local population [11–13,46]. Through citizen science and the participation of local actors in the identification of appropriate criteria and indicators, an original analysis may capture crucial aspects connected to human well-being and environmental issues. These values and indicators were identified in interviews with actors in the chains of pirarucu, açaí, cocoa, and Brazil nuts in the Amazon, specifically in Amazonas State (Brazil). From March 2021 to August 2022, 22 semi-structured online interviews lasting 45–120 min each were conducted with members of producer associations, researchers and staff of research institutes, social organizations, and public sector agents interacting with producers and actors in other chain links such as middlemen, processors, buyers, a tannery, and a restaurant. Finally, in June 2022, field research was conducted in the Mamirauá Sustainable Development Reserve (Amazonas), where we took part in the assembly of the Federation of Pirarucu Management Fishermen and Fisherwomen of Mamirauá (FEMAMPAM, acronym in Portuguese) and applied face-to-face questionnaires to 31 pirarucu fishermen and fisherwomen. Each interview took between 30 and 120 min.

Based on the priorities identified in interviews and field research, we performed a critical reexamination of traditional value chain analysis and their focus on economics skewed towards competitiveness [9]. In order to deal with crucial dimensions such as value distribution among production chain segments and environmental impacts over time, we considered approaches to value chain upgrading [47–50] and polycentric governance [50,51]. Building on the intersection of these theoretical perspectives with research analyses, we present a proposal for mapping and evaluating value chains using metrics consonant with socio-biodiversity bioeconomy principles. This proposal is detailed next.

3.1. Value Chain Mapping

Figure 1 presents our proposed method for analyzing socio-biodiversity bioeconomy value chains, detailing the stages of the chain, the production flow (gray arrows), and the income flow (orange arrows). The orange arrows denote the orientation of the analysis. While in the traditional analysis the objective was to propose competitive improvements (gray arrows), here the focus is to raise the economic and social benefits for communities while preserving the ecosystem. In each box, we exemplify factors to be evaluated and suggest relevant questions for the chain analysis. The value chains are seen as embedded in broader institutional contexts, operating under formal and informal rules [51–53]. Interviews with local actors revealed the decisive role of institutions such as community associations and social organizations in value chains. The analysis of macro- (regulation), meso- (implementation), and micro- (coordination of production arrangements) levels aims to identify bottlenecks at each level and the determining factors for value distribution throughout the chain [54].

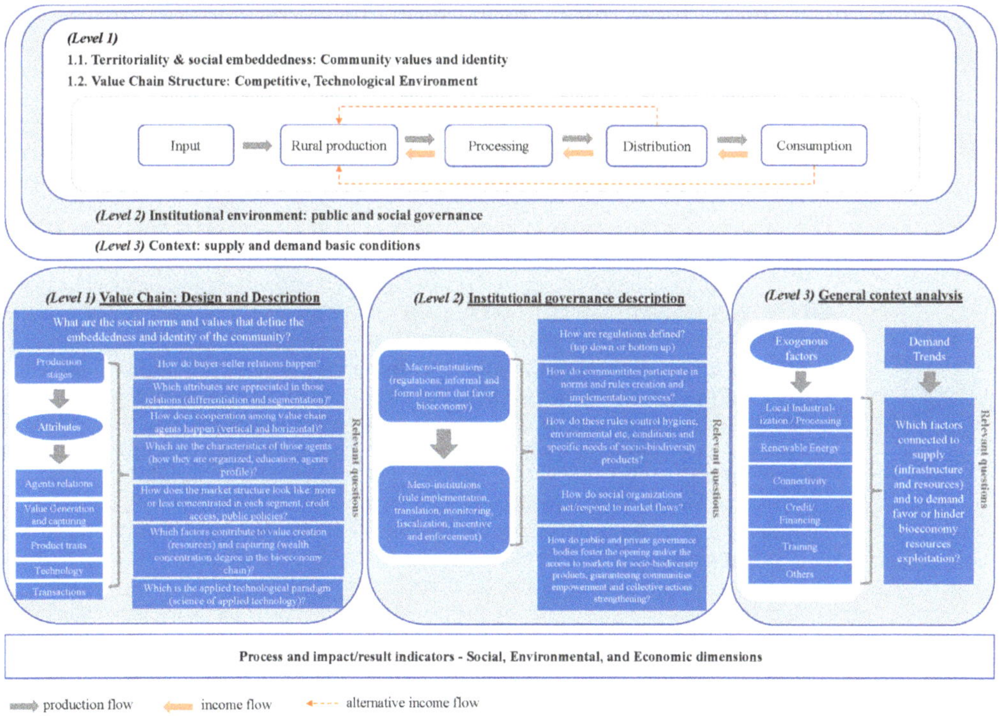

Figure 1. Proposed method for analyzing socio-biodiversity bioeconomy value chains. Source: elaborated by the authors based on [11–13,46,53,54].

The first level addresses (1.1) the characteristics linked to territoriality and identity of communities and (1.2) how the value chain is configured and chain actors relate to one another, regarding (a) social and cultural norms and locality-specific common property management practices, (b) competition (e.g., actors and organizations involved, size of each production segment, growth and competition strategies, product attributes, consumption frequency, and substitutes); (c) technology; and (d) market transactions (e.g., relationships among productive segments, their actors and objectives, leading actors, and degrees of dependence among actors).

The second level addresses institutional arrangements, i.e., the rules of the game at the macro- and meso-levels. Macro-level relates to formal institutions, including regulation and pertaining legislation, as well as informal and cultural rules that may enable or restrict immaterial infrastructure development (e.g., certification and labeling, branding, and declaration of origin). The meso-level regards how organizations implement rules, i.e., public, private, and collective governance.

The third level (general context) regards the general factors linked to supply and demand that favor or hinder the development of value chains. Exogenous trends are identified to analyze demand shifts related to globalization, dietary and lifestyle changes, or environment and climate change, as well as basic supply conditions (e.g., infrastructure, logistics, storage, credit/financing, connectivity and access to digital services, know-how, and natural resources).

3.2. Value Chain Evaluation

In addition to the value chain mapping, metrics were built to assess the coherence of the chain with the socio-biodiversity bioeconomy. The proposed evaluation is based on the concept of upgrading developed by Gary Gereffi and colleagues [47–50,55]. Upgrading refers to increasing the economic, social, and environmental value generated by a chain while benefiting all stakeholders. For each dimension of upgrading—economic, social, and environmental—metrics must be adapted and broadened to reflect the concerns and perspectives of local communities as well as territorial dynamics.

Economic upgrading reflects productivity gains, price improvements for producers, and more equitable gain sharing throughout the value chain. It leads to improvements in (i) products, when moving towards more sophisticated product lines; (ii) processes, by achieving a more efficient transformation of inputs into products through superior technology or better organization; (iii) product/service functionality, adding new uses to a product; and (iv) chain architecture, turning relations among agents more efficient. Social upgrading improves income and employment, empowers individuals and communities, and enhances their autonomy [50,55,56]. Finally, environmental upgrading reflects environmental performance and outlines changes in technology or social and organizational processes that prevent or minimize impacts and strengthen environmental services and biodiversity.

The choice, definition, and interpretation of indicators benefit from Elinor Ostrom's insights into the governance of the commons [51,52]. Indicators were built considering the evidence that polycentric governance reinforces the resilience of eco-systemic services by providing: (i) opportunities for learning and experimenting; (ii) ample stakeholder participation, mobilizing traditional and local knowledge; and (iii) diversity, minimizing and/or correcting errors in decision making. Table 1 presents the socio-biodiversity bioeconomy objectives and potential evaluation criteria to analyze bioeconomy value chains. Indicators reflect communities' priorities and perceptions and allow for community monitoring along with value chain development. It is important to note that the bioeconomy value chain can upgrade the entire locality (territorial scale) or only those who participate in the value chain. Data collection often requires fieldwork and interviews with actors in the value chain.

Table 1. Bioeconomy objectives and potential criteria for evaluation.

Bioeconomy Objectives	Scale of Analysis	Type and Source of Data	References
Social: Strengthening cultural diversity. Integration of S&T knowledge with local community knowledge, aiming at human well-being			
Potential evaluation criteria			
Education and training of human resources	C/T	SD/PD	[38,49–52,55–58]
Health quality	T	SD	
Social governance: community participation in decision making (top down/bottom up). Youth and women participation	C/VC	PD/N	
Collective action, and Respect for local culture and knowledge, Decision-making modal, Community attributes	C/VC	N	
Environmental: Production process of goods and services that safeguard biome resilience and biodiversity conservation			
Potential evaluation criteria			
Biodiversity conservation (Forest area and composition, Lake protection)	T	SD	[59–61]
Water/Soil condition	T	SD	
Waste recovery (Circular economy)	T	SD	
Economic: Improved income generation/well-being; transparent and equitable market access			
Potential evaluation criteria			
Income generation and distribution	C/VC	PD/N	[39,51,62,63]

Note: (T) Territorial; (VC) Value Chain; (C) Community; (SD) Secondary Data; (N) Narratives, (PD) Primary Data. Source: The authors.

4. Results

Participatory pirarucu management in the Mamirauá Sustainable Development Reserve—RDSM (in the Portuguese acronym), Amazonas (Brazil) (Sustainable Development Reserve (SDR)—is defined as a natural area that houses traditional communities, whose existence is based on sustainable systems for natural resource use. RDSM, which was created by the Amazonas Government on 16 July 1996), is an emblematic case for the Amazonian socio-biodiversity bioeconomy, as it seizes economic and nutritional potential while conserving the environment and enabling active community participation to generate income and improve well-being.

4.1. Value Chain Structure

Figure 2 presents the RDSM pirarucu value chain, including key actors and relations among agents. The configuration of the value chain was mapped based on documents from the Mamirauá Sustainable Development Institute (IDSM, Portuguese acronym) [64] and interviews with local actors (Field research, 2021).

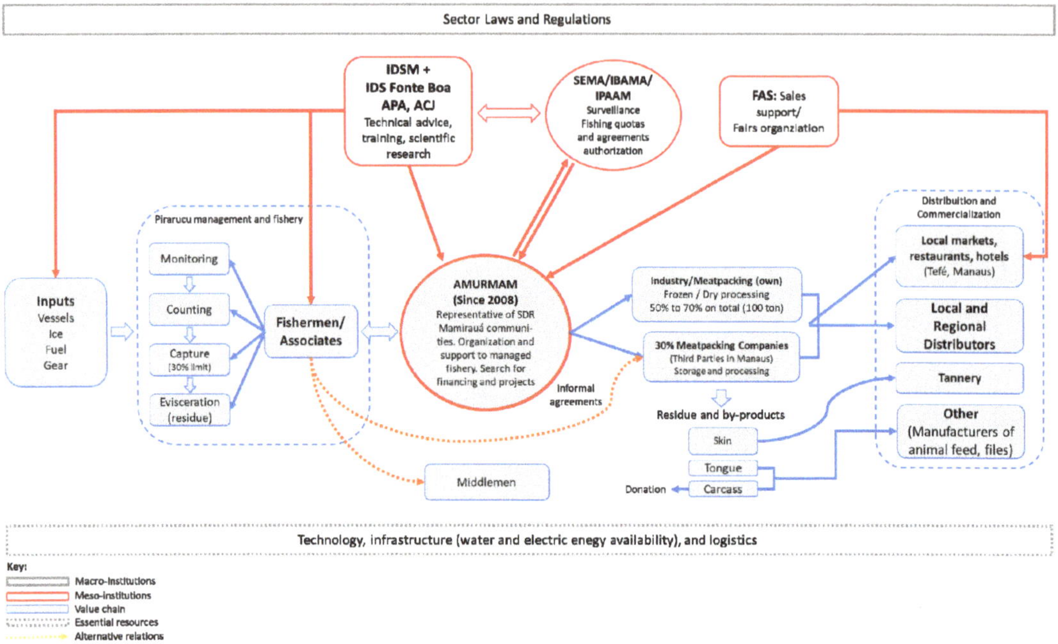

Figure 2. SDR Mamirauá Pirarucu Value Chain. Source: elaborated by the authors.

Pirarucu fishing involves a series of activities: monitoring lakes all year round to curb invaders who practice illegal fishing; planning the fishing schedule, i.e., counting the fish in the lake in order to decide the number of fish to be caught and how the income will be distributed that year; organization; and actual fishing (catching and cleaning the fish, preparing food for the team, etc.).

Fish processing is completed partly by the fishermen's association and partly by slaughterhouses that buy fresh fish. The fish is distributed through different channels, such as local fairs, restaurants, or hotels, as well as local and regional distributors and a tannery that purchases pirarucu skin for leather. Finally, consumption is mainly restricted to the local market. The chain is supported by a set of meso-institutions—public and non-governmental organizations—that help the fishermen and fisherwomen organize collective actions, provide training, and seek to enable their access to markets with better prices.

Two points have risen in the chain mapping: (i) the collective actions of fishermen and fisherwomen in organizing fishing supported by non-governmental organizations, which also play a prominent role in improving pirarucu trade; (ii) the absence of public rule enforcement, with the communities responsible for watching over the lakes to prevent illegal fishing. This represents about 50% of fishing operating costs and is important for communities' perceptions of fishing gains: 74.2% stated that their income was low, and their activity costs were high.

4.2. Institutional Governance

Three key regulations condition the institutional environment in which the pirarucu value chain is embedded: (a) the establishment of environmental conservation units in Amazonas State (Decree no. 12,836, on 9 March 1990), regulating human activity to ensure sustainable exploitation; (b) the imposition of a closure period (*defeso*), during which capture, commercialization, and transportation are prohibited (IBAMA Ordinance no. 480, on 4 March 1991, updated by NI IBAMA no. 34/2004), and during which fishermen and fisherwomen are eligible for a monthly income from unemployment insurance; and (c) criteria and procedures for pirarucu fishing in protected areas (NI IBAMA no. 01/2005). Beyond state regulation, an array of formal and informal rules have been created by local communities that participate in pirarucu management and are enforced through collective action. We highlight fishery agreements that regulate the use of fishing resources as defined by community members (such as quantities that can be fished, equipment allowed, number of vessels authorized to be simultaneously on the lake, and fishing period, among others), including measures and sanctions to be taken against violators.

Interviews showed that meso-institutions support the implementation of macro-institutional rules by creating incentives, enforcing them, or monitoring them. The organizations standing out are: (i) the IDSM, which translates general rules, protocols, and government policies, such as fishery agreements, into specific guidelines adapted to local contexts, aiming to make them more effective; (ii) the Sustainable Amazon Foundation (FAS, in the Portuguese acronym), which coordinates the activity of local actors, fills institutional voids created by the State, and helps to improve commercialization infrastructure; and (iii) the Association of Residents and Users of SDR Mamirauá Antonio Martins (Amurmam), representing local dwellers before governmental, environmental, landholding, and legal institutions. The association defends the rights of communities and organizes decision making in fishing management, also playing a key role in overseeing contractual relationships through formal and informal control mechanisms and sanctions in cases of non-compliance.

4.3. General Context

Pirarucu fish is part of the traditional diet in Northern Brazil but is also consumed in other regions and in international markets. In the 1960s and 1970s, the expansion of the fishing fleet and ice factories stimulated by government policies led to overfishing [64]. As reproduction did not keep up with capture, the pirarucu was classified as an endangered species in 1976. Since then, regulatory measures, such as closures of fisheries for six months every year (October to March), and fishing management in reserve areas, have been put in place aiming at sustainable use [65].

The RDSM was the first to implement sustainable management in Amazonas State. The policy had a clear effect: from 1999 to 2017, fish stocks grew by 427%, and the number of fishermen and fisherwomen who joined management projects jumped from 42 to 1590. In 2017, pirarucu fishing generated an average gross income of R$1739.38 (US$536.85) per fisherman per year, with individual amounts reaching up to R$6533.70 (US$2016.57), with each fisherman or fisherwoman working directly in fishing for a maximum of 50 days throughout the year. In comparison, the Brazilian minimum wage at the time was R$973.00 (US$289.20) per month [64] (p. 88).

Nevertheless, communities still face several bottlenecks: infrastructure (logistics, fish transport, processing, distance from ice factories), financing (to purchase boats and fishing gear), trade (dependence on a few channels and, given that the product is highly perishable, prices are lower than in larger regional markets), bureaucracy (documentation for sale), and lake surveillance. These are partially due to institutional voids left by the state, which contrast with the resolute action of meso-institutions and the collective action of communities.

4.4. Process, Impact, and Results Indicators

To assess coherence with the objectives of the socio-biodiversity bioeconomy, Table 2 presents indicators for the social, economic, and environmental dimensions, including several indicators based on the local communities' priorities and concerns. Despite the bottlenecks found in the context analysis, indicators show that the value chain has evolved over the years, with some upgrading in all three dimensions: social, economic, and environmental.

Table 2. Process and Impact/Result Indicators.

Bioeconomy Targets: Social Dimension Strengthening Cultural Diversity. Integration of S&T Knowledge with Local Community Knowledge, Aiming at Human Well-Being.					
Potential Evaluation Criteria		**Metrics/Indicators**	**Tiers**	**Outputs/Outcomes**	**Source**
Education and training of human resources	1	HDI education	T	Education: average of the Maraã, Fonte Boa and Urani municipalities: 2005—0.308 and 2016—0.498 (rate of growth: 61.69%)	Firjan [1]
	2	% of fishermen trained within the year	T	25.94% of fishermen in 2021 (between men and women) (Trained = 248 people; Total = 956 fishermen)	IDSM [66]
Health	3	HDI health	T	Health: average of the Maraã, Fonte Boa and Urani municipalities: 2005—0.398 and 2016—0.621 (rate of growth: 56.03%)	Firjan [1]
Social governance: communities' participation in decision-making process (top down/bottom up). Youth and women participation.	4	% of women participating in fishing activities	C	Average participation rate of women in fishing = 38.2% in 2021.	IDSM [66]
	5	Participation of women in assemblies	C	It was reported during the conversation circle that women began to have a large participation in assemblies and in the definition of income distribution rules.	(FR)
Collective action, and Respect for local culture and knowledge, Decision making rule, Community attributes	6	Rate of change in the number of communities participating in fishing	T	Number of communities involved in fishing: 1999 = 4 communities, 2017 = 42 communities. Rate of change: 950% (13.95% per year)	IDSM [66]
	7	Generations involved in fishing in the community	C	83.9% of respondents mentioned having started fishing because of family influence, grandparents and parents were fishermen.	FR
	8	Support from social organizations: qualitative, type of organization (local, international, university, church)	VC	Amurnam: local, role: coordination of fishermen. FAS: local coverage, chain coordination, and training. IDSM: Regional coverage, training, and development. Sebrae: National coverage, training, and entrepreneurship initiatives.	FR

Table 2. *Cont.*

Bioeconomy Targets: Social Dimension Strengthening Cultural Diversity. Integration of S&T Knowledge with Local Community Knowledge, Aiming at Human Well-Being.					
Potential Evaluation Criteria		**Metrics/Indicators**	**Tiers**	**Outputs/Outcomes**	**Source**
Bioeconomy targets: Environmental dimension Goods and services production process that safeguards biomes' resilience and biodiversity conservation					
Potential evaluation criteria		Metrics/Indicators	Scale	Outputs/Outcomes	Source
Biodiversity conservation (Forest area, Lake protection)	9	Vegetation cover	T	Vegetation coverage: average of the areas Maraã, Fonte Boa and Urani: 2000—1,165,197 ha; 2020—1,153,268 ha; growth rate: −1.02 (Amazon biome growth rate was—5.81 for the same period)	MapBiomas [67]
	10	Pirarucu population growth rate (average per community)	T	Increase in pirarucu population in lakes: 533%, 10.8% per year (1999 = 627 un; 2017 = 3970 un)	IDSM [66]
Water/Soil management	11	Rate of evolution of the water surface	T	Water surface: average of the areas Maraã, Fonte Boa and Urani (municipalities where the main source of economic activity is pirarucu): 2000—87,263 ha and 2020—92,166 ha; growth rate: +5.62%.	MapBiomas [67]
Bioeconomy targets: Economic dimension Improved income generation/well-being: transparent and equitable market access.					
Potential evaluation criteria		Metrics/Indicators	Scale	Outputs/Outcomes	Source
Income generation and distribution	12	Employment and income	T	Employment and income: average of the Maraã, Fonte Boa and Urani municipalities: 2005—0.272 and 2016—0.247	Firjan [1]
	13	Number of fishermen benefited per year	T	Number of communities involved in fishing: 1999 = 42 fishermen and 2017 = 1590 (growth rate: +3.685%~average 22.37% per year)	IDSM [66]
	14	Other sources of income	VC	Fishermen receive closed season insurance (*defeso*) or *Bolsa Floresta* or *Bolsa Família* (government programs).	FR
	15	Gross average income per fisherman	VC	Average gross earnings per fisher (deflated value, IPCA-1995=100): 1999—R$315.26; 2011—R$575.61 and 2017—R$466.93	IDSM [66]

Table 2. *Cont.*

Bioeconomy Targets: Social Dimension Strengthening Cultural Diversity. Integration of S&T Knowledge with Local Community Knowledge, Aiming at Human Well-Being.					
Potential Evaluation Criteria		**Metrics/Indicators**	**Tiers**	**Outputs/Outcomes**	**Source**
Market access and commercialization	16	Participation in fairs	C	58.1% of respondents sell at fairs. Participation in fairs promoted by FAS in Manaus was reported. FAS mobilizes fishermen to negotiate their production, promoting the practice of better prices.	FR
	17	Number of buyers	C	58.1% of respondents reported having only one buyer.	FR
	18	Market share of commercialization	VC	86.4% state regional market (Manaus, Manacapuru and Parintins) 9.6% interstate market (Santarém/PA, Itapoã and Oeste/RO) 4.0% local regional market (Tefé, Alvarães and Maraã).	IDSM [66]
	19	Participation in institutional programs		No cases were reported where commercialization was carried out for institutional/governmental programs.	FR
	20	How it is traded (whole/processed/salted/leather)	C	The community does not process the pirarucu, they just remove the viscera and sell it, which means that the fish is sold with less added value. Field research data: 87.1% of fishermen sell whole ("charuto") Data from the IDSM report: 97.7% were traded as gutted whole fish and only 2.3% as fresh manta.	FR IDSM [66]
	21	Distribution of income along the chain	VC	Communities 15%; Intermediaries 35%; slaughterhouses 50%.	[68]
Certification	22	Types of certification/collective trademark	VC	Denomination of Origin Mamirauá for managed pirarucu in nine municipalities (Alvarães, Fonte Boa, Japurá, Juruá, Jutaí, Maraã, Tefé, Tonantins and Uarini).	INPI [69]
Production costs	23	Estimated critical production cost	VC	Cost of monitoring the lakes represents about 50% of the total cost	FR

Table 2. *Cont.*

Bioeconomy Targets: Social Dimension Strengthening Cultural Diversity. Integration of S&T Knowledge with Local Community Knowledge, Aiming at Human Well-Being.					
Potential Evaluation Criteria		**Metrics/Indicators**	**Tiers**	**Outputs/Outcomes**	**Source**
Infrastructure	24	Drinking water	C	67.74% have access to piped water.	FR
	25	Electricity	C	90.3% have access to electricity. They use a diesel generator (light engine, in some cases available only from 6 pm to 10 pm)	FR
	26	Internet	C	22.58% of respondents have poor quality internet access.	FR
	27	Basic sanitation	C	Absence of basic sanitation in the visited community	FR
	28	Media	C	93.5% of respondents use cell phones	FR

Note: (T) Territorial; (VC) Value Chain; (C) Community; (SD) Secondary Data; (N) Narratives, (PD) Primary Data; Human Development Index (HDI); Field Research (FR); Associação dos Moradores e Usuários da RDSM Antonio Martins (Amurmam); Fundação Amazonas Sustentável (FAS); Instituto de Desenvolvimento Sustentável Mamirauá (IDSM); Serviço Brasileiro de Apoio às Micro e Pequenas Empresas (Sebrae). Source: The authors. 1 The Firjan index ranges from 0 (minimum) to 1 point (maximum) to classify the level of each location into four categories: low (from 0 to 0.4), regular (0.4 to 0.6), and moderate (from 0.6 to 0.8) and high (0.8 to 1) development. Source: https://www.firjan.com.br/ifdm/ (accessed on 1 August 2022).

By including women in the production process, participatory fisheries management has improved gender equality in the region. Collective actions reinforce the role of fishermen and fisherwomen in performing all the tasks related to fishing and commercialization, as well as in the decision-making process. Moreover, fisheries management has not only improved pirarucu supply in the region but also contributed to the conservation of lakes and other species, as evidenced by strong increases in fish stocks and very low deforestation.

Meso-institutions have enabled the internalization of innovation, such as training, and improved infrastructure for processing and commercialization, contributing to production costs that are more compatible with prices. This is reflected in the relative improvement in income and well-being of communities, albeit timid in several indicators, which suggest room for additional economic and social upgrading.

5. Discussion

Our results show that the assessment of the socio-biodiversity bioeconomy may benefit from indicators that consider the perspectives of individuals and communities and from a qualitative evaluation of bioeconomy value chains. Based on this assessment, the participatory pirarucu management in the RDSM has proved to be a successful case of bioeconomy development. This finding is in line with positive outcomes reported in previous studies on pirarucu management in the Amazon [70–73]. The measured impacts reflect the generally positive perception of communities about fisheries management, although several social and economic aspects have yet to be upgraded, particularly local income generation. In most cases, fishing provides just a small complement to household income and is often considered insufficient. Regarding the positive environmental and social outcomes, the analysis of the broader institutional context, such as the facilitation role of meso-institutions, was important in accessing key drivers of the pirarucu bioeconomy. To point out opportunities and challenges related to advancing the socio-biodiversity bioeconomy, we close this article by discussing some implications of our results.

The value chain mapping indicates that meso-institutions may organize and aggregate local actors in contexts with social and regional disarticulation, as found in many socio-biodiverse regions in the Global South. The literature has shown that polycentric institutions can have a positive role in the governance of common-pool resources when fostering innovation, learning, cooperation between participants, and the achievement of more equitable and sustainable results [51,52]. In participatory pirarucu management, public and non-governmental organizations changed the dynamics of socio-biodiversity chains by increasing transparency and fairness in production chains and by enhancing existing self-organizing initiatives, thereby creating conditions for the bioeconomy to flourish. These organizations met ample community participation, contributing to strengthening economic alternatives adapted to the communities environmental and cultural contexts. This finding is consistent with the evidence that local settings and the active participation of local communities are important to successful outcomes in the management of common-pool resources [52,74].

Still, the deficient economic outcomes in the pirarucu chain contrast with the bioeconomy's promise of win-win solutions and synergies between sustainability and economy, often emphasized in bioeconomy perspectives [2,3,6,40]. The vast literature on non-timber forest products, including several studies in the Amazon, also challenges this emphasis on synergies by showing that NTFP commercialization often implies a trade-off between environmental conservation and economic development [10,43,45,66,75–77].

At the same time, our criteria and indicators based on the perspective of communities help define priorities and the notion of development that are compatible with local livelihoods. For example, indicators related to the role of women in the economic activities of the communities (participation in fishing and in assemblies) in the case analyzed show attention to gender inequality. Similarly, the emphasis on information about lake surveillance indicates concern with security and illegal activity in the region. Positive outcomes related to the empowerment and autonomy of communities, economic stability, and security to carry out their activities may be more important than increases in income alone. Indeed, studies that consider socio-cultural aspects such as reproduction of culture, creation of social capital, and empowerment find more positive outcomes in NTFP trade than analyses focused only on material gains (mainly income) [45].

Strategies and policies to foster a socio-biodiversity bioeconomy should respond to ambiguous results. On the positive side, collective action is not a complication but a critical ingredient in fisheries management. Meso-institutions can play a vital role in filling governance and technical voids in isolated regions. They translate legislation into practices, provide access to appropriate technology, and can improve local value-added retention. On the other hand, the dramatic overfishing following heavily subsidized, over-dimensioned cooling facilities sends a clear warning: politicians should be wary of quick fixes and of 'throwing money at the problem'. Rather, they should adopt a more realistic view of social and environmental challenges and respond to local prospects and concerns regarding development. Policies such as the unemployment insurance paid during fishery closures, for example, are more in sync with the needs of the community and the dynamics of the value chain.

Finally, the proposed methods and indicators have some limitations. Our holistic and systemic approach was instrumental in revealing inter-relations and trade-offs, such as the need for and success of surveillance through collective action, while weighing cost/benefit ratios and community perceptions. The study has identified bottlenecks and potential areas for intervention to enhance the functioning of the value chain, but it has not yet been able to prioritize them and draw up an intervention plan. A more rigorous and action-oriented assessment requires further data, co-validation, and co-construction of solutions with the community. This would also require the construction of solid cause-effect cascades for attribution. Those would guide short- to medium-term action and monitoring, but a reliable diagnosis of sustainability may require long-term follow-up. That, in turn, would be greatly helped by methods and indicators amenable to 'citizen science' by members of the local communities, a question we have not yet been able to explore. Finally, the need to adapt indicators to capture the relevant specificities of each chain may compromise the comparability of bioeconomy cases.

The present assessment of the pirarucu bioeconomy provided a broad picture of the current strengths and challenges of the productive chain, but it could not produce sufficient information about the past trajectory of communities. To evaluate and monitor the evolution of this bioeconomy, the study needs to be replicated over time. Communities should evaluate these results and indicate possible gaps and new criteria to be included. Ideally, communities themselves would perform such continuous monitoring and evaluation. We also believe the method should be applied to more cases to confirm its suitability, especially its ability to generate reliable comparisons and evidence to guide public policies.

Author Contributions: Conceptualization, M.S.M.S., B.M.S., E.R.M.F., P.P., A.L.V. and J.M.; methodology, M.S.M.S. and E.R.M.F.; validation, M.S.M.S., B.M.S., E.R.M.F., P.P., A.L.V. and J.M.; formal analysis, M.S.M.S., B.M.S. and E.R.M.F.; investigation, M.S.M.S., B.M.S., E.R.M.F. and P.P.; resources, M.S.M.S., E.R.M.F., P.P., A.L.V. and J.M.; writing—original draft preparation, M.S.M.S., B.M.S., E.R.M.F. and P.P.; writing—review and editing, M.S.M.S., B.M.S., E.R.M.F., P.P., A.L.V. and J.M.; visualization, M.S.M.S., P.P., A.L.V. and J.M.; project administration, A.L.V. and J.M.; funding acquisition, M.S.M.S., P.P., A.L.V. and J.M. All authors have read and agreed to the published version of the manuscript.

Funding: This research was funded by Fundação de Amparo à Pesquisa do Estado de São Paulo (FAPESP) and Fundação de Amparo à Pesquisa do Estado do Amazonas (FAPEAM) (grant number 2020/08886-1).

Institutional Review Board Statement: Not applicable.

Informed Consent Statement: Informed consent was obtained from all subjects involved in the study.

Data Availability Statement: The data are not publicly available due to ethical and privacy restrictions.

Conflicts of Interest: The authors declare no conflict of interest.

References

1. Bugge, M.M.; Hansen, T.; Klitkou, A. What is the bioeconomy? A review of the literature. *Sustainability* **2016**, *8*, 691. [CrossRef]
2. Vivien, F.D.; Nieddu, M.; Befort, N.; Debref, R.; Giampietro, M. The hijacking of the bioeconomy. *Ecol. Econ.* **2019**, *159*, 189–197 [CrossRef]
3. Giampietro, M. On the circular bioeconomy and decoupling: Implications for sustainable growth. *Ecol. Econ.* **2019**, *162*, 143–156 [CrossRef]
4. Ortega-Pacheco, D.V.; Silva, A.; López, A.; Espinel, R.; Inclán, D.; Mendoza-Jiménez, M.J. Tropicalizing Sustainable Bioeconomy: Initial Lessons from Ecuador. In *Towards a Sustainable Bioeconomy: Principles, Challenges and Perspectives*; Leal Filho, W., Pociovălișteanu, D., Borges de Brito, P., Borges de Lima, I., Eds.; Springer: Cham, Switzerland, 2018. [CrossRef]
5. Rodríguez, A.G.; Rodrigues, M.; Sotomayor, O. Towards a sustainable bioeconomy in Latin America and the Caribbean: Elements for a regional vision. Natural Resources and Development Series; N°191, (LC/TS.2019/25 Santiago, Economic Commission for Latin America and the Caribbean (ECLAC). 2019. Available online: https://repositorio.cepal.org/bitstream/handle/11362/44994/1/S1901014_en.pdf (accessed on 1 April 2022).
6. Nobre, I.; Nobre, C. Projeto 'Amazônia 4.0': Definindo uma Terceira Via para a Amazônia. *Futuribles* **2019**, *2*, 7–20.
7. Abramovay, R.; Ferreira, J.; Costa, F.D.A.; Ehrlich, M.; Castro Euler, A.M.; Young, C.E.F.; Villanova, L. The new bioeconomy in the Amazon: Opportunities and challenges for a healthy standing forest and flowing rivers. *Sci. Panel Amaz.* **2021**. Available online https://www.theamazonwewant.org/wp-content/uploads/2022/05/Chapter-30-Bound-May-16.pdf (accessed on 1 August 2022)
8. Costa, F.; Nobre, C.; Genin, C.; Frasson, C.M.R.; Fernandes, D.A.; Silva, H.; Vicentes, I.; Santos, I.T.; Barbieri, R.F. Ventura Neto, R.; et al. *Uma Bioeconomia Inovadora para a Amazônia: Conceitos, Limites e Tendências para uma Definição Apropriada ao Bioma Floresta Tropical*; Texto para Discussão; WRI Brasil: São Paulo, Brasil, 2022.
9. Ros-Tonen, M.A.; Wiersum, K.F. The scope for improving rural livelihoods through non-timber forest products: An evolving research agenda. *For. Trees Livelihoods* **2005**, *15*, 129–148. [CrossRef]
10. Davis, J.H.; Goldberg, R.A. *A Concept of Agribusiness*; Harvard University Graduate School of Business Administration: Boston MA, USA, 1957.
11. Neves, M.F.; Kalaki, R.B.; Rodrigues, J.M.; Gray, A.W. Strategic planning and management of food and agribusiness chains: The chainplan method (framework). *Rev. Bus. Manag.* **2020**, *21*, 628–646. [CrossRef]
12. Farina, E.M.M.Q.; Zylbersztajn, D. *Competitividade No Agribusiness Brasileiro*; PENSA/FIA/FEA/USP: São Paulo, Brasil, 1998 Volume 1, p. 196.
13. Saes, M.S.M.; Silveira, R.L.F. Novas Formas de Organização das Cadeias Agrícolas Brasileiras: Tendências Recentes. O Mundo Rural no Brasil do Século 21: A Formação de um novo Padrão Agrário e Agrícola. Brasília, Embrapa. 2014. Available online https://repositorio.usp.br/item/002496579 (accessed on 10 July 2022).
14. Georgescu-Roegen, N. *The Entropy Law and the Economic Process*; Harvard University Press: Boston, MA, USA, 1971.
15. Georgescu-Roegen, N. Energy and economic myths. *South. Econ. J.* **1975**, *41*, 347–381. [CrossRef]
16. Gerber, J.F.; Scheidel, A. In search of substantive economics: Comparing today's two major socio-metabolic approaches to the economy–MEFA and MuSIASEM. *Ecol. Econ.* **2018**, *144*, 186–194. [CrossRef]
17. Giampietro, M.; Mayumi, K.; Ramos-Martin, J. Multi-scale integrated analysis of societal and ecosystem metabolism (MuSIASEM) Theoretical concepts and basic rationale. *Energy* **2009**, *34*, 313–322. [CrossRef]
18. Fischer-Kowalski, M. Society's metabolism: The intellectual history of materials flow analysis, Part I, 1860–1970. *J. Ind. Ecol.* **1998** *2*, 61–78. [CrossRef]
19. Krausmann, F.; Fischer-Kowalski, M.; Schandl, H.; Eisenmenger, N. The Global Sociometabolic Transition. *J. Ind. Ecol.* **2008**, *12* 637–656. [CrossRef]
20. Fischer-Kowalski, M.; Haberl, H. *Socioecological Transitions and Global Change: Trajectories of Social Metabolism and Land Use*; Edward Elgar: Cheltenham, UK, 2007.
21. Patermann, C.; Aguilar, A. The origins of the bioeconomy in the European Union. *New Biotechnol.* **2018**, *40*, 20–24. [CrossRef]
22. Bröring, S.; Laibach, N.; Wustmans, M. Innovation types in the bioeconomy. *J. Clean. Prod.* **2020**, *266*, 121939. [CrossRef]
23. OECD. *The Bioeconomy to 2030: Designing a Policy Agenda*; OECD: Paris, France, 2009. [CrossRef]
24. EU-European Commission. Directorate-General for Research and Innovation, a Sustainable Bioeconomy for Europe: Strengthening the Connection between Economy, Society and the Environment: Updated Bioeconomy Strategy. 2018; Publications Office Available online: https://data.europa.eu/doi/10.2777/792130 (accessed on 15 January 2022).
25. Philp, J.; Winickoff, D. *Innovation Ecosystems in the Bioeconomy, OECD Science, Technology and Industry Policy Papers, No. 76*; OECD Publishing: Paris, France, 2019. [CrossRef]
26. EU-European Commission. *Innovating for Sustainable Growth: A Bioeconomy for Europe*; European Commission: Brussels, Belgium, 2012
27. Keegan, D.; Kretschmer, B.; Elbersen, B.; Panoutsou, C. Cascading use: A systematic approach to biomass beyond the energy sector. *Biofuels Bioprod. Biorefining* **2013**, *7*, 193–206. [CrossRef]
28. D'Amato, D.; Gaio, M.; Semenzin, E. A review of LCA assessments of forest-based bioeconomy products and processes under an ecosystem services perspective. *Sci. Total Environ.* **2020**, *706*, 135859. [CrossRef]
29. Talwar, N.; Holden, N.M. The limitations of bioeconomy LCA studies for understanding the transition to sustainable bioeconomy *Int. J. Life Cycle Assess.* **2022**, *27*, 680–703. [CrossRef]

30. Sevigné-Itoiz, E.; Mwabonje, O.; Panoutsou, C.; Woods, J. Life cycle assessment (LCA): Informing the development of a sustainable circular bioeconomy? *Philos. Trans. R. Soc. A* **2021**, *379*, 20200352. [CrossRef]
31. NAS-National Academies of Sciences, Engineering, and Medicine. *Safeguarding the Bioeconomy*; The National Academies Press: Washington, DC, USA, 2020. [CrossRef]
32. Wesseler, J.; von Braun, J. Measuring the Bioeconomy: Economics and Policies. *Annu. Rev. Resour. Econ.* **2017**, *9*, 275–298. [CrossRef]
33. D'Adamo, I.; Falcone, P.M.; Morone, P. A new socio-economic indicator to measure the performance of bioeconomy sectors in Europe. *Ecol. Econ.* **2020**, *176*, 106724. [CrossRef]
34. Heijman, W. How big is the bio-business? Notes on measuring the size of the Dutch bio-economy. *NJAS-Wagening. J. Life Sci.* **2016**, *77*, 5–8. [CrossRef]
35. Becker, B.K. Novas territorialidades na Amazônia: Desafio às políticas públicas. Boletim do Museu Paraense Emílio Goeldi. *Ciências Hum.* **2010**, *5*, 17–23. Available online: http://scielo.iec.gov.br/scielo.php?script=sci_arttext&pid=S1981-81222010000100003&lng=pt&nrm=iso (accessed on 1 March 2022).
36. Bergamo, D.; Zerbini, O.; Pinho, P.; Moutinho, P. The Amazon bioeconomy: Beyond the use of forest products. *Ecol. Econ.* **2022**, *199*, 107448. [CrossRef]
37. WTT&COI. Bioeconomia Amazônica: Uma Navegação Pelas Fronteiras Científicas e Potenciais de Inovação. 2022. Available online: https://wttventures.net/bioeconomia-amazonica-uma-navegacao-pelas-fronteiras-cientificas-e-potenciais-de-inovacao/ (accessed on 8 September 2022).
38. UNSDN. Empowerment: What Does It Mean to You? 2012. Available online: https://www.un.org/esa/socdev/ngo/outreachmaterials/empowerment-booklet.pdf (accessed on 1 March 2022).
39. Petesque, M.; Cipriani, S.; Arriaza, P. *Community Empowerment Manual*; Community Protection Approach. We World; 2020; 59p. Available online: https://reliefweb.int (accessed on 8 April 2022).
40. Amazonas. *Nota Técnica: Um Panorama Sobre a Bioeconomia No Estado do Amazonas*; SEDECTI: Manaus, Brazil, 2021.
41. Pfau, S.F.; Hagens, J.E.; Dankbaar, B.; Smits, A.J. Visions of sustainability in bioeconomy research. *Sustainability* **2014**, *6*, 1222–1249. [CrossRef]
42. Piplani, M.; Smith-Hall, C. Towards a Global Framework for Analysing the Forest-Based Bioeconomy. *Forests* **2021**, *12*, 1673. [CrossRef]
43. Arnold, J.M.; Pérez, M.R. Can non-timber forest products match tropical forest conservation and development objectives? *Ecol. Econ.* **2001**, *39*, 437–447. [CrossRef]
44. Belcher, B.; Ruiz-Pérez, M.; Achdiawan, R. Global patterns and trends in the use and management of commercial NTFPs: Implications for livelihoods and conservation. *World Dev.* **2005**, *33*, 1435–1452. [CrossRef]
45. De Mello, N.G.R.; Gulinck, H.; Van den Broeck, P.; Parra, C. Social-ecological sustainability of non-timber forest products: A review and theoretical considerations for future research. *For. Policy Econ.* **2020**, *112*, 102109. [CrossRef]
46. Neven, D. Developing Sustainable Food Value Chains; FAO: 2014. Available online: https://www.fao.org/3/I3953E/i3953e.pdf (accessed on 10 April 2022).
47. Gereffi, G. Economic upgrading in global value chains. In *Handbook on Global Value Chains*; Edward Elgar Publishing: Cheltenham, UK, 2019. [CrossRef]
48. Gereffi, G.; Lee, J. Economic and Social Upgrading in Global Value Chains and Industrial Clusters: Why Governance Matters. *J. Bus. Ethics* **2016**, *133*, 25–38. [CrossRef]
49. Gereffi, G.; Humphrey, J.; Sturgeon, T. The governance of global value chains. *Rev. Int. Polit. Econ.* **2005**, *12*, 78–104. [CrossRef]
50. Humphrey, J.; Schmitz, H. Chain governance and upgrading: Taking stock. In *Local Enterprises in the Global Economy*; Schmitz, H., Ed.; Edward Elgar: Cheltenham, UK, 2004; pp. 349–382. Available online: https://www.elgaronline.com/view/1843760991.xml#page=361 (accessed on 5 May 2023).
51. Ostrom, E.; Chang, C.; Pennington, M.; Tarko, V. *The Future of the Commons—Beyond Market Failure and Government Regulation* (December 10, 2012). Institute of Economic Affairs Monographs, 2012, Indiana University, Bloomington School of Public & Environmental Affairs Research Paper No. 2012-12-02. Available online: https://ssrn.com/abstract=2267381 (accessed on 5 May 2023).
52. Ostrom, E. Beyond Markets and States: Polycentric Governance of Complex Economic Systems. *Am. Econ. Rev.* **2010**, *100*, 641–672. Available online: http://www.jstor.org/stable/27871226 (accessed on 5 May 2023). [CrossRef]
53. North, D.C. Institutions, Institutional Change and Economic Performance. In *Political Economy of Institutions and Decisions*; Cambridge University Press: Cambridge, UK, 1990. [CrossRef]
54. Ménard, C.; Shirley, M.M. *A Research Agenda for New Institutional Economics*; Edward Elgar Publishing: Cheltenham, UK, 2018. [CrossRef]
55. Barrientos, S.; Gereffi, G.; Rossi, A. Economic and social upgrading in global production networks: A new paradigm for a changing world. *Int. Labour Rev.* **2011**, *150*, 319–340. [CrossRef]
56. Rossi, A. Does economic upgrading lead to social upgrading in global production networks? Evidence from Morocco. *World Dev.* **2013**, *46*, 223–233. [CrossRef]
57. Reinecke, G.; Posthuma, A.C. The link between economic and social upgrading in global supply chains: Experiences from the Southern Cone. *Int. Labour Rev.* **2019**, *158*, 77–703. [CrossRef]

58. Mani, V.; Gunasekaran, A.; Delgado, C. Enhancing supply chain performance through supplier social sustainability: An emerging economy perspective. *Int. J. Prod. Econ.* **2018**, *195*, 259–272. [CrossRef]
59. Achabou, M.A.; Dekhili, S.; Hamdoun, M. Environmental upgrading of developing country firms in global value chains. *Bus. Strat. Environ.* **2017**, *26*, 224–238. [CrossRef]
60. Azevedo-Ramos, C.; Moutinho, P.; Arruda, V.L.D.S.; Stabile, M.C.; Alencar, A.; Castro, I.; Ribeiro, J.P. Lawless land in no man's land: The undesignated public forests in the Brazilian Amazon. *Land Use Policy* **2020**, *99*, 104863. [CrossRef]
61. Peralta, M.E.; Luna, P.; Soltero, V.M. Towards standards-based of circular economy: Knowledge available and sufficient for transition? *Int. J. Sustain. Dev. World Ecol.* **2020**, *27*, 369–386. [CrossRef]
62. Golini, R.; De Marchi, V.; Boffelli, A.; Kalchschmidt, M. Which governance structures drive economic, environmental, and social upgrading? A quantitative analysis in the assembly industries. *Int. J. Prod. Econ.* **2018**, *203*, 13–23. [CrossRef]
63. Jindra, B.; Hatani, F.; Steger, T.; Hiemer, J. Social upgrading and cooperative corporate social responsibility in global value chains: The case of Fairphone in China. *Glob. Netw.* **2019**, *19*, 371–393. [CrossRef]
64. Gonçalves, A.C.T.; Cunha, J.B.C.; Batista, J.S. *O Gigante Amazônico: Manejo Sustentável de Pirarucu*; Instituto de Desenvolvimento Sustentável Mamirauá: Tefé, Brazil, 2018.
65. CONAB. Boletim da Sociobiodiversidade; 2020; Volume 4, pp. 1–52. Available online: https://www.conab.gov.br/info-agro/analises-do-mercado-agropecuario-e-extrativista/boletim-da-sociobiodiversidade/boletim-sociobio (accessed on 15 May 2022).
66. IDSM. Relatório Técnico Anual 2021 dos Projetos de Manejo dos Recursos Pesqueiros nas Reservas de Desenvolvimento Sustentável Mamirauá, Amanã e Entorno. Tefé-AM. 2022. Available online: https://www.mamiraua.org.br/noticias/relat%C3%B3rio-tecnico-aponta-que-manejo-de-pesca-rendeu-mais-de-r$-3,5-milhoes-a-comunidades-do-m%C3%A9dio-solim%C3%B5es-em-2021 (accessed on 1 August 2022).
67. Mapbiomas. MAPBIOMAS—The Brazilian Annual Land Use and Land Cover Mapping Project. 2022. Available online: https://mapbiomas.org/ (accessed on 1 October 2022).
68. Homma, A.K.O. Amazônia: Da bioeconomia das "drogas do sertão" à bioeconomia do século (Cap.1). In *O Brasil rural contemporâneo: Interpretações. Geraldo Sant'Ana de Camargo Barros e Zander Navarro (Org.)*, 1st ed.; Editora Baraúna: São Paulo, Brazil, 2022; p. 385.
69. INPI-Insituto Nacional de Propriedade Industrial. INPI Reconhece a Denominação de Origem Mamirauá para o Pirarucu Manejado. 2021. Available online: https://www.gov.br/inpi/pt-br/central-deconteudo/noticias/inpi-reconhece-a-denominacao-de-origem-mamiraua-para-o-pirarucu-manejado (accessed on 10 May 2022).
70. Campos-Silva, J.V.; Peres, C.A.; Hawes, J.E.; Haugaasen, T.; Freitas, C.T.; Ladle, R.J.; Lopes, P.F. Sustainable-use protected areas catalyze enhanced livelihoods in rural Amazonia. *Proc. Nat. Acad. Sci. USA* **2021**, *118*, e2105480118. [CrossRef]
71. Campos-Silva, J.V.; Peres, C.A. Community-based management induces rapid recovery of a high-value tropical freshwater fishery. *Sci. Rep.* **2016**, *6*, 34745. [CrossRef] [PubMed]
72. Gamarra, N.C.; Hawes, J.E.; Costa, A.C.L.; Vieira, F.A.S.; Rodrigues, A.C.; Ladle, R.J.; Campos-Silva, J.V. Arapaima co-management through the lens of the Protected Areas Assets Framework. *J. Nat. Conserv.* **2022**, *67*, 126161. [CrossRef]
73. Freitas, C.T.; Espírito-Santo, H.M.; Campos-Silva, J.V.; Peres, C.A.; Lopes, P.F. Resource co-management as a step towards gender equity in fisheries. *Ecol. Econ.* **2020**, *176*, 106709. [CrossRef]
74. Cox, M.; Arnold, G.; Villamayor-Tomas, S. *A Review of Design Principles for Community Based Natural Resource Management*; Elinor Ostrom and the Bloomington School of Political Economy: Resource Governance; Lexington Books; Lanham, MD, USA, 2015; pp. 249–280. Available online: http://www.ecologyandsociety.org/vol15/iss4/art38/ (accessed on 2 January 2022).
75. Brites, A.D.; Morsello, C. Efeitos ecológicos da exploração de produtos florestais não madeireiros: Uma revisão sistemática. *Desenvolv. Meio Ambient.* **2016**, *36*, 1–18. [CrossRef]
76. Kusters, K. Non-Timber Forest Product Trade: A Trade-off between Conservation and Development: Assessing the Outcomes of Non-Timber Forest Product Trade on Livelihoods and the Environment, with Special Emphasis on the Damar Agroforests in Sumatra, Indonesia. Ph.D. Thesis, Universiteit van Amsterdam, Amsterdam, The Netherlands, 2009.
77. Bartkus, V.O.; Brooks, W.; Kaboski, J.P.; Pelnik, C. Big fish in thin markets: Competing with the middlemen to increase market access in the Amazon. *J. Dev. Econ.* **2022**, *155*, 102757. [CrossRef]

sustainability

Article

Assessing Circular Economy Opportunities at the Food Supply Chain Level: The Case of Five Piedmont Product Chains

Franco Fassio [1], Isaac Enrique Perez Borda [1], Elisa Talpo [1], Alessandra Savina [1,*], Fabiana Rovera [1,*], Ottavia Pieretto [2] and Davide Zarri [3]

1 Sustainability and Circular Economy Laboratory, University of Gastronomic Sciences, Pollenzo, 12042 Cuneo, Italy
2 Slow Food Piemonte Valle d'Aosta, 10126 Torino, Italy
3 Slow Food Youth Network Italy, Bra, 12042 Cuneo, Italy
* Correspondence: a.savina@unisg.it (A.S.); f.rovera@unisg.it (F.R.); Tel.: +39-389-3435866 (A.S.); +39-377-2698196 (F.R.)

Abstract: The impact of linear agri-food supply chains is progressively leading to ever wider socio-environmental and cultural repercussions, undermining the balance of territories and communities to the point of disrupting the entire planet's health. Therefore, there is an urgent need for a paradigm change involving the agri-food sector, the economic sector, and local public policies, in the direction of a diffused ecological transition. In this scenario, the Circular Economy, supported by the adoption of a Systemic Approach, represents a useful operational tool to respond to complex transversal challenges, to reduce and enhance waste, minimize the use of new raw materials, and strengthen the territorial identity and relations among local stakeholders. This article describes a research project conducted for the Piedmont Region (Italy), through which it was possible to apply these innovative tools and approaches to five typical local agri-food chains (wine, dairy and cheese, rice, water, and bovine beef). Currently, at the Piedmontese level, concrete proposals for public policies capable of supporting the ecological transition of the local agri-food chains have not yet been developed, especially in terms of cyclicality of outputs and territorial co-evolution. For this reason, through the use of a multi-stakeholder approach, participatory mechanisms of local actors, and the analysis of several national and international case studies, the purpose of this research was to evaluate the possible enhancement of food waste and by-products, finally developing specific proposals for good practices and public policies capable of contributing to the achievement of the Piedmontese Regional Strategy for Sustainable Development.

Keywords: circular economy for food; sustainable food supply chains; systemic approach; multi-stakeholder approach; public food policy; local agri-food system; food waste management

Citation: Fassio, F.; Borda, I.E.P.; Talpo, E.; Savina, A.; Rovera, F.; Pieretto, O.; Zarri, D. Assessing Circular Economy Opportunities at the Food Supply Chain Level: The Case of Five Piedmont Product Chains. *Sustainability* **2022**, *14*, 10778. https://doi.org/10.3390/su141710778

Academic Editors: Filippo Vitolla, Nicola Raimo, Ornella Malandrino and Benedetta Esposito

Received: 11 July 2022
Accepted: 24 August 2022
Published: 29 August 2022

1. Introduction

The management and promotion of sustainability is an issue that is currently increasingly addressed within the various human productive activities. Among these, the agri-food system stands out, which plays a central role both in the generation of well-being and social development, and in the environmental balance. Today, such systems are extremely connected to the prevailing economic model, based mainly on the typical approach of a linear economy, "*take-make-dispose*" [1]. The large number of stakeholders involved, the complex network of material and information flows, and the socio-political importance of these systems make the more complex issues connected to them real *wicked problems* [2,3], for which the proposal of any alternatives leads to visible consequences only in the long term, without immediate evidence of the effectiveness of the actions taken [4]. In particular, the management of agri-food supply chains can be unsustainable and not entirely efficient for different reasons, specifically in terms of waste production in the various stages of harvesting, storage, transportation, and distribution of agri-food products which are subject to rapid spoilage due to their natural characteristic [5].

To address this type of problem, it is essential to adopt a systemic vision and perspective, which bring to the fore the notions of complexity and network [6], in which a system is understood as a complex of parties that establish relationships with each other and the behavior of each is characterized by the connection in which it is involved. These components, taken together, do not represent the mere sum of the separate parts, but a single holistic system with a wider value [7]. From this point of view, the application of a systemic approach to the agri-food scenario is a valid design method, to trigger a paradigm shift that involves the transition from linear to circular structures, supported by a collaborative approach [8–10]. However, the concept of circularity is not new: it can be understood as a reinterpretation of rural economies, in which previous generations were accustomed to more sustainable uses of natural resources, based on the reuse and reduction of waste [11]. The novelty is represented by the constitutive elements of the circular economy, which have more articulated roots and derive from a fairly recent evolution of the concept of sustainable development and the economy–environment relationship. In recent years, scientific research and the political landscape are moving toward this new economic paradigm [12]. This is demonstrated by the progressive increase in the quantity of research projects relating to the circular economy in the scientific literature, in conjunction with the publication of the United Nations Sustainable Development Strategy in 2015 [13] and the growing sensitivity toward environmental, social, and economic potentialities of this disciplinary sector.

The energy crisis of the 1970s led to the search for new solutions to address the evident scarcity of resources due to the linear growth model [14], proposing a new regenerative paradigm in which production is based on the precept "*From Cradle to Cradle*", going beyond the simplistic vision of the mere recycling of materials [15] and looking to nature for the formulation of new circular business models [16]. Thus, human production processes can employ regenerative practices and guarantee the cyclical nature of materials, emulating the dynamics of natural systems, with a view to a *Circular Economy* [17].

The latter has been integrated into the environmental policies of the European Union whose current priority is to promote the *European Green Deal* [18]. It proposes specific measures to make the production of energy and resources (including food) and, consequently citizens' lifestyles more sustainable, to initiate an ecological transition that involves the protection of natural capital, the generation of employment, the development of sustainable technologies, and the extension of the life cycle of products, among others. In particular the *Farm-to-Fork Strategy* [19], adopted by the European Commission in 2020, is considered the heart of the European Green Deal and aims to make food systems more equitable healthier, and more sustainable from the economic, social, and environmental point of view. This strategy underlines the priority of acting on the food system to achieve a complete ecological transition, considering food as the basic unit that connects the *17 Sustainable Development Goals* (SDGs) [20]. The ultimate goal is to accelerate the transition to a more resilient food system that is able to cope with sudden contextual changes that would make life on the planet precarious.

Thus, the concept of the circular economy has begun to become part of the agenda of political institutions [21]. Finally, as part of a deep reflection on the dominant economic paradigm and on the need to bend the linear system to reduce its negative impact, in 2015, the European Commission promoted its first *Action Plan for the Circular Economy* [22] an important step financially supported by structural and investment funds such as, for example, the European Fund for Strategic Investments (EuSEF), Horizon 2020, and LIFE programs. In Italy, the Ministry of Ecological Transition followed the European line introducing, in 2017, the *National Strategy for Sustainable Development* (SNSvS) [23]. It outlines a vision of a future centered on sustainability, as a shared and indispensable value for facing global challenges.

Specifically, in the context of Piedmont (Piedmont, Italy), the agri-food scenario does not only represent a strategic economic sector, but also a real key element in the pursuit of social and environmental objectives. Consequently, the need for an approach that ranges from practices to policies emerges, in which public institutions play a direct and concrete role, so that supply chain problems are not faced only by civil society or by individual virtuous producers [24]. Indeed, it is unthinkable to achieve the objectives of European policies without pursuing sustainable strategies on a local scale in the same way, adopting an approach that goes from *micro to macro* [25,26]. From here, the centrality of local authorities and the strategic role of regional policies as an engine of change becomes evident.

One of the most significant challenges for the development and implementation of integrated food policies is the definition of forms of governance that can support the active participation of stakeholders [27]. In fact, in the agri-food scenario, a multi-stakeholder approach based on a public–private–third sector partnership is fundamental, involving the various institutional levels in the same way [28]. The benefits of an ecological transition also implemented through the adoption of a circular economy go far beyond environmental protection and resource saving, leading to new opportunities and the design of new, systemic, highly innovative business models. This transition must be mainly driven by companies, through the support of regional authorities, stricter regulatory systems, and more demanding and aware consumers [29].

Going into detail, the objectives of European and international policies cannot be achieved if—*in cascade at each level*—a strategy that allows the achievement of global targets is not implemented territorially. For example, in the case of Italy, each region differs profoundly in terms of priority areas of action and internal organization. For this reason, it is necessary that public food policies take into account local specificities and be interpreted in different ways according to the context and the nature of the problems [30–32]. This shows the centrality of local authorities and the strategic role of regional policies as a driver of change, in contrast to the "*by nation*" approach that has characterized the past decades [33]. The development of effective public policies, therefore, requires in-depth research and analysis relating to the territory in which action is to be taken, to ensure objective action based on shared criticalities and priorities, not dictated by political preferences [34,35]. Scientific research is found to be fundamental for informing multi-sectoral policies that address trade-offs and synergies, also for enabling policy makers to orient themselves in complexity, through an objective knowledge support [36,37]. Along with this, as argued by den Boer et al. [38], it is fundamental to deepen R&I through the development of "transdisciplinary research approached by investing in the creation of meaningful interactions between researchers, societal actors, and policymakers."

Acting concretely from the local level is essential: half of the human population—3.5 billion people according to the United Nations—lives in urban areas and this number is destined to increase significantly by the end of the decade [39]. With population growth, resource demand in urban areas increases as well, and environmental criticalities and socio-economic differences among citizens emerge and deepen. A new food uncertainty enters cities all over the world, not only linked with the issue of lack of food and nutrients, but also with the issue of food excesses.

In 2030, the deadline for reaching the 17 Sustainable Development Goals is also sanctioned and, given the transversality of food with respect to each of them, this must be the strategy adopted to address the criticalities of our food systems. Food, a basic unit of connection among all SDGs, plays an essential role in the transition toward a sustainable development, as shown in the *Wedding Cake model* by Rockstrom and Sukhdev [20]; therefore, the role that the food system can and must play in the ecological transition toward a sustainable development paradigm is essential. In the food sector, the circular economy is presented as a sustainable practice to remedy some of today's greatest challenges, including population growth, inefficient use of resources, environmental impacts on climate, soil and oceans, and food waste [40].

Therefore, there is a growing consensus on the need for public institutions to adopt a systemic approach to food policies to successfully address and solve complex, persistent, and interrelated problems such as food insecurity, climate change, use of resources, poverty, and public health [27]. A systemic approach to food systems allows us to broaden the view, promoting integrated and coherent policies to align different political agendas and transversal issues (e.g., agriculture, environment, trade, health, food safety) to better meet the needs of the actors involved and support multiple objectives (environmental, socio-economic, and healthcare) [41,42]. Therefore, improving the quality and sustainability of the food system means increasing the sustainability of all areas of the territory as a whole. A higher level of design for sustainability is reached through innovation at a systems level, a more radical and strategic approach that involves many stakeholders, such as communities, governments, companies, and customers [43,44]. Within this context, the circular economy applied to the agri-food sector, or Circular Economy for Food, is of considerable importance because it recognizes the mutually influential relationship between food and the circular economic model, the principles of which are taking the first but decisive steps towards a route change [45]. Thus, the need to achieve food systems that are sustainable is seen, recognizing the importance of the indivisible links that exist between healthy people, healthy societies, and a healthy planet (*One Health approach*). One Health is an integrated, unifying approach that aims to sustainably balance and optimize the health of people, animals, and ecosystems. It recognizes that the health of humans, domestic and wild animals, plants, and the wider environment (including ecosystems) are closely linked and interdependent [46].

Within this framework, the 3 Cs of the circular economy for Food [47]—Capital, Cyclicality, and Co-evolution—were taken as a theoretical-practical reference to systematically analyze the supply chains being researched and to arrive at the proposal of public policies that could be transversal as relational bridges between several MAS of the SRSvS of the Piedmont Region and the SDGs. Furthermore, the coexistence of the three criteria of interpretation of the system applied to food allows the transition to a paradigm—*Circular Economy for Food*—that is impactful and meaningful for a development that is sustainable. In particular, by designing the flows of matter, energy, and information that cyclically condition the food system, it is possible to have a positive impact on the 17 SDGs and contribute to regenerating the natural, cultural, and economic capital that supports coevolution between species [48].

This article aims to investigate some aspects of the Piedmontese Regional Strategy for Sustainable Development (in Italian, abbreviated as SRSvS) [49], paying particular attention to the issues of the circular economy applied to the food system. This action focuses on involving the agri-food system in an operational way, creating cross-cutting worktables, to share and initiate integrated policies on circular economy issues that are directly attributable to the objectives of SRSvS. With the aim of making the research as representative and relevant as possible for the regional food system, five different typical food supply chains were selected as a priority object of study. The attempt was to incorporate and restore the diversity and variety of the Piedmontese food production landscape in the best possible way, seeking the involvement of companies with different sizes and positions at the level of the supply chain. Addressing "real-life case studies" makes our research an important contribution to the existing academic literature, as a result of the strong interaction between researchers and key stakeholders aiming to highlight concrete evidence-based solutions to local existing criticalities [5].

In conclusion, to better contextualize the path of the research conducted and the narrative vein used, the contents of the following sections are reported below:

1. Section 2 shows the objectives and the design methodology that characterized the research, as well as the instrument and tools that have made it possible to carry out the investigation;
2. Section 3, in accordance with the stages of the methodological structure presented in Section 2, reports a particular emphasis on the outcome of *Desk-based Research*,

in particular describing the *Academic and Sectoral Document Discovery* (within which the scientific literature was analyzed), the *Quantitative and Qualitative Data Analysis* and the *Supply Chain Mapping* developed; furthermore, the results of the *Stakeholder Engagement* are summarized inside this section;

3. Section 4 shows the meaningful challenges for the development and the expansion of integrated food policies through the definition of a series of cross-cutting solutions;
4. Section 5 describes implications, limits, hypotheses, and future research directions about the *Research Proposal* developed.

2. Materials and Methods

2.1. Objectives of the Project

The process that led to the development of the "*Circular Economy in the Agri-food piedmontese sector*" project is based particularly on two of the seven strategic macro-areas (MAS) defined by the Piedmontese regional government (*MAS 1, accompanying the transition of the Piedmont production system toward a model capable of combining competitiveness and sustainability* and *MAS 3: taking care of cultural heritage and environmental heritage and the resilience of territories*) [49]. These deal with the promotion of an ecological transition of the regional productive system, and are aimed at understanding how the dynamics of the circular economy can be applied in practice to five supply chains of the Piedmontese food system, which coincide with those considered priorities by both the Piedmont Region and the University of Gastronomic Sciences of Pollenzo. This research focused on the agri-food supply chains that play an important and decisive role for the Piedmontese economy in terms of production and turnover, chosen in such a way as to include the agri-food production diversity of the region: wine, dairy and cheese, rice, bottled water, and beef supply chains (Figure 1).

Figure 1. The five local food supply chains were taken into consideration for the project.

The Piedmont region has promoted this research in order to focus on the regional agri-food system, with the aim of supporting the development of actions in line with the objectives of the SRSvS [49]. It stems from the need to realize the SNSvS on the circular economy front in the agri-food sector and, therefore, the regional government collaborated with UNISG to ensure that it would work to bring the sustainability targets set at national and then regional level to the ground.

Through the operational involvement of the actors of the agri-food system, it promoted the sharing of issues concerning the circular economy, identified as the key to the SRSvS. There are essentially two main objectives on which the *Research Proposal* is based:

1. to identify priority issues for the agri-food system regarding the possibility of transition toward a model based on a better use of renewable resources, reuse of raw materials, and waste valorization;
2. to identify and suggest regional system policies according to sustainability objectives in relation to the priority areas.

More specifically, the project objectives required the identification, through the involvement of stakeholders, of the main challenges and potentialities concerning the Piedmontese agri-food system in relation to the transition toward a circular economic model that focuses on the use, recycling, and recovery of waste or by-products from the processing phases of each supply chain. After that, the identification of those best practices, technologies, and services that could help in the development of alternative and innovative paths involving the reduction, but above all the valorization, of waste present in the agri-food chains was requested.

2.2. Instruments and Tools

As already mentioned in the Introduction, the majority of current scientific research about the opportunities for circular economy applications in the agri-food sector seeks to widen the focus of their control volume to include an entire food product's supply chain. In this way, it is possible to better address the complexity and wide-ranging impact of agricultural supply chains, whose analysis benefits from a systemic perspective, within which most of the inputs and outputs can be accounted for as an interconnected set of flows of matter, energy, and information, rather than as separate units.

In order to provide sufficient and accurate information for the development of public policy proposals in the context of the promotion of circular economy practices and the ecological transition of the chosen supply chains, the expansion of this focus has been translated into the research project described inside of this article. Such an approach has further allowed for an in-depth understanding of the current circularity and baseline of ecological transition in regional food supply chains, as well as the possible obstacles and potentialities for further adoption. Specifically, a total of 72 people were involved in the development of the research project, nine of whom are members of the University's research team, while the remaining number can be traced back to 21 companies in the agri-food sector, three innovation poles in Piedmont, and 39 students following the course in *Systemic Design for Circular Economy* of the master's degree in *Food Innovation & Management* at UNISG, who contributed to the project by offering innovative solutions to what they perceived as the main challenges being faced by the supply chains.

Although there is a wide field of application for the research conducted within the discipline of Circular Economy in the agri-food sector, in terms of supply chain stages, from the scientific literature it has emerged that a singularly predominant tool (e.g., LCA, SLCA, WFA, LCIA) or variable chosen to measure impact is not generally used (e.g., carbon and water footprints, energy consumption, etc.), leading to a lack of comparability and reproducibility among studies [13]. Therefore, it seems preferable to employ an approach which gathers the results of more specific previous scientific research and applies it to the food supply chain system as a whole, generating a holistic understanding of the priorities to be defined and the challenges to be faced.

For this reason, on the basis of the methodology of the systemic approach adopted in the research path discusses in this paper, specific phases were followed (Figure 2): firstly, desk research allowed the discovery of academic and sectoral documents for the review of a scientific literature consistent with the research topic, after which it was necessary to carry out a qualitative and quantitative analysis of the five selected Piedmontese agri-food chains and map the main process steps and the actors involved in them, supported by an investigation of several innovative case studies, which made it possible to identify the best practices to be exploited. Finally, the organization and implementation of participatory mechanisms made it possible to finalize a shared research proposal regarding possible public policies applicable to the regional context, responding to the feasibility criteria and territorial requirements.

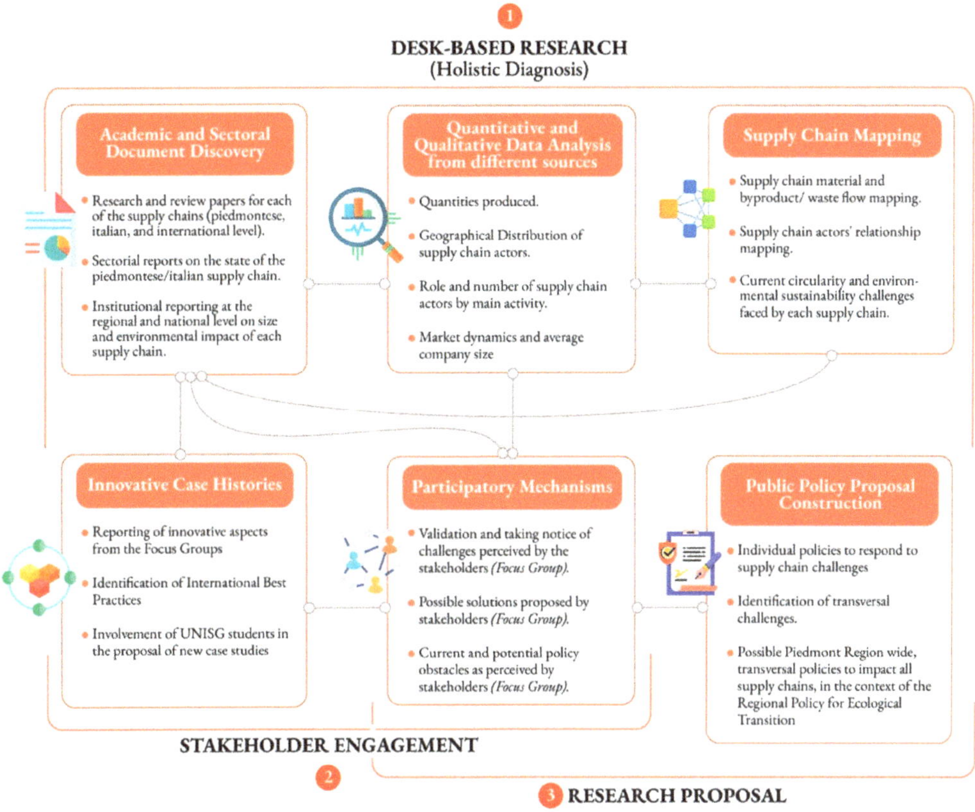

Figure 2. Design methodology: the representation of the single steps addressed and the links among them.

As follows, a description of each of the steps of the project are included, along with their relevance to the objectives stated by its definition:

- *Academic and sectoral document discovery*: as a primary step, and to better define the extent of each supply chain, relevant research papers and sectoral documents were researched. Using the keywords "*waste*", "*byproduct*", "*waste management*" and "*circular economy*", followed by each of the individual supply chain names, in an academic search engine (Google Scholar) and in a scientific article database (Scopus), with a total number of 75 selected articles, ranging from scientific articles reporting on similar initiatives on a smaller or similar scale, to sectoral documents, used to better delineate the size and boundaries of each supply chain. In addition to supply chain structure and boundaries, the documents were also analyzed for information regarding the current practices of waste or by-product disposal, valorization and/or and treatment, as well as for the private and public sectors' interests and understanding of the challenges currently present in the Piedmontese regional context. This step, using other terms, could also be called a *literature review* because it was performed to understand the scenario of the research proposal better and to analyze—*at a micro level*—the research gap identified. In fact, it consists of a thorough study review process of papers about the *circular economy, food waste management, sustainable food supply chains, and local agri-food system*. All the considerations developed during this phase have been inserted in the aforementioned text, in particular, in Sections 1, 3 and 4.

- *Quantitative and Qualitative Data Analysis*: once the current state and main challenges for each supply chain were taken into account, information regarding the distribution of actors in each step of the chain was procured by employing two datasets: one, provided by the Piedmont Region joint Chambers of Commerce (*Unioncamere Piemonte*) [50], containing statistical and location data for all the companies registered within the region, and the other by the University of Gastronomic Sciences' Food Industry Monitor, a performance observatory containing historical financial data for the most relevant companies in the Italian food sector. Both datasets were filtered by only including the companies whose economic activity code (*Codice ATECO*) was within those belonging to the mapped supply chain, with a resulting total of 3261 records analyzed. In addition to individual company data, other statistical databases for each of the supply chains analyzed were taken into consideration, in order to better understand the complexity and scale of each one. Namely, the regional and national agricultural registries were consulted for a region-wide perspective on the data that is reported by each supply chain's multiplicity of actors to the regional and national governments.
- *Supply chain stage mapping*: based on the information gathered in the document discovery stage, as well as the quantitative data gathered and analyzed, a simplified "map" of each supply chain was constructed; said map would feature the main actors in the supply chain as system nodes, as well as the financial, material, and information flows among them. The mapping covered the flow from direct material suppliers all the way to retail channels, including, in some cases, the actors involved in the treatment and valorization of waste and by-products, based on the level of connection to the main supply chain, and to the available information gathered. Following the development of each supply chain map, the research team validated their contents with actors present at different stages of each supply chain, who either attested to the accuracy of the mapping, as well as pointing out blind spots or missing nodes/connections. Having received this feedback, a definitive map for each supply chain, including the actors' feedback, was constructed.
- *Innovative case histories research*: in parallel, in keeping with the practical, implementation-centered intention and goals of the project, a set of 28 innovative case studies, relevant to each of the five supply chains were studied and summarized, intended to be used as input for both the proposed solutions to challenges, and to foster conversation in the participatory mechanism sessions further ahead.
- *Participatory mechanisms:* while a better understanding of each supply chain was established by mapping and quantifying their dimension, complexity, and economic relevance at a regional level, the next step of the project was designed to include the personal experience of several actors from each supply chain, as well as their interaction with institutional representatives. This phase was conducted, in line with the scientific literature analyzed, on the topic of the circular economy, which revealed the need to involve stakeholders at the level of the supply chain in the development and evaluation of possible directions for the transition from a linear to a circular and more sustainable system, in order to understand potential, dormant assets and possible barriers [13]. The process followed therefore placed emphasis on the engagement of relevant stakeholders for contemporary agri-food circular economy research. Taking the project's final objective of public policy development into account, it was decided that it was best to employ a participatory, deliberative approach to the validation of the quantitative analysis, as well as to the understanding of challenges faced by each of the supply chains. Therefore, a series of supply chain circularity and ecological transition-focused focus groups were designed and implemented.
- *Data analysis and drafting of public policy proposals*: as all necessary inputs were gathered, the following and final step was to synthesize and process the data gathered along every step of the research process and translate them into actionable recommendations for the Piedmontese regional government.

In total, five focus groups were performed, in which a total of 35 people participated actively and collaboratively. Of that number, 26 represented the rich diversity of the agribusiness sector: 21 companies, including agricultural producers, processors, and distributors, with participation from both family-owned businesses and some of the largest food and agriculture companies in the region.

In order to best represent the multiplicity of actors of each supply chain, at least five companies were selected per supply chain, considering the different production phases, the size of the company, the innovations implemented within some companies, and their size in terms of turnover and production. The need to involve the private sector is often cited in the scientific literature regarding the implementation and improvement of circular economy practices, given the key role that private companies play, as they possess greater capabilities and resources than other stakeholders [12]. Nevertheless, in addition to the private sector, all the regional innovation poles, regional government representatives, and the main regional research institutions (*Ires Piemonte* and *Environment Park Torino*) were invited to participate in these transversal worktables.

The focus groups were developed with the following structure:

- Introduction to the University's project and its objectives;
- Explanation of the context of the project at the European (*Green Deal*), national (*SNSvS*), and regional (*SRSvS*) levels;
- Presentation of the UNISG vision of the Circular Economy for Food (3C);
- Validation of the quantitative, qualitative, and financial representation of the specific supply chain;
- Illustration of the main waste products and by-products of the supply chain under analysis;
- Emerging issues related to the application of circular activities or the implementation of circular practices that also address the management and valorization of the supply chain's outputs;
- Presentation and identification of relevant case studies and virtuous business models of the circular economy applied to the agri-food sector, leaving a space for discussion among participants for their assessment of the feasibility of applying such examples in the Piedmont context;
- Further insight and discussion on the scenario of the circular economy in Piedmont.

In order to engage people and encourage dialogue among different actors (business representatives, innovation poles and invited researchers, and regional government representatives), a number of questions were posed to each of the focus group participants whose subject was the critical issues in the circular economy. Through answering these questions, as well as helped by the context given by the previous interventions with regard to the current situation and the opportunities that a circular economy and policy change brings about, all participants of the focus groups gave interventions on the circular economy potentialities and proposals that individual businesses, the supply chain, and the regional context as a whole held, and the steps and concerns about the implementation of an ecological transition agenda.

These interventions were recorded and furthermore analyzed for commonalities and differences, as well as to their relevance to the SRSvS and SNSvS. After recording and summarizing all interventions, a brief report for each supply chain's focus group was constructed and shared with the participants, in order to validate that their positions and opinions were accurately represented and that no further inputs were missing from the research.

2.3. *Construction of Public Policy Proposals from the Information Gathered and Processed*

In order to help achieve the objectives of the SRSvS, a set of Public Policy and Best Practices proposals was given to the Piedmontese regional government. These proposals, outlined with the purpose of being put into practice by the institutional body, were the result of a careful analysis, based on the understanding of all those problems and opportunities in a circular and systemic key that characterize the agri-food sector of the Piedmontese territory. In this way, with the participation of Slow Food, a non-profit international association committed to restoring value to food, sectoral and transversal recommendations were identified among the five supply chains that are consistent with the SRSvS in order to boost Piedmont's ecological transition, by the combination of desk-based research the intervention of citizens represented by the Slow Food Piedmont and Aosta Valley Association [51], the focus groups and the proposals for innovative solutions suggested by the innovation poles, as well as by the students of the master's degree in Food Innovation and Management at UNISG.

In fact, a full understanding of the opportunities and problems that characterize a certain territory allows the political decision maker in possession of these elements to prepare adequate responses, giving impetus to innovation and social experimentation, in line with the objectives of the SRSvS and aligned with the Ecological Transition guidelines [52].

As a result of the gathered information, a total of 45 proposals for actions impacting the SRSvS divided by supply chain and nine transversal recommendations addressed to regional policy makers were listed. In the following Figure 3, it is possible to read a graphic synthesis of the process that led to the generation of the specific public policy proposals.

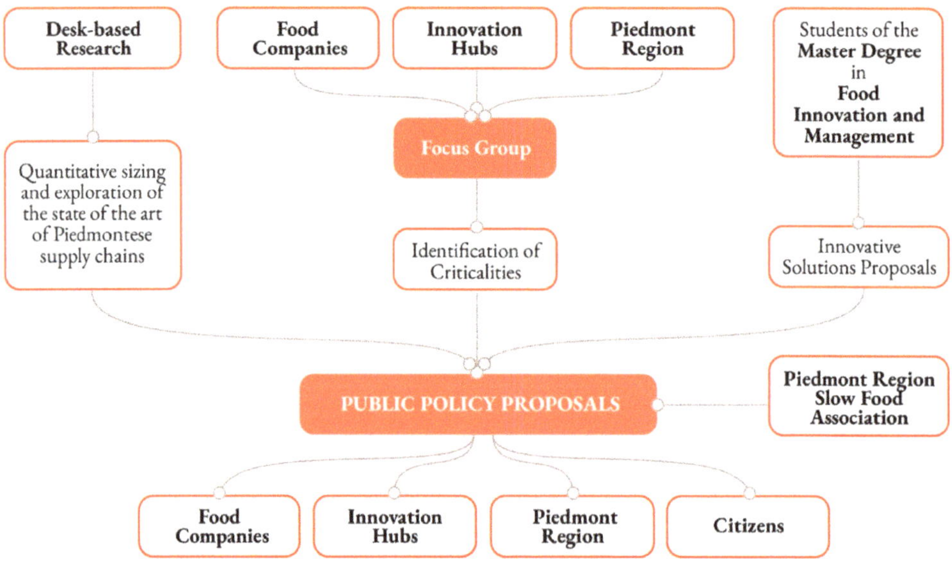

Figure 3. Process followed for public policy proposal generation.

3. Results

The entirety of the results and proposals produced by the *Circular Economy for the Piedmontese Food System Project* were delivered in a detailed report to the regional government of the Piedmont region, in order to support the regional government's ecological transition goals and implementation. However, in the following section, some illustrative examples with the kind of results reported, their mode of presentation, and their relevance to the overarching project objective, are shown, in accordance with the stages of the methodological structure presented in the previous section, with a particular emphasis on the results of the stages of *Academic and Sectoral Document Discovery*, *Quantitative and Qualitative Data Analysis* from different sources, and *Supply Chain Mapping*.

Furthermore, the results from the *Participatory Mechanisms* stage are summarized in the following sections, as the transversal challenges found to be common to all of the five supply chains under consideration are included in the results.

3.1. Examples of Project Outputs

3.1.1. Academic and Sectoral Document Discovery

As the first product of the research, and in order to better understand the complexities of each supply chain at a glance, what were considered relevant production and actor numbers were reported for the latest recorded number, which, in most cases, was 2019 (Figure 4). These aggregate numbers, along with any relevant information were also shown during the focus group sessions, allowing for their validation and for a contextualization of the entire supply chain's scale available to all participants.

Figure 4. The production and actor quantitative, aggregate summary for the dairy supply chain in the Piedmont region.

3.1.2. Quantitative and Qualitative Data Analysis

Furthermore, employing the data from UNISG's Food Industry Monitor, it was possible to create a generalized representation for the aggregate economic performance for the main players in each supply chain, further contextualizing the reality of the supply chain in terms of its economic potential, as well as any possible investment or income challenges that might hinder a smoother ecological transition within the leading companies of the supply chain, and, by extension, smaller actors without the same access to financial or investment opportunities. These results are summarized in quick overview images (Figure 5).

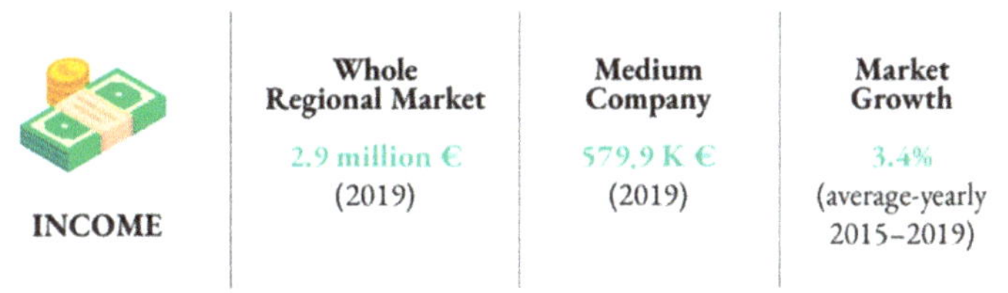

Figure 5. Cumulative supply chain income summary, as presented in the final project report.

In order to better place each supply chain's actors in the Piedmont region's territory, their distribution and possible sub-regional areas to be considered for grouping in future interventions, geographical heat maps were constructed (Figure 6), employing the information from the regional company registry. Furthermore, the geographical representation of the presence of the supply chain's actors in the regional territory helped view areas of overlap among supply chains, where possible transversal interventions could benefit two or more of them at the same time, further guiding policy suggestions by placing the supply chains in a geographical and territorial context.

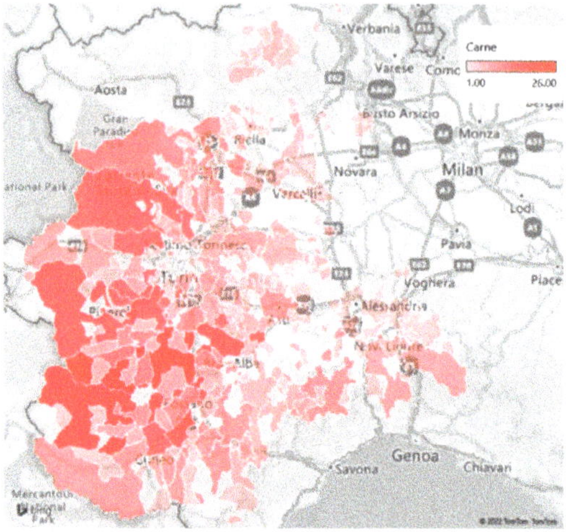

Figure 6. Heatmap of the concentration of businesses classified as participating in the Bovine Beef supply chain according to Unioncamere Piemonte's 2020 database [50]; as presented in the final project report.

3.1.3. Supply Chain Stage Mapping

For each of the supply chains, its corresponding supply chain steps, the number of actors counted as registered in the regional Chambers of Commerce database in each the steps, and the flows of information and material among them, were summarized in simplified supply chain diagrams (Figure 7), intended for a top-level view of supply chain dynamics. These diagrams served for both validation and understanding by each supply chain's selection of stakeholders participating in the focus groups, such that the level of understanding of these sessions, as well as the following steps of the research, were agreed among those involved.

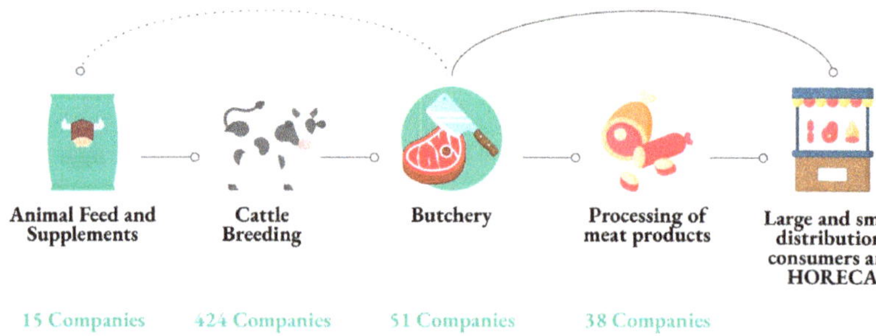

Figure 7. Simplified supply chain map diagram, as validated in the focus group session for the Bovine Beef supply chain.

Additionally, the information from previous diagrams was expanded, based on the information gained during the document and discovery steps of research. This is a more detailed flow diagram, in Figure 8, illustrating the standard steps of the production and supply chain for each product, as well as listing the main inputs, waste, by-products, and outputs for each.

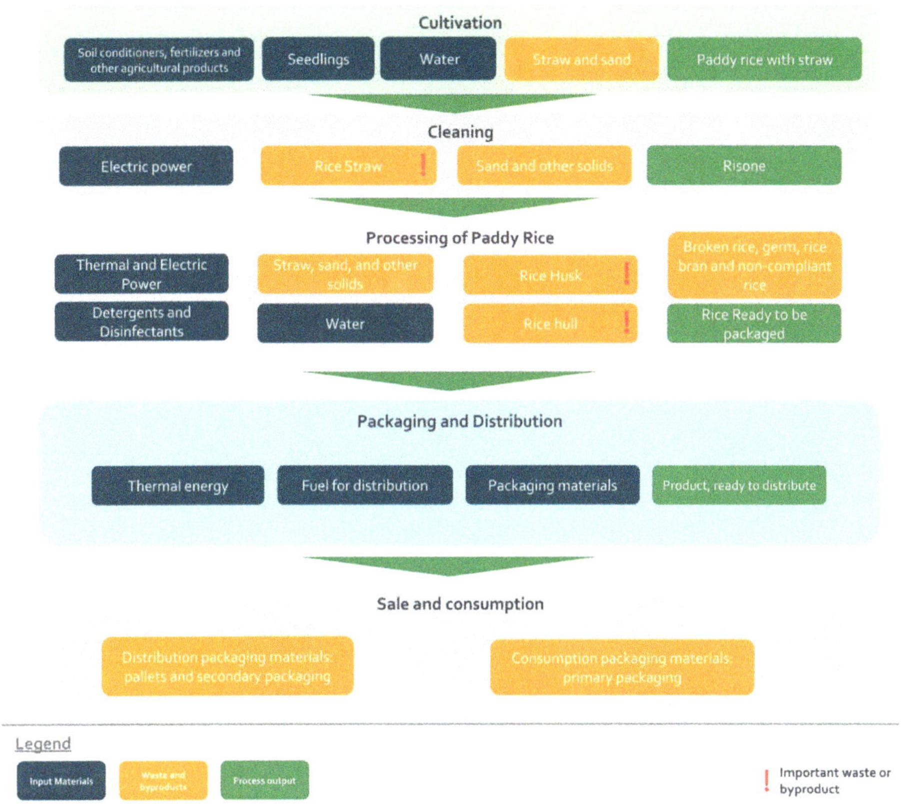

Figure 8. Flow diagram of the main supply chain steps with the main inputs (in blue), waste and by-products (in yellow), and products (in green), as reported both in the focus group sessions, as well as the project report for the Rice supply chain [53].

In addition to this information, an emphasis (marked as a red exclamation point) could be placed on any of the waste/by-products, denoting their importance and/or challenge for the supply chain as a whole, be it in terms of environmental impact on disposal, or volume produced per unit of finished product.

The final diagram produced to illustrate each supply chain's structure and dynamics was an actor categories map (Figure 9), in which all possible independent actor categories were identified and placed within a grouping according to their position within the supply chain. In this case, connections were also drawn according to the information gathered during the steps of document discovery and supply chain mapping, using both sectorial and academic sources in order to list the main actors of each supply chain, as well as their relationships. In it, categories of actors, grouped by their most relevant activity or input to the supply chain were presented, and then ordered into greater categories, representing their role in the supply chain overall. Finally, the direction (uni or bidirectional), intensity (direct/indirect or constant/sporadic) of their relationship were shown using arrows. The criteria employed for defining a relationship were the existence of either financial (e.g., payments for inputs or services), information (e.g., sales numbers, production practices, or marketing information) and material (inputs, semi-finished and finished products) flows.

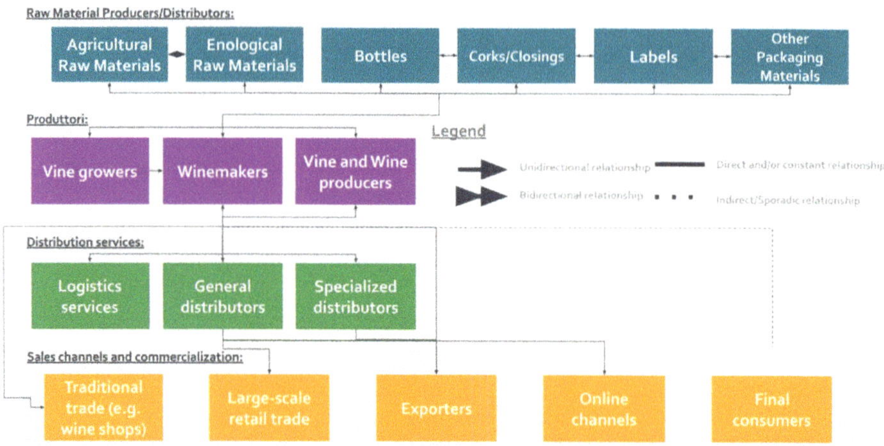

Figure 9. Mapping of the main supply chain actor categories and their groupings for the Wine supply chain, along with their commercial connections.

3.2. Main Waste Products and By-Products Found by the Project

After having carried out a thorough analysis of the material flows for each supply chain under consideration (Figure 8), a significant part of the investigation was dedicated to the current disposal methods for waste and by-products of the production chain and to the main problems related to their management. Thanks to the discussion fostered within the focus groups with the various stakeholders, both critical issues in the standard disposal method and possible solutions for their valorization in a circular way emerged, so that a solution could be found to maintain the value in a production chain of what would otherwise be thrown away.

Different types of packaging waste were indicated in the five supply chains and, through focus group comparisons, some sectors were found to be problematic (as in the case of meat, wine, and water, see Appendix A). However, they are to be considered as part of municipal solid waste management.

It is important to emphasize that those analyzed are five agri-food chains that are intrinsically different from each other and, for this reason, the disposal method for by-products and waste is different in each chain. This is evident in the case of the beef supply chain, where the circular economy is more mature, with a purpose being applied for almost all the outputs from the production process [54] and, therefore, few difficulties are found in the disposal of these, as their value is maintained by using them as inputs in a new production process. In the beef sector, in fact, many animal by-products are commonly used in important production sectors, such as the pharmaceutical, feed, and leather industries, where the by-products are employed to extract bioactive compounds, nutrients, or raw materials for the production of leather, respectively

Innovative Case Study Research

In conclusion, in order to better support the participative stages of the project, a series of 28 relevant case studies for each supply chain were gathered, specifically looking for already implemented, regional-to-multinational-scale initiatives that included circular economy practices or business models as their main feature. These were employed to foster discussion among focus groups participants, as well as included as examples for ideas or practices that could be applied to each of the regional supply chains.

Each case study was summarized on a quick-reference card (Figure 10), in which the main points of interest for their use as a basis for regional proposals (i.e., supply chain stages, principal resources, and reported benefits) were included.

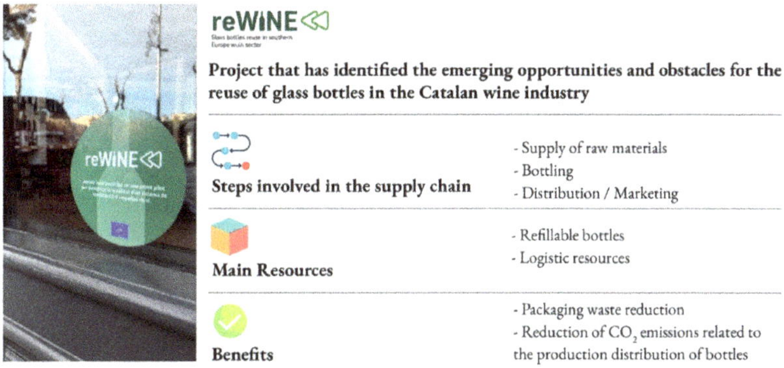

Figure 10. Example of one of the quick-reference cards for a case study regarding the Wine supply chain.

All of the 28 circular economy business model and practice case studies included in the detailed report are shown in Table 1. They represent a relevant overview of the possibilities for circular economy developments in each supply chain, serving as reference for the construction of the proposals tailored to the Piedmont region's reality and current issues in each supply chain.

Table 1. A summary of the case studies presented and validated in the focus group sessions, and included in the final project report.

Supply Chain	Name of the Case History	Circular Business Model/ Practice Shown
Wine	reWINE	System for the return and reuse of glass wine bottles at a local level
	VEGEA	Valorization of grape pomace for manufacturing of leather-like fabrics
	Caviro group	Production of bioenergy and compost from waste from pruning and destemming
	Poliphenolia	Extraction of grape pomace polyphenols for cosmetics production
	NOMACORC	Adoption of cork substitutes from recycled and bio-based alternatives to cork wood
Milk and Cheese	KRINGLOOP WIJZER	Monitoring soil nutrient cycle
	FrieslandCampina	Tool for assessment of biodiversity improvement in the dairy sector
	BIOCOSI'	Production of bioplastics from wastewater
	ORIGAMI Organics	Production of fabrics based on milk waste
	fluence	Biogas production from ricotta whey, buttermilk, and wastewater
	UNTER/EGGER	Production of cosmetics based on nutrients extracted from whey
	Milk Brick	Production of building materials based on milk waste
Rice	VIPOT	Production of biodegradable pots based on rice husk
	GENIA BIOENERGY	Biogas production from rice straw
	IKEA India	Production of furniture items from rice straw
	RICE HOUSE	Production of building materials based on rice waste (husk, chaff, straw)

Table 1. *Cont.*

Supply Chain	Name of the Case History	Circular Business Model/ Practice Shown
Water	PABOCO	Bottles made from sustainably sourced wood fibers
	Carlsberg	Reduction in plastic use in beverage multipacks
	E6PR	Biodegradable and compostable secondary packaging for beverages (multi-pack rings)
	Ferrarelle	Bottles composed of 100 percent R-PET
	VERITAS	Returnable glass bottle service in partnership with retailer
	Pfand System-Germany	Public bottle return system with incentives for adoption
Bovine Meat	BovINE	Reward systems to farmers who practice regenerative agriculture
	Water2Return/Bioazul	Extraction of nutrients in slaughterhouse wastewater
	Circ4Life/Alia	Co-creation of circular synergies among various actors in the supply chain
	La Granda	'Symbiotic farming' for animal husbandry
	BTS	Biogas production through wastewater
	Fileni	Biodegradable and recyclable packaging for meat products

In addition to the analysis of these real case studies, which made it possible to examine the existing circular realities in the panorama of the five selected agri-food chains, 39 students of the graduate degree in Food Innovation and Management at the University of Gastronomic Sciences of Pollenzo were encouraged to respond to the challenges that emerged in the food sectors involved and to draw up hopefully applicable circular project proposals. A total of 20 potentially innovative food projects have emerged, involving the point of view, sensibilities, and ideas of the future generation of professionals in the food sector in the design.

3.3. Challenges and Obstacles to a Regional, Supply Chain-Level Transition

To contextualize the challenges and issues raised during the desk-based research, as well as during the participatory mechanism stages of the project, a series of *transversal issues* (i.e., pertaining to two or more supply chains) (Appendix B) were identified, as well as their connection to the theoretical framework chosen for the research, and their potential impact on the goals set by the regional ecological transition plans.

In this scenario, it is preferable to make a brief focus on the 3 Cs of the Circular Economy for Food, presented in the Figure 11 [48], to better understand why it was decided to analyze the transversal problems also from that point of view.

In short, the new circular economic paradigm, when applied to food, takes as its starting point the preservation and regeneration at a local level of *Capital*, the first C, the natural Capital that contributes to providing goods and ecosystem services for humanity and which are necessary for the survival of the environment from which they are generated. In particular, natural, cultural, and economic Capital are therefore inseparable factors, supported and in dialogue thanks to relational Capital. The second C is *Cyclicality*, which invites us to think in regenerative terms, which comprises the three fundamental concepts of extension, metabolization, and renewability. Extension refers to an expansion of the responsibility of a business; metabolization means the final adding of value, or upcycling [55], with the goal of generating only resources for the same system or another system (biological and technical metabolization cycle) and not waste. In these terms, the emphasis is on renewability, because every action must be in harmony with the regenerative cycles found in nature.

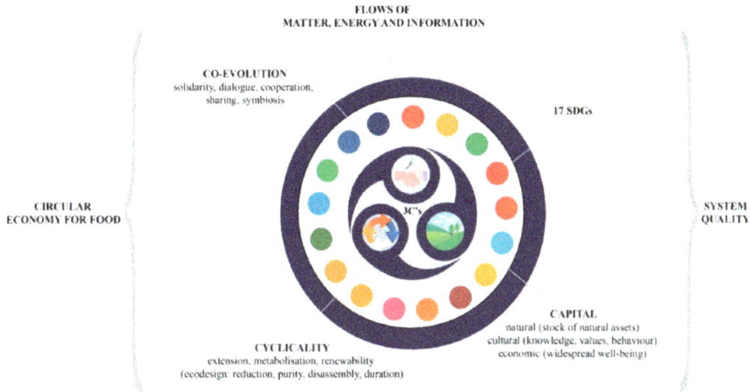

Figure 11. The 3 Cs of the Circular Economy for Food [48].

The last C is *Co-evolution*, inspired by mutualistic symbiosis in nature. It develops thanks to a collaborative paradigm, which through the application of a win–win logic generates a solution that is advantageous for all, among which the environment is included.

Solidarity (between individuals and peoples to reduce social inequality and increase access to quality food), dialogue (between natural and artificial ecosystems, to eliminate the asynchronicity of the human economic model with natural cycles), cooperation (between communities that share values and objectives), sharing (of materials, energy, and information to accelerate the transition and facilitate evolution) and symbiosis (between businesses and between businesses and the community, the local area and the five natural kingdoms) are the priorities on which we must work to give resilience to the circular economic paradigm.

4. Discussion

Once all individual and transversal issues about the five supply chains considered were analyzed and listed, a series of transversal solutions were identified (Table 2), to give their potential for inter-supply chain efficiencies and to generate a higher impact on the food supply chains analyzed, naturally considering the interconnectedness of issues and solutions in the scenario of *Ecological Transition* and *Circular Economy*. In fact, one of the most meaningful challenges for the development and the expansion of integrated food policies was the definition of a series of cross-cutting solutions.

Table 2. The relevant transversal opportunities.

N°	Relevant Transversal Opportunities	Affected Supply Chains	SDGs Affected	3 Cs of CEFF	Relationship to Issues
1	Biogas production through shared plants for the recovery of waste and by-products of organic nature				3—Air pollution 7—The volatility of energy prices 9—Management and treatment of wastewater from production processes 11—The low valorization of organic waste/subproducts 13—Individual and little collective effort in executing and promoting circular initiatives

Table 2. *Cont.*

N°	Relevant Transversal Opportunities	Affected Supply Chains	SDGs Affected	3 Cs of CEFF	Relationship to Issues
2	Use of regenerative cultural practices and symbiotic agriculture for biodiversity presentation and resilience				1—Climate Change 3—Air pollution 4—Biodiversity loss 8—The lack of attention to animal welfare 9—Management and treatment of wastewater from production processes 13—Individual and little collective effort in executing and promoting circular initiatives
3	Use of by-products for green building, to propose an alternative to traditional materials made from non-renewable resources				3—Air pollution 7—The volatility of energy prices 10—Logistics for the management of waste/organic by-products 11—The low valorization of organic waste/subproducts 12—Lack of communication of information or best practices among the various players in the same supply chain 13—Individual and little collective effort in executing and promoting circular initiatives
4	Cross-sectional research on the production of bioplastics from organic packaging waste				1—Climate Change 3—Air pollution 5—Exploitation of non-renewable resources for packaging production 6—The end of life of after-consumer packaging 9—Management and treatment of wastewater from production processes 11—The low valorization of organic waste/subproducts 12—Lack of communication of information or best practices among the various players in the same supply chain 13—Individual and little collective effort in executing and promoting circular initiatives 14—Difficulty of dialogue and confrontation of companies/innovation poles with regional institutions

Table 2. *Cont.*

N°	Relevant Transversal Opportunities	Affected Supply Chains	SDGs Affected	3 Cs of CEFF	Relationship to Issues
5	Creation of joint participatory tables for the common achievement of competitiveness and sustainability objectives		17		12—Lack of communication of information or best practices among the various players in the same supply chain 13—Individual and little collective effort in executing and promoting circular initiatives 14—Difficulty of dialogue and confrontation of companies/innovation poles with regional institutions
6	Using R-PET for water bottles		9 12		1—Climate Change 3—Air pollution 5—Exploitation of non-renewable resources for packaging production 6—The end of life of after-consumer packaging 14—Difficulty of dialogue and confrontation of companies/innovation poles with regional institutions
7	Extraction of nutrients from processing and manufacturing waste and by-products for pharmaceutical and cosmetic production		9 12		9—Management and treatment of wastewater from production processes 10—Logistics for the management of waste/organic by-products 11—The low valorization of organic waste/subproducts 12—Lack of communication of information or best practices among the various players in the same supply chain 13—Individual and little collective effort in executing and promoting circular initiatives
8	Realization of spaces that exploit the sharing economy for the creation of economies of scale		9 8 17		7—The volatility of energy prices 9—Management and treatment of wastewater from production processes 10—Logistics for the management of waste/organic by-products 12—Lack of communication of information or best practices among the various players in the same supply chain 13—Individual and little collective effort in executing and promoting circular initiatives 14—Difficulty of dialogue and confrontation of companies/innovation poles with regional institutions

The *Transversal Opportunities* were analyzed through various aspects, such as:

- *Affected supply chain*: in particular, describing which supply chain they refer to;
- *SDGs affected:* describing which of the SDGs are affected by the opportunity;
- *3 Cs of Circular Economy for Food:* indicating which 3 C they concern—Capital, Cyclicality, Coevolution;
- *Related issues:* describing which other transversal problems are referred to. In particular, reporting the content connection between the transversal opportunities and the individual issues. In particular, they refer to the contents present in the regional MAS.

As a final result of the research, a series of *Public Policy Proposals* were presented to the regional institutional bodies. In particular, they were based on the evidence gathered and analyzed along all stages of the project. The proposals were designed in such a way that they could be indirect to the Piedmontese territory and its supply chains, as well as being consistent with the goals for Ecological Transition and the CEFF framework [48].

Especially, the analysis of the five chains allowed the adoption of a systemic approach based on the needs of the territory, suggesting integrated and coherent recommendations with the SRSvS to align different political agendas and cross-cutting issues. From the synthesis of these, some *Transversal Public Policy Proposals* have been developed and collected in Table 3, which suggest interventions, tools, and strategies common to all research focuses, to be implemented according to different contexts.

Table 3. The Transversal Public Policy Proposals.

N°	Transversal Public Policy Proposals	Affected SDGs	3C of CEFF
1	Promote business participation in specific funding programs aimed at boosting agribusiness investments to improve the competitiveness of the agricultural sector, ensure sustainable management of natural resources, and promote climate actions, achieve balanced territorial development of rural economies and communities (supply chain and district contracts, RDP 2014–2022)		
2	Promote new forms of territorial aggregation between enterprises aimed at fostering innovation in agribusiness and the integration of activities characterized by territorial proximity (food district contracts)		
3	Involvement of stakeholders through the establishment of specific working tables to identify shared solutions with impacts on the community		

Table 3. *Cont.*

N°	Transversal Public Policy Proposals	Affected SDGs	3C of CEFF
4	Advisory activities and accompaniment of economic operators toward the development of a by-product and waste supply chain and to the development of new sustainable materials from them.		
5	Connecting the regional economic tissue with research and development facilities (Piedmont universities, research institutions, innovation hubs, etc.) by enhancing their expertise for positive spillover to their local area		
6	Promoting the digitization of bureaucracy and its simplification/streamlining		
7	Mapping the supply chain for by-products and waste in the Piedmont region		
8	Promote the provision of training courses aimed at practitioners focused on sustainable production systems, facilitating access to resources, best practices, and useful tools		
9	Involvement of stakeholders (municipalities, farmers, restaurateurs, teachers, consumers, recycling operators) by setting up roundtables to promote conscious and responsible consumption by the citizenry		

The Transversal Public Policy Proposals were proposed and analyzed through two of four aspects already mentioned for the Transversal Opportunities: SDGs affected and the 3 Cs of the Circular Economy for Food.

These proposals are not only transversal to all supply chains, but points of contact between them can also be highlighted. For example, the relaunch of investments in the agri-food sector requires a combination of interventions that promote synergy between companies and, at the same time, support their financial commitments.

There is also a major issue of capacity development, namely, the need to create skills within companies and the public administration. In fact, during the analysis of many supply chains, training gaps were found in relation to the impact (especially in the long term) of good/bad agricultural practices, as well as technological and *information technology* gaps. Digitization is now a necessary step to remain competitive in the current context, allowing companies that decide to invest to work more productively, and plan and organize processes more efficiently [56].

In addition to this, the Piedmont area has a specific need, as emerged during the phases of the holistic diagnosis: to strengthen the network at several levels, promoting positive synergies both between operators in the same sector and between players in the same supply chain but who work in different areas.

From this point of view, even those network nodes not included in this research could be included in subsequent analyses (such as restaurateurs, teachers, but also consumers in general) whose awareness and personal decision-making capacity certainly plays a fundamental role in transition to a more sustainable and shared Piedmontese food system

The information obtained through the research and design path addressed was useful in outlining which circular *Public Policy Proposals* should be indicated to the institutions in order to improve the performance of the circular economy in the region.

With a view to analyzing the potential, already existing circular aspects within the supply chains taken into consideration, during the research process, there was a need to further explore the presence or absence of a waste and by-product recovery chain of the five supply chains, as part of the mapping stage, as well as including any potential actors involved in these activities, in further stages.

This need represented a more solid information base for the phase of elaboration of local policy proposals, expanding the range of analysis of the individual supply chains and allowing us to consider also the phases subsequent to those of the sale of a food product, or those connected to the disposal and to the correct management of the outputs Extending the analysis to this phase, not always considered in the evaluation processes of a supply chain, allows a more effective integration of further waste valorization chains in the regional economic landscape, triggering high-value systemic economic and production networks, in terms of sustainability. At this point, the phase of participatory processes could be extended to include these additional stakeholders, beginning the construction of a common perspective.

In fact, the participatory mechanisms put in place along this research path represented an opportunity for sharing knowledge, for networking among professionals and companies that, albeit in the same territory and in the same agri-food sector, use different production methods.

In brief, the strengths of the research developed were already highlighted and guaranteed by a broad, evidence-based, and constructed picture of the current state of art of the five Piedmontese agri-food supply chains—*wine, milk and cheese, rice, water, and bovine meat*—in which over 3200 companies were analyzed, and by their hypothetical future scenario, which emerged during the focus groups. In detail, as reported above, the scientific research enclosed the results of the five focus groups, in which a total of 35 people participated in an active and collaborative way, in particular, in a free and spontaneous discussion about the sustainability perspective of the food supply chain. Of this number, 26 represented the rich variety of the agri-food sector: 21 companies, including agricultural producers, processors, and distributors, with the participation of both family-run businesses and some of the largest agri-food companies in the Piedmont region. In this context, the information collected, the points of view, and the concerns expressed during the focus groups, the inspiring case studies and the priorities of the SRSvS, were summarized in:

- 45 recommendations/proposals for impacting actions on SRSvS divided among the five sectors most deeply involved in research innovation poles and research centers;

- 11 recommendations with a transversal approach (on the fivesupply chains);
- 20 indications on areas that should be explored through specific applied research (on the five supply chains).

5. Conclusions

Food has always been an element of connection between the environment, culture, society, and economy. For this reason, nowadays, it can be the key to addressing some of the many and complex contemporary challenges [9]. From this point of view, public policies, in conjunction with a socially and environmentally responsible design approach, become an indispensable tool and aid to act concretely, to respond to the needs of territories and communities, and to support their identity, acting on a local level, but leading to wider positive spillovers [26,57–59]. In more detail, the systemic approach adopted in this research path, located in the Piedmont area, plays a significant role in the design of adequate articulated solutions in response to complex and interconnected problems, such as those connected to a local linear agri-food system that requires a sustainable transition in harmony with the different peculiarities of the territory. To achieve this objective, which started from the analysis of the specific agri-food chains of the region to direct the development of targeted food policies, the active participation of the interested parties was fundamental, through the creation of real focus groups, whose objective knowledge supports policy makers in the formulation and implementation of more sustainable and identity-based decision-making and operational plans. Certainly, this research will be able to undertake further future developments, expanding the field of action not only to other Piedmontese companies belonging to the agri-food chains already mentioned, but also to entrepreneurial realities outside the identified territory. Furthermore, continuing to exploit a multi-stakeholder approach, it will be interesting to implement the network of territorial connections, among companies, research institutions, universities, and agri-food professionals, more generally, in order to find new and more interesting opportunities for the enhancement of agri-food sector waste, supporting the levels of territorial resilience. This potential direction necessarily implies the continuation of a transdisciplinary research, which aspires to innovative results through the strategic interaction among different disciplines and areas of study, involving not only agricultural and gastronomic sciences, but also the branches of chemistry–energy, systemic design, eco-design of processes, products, services, biological and environmental sciences, and much more. In this scenario, the role of an enlarged and permeable scientific community is essential to contribute to the evaluation of the opportunities of the circular economy at the level of the food supply chain, first of all to spread awareness of the existence of good practices and innovative processes capable of making organic and non-organic waste of the agri-food chain new and precious resources for other production possibilities; secondly, to decisively support political decision makers in undertaking sustainable long-term solutions. In fact, it is good to remember that without a collaborative and co-evolutionary effort, scientific research is not enough. At the same time, the political class cannot undertake sustainable pragmatic solutions without interaction with the scientific community, with the network of local businesses, citizens, and the multitude of local actors. The natural, social, and cultural capital of the planet requires urgent actions, in terms of protection and regeneration; for this reason, it is right to ask how to accelerate the ecological transition. Circular solutions close to those proposed in this contribution are progressively developing, in the current panorama; however, for a paradigm change to really take hold, it is necessary to guarantee an economic feasibility of the solutions found, which today represents perhaps one of the most complex limits to be overcome in research projects in the sector of the circular economy and environmental sustainability. For this reason, it is necessary to develop transparent support methods for companies and organizations that decide to undertake circular and sustainable solutions, through a system of benefits and concessions, which can reduce the economic challenges to be undertaken along this transition path. This would consequently make it possible to make more accessible products and services that derive from sustainable and circular choices, which today, as in a paradox, remain the prerogative of a high-spending public, making sustainability an economic, as well as an environmental, challenge.

Author Contributions: Conceptualization, F.F., I.E.P.B. and E.T.; methodology, F.F., A.S. and F.R.; validation, F.F., F.R. and A.S.; formal analysis, F.F., I.E.P.B., E.T., O.P. and D.Z.; investigation, F.F., A.S., F.R., I.E.P.B., E.T., O.P. and D.Z.; resources, F.F., I.E.P.B., E.T., O.P. and D.Z.; data curation, F.F., F.R., A.S., I.E.P.B., E.T., O.P. and D.Z.; writing—original draft preparation, F.F., E.T. and I.E.P.B.; writing—review and editing, F.F., F.R. and A.S.; visualization, F.F., A.S. and F.R.; supervision, F.F., A.S. and F.R.; project administration, F.F.; funding acquisition, F.F. All authors have read and agreed to the published version of the manuscript.

Funding: This research was funded by Regione Piemonte. The APC was funded by the University of Gastronomic Sciences of Pollenzo.

Institutional Review Board Statement: Not applicable for studies not involving humans or animals.

Informed Consent Statement: Not applicable for studies not involving humans.

Data Availability Statement: A part of the data was obtained from Unioncamere Piemonte and is available with the permission of Unioncamere Piemonte.

Acknowledgments: We acknowledge the support of the Regione Piemonte and the local agri-food companies that took part in the focus groups, the strategic partners and supporting members of the University of Gastronomic Sciences of Pollenzo and, finally, its Fundraising and External Relations Office.

Conflicts of Interest: The funders took part in the organized focus groups and contributed to the data collection essential to the realization of the research project described in this paper, making their databases available.

Appendix A

This section contains details and supplementary data on the main issues identified within the individual supply chains.

Table A1. A Summary of the main issues with standard disposal, and circular opportunities for waste and by-product valorization by the supply chain.

Supply Chain	Waste or By-Product	Standard Disposal Methods	Disposal-Related Issues	Circular Opportunity of Valorization
Wine	Prunings and Stalks	• Livestock Feeding • Burning in the field • Soil Fertilizer	• Poorly digestible for animals • Creation of particulate matter in the atmosphere	• Biogas production • Recovery of antioxidant compounds (polyphenols) • Activated carbon or lignocellulosic fractionation from stalks • Bio-based packaging from prunings
	Grape Pomace	• Production of distillates (i.e., grappa) • Spread on the ground • Compost	• Release of carbon substances	• Biogas production • Extraction of polyphenols for cosmetics/pharmaceutical purposes • Leather-like fabric production • Natural textile dye
	Wine lees	• Used for wine aging • Distillery	• Can give an unpleasant taste	• Biogas production • Compost • Natural textile dye

Table A1. *Cont.*

Supply Chain	Waste or By-Product	Standard Disposal Methods	Disposal-Related Issues	Circular Opportunity of Valorization
	Waste water and filter cakes	Regular disposal	Difficult to manage	• Biogas production • Extraction of components
Dairy and Cheese	Whey	• Ricotta production • Additive • Livestock Feeding	• Large quantity produced	• Biogas production • Extraction of components for cosmet-ics/pharmaceutical purposes • Bioplastics production • Milk fiber production
	Packaged expired and non-compliant milk at the factory	• Thrown away according to the rules of recycling	• Food waste	• Livestock feeding • Materials for bio-building • Bioplastic production • Biogas production • Milk fiber production
Dairy and Cheese and Bovine Meat	Waste waters and Sludge	• Regular disposal	• Difficult to manage	• Bioplastic production from waste water • Filtered sludge for biogas production, using the digestate as litter for dairy cow breeding
	Manure	• Soil fertilizer	• If not properly handled, it can create pollutants	• Biogas production
Rice	Rice Straw	• Livestock Feeding • Soil fertilizer • Burning in the field	• Poorly digestible for animals • Deterioration of the growth of rice plants and increased methane emissions • Creation of particulate matter in the atmosphere	• Materials for bio-building • Biogas production • Furnishing objects
	Rice Husk (lolla)	• Disposed as special waste • Animal litter • Electricity production	• Difficult to dispose for high percentage of silica	• Materials for bio-building • Biodegradable pots for floriculture • Extraction of components for cosmetics/pharmaceu tical purposes • Production of wood-like material
	Rice Hull (pula)	• Livestock breeding	• Large quantity produced	• Materials for bio-building • Extraction of components for cosmetics/pharmaceu tical purposes
Bottled Water	Plastic water bottles	• Disposed of according to the regulations of recycling collection	• Large presence of plastic material • Often dispersed in the environment by consumers • Not designed for repeated use	• R-PET production • Bottles made from sustainable wood fiber • Returnable glass bottle to the shop • Deposit return model

Table A1. *Cont.*

Supply Chain	Waste or By-Product	Standard Disposal Methods	Disposal-Related Issues	Circular Opportunity of Valorization
Bovine Meat	Blood	• Soil fertilization	• If blood is mixed at the slaughterhouse, it is not legally possible to use it as fertilizer	• Creating a bovine-only slaughterhouse for soil fertilization • Biogas production
Bovine Meat Rice Dairy and Cheese	Final packaging	• Disposed of according to the regulations of recycling collection	• Large presence of plastic material • Often dispersed in the environment by consumers	• Use of biodegradable and recyclable packaging • Use of recycled PET trays

Appendix B

In this section, the *Transversal issues* were analyzed through various aspects, such as:

- *Supply chain:* in particular, describing which supply chain they relate to;
- *SDGs affected:* describing which SDGs they refer to;
- *3 Cs of the Circular Economy For Food:* describing which 3 C they pertain to—Capital, Cyclicality, Coevolution;
- *Related issues:* describing other transversal problems concerned. In particular, reporting the content connection between the individual problems.

Table A2. The Transversal issues.

N°	Transversal Issues	Supply Chains	SDGs Affected	3 Cs of CEFF	Relationship to Issues
1	Climate Change				2—Soil acidification 3—Air pollution 4—Biodiversity loss 5—Exploitation of non-renewable resources for packaging production 9—Management and treatment of wastewater from production processes 12—Lack of communication of information or best practices among the various players in the same supply chain

Table A2. *Cont.*

N°	Transversal Issues	Supply Chains	SDGs Affected	3 Cs of CEFF	Relationship to Issues
2	Soil acidification				1—Climate Change 3—Air pollution 4—Biodiversity loss 8—The lack of attention to animal welfare 9—Management and treatment of wastewater from production processes 11—The low valorization of organic waste/subproducts 12—Lack of communication of information or best practices among the various players in the same supply chain
3	Air pollution				1—Climate Change 2—Soil acidification 4—Biodiversity loss 5—Exploitation of non-renewable resources for packaging production 6—The end of life of after-consumer packaging 9—Management and treatment of wastewater from production processes 10—Logistics for the management of waste/organic by-products 11—The low valorization of organic waste/subproducts 12—Lack of communication of information or best practices among the various players in the same supply chain 13—Individual and little collective effort in executing and promoting circular initiatives
4	Biodiversity loss				1—Climate Change 2—Soil acidification 3—Air pollution 5—Exploitation of non-renewable resources for packaging production 8—The lack of attention to animal welfare 12—Lack of communication of information or best practices among the various players in the same supply chain

Table A2. *Cont.*

N°	Transversal Issues	Supply Chains	SDGs Affected	3 Cs of CEFF	Relationship to Issues
5	Exploitation of non-renewable resources for packaging production				1—Climate Change 3—Air pollution 4—Biodiversity loss 7—The volatility of energy prices 13—Individual and little collective effort in executing and promoting circular initiatives
6	The end of life of after-consumer packaging				1—Climate Change 3—Air pollution 5—Exploitation of non-renewable resources for packaging production
7	The volatility of energy prices			Social and economic Capital	1—Climate Change 5—Exploitation of non-renewable resources for packaging production
8	The lack of attention to animal welfare				2—Soil acidification 3—Air pollution 4—Biodiversity loss

Table A2. *Cont.*

N°	Transversal Issues	Supply Chains	SDGs Affected	3 Cs of CEFF	Relationship to Issues
9	Management and treatment of wastewater from production processes		6, 13, 14, 15		2—Soil acidification 3—Air pollution 10—Logistics for the management of waste/organic by-products 11—The low valorization of organic waste/subproducts 12—Lack of communication of information or best practices among the various players in the same supply chain 13—Individual and little collective effort in executing and promoting circular initiatives
10	Logistics for the management of waste/organic by-products		9		9—Management and treatment of wastewater from production processes 11—The low valorization of organic waste/subproducts 12—Lack of communication of information or best practices among the various players in the same supply chain 13—Individual and little collective effort in executing and promoting circular initiatives
11	The low valorization of organic waste/subproducts		9, 12		9—Management and treatment of wastewater from production processes 10—Logistics for the management of waste/organic by-products 12—Lack of communication of information or best practices among the various players in the same supply chain 13—Individual and little collective effort in executing and promoting circular initiatives
12	Lack of communication of information or best practices among the various players in the same supply chain		4, 9, 11, 17		1—Climate Change 2—Soil acidification 3—Air pollution 4—Biodiversity loss 8—The lack of attention to animal welfare 9—Management and treatment of wastewater from production processes 10—Logistics for the management of waste/organic by-products 11—The low valorization of organic waste/subproducts 13—Individual and little collective effort in executing and promoting circular initiatives 14—Difficulty of dialogue and confrontation of companies/innovation poles with regional institutions

Table A2. *Cont.*

N°	Transversal Issues	Supply Chains	SDGs Affected	3 Cs of CEFF	Relationship to Issues
13	Individual and little collective effort in executing and promoting circular initiatives				5—Exploitation of non-renewable resources for packaging production 6—The end of life of after-consumer packaging 8—The lack of attention to animal welfare 9—Management and treatment of wastewater from production processes 10—Logistics for the management of waste/organic by-products
14	Difficulty of dialogue and confrontation of compa-nies/innovation poles with regional institutions				9—Management and treatment of wastewater from production processes 10—Logistics for the management of waste/organic by-products 11—The low valorization of organic waste/subproducts 13—Individual and little collective effort in executing and promoting circular initiatives

References

1. Ellen Macarthur Foundation (FME). Towards the Circular Economy: Economic and Business Rationale for an Accelerated Transition. Available online: https://emf.thirdlight.com/link/x8ay372a3r11-k6775n/@/preview/1?o (accessed on 27 May 2022).
2. Buchanan, R. Wicked Problems in Design Thinking. *Des. Issues* **1992**, *8*, 5–21. [CrossRef]
3. Rittel, H.W.J.; Webber, M.M. Dilemmas in a general theory of planning. *Policy Sci.* **1973**, *4*, 155–169. [CrossRef]
4. Jones, P.H. Systemic Design Principles for Complex Social Systems. *Soc. Syst. Des.* **2014**, *1*, 91–128.
5. Mirabelli, G.; Solina, V. Optimization strategies for the integrated management of perishable supply chains: A literature review. *J. Ind. Eng. Manag.* **2022**, *15*, 58–91. [CrossRef]
6. Capra, F.; Luisi, P.L. *Vita e Natura—Una Visione Sistemica*; Aboca: Sansepolcro, Italy, 2014.
7. Bertalanffy, L.V. *General Systems Theory*; George Braziller: New York, NY, USA, 1968.
8. Peruccio, P.; Savina, A. The planning of a territorial network among food, health and design. *Agathón | Int. J. Archit. Art Des.* **2020**, *8*, 262–271.
9. Bistagnino, L.; Cantino, V.; Gibello, P.; Puddu, E.; Zaccone, D. *Il Settore Agroalimentare. Scenari e Percorsi di Crescita Sostenibile*; Slow Food Editore: Bra, Italy; Deloitte & Touche S.p.A: Milan, Italy, 2015.
10. Bistagnino, L. *Systemic Design—Designing the Productive and Environmental Sustainability*; Slow Food Editore: Bra, Italy, 2011.
11. Grimaldi, P. La cultura contadina: Le radici del pensiero circolare. In *Circular Economy for Food. Materia, Energia e Conoscenza in Circolo*; Fassio, F., Tecco, N., Eds.; Edizioni Ambiente: Milano, Italy, 2018; pp. 13–18.
12. Geissdoerfer, M.; Savaget, P.; Bocken, N.M.; Hultink, E.J. The Circular Economy—A New Sustainability Paradigm? *J. Clean. Prod.* **2017**, *143*, 757–768. [CrossRef]
13. Esposito, B.; Sessa, M.; Sica, D.; Malandrino, O. Towards Circular Economy in the Agri-Food Sector. A Systematic Literature Review. *Sustainability* **2020**, *12*, 7401. [CrossRef]

14. Giarini, O.; Stahel, W.R. *The Limits to Certainty: Facing Risks in the New Service Economy*; Kluwer Academic: Alphen aan den Rijn, The Netherlands, 1989.
15. McDonough, W.; Braungart, M. *Cradle to Cradle: Remaking the Way We Make Things*; North Point Press: Milano, Italy, 2002.
16. Bonacina, R. Economia Circolare: Il Nuovo Paradigma Della Sostenibilità. Master's Thesis, LUISS—Libera Università Internazionale degli Studi Sociali Guido Carli, Rome, Italy, 2020.
17. Bompan, E.; Brambilla, I.N. *Che Cosa è l'Economia Circolare?* Edizioni Ambiente: Milan, Italy, 2016.
18. Un Green Deal Europeo. Available online: https://ec.europa.eu/info/strategy/priorities-2019-2024/european-green-deal_it (accessed on 30 May 2022).
19. European Commission. Farm to Fork Strategy. Available online: https://ec.europa.eu/food/horizontal-topics/farm-fork-strategy_it (accessed on 30 May 2022).
20. Rockström, J.; Sukhdev, P. *How Food Connects all the SDGs; Opening KeyNote Speech at the 2016 EAT Forum*; Stockholm Resilience Center: Stockholm, Sweden, 2016.
21. Brennan, G.; Tennant, M.; Blomsma, F. Business and production solutions: Closing the Loop and the circular economy. In *Sustainability: Key Issues*; Kopnina, H., Shoreman-Ouimet, E., Eds.; Routledge: New York, NY, USA, 2015; pp. 219–239.
22. European Commission. Circular Economy Action Plan. Available online: https://ec.europa.eu/commission/presscorner/detail/en/fs_20_437 (accessed on 25 May 2022).
23. La Strategia Nazionale per lo Sviluppo Sostenibile. Available online: https://www.mite.gov.it/pagina/la-strategia-nazionale-lo-sviluppo-sostenibile (accessed on 30 May 2022).
24. Pereira, L.M.; Drimie, S.; Maciejewski, K.; Tonissen, P.B.; Biggs, R. Food system transformation: Integrating a political–economy and social–ecological approach to regime shifts. *Int. J. Environ. Res. Public Health* **2020**, *17*, 1313. [CrossRef] [PubMed]
25. Bistagnino, L. *MicroMACRO. Micro Relazioni Come Rete Vitale Del Sistema Economico e Produttivo*; Edizioni Ambiente: Milan, Italy, 2016.
26. Sonnino, R.; Milbourne, P. Food system transformation: A progressive place-based approach. *Local Environ.* **2022**, *27*, 915–926. [CrossRef]
27. Food and Agriculture Organization of the United Nations (FAO). Transforming Food and Agriculture to Achieve the SDGs: 20 Interconnected Actions to Guide Decision-Makers. Available online: https://www.fao.org/3/I9900EN/i9900en.pdf (accessed on 25 May 2022).
28. Nutrition Connect. Available online: https://symposium.bayes.city.ac.uk/__data/assets/pdf_file/0018/504621/7643_Brief-5_Policy_coherence_in_food_systems_2021_SP_AW.pdf (accessed on 28 May 2022).
29. Barbero, S. *Systemic Design Method Guide for Policymaking: A Circular Europe on the Way*; Allemandi: Turin, Italy, 2017.
30. Eraydin, A. The role of regional policies along with the external and endogenous factors in the resilience of Regions. *Camb. J. Reg. Econ. Soc.* **2015**, *9*, 217–234. [CrossRef]
31. European Commission. The Role of Regional Policy in the Future of Europe. European Commission. Available online: https://ec.europa.eu/regional_policy/sources/docgener/panorama/pdf/mag39/mag39_en.pdf (accessed on 26 May 2022).
32. Ciaffi, A.; Odone, C. *Le Politiche Europee e Il Ruolo Delle Regioni: Dalla Teoria Alla Pratica*; Centro Interregionale Studi e Documentazione: Rome, Italy, 2017.
33. Calori, A.; Magarini, A. *Food and the Cities—Politiche Del Cibo Per Città Sostenibili*; Edizioni Ambiente: Milan, Italy, 2015.
34. Global Panel on Agriculture and Food Systems for Nutrition. Future Food Systems: For People, Our Planet, and Prosperity. London, UK. Available online: https://www.glopan.org/wp-content/uploads/2020/09/Foresight-2.0_Future-Food-Systems_For-people-our-planet-and-prosperity.pdf (accessed on 10 May 2022).
35. Deakin, M.; Borrelli, N.; Diamantini, D. *The Governance of City Food Systems: Case Studies from Around the World*; Feltrinelli: Milano, Italy, 2016.
36. Mayne, R.; Green, D.; Guijt, I.; Walsh, M.; English, R.; Cairney, P. Using evidence to influence policy: Oxfam's experience. *Palgrave Commun.* **2018**, *4*, 122. [CrossRef]
37. Cairney, P. The Barriers to Evidence Based Policymaking in Environmental Policy. 2015. Available online: https://paulcairney.wordpress.com/2015/08/21/the-barriers-to-evidence-based-policymaking-in-environmental-policy/ (accessed on 15 May 2022).
38. Den Boer, A.C.; Kok, K.P.; Gill, M.; Breda, J.; Cahill, J.; Callenius, C.; Caron, P.; Damianova, Z.; Gurinovic, M.; Lähteenmäki, L.; et al. Research and innovation as a catalyst for food system transformation. Trends. *Food Sci. Technol.* **2021**, *107*, 154. [CrossRef] [PubMed]
39. United Nations. Cities—United Nations Sustainable Development Action 2015. Available online: https://www.un.org/sustainabledevelopment/cities/ (accessed on 28 March 2021).
40. Jurgilevich, A.; Birge, T.; Kentala-Lehtonen, J.; Korhonen-Kurki, K.; Pietikainen, J.; Saikku, L.; Schosler, H. Transition towards Circular Economy in the Food System. *Sustainability* **2016**, *8*, 69. [CrossRef]
41. IPES FOOD. Verso Una POLITICA Alimentare Comune Per l'Unione Europea. Le Riforme Politiche e Gli Aggiustamenti Necessari Alla Creazione di Sistemi Alimentari Sostenibili in Europa. Available online: https://www.ipes-food.org/_img/upload/files/CFP_ExecSummary_IT.pdf (accessed on 25 May 2022).
42. Hawkes, C.; Parsons, K. Brief 1: Tackling Food Systems Challenges: The Role of Food Policy. In *Rethinking Food Policy: A Fresh Approach to Policy and Practice*; Centre for Food Policy: London, UK, 2019.

43. Lever, J.; Sonnino, R. Food system transformation for sustainable city-regions: Exploring the potential of circular economies. *Reg. Stud.* **2022**, 1–13. [CrossRef]
44. Bhamra, T.; Hernandez, R.; Mawle, R. Sustainability: Methods and practices. In *The Handbook of Design for Sustainability*; Walker, S., Giard, J., Eds.; Bloomsbury: London, UK, 2013; pp. 100–120.
45. Fassio, F.; Tecco, N. *Circular Economy for Food: Materia, Energia e Conoscenza in Circolo*; Edizioni Ambiente: Milan, Italy, 2018.
46. World Health Organization. One Health. Available online: https://www.who.int/health-topics/one-health#tab=tab_1 (accessed on 20 May 2022).
47. Fassio, F.; Cionchi, E.; Tondella, A. The Circular Economy for Food in Future Cities. Good Practices That Define Smart Food. *Agathón | Int. J. Archit. Art Des.* **2020**, *8*, 244–253.
48. Fassio, F. The 3 C's of the Circular Economy for Food. A conceptual framework for circular design in the food system. *Diid Disegno Ind. Ind. Des.* **2021**, *73*, 8. [CrossRef]
49. La Strategia di Sviluppo Sostenibile. Verso un Presente Sostenibile. Available online: https://www.regione.piemonte.it/web/temi/strategia-sviluppo-sostenibile (accessed on 30 May 2022).
50. TELEMACO-INFOCAMERE. Available online: https://www.registroimprese.it/area-utente#elenchi (accessed on 20 May 2022).
51. Slowfood.It. Available online: https://www.slowfood.it (accessed on 20 May 2022).
52. Ministero della Transizione Ecologica. Piano Per La Transizione Ecologica. Available online: https://www.mite.gov.it/pagina/piano-la-transizione-ecologica (accessed on 15 June 2022).
53. Fiore, E.; Stabellini, B.; Tamborrini, P. A Systemic Design Approach Applied to Rice and Wine Value Chains. The Case of the InnovaEcoFood Project in Piedmont (Italy). *Sustainability* **2020**, *12*, 9272. [CrossRef]
54. Bernardi, E.; Capri, E.; Pulina, G. *La Sostenibilità Delle Carni e Dei Salumi in Italia: Salute, Sicurezza, Ambiente, Benessere Animale, Economia Circolare e Lotta Allo Spreco*; FrancoAngeli: Milan, Italy, 2018.
55. McDonough, W.; Braungart, M.; Clinton, B. *The Upcycle: Beyond sustainability—Designing for Abundance*; Melcher Media, North Point Press: New York, NY, USA, 2013.
56. Christian, R.D.; Jason, C. Agriculture 4.0: Broadening Responsible Innovation in an Era of Smart Farming Frontiers. *Sustain. Food Syst.* **2018**, *2*, 87.
57. Parente, M.; Sedini, C. *D4T—Design Per i Territori—Approcci, Metodi, Esperienze*; List Editore: Trento, Italy, 2019.
58. Boehnert, J. *Design, Ecology, Politics. Towards the Ecocene*; Bloomsbury Academic: London, UK, 2018.
59. Soma, T.; Wakefield, S. The emerging role of a food system planner: Integrating food considerations into planning. *J. Agric. Food Syst. Community Dev.* **2011**, *2*, 53–64. [CrossRef]

sustainability

MDPI

Article

Does Contextual Factor Influence Travelers' Towel Reuse Behavior? Insights from Circular Economy

Farrukh Rafiq [1], Mohd Adil [2] and Mohd Sadiq [3,*]

[1] Department of Business Administration, College of Administrative and Financial Sciences, Jeddah-M Campus, Saudi Electronic University, Riyadh 11673, Saudi Arabia; f.ahmad@seu.edu.sa
[2] Department of Management Studies, NIT Hamirpur, Hamirpur 177005, India; adil.dms@nith.ac.in
[3] Department of Marketing, University of Otago, Dunedin 9016, New Zealand
* Correspondence: sadiq.otago@gmail.com

Abstract: This study examines the role of environmental knowledge, perceived consumer effectiveness, and willingness to sacrifice on travelers' towel reuse behavior. Additionally, it tests whether environmental consciousness moderates the effect of environmental knowledge, perceived consumer effectiveness, willingness to sacrifice, and attitude towards towel reuse on actual towel reuse behavior. The proposed research model was tested using Amazon Mechanical Turk data. Using partial least square structural equation modelling, we analyzed the hypotheses. The results demonstrate that environmental knowledge, perceived consumer effectiveness, and willingness to sacrifice have significant influences on travelers' towel reuse behavior. Additionally, findings about environmental consciousness indicate that high levels of environmental consciousness can help bridge the attitude-behavior gap. To bridge the gap between travelers' attitude and behavior, managers should develop communication strategies to raise awareness and a sense of responsibility among them.

Keywords: perceived consumer effectiveness; circular economy; towel reuse; willingness to sacrifice; environmental consciousness; environmental knowledge

Citation: Rafiq, F.; Adil, M.; Sadiq, M. Does Contextual Factor Influence Travelers' Towel Reuse Behavior? Insights from Circular Economy. *Sustainability* **2022**, *14*, 6155. https://doi.org/10.3390/su14106155

Academic Editors: Nicola Raimo

Received: 4 May 2022
Accepted: 17 May 2022
Published: 19 May 2022

Publisher's Note: MDPI stays neutral with regard to jurisdictional claims in published maps and institutional affiliations.

1. Introduction

A significant contribution to global warming is attributed to the hotel industry [1]. About 75% of the negative environmental impacts caused by hotels come from the excessive consumption of energy and natural resources [2]. One such example could be the release of pollutants directly into the air, water, and soil, contributing to environmental pollution [2]. However, travelers tend to reflect their concern for the environment in a variety of ways, one of which might be choosing to stay at a green hotel while traveling [3]. Even though travelers are becoming more aware of environmental concerns and having a positive attitude toward reusing towels during their hotel stays, their actions do not reflect these attitudes [4]. This results into an attitude-behavior gap [5], causing researchers to focus on this concern.

This research responds to the call by examining how individual and contextual factors such as perceived effectiveness, environmental knowledge, and willingness to sacrifice are related to travelers' intention to reuse towels. Notably, towel reuse is a part of the circular economy, as it reduces waste generation by reusing products [6]. We thus argue that understanding travelers' towel reuse behavior would not only aid in bridging the gap between their attitude and behavior, but would also help in achieving sustainability at hotels through the concept of circular economy. Although the circular economy literature indicates that consumers' willingness to sacrifice is studied [7], but not in the context of towel reuse programs. Since 'willingness to sacrifice' is a key indicator of travelers' attitude and actual behavior [8], we decided to include it as a driver of attitude and towel reuse behavior.

According to the current literature, the research on the towel reuse program or circular economy has grown considerably in developed nations, while in developing nations such as India, consumer awareness of environmental protection has seen a slow growth Interesting to note that Indians place a high value on the environment, which results in them having a positive outlook toward environmental safety [9]. Further, the hospitality industry contributes approximately 5% to 7% [3] of India's GDP, providing a significant growth opportunity. As a result, we chose India as the context for our study.

In light of the above discussion, this study seeks to answer the following research questions:

(RQ1) Do contextual factors (perceived consumer effectiveness, and willingness to sacrifice) influence travelers' attitude and their behavior?
(RQ2) How does environmental knowledge influence travelers' attitude and their behavior?
(RQ3) What role does environmental consciousness play in bridging the attitude-behavior gap?

Using the attitude-behavior-contextual (ABC) theory [10] and knowledge-attitude-behavior (KAB) model [11], this study aims to answer the above-mentioned research questions. Due to the fact that both theoretical frameworks (i.e., ABC and KAB) discuss disparity between attitude and behavior, both are relevant to this study. Since attitude does not always translate into behavior, additional factors are required to bridge the most debatable gap. To answer our research questions and support the translation of attitude into behavior, we rely on environmental knowledge, perceived consumer effectiveness willingness to sacrifice, and environmental consciousness.

Academicians and practitioners can gain three significant insights from our study First of all, this is the first study to integrate both KAB and ABC theory in the context of circular economy. Secondly, ours is the first study to consider willingness to sacrifice as a significant factor in the circular economy context. Third, since the towel reuse program is a major factor in achieving sustainability, managers can adopt this program and use our findings to bridge the gap between travelers' attitude and behavior.

Further, the study offers a brief overview of circular economy and towel reuse programs. This was followed by the development of a research model. Sections 3 and 4 present the adopted methodology and the analysis of the data collected. In Section 5, the findings are discussed in light of existing literature. Lastly, Section 6 discusses conclusions and limitations, as well as future directions.

2. Materials

2.1. Circular Economy

Of late, the concept of circular economy is becoming more popular as a way to achieve sustainability. In essence, circular economy focuses on minimizing wastage of natural resources, leading to sustainability. While, due to the rise of sustainability issues, the circular economy is being considered by policy makers in a more critical manner [12], yet there are still questions about an acceptable definition of circular economy [6]. Previously Ellen MacArthur Foundation [13] defined circular economy as "an industrial system that is restorative or regenerative by intention and design. It replaces the end-of-life concept with restoration, shifts towards the use of renewable energy, eliminates the use of toxic chemicals, which impair reuse, and aims for the elimination of waste through the superior design of materials, products, systems, and, within this, business models". Moreover Kirchherr et al. [14] have analyzed over 100 definitions of circular economy in order to create a comprehensive definition and proposed it as an economic system that replaces the "end-of life" concept with reducing, alternatively reusing, recycling, and recovering materials in production/distribution and consumption processes". The above two definitions highlight that circular economy focuses on 'zero waste and emission' and 'design to reduce the waste'. Lastly, the most acceptable definition of circular economy is given by European Commission [15] as "an economic system that keep the added value in products for as long as possible and eliminate waste". Despite the diverse definitions and divergent approaches to circular economy, academic research has been able to produce an impressive number of studies. While there are numerous studies examining circular economy in the hospitality industry, there are none that investigate how circular economy is manifested in towel reuse behaviors (see Table 1). Therefore, this study is the first to investigate role of towel reuse program in achieving sustainability.

Table 1. Circular economy and hospitality.

Authors	Method	Study Focus
Rodríguez-Antón and Mar Alonso-Almeida [16]	The qualitative multicase	Circular economy in hospitality industry
Jones and Wynn [17]	Review study	Role of resilience and natural capital in circular economy in hospitality domain.
Khodaiji and Christopoulou [18]	Case Study	Circular economy's role in achieving sustainability in hospitality industry
Sorin and Sivarajah [19]	Interviews	Circular economy in hotels' supply chain.
Manniche et al. [20]	Review study	Circular economy in hospitality industry
van Keulen and Kirchherr [21]	Interviews	Barriers and enablers of circular economy in hospitality industry
Khan et al. [22]	Survey	Circular economy in hospitality industry
Aguilar et al. [23]	Review study	Role of quality of life and work-life balance in circular economy in hospitality industry
Vecchio et al. [24]	Interviews	Circular economy in hospitality industry

Note: The list includes articles published only in journals that are indexed either in the Australian Business Dean Council journal quality list or Web of Science.

2.2. Knowledge-Attitude-Behavior Model

KAB model was proposed by Kallgren and Woods [11] to study the influence of consumer knowledge on attitude and behavior. According to the theory, consumers develop an attitude based on knowledge, either a positive one or a negative one, in order to behave in the way they desire. The current study adopts KAB model since; (1) this theory has been applied to discover sustainable consumption patterns behavior. For example, Dhir et al. [25] and Taufique et al. [26] examined the role of environmental knowledge in the development of attitude and consumption behavior; (2) it helps in examining the knowledge-behavior gap and the attitude-behavior gap, which are significant barriers for achieving sustainability behavior [25].

2.3. Attitude-Behavior-Context Theory

ABC theory was put forth by Guagnano et al. [10] to study the most debatable gap, namely the attitude-behavior gap. ABC theory states that consumer behavior is an outcome of attitude and contextual factors. Further, scholars such as Dhir et al. [25] and Zhang et al. [27] suggest that a consumer acts in a specific way to obtain a certain outcome behavior. The manifestation of such behavior requires consumers to develop a positive attitude toward it, combined with the significant influence of contextual factors [27]. By increasing the predictability of consumers' behavior in a given situation, such contextual factors increase the uniqueness of a given situation [28]. Hence, we adopted the ABC theory in this study. The reasons are three-fold (1) ABC theory has been used to investigate the influence of environmental attitudes on target behavior in the past [29], (2) Researchers use the ABC theory to understand the gap between attitude and behavior [25], (3) Lastly, this theory has third constituent as a contextual factor, which is significantly related to consumers' decision making [30].

3. Model and Hypotheses Development

3.1. Attitude towards Towel Reuse and Towel Reuse Behavior

Previously, many scholars in the area of environmental sustainability claim that attitude is among the strongest predictors of actual behavior [25,26,31]. In the case of the towel reuse program, attitude is a significant and strong predictor of behavior [32], while Budovska et al. [5] observed a weak relationship between attitude and behavior. Inferring that attitude may not always results in towel reuse behavior. We therefore examined the relationship between attitude and behavior in order to explain the paradoxical observation of the attitude-behavior gap. Therefore, we anticipate the following.

Hypothesis 1 (H1). *Attitude towards towel reuse has a significant influence on towel reuse behavior.*

3.2. Environmental Knowledge, Attitude towards Towel Reuse, and Towel Reuse Behavior

Environmental knowledge refers to "an individual's general knowledge of facts, concepts and relationships related to environmental protection and its major ecosystems" [4] (p. 69). Extant research on environmental sustainability highlights that mapping environmental knowledge is important as it helps to understand the 'sustainability movement' of any country [33]. Further, as travelers become more aware of sustainability issues, their attitude towards sustainability issues changes as well. For example, Taufique et al. [26] observed that environmental knowledge develops a positive attitude towards sustainability. On the contrary, Dhir et al. [25] and Han et al. [32] suggests that environmental knowledge may not always lead to a positive attitude in behavior and hence, there exists a knowledge-attitude-behavior gap. Yet, scholars such as Teng, Lu, and Huang [34], Wang et al. [35] suggest that increasing environmental literacy leads to changes in behavior. Accordingly, we argue that knowledge motivates travelers to adopt a positive attitude toward towel reuse and adoption. Thus, we hypothesize.

Hypothesis 2 (H2). *Environmental knowledge has a positive influence on attitude towards towels reuse.*

Hypothesis 3 (H3). *Environmental knowledge has a positive influence on towel reuse behavior.*

3.3. Perceived Consumer Effectiveness, Attitude towards Towel Reuse and Towel Reuse Behavior

Perceived consumer effectiveness refers to "as the degree of the consumers' domain-specific self-belief that they will be able to solve the problem at hand" [36] (p. 3593). In a situation where consumers believe their actions could significantly impact the outcome, they are more likely to take steps towards making a positive difference [37]. Indeed, prior research highlights that consumers with positive perceptions of their effectiveness in resolving sustainability issues are more likely to develop positive attitude [9,38]. Similarly, Dhir et al. [36] noted that consumers with high perceived consumer effectiveness in sustainable consumption were more likely to adopt sustainable consumption to address sustainability issues. In accordance with the existing literature, this study assumes that travelers with a high perception of consumer effectiveness are more likely to express positive attitudes about towel reuse and actual towel reuse behavior. Thus, we hypothesize.

Hypothesis 4 (H4). *Perceived consumer effectiveness has a positive influence on attitude towards towels reuse.*

Hypothesis 5 (H5). *Perceived consumer effectiveness has a positive influence on towel reuse behavior.*

3.4. Willingness to Sacrifice and Towel Reuse Behavior

Researchers have found that willingness to sacrifice is an important determinant of travelers' decision making [39]. Travelers who are committed to creating a sustainable world are willing to make sacrifices to reach that goal [40]. Such 'sacrifice' is defined as "the degree to which one's behavior prioritises the benefit of the environment despite the expense/loss of his/her immediate costs, self-interest, and time/effort" [41] (p. 2121). If travelers show a strong desire to solve sustainability issues, they are more likely to pay premium or step outside of their comfort zone to protect the planet [42]. For example, Burhanudin and Unnithan [43] found that travelers who were willing to sacrifice did so in an environmentally-friendly manner behavior. Indeed, in case of towel reuse, travelers who have strong willingness to sacrifice are more likely to perform action. Thus, we hypothesize.

Hypothesis 6 (H6). *Willingness to sacrifice has a positive influence on towel reuse behavior.*

3.5. Indirect Effect through Attitude towards Towel Reuse

Attitude towards towel reuse refers to as "the degree of favourable or unfavourable predisposition a person has to respond to a behavior in question (in our case towel reuse)" [5] (p. 107). Attitude is considered as an important factor that drives travelers' actual behavior, yet the literature on sustainability issues remedy program such as the attitude-behavior gap exists [5,32]. By examining the indirect effect of travelers' attitude on the relationship of towel reuse behavior with environmental knowledge and consumer perceptions, the authors argue that the gap between attitude and actual behavior can be better understood. The KAB theoretical framework proposed by Kallgren and Woods [11] postulates attitude as the mediating influence between knowledge and behavior. This has been tested in literature related to sustainability. For example, Dhir et al. [25] determined that knowledge significantly influences the development of attitudes, which lead to the purchase of green apparel. Similarly, Taufique et al. [26] argued that individuals with a high level of knowledge about the environmental issues are likely to perform sustainable behavior. The possible justification is derived from the Kargren and Woods [11] model that environmental knowledge leads to sustainable behavior as knowledge nurtures a positive attitude, which is then translated into actual behavior. Therefore, we argue that environmental knowledge indirectly influences towel reuse programs through attitude. Hence, we hypothesize.

Hypothesis 7a (H7a). *Attitude towards towel reuse significantly mediates the association of environmental knowledge with towel reuse program.*

Research on perceived consumer effectiveness in sustainability has been well documented [9,37]. It is considered as one of the significant factors that derive actual sustainable behavior [38]. We argue that travelers with high perceived consumer effectiveness would adopt sustainable behavior such as towel reuse behavior because travelers' effectiveness to solve sustainability issues lead to development of their attitude towards remedy action, which in turn results in towel reuse behavior. Therefore, we hypothesize.

Hypothesis 7b (H7b). *Attitude towards towel reuse significantly mediates the association of perceived consumer effectiveness with towel reuse program.*

3.6. The Moderating Effect of Environmental Consciousness

Environmental consciousness refers to as "an element of the belief system that denotes to specific psychological influences related to individuals' propensity to join pro-environmental behaviors" [44] (p. 200). Policy makers suggest that measuring a country's environmental consciousness is relevant for understanding that country's 'sustainability activities' [45]. One of the major factors that affect consumers' sustainable consumption decisions is their environmental consciousness [46]. As a result, travelers who are aware of environmental issues are more likely to adopt sustainable behavior to mitigate the effects [47].

Further, extant research on sustainability suggests that environmental consciousness has been studied as one of the significant predictors of sustainable behavior [45,46]; while some researchers have considered it as a moderator on the relationship of sustainable behavior with its drivers [48,49]. Interestingly, Kautish et al. [48] claimed that individuals having high environmental consciousness are more likely to increase the conversion of perceived consumer effectiveness into actual behavior, towel reuse behavior in our case. Similarly rise in environmental consciousness among travelers results into rise in awareness about environmental/sustainability issues [50], resulting into adoption of sustainable behavior [26]. In the same vein, according to Kautish et al. [48] travelers with high environmental consciousness are likely to be more willing to contribute to preserving the environment or sacrifice for it, which results into higher chances of sustainable behavior adoption. Further travelers exhibiting strong attitude towards towel reuse are less likely to go for towel reuse behavior [5], reflecting a disparity between attitude and actual behavior. Therefore, extant literature [45,51] suggests that high environmental consciousness among travelers develop positive attitude towards sustainability, which in turn leads to adoption of sustainable behavior. Therefore, we assume that environmental consciousness is significantly bridge the attitude-behavior gap. Thus, we hypothesize.

Hypothesis 8a (H8a). *Environmental consciousness positively moderates the association of environmental knowledge with towel reuse behavior.*

Hypothesis 8b (H8b). *Environmental consciousness positively moderates the association of perceived consumer effectiveness with towel reuse behavior.*

Hypothesis 8c (H8c). *Environmental consciousness positively moderates the association of willingness to sacrifice with towel reuse behavior.*

Hypothesis 8d (H8d). *Environmental consciousness positively moderates the association of attitude towards towel reuse with towel reuse behavior.*

As a result of the foregoing arguments, we developed a conceptual model (see Figure 1).

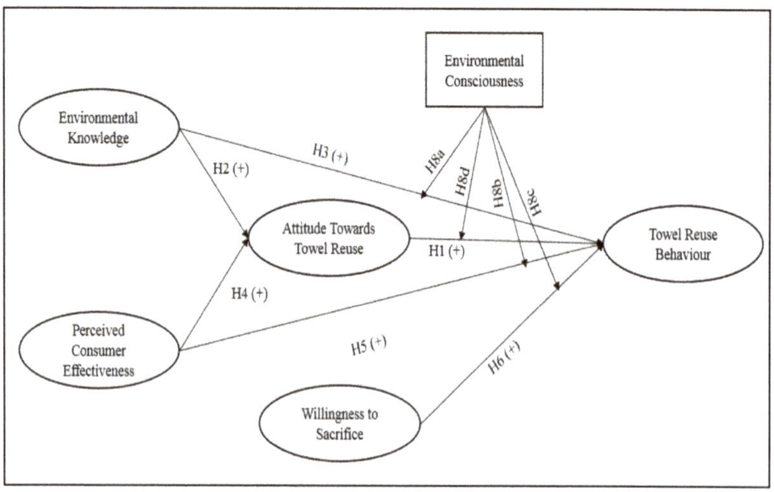

Figure 1. Conceptual model.

4. Methods

4.1. Measure

This research aims at understanding the role of environmental knowledge, perceived consumer effectiveness, willingness to sacrifice, and environmental consciousness in adoption of towel reuse behavior. As part of our research, we developed a questionnaire containing 20 items to test our hypotheses. All the items were measured on a 5-point Likert scale. To measure attitude towards towel reuse, we used the 5-items scale of Budovska et al. [5]. The sample items were "For me, reusing a towel at a hotel when travelling is—(a) Extremely undesirable (1)/Extremely desirable (5); (b) Extremely unpleasant (1)/Extremely pleasant (5); (c) Extremely unfavorable (1)/Extremely favorable (5)". Three items were adapted from Wang et al. [38] and Sadiq et al. [33] to measure perceived consumer effectiveness ("An individual's behavior has a positive impact on society if he/she reuses towels during hotel stay") and environmental knowledge ("Reusing towels is a substantial way to reduce wasteful use of natural resources"; "Reusing towels is a great way to conserve natural resources"), respectively. To measure willingness to sacrifice, we adopted the two-items scale from Han et al. [41]. The items were "Even when it is inconvenient to me, I am willing to reuse towel as I think it is best for the environment"; "I am willing to reuse towel for the environment, even if I'm not thanked for my efforts". Similarly, four items [38] were used to measure environmental consciousness ("I am extremely worried about the situation of the world's environment and what it will mean for my future"; "Mankind is severely abusing the environment"). Lastly, we adapted the 3-items ("I feel that I have played a great part in helping the environment when I reuse towel during my hotel stay"; "I feel more comfortable when I use fresh towels rather than reuse the same towel during my hotel stay") to measure towel reuse behavior from Kautish et al. [48].

4.2. Data Collection

The sample size for this study was calculated based on the recommendation of Hair et al. [52], sample size should be ten to fifteen times of latent variables that are there in the research model i.e., $15 \times 20 = 300$. In addition, extant literature on sustainable behavior in India highlights that studies with sample size in between 200 to 500 [29,53,54] provide reliable and valid results. Therefore, this research made available the questionnaire on M-Turk on 21 January 2022 for 350 respondents. We have terminated the survey on 24 January 2022 after receiving 350 responses. We have selected MTurk because it is known for providing reliable and valid data with a few errors [55]. Yet, we have taken precautions to reduce the probability of getting unreliable data. For this, the current study has set three inclusion criteria: (1) Screening question: Do you have any recent experience (six months) with hotels that offer towel reuse programs? (2) only those respondents can take part in the survey who has an acceptance rate of 98% or above; and (3) respondents should be a resident of India.

Of 350 respondents, 203 (58%) were male and 147 (42%) were female. Majority of the respondents (70.8%) hold graduation degree and are single (67.4%). Approximately half (51.1%) of the respondents belong to income group of 100,000 INR or above.

4.3. Statistical Tools

To test our research model, we have used Partial Least Square based Structural Equation Modelling in ADANCO software. Further to test the moderation effect, we have used process macro for SPSS.

4.4. Common Method Bias

In social science research, common method bias is a major problem [36]. Therefore, to tackle the CMB problem, we have taken three steps. First, to test respondents' attention, we have reversed a few items in the questionnaire and informed them that there is no right and wrong answer. Second, we check CMB of the collected data through Harman's single factor test. According to the result, 28.3% of variance was attributed to a single factor, which is within the threshold limit of 50%. Hence, it confirms that there exist no CMB problem in the data. Lastly, to re-confirm the absence of CMB problem in our data, we have used Marker Variable Test. The result re-confirms that our data is free from any CMB issue.

5. Results

To test the research model, we have adopted two-step structural equation modelling. The first step is to test the reliability and validity of the research model. Next, is to assess the hypotheses. The study tested the validity, reliability, and model fitness of the proposed research model. Henseler [56] suggested three criteria to test model fitness; "unweighted least squares discrepancy (dULS)"; "geodesic discrepancy (dG)"; and "standardized root mean square residual (SRMR)" (p. 2223). All these values should be lower than threshold values i.e., HI99 [56]. According to the results (see Table 2), the values obtained for three criteria are below the threshold levels, which is suggestive of good fit for the proposed research model.

Table 2. Model fit.

	SRMR	dULS	dG
Observed values	0.057	0.738	0.416
HI95	0.061	0.791	0.483
HI99	0.072	0.949	0.617

Further, our research model has also been tested for reliability and validity. "Cronbach's alpha" (α) and "Joreskog's rho" (ρ) were used to determine the reliability. Similarly, convergent and discriminant validity of the research model was tested through average variance extracted (AVE) and "Heterotrait–Monotrait Ratio of Correlation" (HTMT). The findings reveal that the research model is reliable as value of α and ρ are higher than the minimum cut-off value of 0.70 for each employed variable (see Table 3).

Table 3. Measurement model results.

Variable	Item Code	λ	AVE	α	ρ
Perceived consumer effectiveness	PCE2	0.82	0.65	0.81	0.79
	PCE3	0.79			
Environmental knowledge	EK1	0.83	0.72	0.89	0.88
	EK2	0.87			
	EK3	0.84			
Willingness to sacrifice	WtS1	0.71	0.54	0.88	0.86
	WtS2	0.76			
Attitude towards towel reuse	A1	0.78	0.60	0.91	0.88
	A2	0.71			
	A3	0.83			
	A4	0.81			
	A5	0.75			
Towel reuse behavior	TRB1	0.84	0.64	0.85	0.84
	TRB2	0.77			
	TRB3	0.79			

Key: λ = factor loadings; AVE = average variance extracted; α = Cronbach's alpha; ρ = Joreskog's rho.

Additionally, the AVE value for each variable exceeds the threshold of 0.50, hence, convergent validity of the research model is ascertained. Similarly, values of all variables in the HTMT analysis are less than 0.90, indicating discriminant validity (see Table 4).

Table 4. HTMT analysis.

	PCE	EK	WtS	A	TRB
PCE	1				
EK	0.51	1			
WtS	0.58	0.64	1		
A	0.45	0.67	0.48	1	
TRB	0.37	0.51	0.53	0.31	1

5.1. Hypotheses Testing

Results indicate that attitude influences towel reuse behavior both significantly but weakly ($\beta = 0.21$, $p < 0.05$; $f^2 = 0.11$), therefore, H1 was supported (see Figure 2).

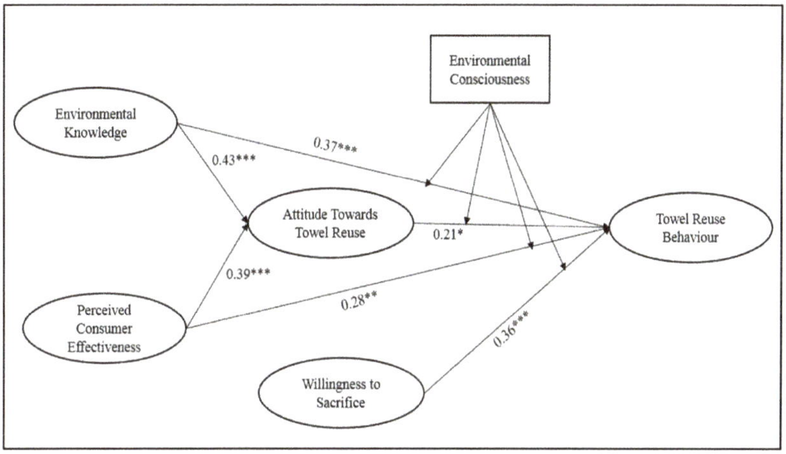

Figure 2. Structural model. * $p < 0.5$; ** $p < 0.01$; *** $p < 0.001$.

Further, environmental knowledge influences attitude both significantly and moderately ($\beta = 0.43$, $p < 0.001$; $f^2 = 0.32$) and towel reuse behavior ($\beta = 0.37$, $p < 0.001$; $f^2 = 0.18$). Therefore, H2 and H3 were supported. Perceived consumer effectiveness significantly and moderately influences travelers' attitude towards towel reuse ($\beta = 0.39$, $p < 0.001$; $f^2 = 0.17$). Similarly, perceived consumer effectiveness significantly but weakly influences towel reuse behavior ($\beta = 0.28$ $p < 0.01$; $f^2 = 0.12$). Therefore, H4 and H5 were supported. Lastly, willingness to sacrifice significantly and moderately associated with towel reuse behavior ($\beta = 0.36$ $p < 0.001$; $f^2 = 0.24$), therefore, H6 was supported. Table 5 shows that all the direct hypotheses were supported.

Table 5. Direct hypotheses results.

Hypothesis	Path	β	Supported?
H1	A → TRB	0.21 *	Yes
H2	EK → A	0.43 ***	Yes
H3	EK → TRB	0.37 ***	Yes
H4	PCE → A	0.39 ***	Yes
H5	PCE → TRB	0.28 **	Yes
H6	WtS → TRB	0.36 ***	Yes

R^2 value of Attitude towards towel reuse = 41.3%. R^2 value of Towel reuse behavior = 44.7%. * $p < 0.5$; ** $p < 0.01$; *** $p < 0.001$.

Next, we tested the indirect effects of environmental knowledge (H7a) and perceived consumer effectiveness (H7b) on towel reuse behavior through attitude towards towel reuse. Table 6 shows that H7a and H7b were supported as the values between ULCI and LLCI do not have zero. In addition, attitude has a partial mediating effect as direct links of towel reuse behavior with perceived consumer effectiveness and environmental knowledge remained significant.

Table 6. Results of Indirect effect.

Hypothesised Pathway	β	ULCI	LLCI	Mediation Type
EK → A → TRB (H7a)	0.09 *	0.194	0.073	Partial
PCE → A → TRB (H7b)	0.08 *	0.158	0.062	Partial

* $p < 0.5$.

5.2. Moderating Analysis

We test the hypotheses related to moderation effect of environmental consciousness using the 'Model 1' in process macro. Table 7 shows that environmental consciousness significantly and positively moderates the relationship of towel reuse behavior with environmental knowledge, perceived consumer effectiveness, willingness to sacrifice, and attitude towards towel reuse. Thus, H8a, H8b, H8c, and H8d were supported. For a graphical view, see Figures 3–6.

Table 7. Moderation result.

	Moderator: Environmental Consciousness			
Path	β	ULCI	LLCI	Moderation
EK → TRB	0.13 **	0.473	0.097	Yes
PCE → TRB	0.05 *	0.118	0.039	Yes
WtS → TRB	0.11 **	0.419	0.086	Yes
A → TRB	0.07 *	0.184	0.091	Yes

* $p < 0.5$; ** $p < 0.01$.

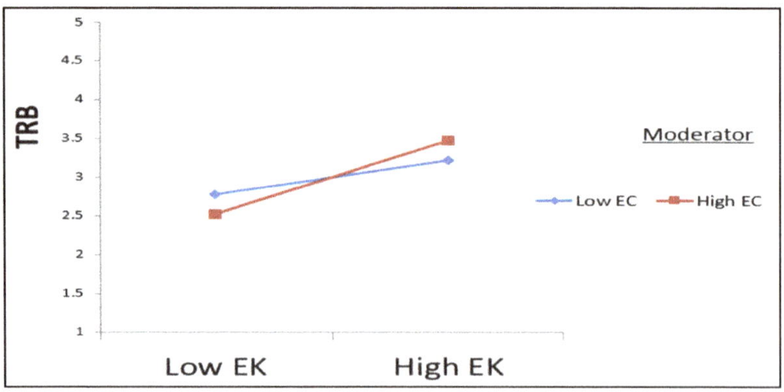

Figure 3. The relationship between environmental knowledge (EK) and towel reuse behavior (TRB) is significantly moderated by environmental consciousness (EC).

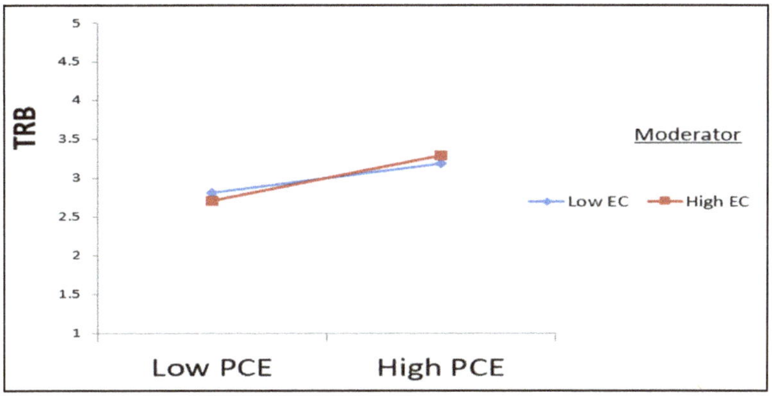

Figure 4. The relationship between perceived consumer effectiveness (PCE) and towel reuse behavior (TRB) is significantly moderated by environmental consciousness (EC).

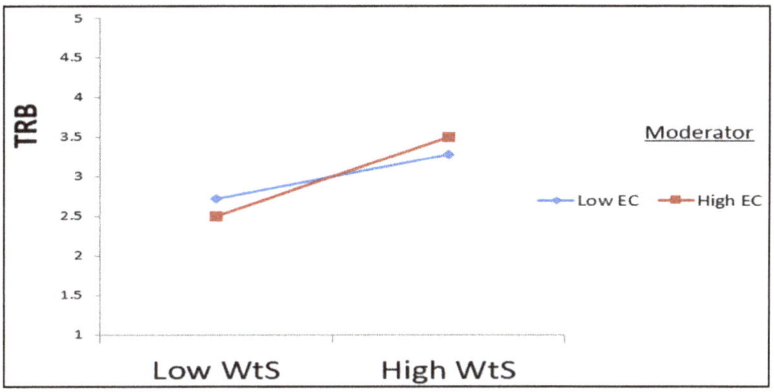

Figure 5. The relationship between willingness to sacrifice (WtS) and towel reuse behavior (TRB) is significantly moderated by environmental consciousness (EC).

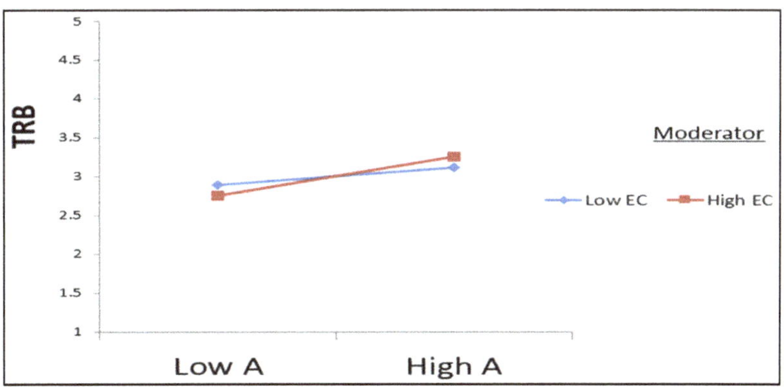

Figure 6. The relationship between attitude towards towel reuse (A) and towel reuse behavior (TRB) is significantly moderated by environmental consciousness (EC).

6. Discussion

The aim of this research was to examine the influence of environmental knowledge, perceived consumer effectiveness, willingness to sacrifice, and environmental consciousness on the development of travelers' attitude and towel reuse behavior. For this, we have proposed six direct, two indirect, and four moderating hypotheses using the theoretical lens of KAB model and ABC theory.

The study supports H1, wherein it was claimed that attitude towards towel reuse has a significant and positive influence on towel reuse behavior. This finding is in line with the observation of Budovska et al. [5] and Cvelbar, Grün, and Dolnicar [57]. Further, our finding of H1 also supports the existence of attitude-behavior gap as the influence of attitude on actual behavior is weak ($f^2 = 0.11$). The possible reason is that travelers are health conscious, which leads them to avoid reusing the same towel multiple times. In addition, health consciousness among travelers might have increased due to COVID-19 [31]. Further, lack of internal locus of control to perform certain actions is another possible reason for attitude-behavior gap. For example, Trivedi et al. [58] suggests that individuals who exhibit no interest in accomplishing the task because they are low on internal locus of control, result into creating attitude-behavior gap.

Our study extends support to H2 and H3 that state environmental knowledge has a significant and positive association with attitude towards towel reuse and towel reuse behavior. These findings are in line with Dhir et al. [25], and Yadav and Pathak [51]. In their research, they have found that individuals with high environmental knowledge have a strong attitude, resulting in actual environmental-friendly behavior. These findings imply that travelers have awareness about the towel reuse program and its capability to resolve sustainability issues. These relationships offer a significant implication, which is that environmental knowledge can help in bridging the disparity between attitude and behavior. Similarly, H4 and H5 were supported in our research, wherein we argue that perceived consumer effectiveness has a positive and significant relationship with attitude and towel reuse behavior. These findings concur with Jaiswal and Kant [9], Taufique and Vaithianathan [59], and Wang et al. [38], while H5 contradicts with the observation of Dhir et al. [36]. The possible justification of H4 and H5 is that sustainable behavior is considered to be more driven by internally controlled factors such as traveler's effectiveness rather than effectiveness of a group [60]. Therefore, travelers with high perceived consumer effectiveness tend to have a positive attitude towards saving the environment, resulting in adoption of towel reuse behavior. In addition, our study established that willingness to sacrifice has a positive and significant association with towel reuse behavior (H6). This finding is in line with Burhanudin and Unnithan [43] and Rahman and Reynolds [40], while contradicts those of Han et al. [41]. This finding implies that travelers are ready to willingly sacrifice the luxury of using a new towel every day. Therefore, such travelers are ready to exhibit towel reuse behavior during their hotel stay.

We further tested the mediating effect of attitude towards towel reuse between the association of towel reuse behavior with environmental knowledge (H7a) and perceived consumer effectiveness (H7b). The hypotheses H7a and H7b are partially supported as the direct relationship is significant in both the cases. The finding of H7a is in line with Taufique et al. [26], wherein it was found that knowledge about sustainability issues leads to the development of a positive attitude and actual behavior. The possible justification is taken from the knowledge-attitude-behavior model [11], i.e., travelers with high environmental knowledge tend to have a strong and positive attitude towards towel reuse, which in turn results into towel reuse behavior. Similarly, the finding of H7b is in line with Wang et al. [38], wherein travelers with high perceived consumer effectiveness exhibit strong attitude and likelihood of performing behavior. This finding implies that travelers with high perceived consumer effectiveness (i.e., they believe that their efficacy to solve the problem) tend to have a strong and positive attitude, resulting in actual towel reuse behavior.

Next, we tested the moderation effect of environmental consciousness between the association of towel reuse behavior with environmental knowledge (H8a), perceived consumer effectiveness (H8b), willingness to sacrifice (H8c), and attitude towards towel reuse (H8d). All the hypotheses were supported in the current study. To the best of our knowledge, it is the first study to examine the moderating influence of environmental consciousness between the relationship of towel reuse behavior with environmental knowledge and attitude towards towel reuse. Therefore, we cannot correlate our findings with the extant literature on sustainability.

6.1. Implications

Our study offers significant theoretical implications that contribute to the literature on towel reuse programs and the circular economy. First, there are numerous research studies that have used KAB and ABC in general green consumer behavior literature (for e.g., see [17,29]), but no study has proposed a comprehensive research model of towel reuse by merging KAB and ABC theory in the circular economy literature. Second, after a critical review of sustainable tourism literature, we have found that willingness to sacrifice has been studied in a few studies (see, [40,41,43]), while no study has used it in the circular economy literature. Further, Rahman and Reynolds [40] suggest that travelers' willingness to sacrifice is one of the significant factors to achieve sustainability in the tourism and hospitality industry. Therefore, our study extends the circular economy literature by examining the role of willingness to sacrifice. Lastly, we have responded to the call of Budvsko et al. [5] to conduct empirical research to bridge the attitude-behavior gap. For this, we have examined the moderating role of environmental consciousness between the attitude and behavior. Our results suggest that environmental consciousness significantly bridges the attitude-behavior gap. Further, to the best of our knowledge, this study is the first to examine the moderating role of environmental consciousness to bridge the attitude-behavior gap in the sustainability/circular economy literature.

Besides the theoretical, our study also offers significant implications for managers. First, the current research shows that knowledge about sustainability issues stimulates the development of a positive and strong attitude, which results in towel reuse behavior. This highlights that increasing environmental knowledge leads to increased adoption of sustainable behavior. In the light of this finding, we suggest managers draft policies to bridge the attitude-behavior gap. They should aim at disseminating environmental knowledge. For this, managers can use social media platforms to teach travelers about sustainability issues. Second, this study reveals that travelers' efficacy to solve the sustainability problem leads to development of a positive attitude, resulting in actual tower reuse behavior. This reflects that increasing travelers' perception about effectiveness of their action helps in reducing the disparity between attitude and behavior. Therefore, we suggest managers to draft policies to enhance perceived consumer effectiveness to bridge the attitude-behavior gap. Further, managers should also communicate the societal and community responsibility to boost up the sense of their responsibility towards solving the sustainability issues. Third, our study shows that willingness to sacrifice significantly influence the towel reuse behavior. This indicates that increasing the sense of sacrifice among travelers results in higher adoption of towel reuse behavior. We suggest managers do effective marketing strategies to develop a sense of willingness to sacrifice for the environment among travelers. Lastly, our study reflects that environmental consciousness plays an important role in bridging the attitude-behavior gap. Therefore, we suggest managers to create concern among travelers they can develop communication strategies to highlight the initiatives that they have taken to solve the sustainability issues. Additionally, they can also communicate that if individuals do not participate in solving sustainability issues, it will affect the sustainability of the planet.

6.2. Limitations and Future Research Avenues

The current study has a few limitations and offers some important future avenues for researchers. First, we have collected the data using a 'single cross-sectional approach' which may restrict the generalisability of our findings as consumer behavior changes with time. Therefore, future academicians are suggested to conduct longitudinal research to better gauge the towel reuse behavior. Second, our stdy was conducted in India, which has a different culture in comparison to ASEAN nations and western world such as Vietnam and the USA. This may restrict the generalisability of our findings to other nations with different cultural setups. Therefore, we suggest future researchers should replicate our research model in different cultural systems. Third, we suggest future scholars to extend our research model by considering other contextual factors such as price sensitivity, and environmental commitment that may help in bridging the attitude-behavior gap in circular economy literature. Lastly, Mturk as a method of collecting data raises a number of concerns, so future researchers should seek to collect data through more traditional methods, such as conducting field surveys.

Author Contributions: Conceptualization, F.R., M.A. and M.S.; methodology, M.A. and M.S.; software, M.S.; validation, F.R.; formal analysis, M.S.; investigation, M.S.; resources, F.R.; data curation, M.S.; writing—F.R. and M.S.; writing—review and editing, M.A.; visualization, M.A.; supervision, M.A.; project administration, F.R., M.A. and M.S.; funding acquisition, F.R. All authors have read and agreed to the published version of the manuscript.

Funding: This research received no external funding.

Institutional Review Board Statement: Not applicable.

Informed Consent Statement: Not applicable.

Data Availability Statement: The datasets analyzed during the current study are available on request from the corresponding author. The data are not publicly available due to confidentiality and privacy issues.

Conflicts of Interest: The authors declare no conflict of interest.

References

1. Chen, H.; Bernard, S.; Rahman, I. Greenwashing in hotels: A structural model of trust and behavioral intentions. *J. Clean. Prod.* **2019**, *206*, 326–335. [CrossRef]
2. Aboelmaged, M. Direct and indirect effects of eco-innovation, environmental orientation and supplier collaboration on hotel performance: An empirical study. *J. Clean. Prod.* **2018**, *184*, 537–549. [CrossRef]
3. Verma, V.K.; Chandra, B.; Kumar, S. Values and ascribed responsibility to predict consumers' attitude and concern towards green hotel visit intention. *J. Bus. Res.* **2019**, *96*, 206–216. [CrossRef]
4. Wang, L.; Wong, P.P.; Alagas, E.N. Antecedents of green purchase behaviour: An examination of altruism and environmental knowledge. *Int. J. Cult. Tour. Hosp. Res.* **2020**, *14*, 63–82. [CrossRef]
5. Budovska, V.; Torres Delgado, A.; Øgaard, T. Pro-environmental behaviour of hotel guests: Application of the Theory of Planned Behaviour and social norms to towel reuse. *Tour. Hosp. Res.* **2020**, *20*, 105–116. [CrossRef]
6. Camacho-Otero, J.; Boks, C.; Pettersen, I.N. Consumption in the circular economy: A literature review. *Sustainability* **2018**, *10*, 2758 [CrossRef]
7. Hao, Y.; Wang, Y.; Wu, Q.; Sun, S.; Wang, W.; Cui, M. What affects residents' participation in the circular economy for sustainable development? Evidence from China. *Sustain. Dev.* **2020**, *28*, 1251–1268. [CrossRef]
8. Torres-Moraga, E.; Rodriguez-Sanchez, C.; Sancho-Esper, F. Understanding tourist citizenship behavior at the destination level *J. Hosp. Tour. Manag.* **2021**, *49*, 592–600. [CrossRef]
9. Jaiswal, D.; Kant, R. Green purchasing behaviour: A conceptual framework and empirical investigation of Indian consumers *J. Retail. Consum. Serv.* **2018**, *41*, 60–69. [CrossRef]
10. Guagnano, G.A.; Stern, P.C.; Dietz, T. Influences on attitude-behavior relationships: A natural experiment with curbside recycling *Environ. Behav.* **1995**, *27*, 699–718. [CrossRef]
11. Kallgren, C.A.; Wood, W. Access to attitude-relevant information in memory as a determinant of attitude-behavior consistency *J. Exp. Soc. Psychol.* **1986**, *22*, 328–338. [CrossRef]
12. Marrucci, L.; Daddi, T.; Iraldo, F. The integration of circular economy with sustainable consumption and production tools Systematic review and future research agenda. *J. Clean. Prod.* **2019**, *240*, 118268. [CrossRef]
13. MacArthur, E. Towards the circular economy. *J. Ind. Ecol.* **2013**, *2*, 23–44.
14. Kirchherr, J.; Reike, D.; Hekkert, M. Conceptualizing the circular economy: An analysis of 114 definitions. *Resour. Conserv. Recycl* **2017**, *27*, 221–232. [CrossRef]
15. European Commission. Communication from the Commission to the European Parliament, the Council, the European Economic and Social Committee and the Committee of the Regions. In *A Roadmap for Moving to a Competitive Low Carbon Economy in 2050*, European Commission: Brussels, Belgium, 2011; pp. 1–16.

16. Rodríguez-Antón, J.M.; Alonso-Almeida, M.D. The circular economy strategy in hospitality: A multicase approach. *Sustainability* **2019**, *11*, 5665. [CrossRef]
17. Jones, P.; Wynn, M.G. The circular economy, natural capital and resilience in tourism and hospitality. *Int. J. Contemp. Hosp. Manag.* **2019**, *31*, 2544–2563. [CrossRef]
18. Khodaiji, J.D.; Christopoulou, D. Sustainable development and the circular economy in Greece: Case examples from Costa Navarino and Grecotel. *Worldw. Hosp. Tour. Themes* **2020**, *12*, 609–621. [CrossRef]
19. Sorin, F.; Sivarajah, U. Exploring Circular economy in the hospitality industry: Empirical evidence from Scandinavian hotel operators. *Scand. J. Hosp. Tour.* **2021**, *21*, 265–285. [CrossRef]
20. Manniche, J.; Larsen, K.T.; Broegaard, R.B. The circular economy in tourism: Transition perspectives for business and research. *Scand. J. Hosp. Tour.* **2021**, *21*, 247–264. [CrossRef]
21. van Keulen, M.; Kirchherr, J. The implementation of the Circular Economy: Barriers and enablers in the coffee value chain. *J. Clean. Prod.* **2021**, *281*, 125033. [CrossRef]
22. Khan, O.; Bellini, N.; Daddi, T.; Iraldo, F. Effects of behavioral intention and dynamic capabilities on circular economy adoption and performance of tourism SMEs. *J. Sustain. Tour.* **2022**, *14*, 1–20. [CrossRef]
23. Aguilar, J.E.; Zang, L.; Fushimi, S. Towards sustainability in hospitality operations: How is quality of life and work–life balance related? *Worldw. Hosp. Tour. Themes* **2022**, *14*, 274–285.
24. del Vecchio, P.; Malandugno, C.; Passiante, G.; Sakka, G. Circular economy business model for smart tourism: The case of Ecobnb. *EuroMed J. Bus.* **2021**, *17*, 88–104. [CrossRef]
25. Dhir, A.; Sadiq, M.; Talwar, S.; Sakashita, M.; Kaur, P. Why do retail consumers buy green apparel? A knowledge-attitude-behaviour-context perspective. *J. Retail. Consum. Serv.* **2021**, *59*, 102398. [CrossRef]
26. Taufique, K.M.; Vocino, A.; Polonsky, M.J. The influence of eco-label knowledge and trust on pro-environmental consumer behaviour in an emerging market. *J. Strateg. Mark.* **2017**, *25*, 511–529. [CrossRef]
27. Zhang, L.; Li, D.; Cao, C.; Huang, S. The influence of greenwashing perception on green purchasing intentions: The mediating role of green word-of-mouth and moderating role of green concern. *J. Clean. Prod.* **2018**, *187*, 740–750. [CrossRef]
28. Xu, X.; Maki, A.; Chen, C.F.; Dong, B.; Day, J.K. Investigating willingness to save energy and communication about energy use in the American workplace with the attitude-behavior-context model. *Energy Res. Soc. Sci.* **2017**, *32*, 13–22. [CrossRef]
29. Goh, S.K.; Balaji, M.S. Linking green skepticism to green purchase behavior. *J. Clean. Prod.* **2016**, *131*, 629–638. [CrossRef]
30. Grimmer, M.; Kilburn, A.P.; Miles, M.P. The effect of purchase situation on realized pro-environmental consumer behavior. *J. Bus. Res.* **2016**, *69*, 1582–1586. [CrossRef]
31. Sadiq, M.; Adil, M.; Paul, J. Eco-friendly hotel stay and environmental attitude: A value-attitude-behaviour perspective. *Int. J. Hosp. Manag.* **2022**, *100*, 103094. [CrossRef]
32. Han, H.; Lee, M.J.; Kim, W. Promoting towel reuse behaviour in guests: A water conservation management and environmental policy in the hotel industry. *Bus. Strategy Environ.* **2018**, *27*, 1302–1312. [CrossRef]
33. Sadiq, M.; Bharti, K.; Adil, M.; Singh, R. Why do consumers buy green apparel? The role of dispositional traits, environmental orientation, environmental knowledge, and monetary incentive. *J. Retail. Consum. Serv.* **2021**, *62*, 102643. [CrossRef]
34. Teng, C.C.; Lu, A.C.; Huang, T.T. Drivers of consumers' behavioral intention toward green hotels. *Int. J. Contemp. Hosp. Manag.* **2018**, *30*, 1134–1151. [CrossRef]
35. Wang, S.; Wang, J.; Wang, Y.; Yan, J.; Li, J. Environmental knowledge and consumers' intentions to visit green hotels: The mediating role of consumption values. *J. Travel Tour. Mark.* **2018**, *35*, 1261–1271. [CrossRef]
36. Dhir, A.; Talwar, S.; Sadiq, M.; Sakashita, M.; Kaur, P. Green apparel buying behaviour: A Stimulus–Organism–Behaviour–Consequence (SOBC) perspective on sustainability-oriented consumption in Japan. *Bus. Strategy Environ.* **2021**, *30*, 3589–3605. [CrossRef]
37. Sharma, R.; Jha, M. Values influencing sustainable consumption behaviour: Exploring the contextual relationship. *J. Bus. Res.* **2017**, *76*, 77–88. [CrossRef]
38. Wang, J.; Wang, S.; Wang, Y.; Li, J.; Zhao, D. Extending the theory of planned behavior to understand consumers' intentions to visit green hotels in the Chinese context. *Int. J. Contemp. Hosp. Manag.* **2018**, *30*, 2810–2825. [CrossRef]
39. Han, H.; Lee, S.; Al-Ansi, A.; Kim, H.C.; Ryu, H.B.; Kim, J.J.; Kim, W. Convention tourism and sustainability: Exploring influencing factors on delegate green behavior that reduce environmental impacts. *Sustainability* **2019**, *11*, 3903. [CrossRef]
40. Rahman, I.; Reynolds, D. Predicting green hotel behavioral intentions using a theory of environmental commitment and sacrifice for the environment. *Int. J. Hosp. Manag.* **2016**, *52*, 107–116. [CrossRef]
41. Han, H.; Yu, J.; Kim, H.C.; Kim, W. Impact of social/personal norms and willingness to sacrifice on young vacationers' pro-environmental intentions for waste reduction and recycling. *J. Sustain. Tour.* **2018**, *26*, 2117–2133. [CrossRef]
42. Landon, A.C.; Woosnam, K.M.; Boley, B.B. Modeling the psychological antecedents to tourists' pro-sustainable behaviors: An application of the value-belief-norm model. *J. Sustain. Tour.* **2018**, *26*, 957–972. [CrossRef]
43. Burhanudin, B.; Unnithan, A.B. The determinants of eco-friendly tourist behaviour: Perspectives from Indian and Indonesian tourists travelling abroad. *Anatolia* **2022**, *33*, 48–64. [CrossRef]
44. Sharma, K.; Bansal, M. Environmental consciousness, its antecedents and behavioural outcomes. *J. Indian Bus. Res.* **2013**, *5*, 198–214. [CrossRef]

45. Paul, J.; Modi, A.; Patel, J. Predicting green product consumption using theory of planned behavior and reasoned action. *J. Retail. Consum. Serv.* **2016**, *29*, 123–134. [CrossRef]
46. Sadiq, M.; Paul, J.; Bharti, K. Dispositional traits and organic food consumption. *J. Clean. Prod.* **2020**, *266*, 121961. [CrossRef]
47. Sultan, M.T.; Sharmin, F.; Badulescu, A.; Stiubea, E.; Xue, K. Travelers' responsible environmental behavior towards sustainable coastal tourism: An empirical investigation on social media user-generated content. *Sustainability* **2020**, *13*, 56. [CrossRef]
48. Kautish, P.; Paul, J.; Sharma, R. The moderating influence of environmental consciousness and recycling intentions on green purchase behavior. *J. Clean. Prod.* **2019**, *228*, 1425–1436. [CrossRef]
49. Rafiq, F.; Chishty, S.K.; Adil, M. Explanatory or dispositional optimism: Which trait predicts eco-friendly tourist behavior? *Sustainability* **2022**, *14*, 2994. [CrossRef]
50. Pagiaslis, A.; Krontalis, A.K. Green consumption behavior antecedents: Environmental concern, knowledge, and beliefs. *Psychol. Mark.* **2014**, *31*, 335–348. [CrossRef]
51. Yadav, R.; Pathak, G.S. Young consumers' intention towards buying green products in a developing nation: Extending the theory of planned behavior. *J. Clean. Prod.* **2016**, *135*, 732–739. [CrossRef]
52. Hair, J.F.; Gabriel, M.; Patel, V. AMOS covariance-based structural equation modeling (CB-SEM): Guidelines on its application as a marketing research tool. *Braz. J. Mark.* **2014**, *13*, 44–55.
53. Khare, A. Consumers' susceptibility to interpersonal influence as a determining factor of ecologically conscious behavior. *Mark. Intell. Plan.* **2014**, *32*, 2–20. [CrossRef]
54. Sadiq, M.; Adil, M. Ecotourism related search for information over the internet: A technology acceptance model perspective. *J. Ecotourism* **2021**, *20*, 70–88. [CrossRef]
55. Sadiq, M.; Adil, M.; Paul, J. An innovation resistance theory perspective on purchase of eco-friendly cosmetics. *J. Retail. Consum. Serv.* **2021**, *59*, 102369. [CrossRef]
56. Henseler, J. *User Manual Adanco 2.0. 1*; Composite Modeling GmbH & Co.: Kleve, Germany, 2017; pp. 1–47.
57. Cvelbar, L.K.; Grün, B.; Dolnicar, S. Which hotel guest segments reuse towels? Selling sustainable tourism services through target marketing. *J. Sustain. Tour.* **2017**, *25*, 921–934. [CrossRef]
58. Trivedi, R.H.; Patel, J.D.; Savalia, J.R. Pro-environmental behavior, locus of control and willingness to pay for environmental friendly products. *Mark. Intell. Plan.* **2015**, *33*, 67–89. [CrossRef]
59. Taufique, K.M.; Vaithianathan, S. A fresh look at understanding green consumer behavior among young urban Indian consumers through the lens of Theory of Planned Behavior. *J. Clean. Prod.* **2018**, *183*, 46–55. [CrossRef]
60. Mostafa, M.M. Antecedents of Egyptian consumers' green purchase intentions: A hierarchical multivariate regression model. *J. Int. Consum. Mark.* **2006**, *19*, 97–126. [CrossRef]

sustainability

Review

Barriers to Implementing the Circular Economy in the Construction Industry: A Critical Review

Rabia Charef [1], Jean-Claude Morel [2,*] and Kambiz Rakhshan [3]

1 Faculty of Engineering, Environment and Computing, Coventry University, Coventry CV1 5FB, UK; r.charef69@gmail.com
2 Laboratory of Tribology and System Dynamics, National School of State Public Works (ENTPE), 69120 Lyon, France
3 Leeds Sustainability Institute, Leeds Beckett University, Woodhouse Lane, Leeds LS2 8AG, UK; rakhshak@uni.coventry.ac.uk
* Correspondence: jeanclaude.morel@entpe.fr

Abstract: To facilitate the adoption of the circular economy (CE) in the architecture, engineering and construction (AEC) sector, some authors have demonstrated the potential of recent designs that take into account the sustainable management of an asset's end-of-life (EOL), providing an alternative to the dominant designs that end with demolition. However, there is no review of the literature that encompasses a large range of sustainable designs in the current CE context. This paper provides a critical review of journal papers that deal with the barriers to implementing sustainable designs and approaches to the EOL management of assets that have the potential to fulfil the principles of the CE. Eighteen approaches related to prefabrication, design for change, design for deconstruction, reverse logistics, waste management and closed-loop systems were found. Through an analysis of the barriers that are common among these 18 approaches, we classified them into six different categories (organisational, economical, technical, social, political and environmental). Two Sankey diagrams illustrate the interrelation between the barriers, their categories and the 18 approaches. The diagrams clearly show that most of the barriers are common to multiple approaches and that most of the barriers relate to organisational concerns. The study gives a detailed map of the barriers that would help stakeholders from the AEC sector develop strategies to overcome the current obstacles in the shift to a CE.

Keywords: sustainable asset lifecycle; circular economy; design for deconstruction; barriers; sustainable approach; critical literature review

Citation: Charef, R.; Morel, J.-C.; Rakhshan, K. Barriers to Implementing the Circular Economy in the Construction Industry: A Critical Review. *Sustainability* **2021**, 13, 12989. https://doi.org/10.3390/su132312989

Academic Editors: Nicola Raimo, Filippo Vitolla, Ornella Malandrino, Benedetta Esposito and Paris Fokaides

Received: 25 October 2021
Accepted: 17 November 2021
Published: 24 November 2021

1. Introduction

Since the 1973 oil embargo, the architecture, engineering and construction (AEC) sector has had to reduce the energy it uses through various active and passive measures, including increasing the insulation of buildings [1,2]. To reduce heat loss from the envelope, builders add insulation to the full perimeter of buildings at the cost of consuming more primary resources to produce insulation materials. Similar processes occur with double/triple glazing or building-integrated photovoltaics, resulting in more pressure on building component production and its corollary consumption of natural resources. Inevitably, stakeholders in this sector have to tackle various related challenges, including the scarcity of raw material resources and the production of megatonnes of waste [3].

Waste is generated throughout the lifecycle of an asset. However, the end-of-life (EOL) phase of assets is the least sustainable phase, given the amount of waste generated by the demolition process. Demolition is responsible for 50% of all waste produced by the AEC sector worldwide [4]. In the UK, 30% of the total waste sent to landfills comes from the demolition of buildings, costing GBP 200 million annually [5]. That is why the EOL management of assets and the construction and demolition waste (CDW) that is often produced at this stage are of increasing interest to academic researchers [6], who seek to shift the sector into a new paradigm with the advent of the circular economy (CE) [7] Recently, the Ellen MacArthur Foundation [8] has contributed to the shift from the linear system to the CE for the manufacturing industry [9,10]. As an example, at the European Union level, the CE implementation action plan was initiated in 2015 [11]. After five years of implementation in the industry, some authors have noticed some shortcomings. First recycling has been developed most substantially [11], despite the fact that it is the worst option on the scale of the "10Rs", just before the "R" of "Recover" [12]. Secondly, EU policies focus on "end of pipe" solutions without addressing "the many socio-ecological implications of a circularity transition" [13].

In the specific case of the AEC sector, the linear approach is the dominant method in the "design for construction", with the EOL managed by demolition [14]. There is no consensus on a definition of the circularity of a building. The reason might be due to a lack of familiarity with what the notion of the CE means for the AEC sector compared with the manufacturing industry. Moreover, the nature of a building is very different from a given manufactured product, in terms of its use and fabrication process. Indeed in the construction industry, each owner (private, public or company) wishes to have a bespoke building and not a standard one. Moreover, the management of a building from its construction phase to its demolition involves a wide range of stakeholders with different skills and stakes, as illustrated by some authors [15]. Lastly, the timescale of the different phases of a building's lifecycle varies drastically and is considerably long during its operation (typically lasting a minimum of 30 years).

Although the implementation of the CE in the AEC sector is relatively new [16] some authors have already studied alternative approaches that have the potential to fulfil the CE requirements. For example, Ghisellini et al. [17] have linked sustainable buildings and cleaner production to the CE. Some authors have associated the design for deconstruction (DfD) approach to the CE [18], whereas this concept of design for X is already developed in other sectors of the industry [19]. Another example is the extensive focus on resource use and CDW management (e.g., [20]) at the expense of the investigation into alternative approaches, such as supply chain integration, building designs, offsite manufacturing, among others [21]. Moreover, as the AEC sector is currently experiencing the digital revolution (mainly through building information modelling, or BIM), some authors have developed specific BIM uses for overcoming the barriers to the adoption of a CE [22]. However, this research area is still in its infancy, and as such it faces a lack of common understanding which leads to the inconsistent application of terminologies, especially among architects and contractors [23]. More recently, in response to the lack of consensus and to clarify the processes along the asset lifecycle some authors have developed a classification of the current alternative design approaches that have the potential to fulfil the CE requirements [24]. This classification clarifies and illustrates the current diversity of existing alternative approaches with five central categories (prefabrication, design for change, design for deconstruction, reverse logistics and closed-loop systems). Although this current research has fueled the quest for clarity there are still no studies that provide an overview of the barriers facing sustainable design approaches. Indeed, studies found in the literature typically explore barriers associated with a particular sustainable approach, such as the implementation of reverse logistics [25] construction waste reduction, reuse and recycling [26], among others. More recently, the impediments associated with the CE approach are at the centre of the concerns of some authors [27]. However, none of these studies have targeted the overall view of the barriers associated with sustainable approaches, as defined by [24].

This article aims to analyse the barriers found in the literature that are preventing the AEC sector from shifting to a CE. To fulfil this aim, the following objectives are defined:

1. To extract from journal papers the barriers to the implementation of approaches related to prefabrication, design for change, design for deconstruction, reverse logistics waste management and closed-loop systems.

2. To classify the barriers into categories and extract the interrelation between the barriers, their categories and the selected approaches.

2. Method

Figure 1 presents the three stages of the overall method to establish an understanding of the barriers identified in the literature that are hindering the development of the CE in the AEC sector. At Stage 1, we selected the journal articles for the review. First, the approaches that have the potential to help implement the CE in the AEC sector are selected (Points 1 and 2, developed in Section 2.1). Second, the journal articles dealing with the barriers to those approaches are searched for (Section 2.2). Stage 2 is dedicated to the analysis. First, how the barriers could be classified into categories was investigated (Section 2.3). Second, the results are eventually illustrated in Stage 3 by the two Sankey diagrams.

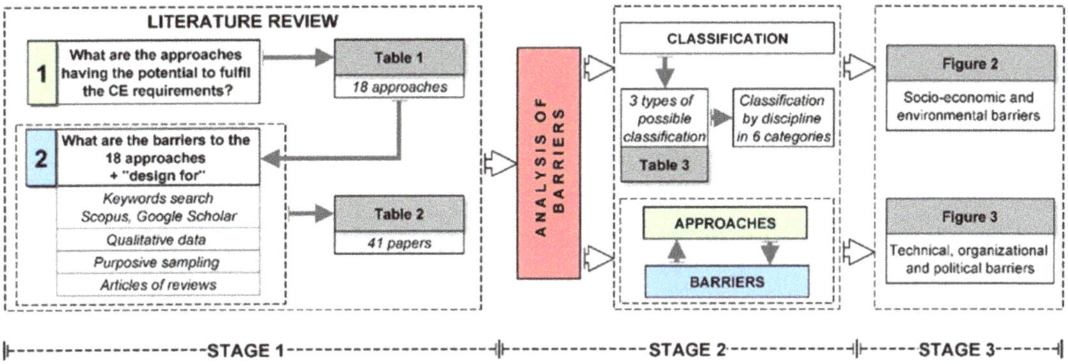

Figure 1. Graphical image of the methodological framework with three stages.

2.1. The Studied Approaches

Eighteen approaches, listed in Table 1, were extracted from previous work conducted on the identification, definition, and classification of asset lifecycle strategies for the CE [24]. In this previous work, the authors provided a synthesis of the literature to clarify the approaches that are related to the asset lifecycle in the context of the CE. The definitions of sets of approaches were analysed to clarify whether they were compliant with a CE or recycling and reuse economy CE [24]. Most of those approaches are alternative design approaches that have the potential to fulfil the CE requirements. However, some approaches deal with CDW management, given that it is a crucial stage in the CE that must be brought down to zero waste. The barriers linked to life cycle assessment (LCA) in the AEC sector were also investigated, because it is a method that is key in the context of the CE where circularity indicators are yet to be properly developed in the AEC sector. Most of the approaches are dual, with the approach linked to the design for the approach (e.g., for reverse logistics (RL) and design for reverse logistics (DfRL)—see Table 1). In the following, to simplify the reading, the comments will specify only the approach and not its associated "design for" approach (e.g., simply RL, not RL-DfRL).

2.2. The Search Method for Papers Dealing with Barriers to the 18 Approaches

This research has targeted only qualitative data in peer-reviewed journals to ensure the quality of the secondary data [28]. Knowing the 18 approaches has guided the extraction of appropriate keywords. For example, in the case of DfD, the keywords would be "barriers" and "deconstruction" (or "dismantling", or "disassembly", or "demountable", among others). The papers were searched for in two databases (Scopus and Google Scholar) and purposive sampling was used as defined by Grant and Booth [29] to select the outputs. Articles of review on a specific approach were privileged to ensure that we obtain the whole picture of the approach and its barriers, Table 2.

Table 1. Acronyms of the 18 studied approaches.

Acronyms	Approaches
3Rs-Df3Rs	Reduce, Reuse, Recycle/Design for Reduce, Reuse, Recycle
AR-DfAR	Adaptive Reuse/Design for Adaptive Reuse
CDW Mana	Construction and Demolition Waste Management
CDW Mini	Construction and Demolition Waste Minimisation
CE-DfCE	Circular Economy/Design for Circular Economy
CL-DfCL	Closed loops/Design for Closed loops
Cy	Constructability
De-DfD	Deconstruction/Design for Deconstruction
Disa-DfDisa	Disassembly/Design for Disassembly
Dis-DfDis	Dismantle/Design for Dismantle
IFD-DfIFD	Industrialised, Flexible and Demountable/Design for Industrialised, Flexible and Demountable buildings
LCA	Lifecycle Assessment
MA	Manufacture and Assembly/Design for Manufacture and Assembly
PFA-DfPFA	Prefabrication/Design for Prefabrication
RL-DfRL	Reverse Logistics/Design for Reverse Logistics
SB-DfSB	Sustainable Building/Design for Sustainable Building
SD	Selective Demolition
TB-DfTB	Transformable Building/Design for Transformable Building

2.3. *The Classification of the Barriers*

Given the number of barriers in the literature, we sought to classify them into categories and subcategories to facilitate the analysis and the discussion. According to the literature, outlined in Table 3, three main types of categories were used, namely those linked to:

1. A discipline (economy, environment, culture, among others).
2. A specific stakeholder (owners, designers, among others).
3. A construction phase (design, in use, etc.).

One of the weaknesses of the construction industry is its siloed working methods where the phases and stakeholders are divided, with poor communications in between. In the particular context of the CE, it is crucial to implement a more holistic approach, involving all of the stakeholders and therefore all of the asset phases as described by [15]. The choice of the disciplines as macro-categories is motivated by the need to cover the whole of the asset lifecycle and all of the stakeholders involved in the asset lifecycle in line with the holistic approach required to implement a CE [15]. Moreover, a classification by discipline gives us the ability to highlight that the technical barriers are only a part of the picture, despite the fact that there is a huge scientific literature focused only on technical barriers. The label of the categories is derived from the literature (see Table 3) especially Huuhka and Hakanen (2015), who used four categories for their barriers: (i) economic, (ii) social, (iii) ecological and (iv) technological. Following this, two categories (organisational and political) were added by Rakhshan et al. (2020). For this study, six main categories are therefore used (economic, sociological, political, organizational, technological and environmental). Within these categories, some subcategories related to specific stakeholders or phases are added when necessary.

Table 2. List of the articles analysed and used to build the Sankey diagrams.

Authors	Article Title	Type of Data (LR = Literature Review)
Ajayi et al. (2015) [30]	Waste effectiveness of the construction industry: Understanding the impediments and requisites for improvements	LR
Akanbi et al. (2018) [31]	Salvaging building materials in a circular economy: A BIM-based whole-life performance estimator	Interviews + questionnaire
Akinade et al. (2020) [32]	Design for deconstruction using a circular economy approach: Barriers and strategies for improvement	Questionnaire + literature review
Bouzon et al. (2015) [33]	Reverse logistics drivers: empirical evidence from a case study in an emerging economy	LR
Brancart et al. (2017) [34]	Transformable structures: Materialising design for change	LR
Carvalho Machado et al. (2018) [35]	Analysis of Guidelines and Identification of Characteristics Influencing the Deconstruction Potential of Buildings	Case study
Chileshe et al. (2015) [25]	Barriers to implementing reverse logistics in South Australian construction organisations	LR
Chileshe et al. (2016) [36]	Drivers for adopting reverse logistics in the construction industry: A qualitative study	Case study
Couto and Couto (2010) [37]	Analysis of Barriers and the Potential for Exploration of Deconstruction Techniques in Portuguese Construction Sites	LR + case studies
Crowther, P. (2002) [38]	Design for buildability and the deconstruction consequences	LR
Crowther, P. (2005) [39]	Design for Disassembly—Themes and principles	Case studies
Cruz Rios et al. (2015) [40]	Design for Disassembly and Deconstruction—Challenges and Opportunities	Case study/modelling
Diyamandoglu and Fortuna (2015) [41]	Deconstruction of wood-framed houses: Material recovery and environmental impact	Questionnaire + case study
Ferreira Correia et al. (2021) [42]	Plan to Overcome Barriers to Reverse Logistics in CDW: Survey of the Construction Industry	LR+Questionnaire
Forsythe, P. (2011) [43]	Drivers of Housing Demolition Decision Making and the Impact on Timber Waste Management	LR + focus groups
Gorgolewski, M. (2008) [44]	Designing with reused building components: Some challenges	Case studies
Leigh and Patterson (2006) [45]	Deconstructing to Redevelop	LR + case study + model
Hakkinen and Belloni (2011) [46]	Barriers and drivers for sustainable building	Semi-structured interviews
Hosseini et al. (2014) [47]	Reverse Logistics for the Construction Industry: Lessons from the Manufacturing Context	LR+ focus group + model
Hosseini et al. (2015) [48]	Reverse logistics in the construction industry	Questionnaire + case study
Huuhka and Hakanen (2015) [49]	Potential and barriers for reusing load-bearing building components in Finland	Case study
Inglis, M. (2007) [50]	Construction and Demolition waste—Best practice and cost-saving	Case study
Jaillon and Poon (2010) [51]	Design issues of using prefabrication in Hong Kong building construction	LR

Table 2. *Cont.*

Authors	Article Title	Type of Data (LR = Literature Review)
Jaillon and Poon (2014) [52]	Lifecycle design and prefabrication in buildings: A review and case studies in Hong Kong	LR
Kibert, C. J. (2003) [53]	Deconstruction: The start of a sustainable materials strategy for the built environment	Case study
Kifokeris and Xenidis (2017) [54]	Constructability: Outline of Past, Present, and Future Research	LR + case studies
Kim et al. (2017) [55]	An estimation framework for building information modelling (BIM)-based demolition waste by type	LR + case studies
Knecht B. (2004) [56]	Designing for Disassembly and Deconstruction	LR
Kohler, N., and Yang, W. (2007) [57]	Long-term management of building stocks.	Case study
Nisbet et al. (2012) [58]	Demolition and Deconstruction: Review of the Current Status of Reuse and Recycling of Building Materials.	LR
Pulaski et al. (2004) [59]	Design for Deconstruction	LR + model
Río Merino and Gracia (2010) [60]	Sustainable construction: Construction and demolition waste reconsidered	LR
Sanchez and Haas (2018) [61]	Capital project planning for a circular economy	LR
Sassi, P. (2008) [62]	Defining closed-loop material cycle construction	LR
Tingley and Davison (2012) [63]	Developing an LCA methodology to account for the environmental benefits of design for deconstruction	LR+model
Xanthopoulos et al. (2009) [64]	Reverse logistics processes of multi-type end-of-life buildings/construction sites: An integrated optimization framework	LR
Yuan et al. (2018) [65]	Design for Manufacture and Assembly-oriented parametric design of prefabricated buildings	Case study
Zaman et al. (2018) [66]	Resource Harvesting through a Systematic Deconstruction of the Residential House: A Case Study of the 'Whole House Reuse' Project in Christchurch, New Zealand	Survey

Table 3. The different types of barrier categories found in the literature.

Authors	Categories
(Hosseini et al., 2014) [47] (Chileshe et al., 2016) [36]	1. External. 2. Internal (to the manufacture and building industry).
(Chileshe et al., 2015) [25]	The authors split the external barriers into two subcategories: 1. "Environmental barriers", referring to barriers imposed by the prevailing business environment in the industry. 2. Barriers due to the nature of construction products (e.g., buildings) and their activities.
(Abdulrahman et al., 2014) [67]	Authors grouped the RL implementation barriers relevant to the Chinese context into four categories: (i) management, (ii) financial, (iii) policy and (iv) infrastructure.
(Bouzon, Govindan, and Rodriguez, 2018) [68]	Authors have classified the barriers to their multi-perspective framework for RL into three categories: (i) governmental perspective, (ii) organisational perspective and (iii) consumers' perspective.
(Häkkinen and Belloni, 2011) [46]	(i) Steering mechanisms: informative regulatory instruments (mandatory labelling)—economic and market-based instruments (certificate schemes), fiscal instruments and incentives (taxation and support)—voluntary action (public leadership programmes); (ii) Economics; (iii) Client understanding; (iv) Process: procurement and tendering, timing, cooperation and networking); (v) Underpinning knowledge: knowledge and common language, availability of methods and tools, innovation (including for normative regulatory instruments, e.g., building codes).
(Huuhka and Hakanen, 2015) [49]	Barriers to reusing load-bearing building components in Finland in four categories: (i) economic, (ii) social, (iii) ecological and (iv) technological barriers.
(Rakhshan et al., 2020) [3]	Barriers to reusing load-bearing building components in six categories: (i) economics, (ii) environmental, (iii) organisational, (iv) social, (v) technological and (vi) regulatory barriers.
(Kifokeris and Xenidis, 2017) [54]	Barriers to constructability implementation in five categories: (i) general barriers, (ii) owner barriers, (iii) designer barriers, (iv) contractor barriers and (v) project-specific barriers.
(Park and Tucker, 2016) [69]	Barriers raised by stakeholders (end users, developers, etc.).

3. Results and Discussion

Sankey diagrams (Figures 2 and 3) are used as a visualisation tool to analyze the interplay between the barriers (structured in different categories and subcategories) and the 18 approaches. Figure 2 presents the sociological, economic and environmental barriers, given that those three categories are the basis of sustainable development.

3.1. Environmental Barriers

The environmental barriers are less prominent than the socio-economic ones (Figure 2). They are mainly limited to the EOL phase of the asset (CDW management and selective demolition). In the cases of deconstruction (De), RL and adaptive reuse (AR), the issue of space for storage is cited by [25]. The site access limitation is also cited by [51]. Some authors have also noticed the existence of lead and asbestos in old buildings when dealing with RL, the 3Rs and CDW management [56,58,60]. Those authors are not linking the high pollutant materials to environmental issues, but to the additional costs of processing such materials. However, the process leads to environmental issues because waste ends up in landfills. The fact that some authors and stakeholders do not directly link pollution to environmental issues but to cost may be related to the lack of awareness of the impact that these processes have on the environment. This is made more apparent with prominence of the economic barriers (cost and market) (Figure 2).

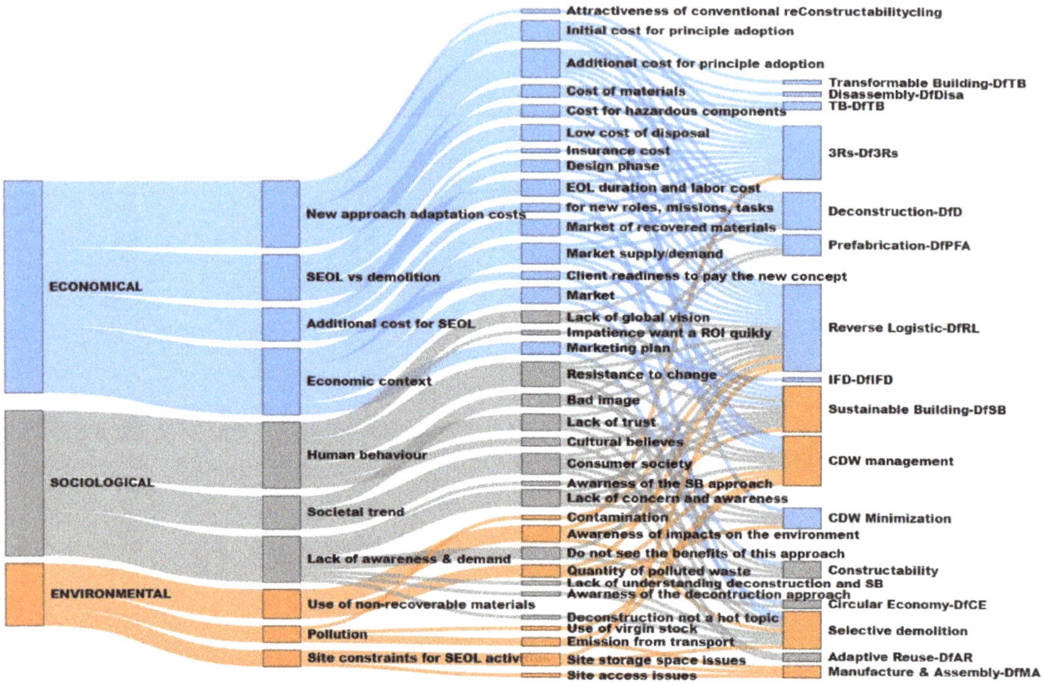

Figure 2. Outline of the sociological, economic and environmental barriers to alternative approaches identified in the literature.

Many studies have identified the lack of strong evidence of the environmental and sustainability benefits of using well-designed approaches, such as De, SB (sustainable buildings), the 3Rs (reduce, reuse, recycle), AR and CDW minimisation [46]. Other authors raised the exposure to health and safety risks from encountering contaminated materials as an essential barrier [25,39]. However, this topic is also linked to a social issue because health and safety are determined by the appraisal of the risks by society.

Direct, specific impacts have been identified by Sassi [62] in the recycling process, where the loss of material mass required additional virgin feedstock to be added. Other authors have identified the emissions from transport and reconditioning for the 3Rs and PFA (prefabrication) approaches [40,51].

3.2. Economic Barriers

The obstacles arising from factors related to the economic context (Figure 2) demonstrate that the markets built around sustainable buildings are, in many cases, insufficient. For example, the marketing plan of the industrial, flexible and demountable approach (IFD) is found to be inappropriate [52]. For SB, the lack of client demand is a barrier [46]. These shortcomings include the lack of demand for second-hand, reused and recycled materials and products [47]. The other factor is due to the constrictive search for profitability, which does not permit any risks to arise from innovation or changes to the current processes [54]. Deconstruction faces economic barriers because it is more expensive than demolition, demanding more time and labour to recover comparably low-cost construction materials [44]. Several authors have argued that the low cost of construction materials compared to recovered/recycled materials is the problem in the cases of De and the 3Rs [43]. Moreover, the standard construction and demolition practices are focused on the fastest and most economical way to finish the job in the case of the 3Rs [44]. Lastly, obstacles are caused by the shortcomings of the recycling processes of conventional materials. For example, a building's aluminium scraps are challenging to recover economically [58], and the recycled aggregates have a considerably lower price than that of the natural materials, due to their poorer quality. The relatively low cost of disposing CDW materials in landfills is also cited as a barrier by many authors, in the cases of De, RL, the 3Rs and PFA [58]. Some authors have also noticed the labour-intensive nature of the deconstruction and reuse processes [33]. Moreover, additional time is necessary in the cases of SD, De, and the 3Rs compared to conventional processes. This extra time results in extra costs [33].

The main economic subcategory relates to the costs attributed to the approach ("new approach adoption cost" in Figure 2). The design phase embodies additional costs due to more work needing to be performed in the case of RL, the 3Rs and PFA [52]. Although no additional costs attributed to the construction phase were found, some costs related to the necessary adaptations to a new approach (or to adopt new approaches) were noted. Several authors have identified such additional costs for adopting the following approaches: De, RL, TB (transformable buildings), Dis (dismantle), the 3Rs, CDWmini (CDW minimisation) and PFA [25]. The additional cost is also due to the management of hazardous components. For example, the existence of lead and asbestos in old buildings makes the process of deconstruction costly and time consuming because the cost of separating the materials to be recycled from contaminating materials is high [58,60]. Another cost of adopting a new approach is the additional initial cost (i.e., the higher cost of the initial investment in the project) cited by many authors, concerning the following approaches: De, RL and PFA [25,30,36,37,51,52,66]. Lastly, the additional costs due to higher insurance fees are a barrier reported by some authors for the 3Rs approach [70].

Some authors have spotted barriers linked to the quantification and sales involved in the approaches. In the case of De, Jaillon and Poon [52] have noticed that the economic benefits are not well established. Similarly, Xanthopoulos et al. [64] highlighted the lack of establishment for the economic and environmental benefits of CDW management. In the case of SB, Häkkinen and Belloni [46] have noticed the lack of understanding of business cases. Finally, in the case of AR, Chileshe et al. [25] have noted the significant differences in the distribution of the construction budget.

Another type of obstacle valid for the 3Rs approach is the necessity of planning and paying upfront early in the asset's lifecycle, which is impossible without the willingness of the client [44]. At this stage, it is often the case that the contractor has not been appointed yet, so the client has to spend money upfront purchasing materials, which many clients will not be willing to do.

3.3. Sociological Barriers

Sociological barriers deal with social issues, focusing on cultural (societal trends), psychological (human behaviour) and personal characteristics (lack of awareness and demand). The main subcategory is "human behaviour" and the most cited barrier is "resistance to change" (Figure 2). Cultural beliefs are involved in the case of sustainable buildings, which notably face the barrier of a low-risk culture [46], as does the case of CDW minimisation [30]. The lack of global vision has also been mentioned in the cases of RL, SB, Disa and CE [39,46]. These authors found that thinking was more linear than circular, with a lack of lateral thinking and an ignorance of life-cycle thinking. The lack of trust in De, RL, the 3Rs, and CDW management is described by [25,70]. Additionally, these authors noticed a lack of acceptance of reclaimed materials. In particular, there is an impatience to get a return on investment quickly, which creates an unfavourable business culture in the case of RL [25]. The last social barrier is the resistance to change, spotted by many authors who focused on the use of Cy (constructability), De, RL, SB, the 3Rs and PFA [37,44,46,51,52,54,61]. Moreover, these authors highlighted scepticism and a preference for traditional methods within the industry, leading to a natural resistance to change from the manufacturers, builders and owners. This resistance to change, common to six approaches, is also seen within the organisations that lack the effort necessary to innovate. Unsurprisingly, the lack of experienced, skilled workers and insufficient knowledge is common to five approaches (Figure 2).

Some barriers are related to "consumer society behaviours". These include, for example, the fear of the additional costs of better waste management [46], or the belief that waste is inevitable [30], or the disbelief in the potential utility of a constructability program [54] Some authors have added that the consumer culture and attitudes towards the quality of salvaged and used items are also an obstacle [25]. For the RL, 3Rs and PFA approaches, the bad image of salvaged materials was reported to be an important barrier [51]. At a broader level, many authors have noticed a lack of awareness for several approaches, such as SB, the 3Rs and CDW minimisation [46,60]. Other authors have noticed the lack of awareness of the benefits of using Cy, RL and CE [59,61]. A lack of concern was raised by several authors for the De, RL and SB approaches [46,52,66,68]. Moreover, the lack of understanding of deconstruction and SB was reported by Zaman et al. (2018) [66], as well as [46]. In addition, and specifically for deconstruction, the demolition contractor's culture was highlighted as a critical issue [37].

3.4. Organisational Barriers

Organisational barriers refer to the hindrance to the flow of information between stakeholders and between construction phases that negatively impact the efficiency of a project. Organisational barriers include the extra time, resources and effort necessary for the consideration of sustainability and circularity throughout the asset lifecycle. They are the most documented barriers in the articles studied and are found to be attributed to 14 of the 18 approaches (Figure 3).

3.4.1. Working Methods and the New Approach

Approximately fifty statements from different authors that deal with issues associated with the current linear approach have been found. The barriers that they detail are raised by the fragmentation of the sector and its inappropriate organisation (Figure 3). The factors cited are the lack of a holistic approach, safety in the deconstruction process, innovation effective methods, and lifecycle performance focus (Figure 3). When dealing with RL, some authors have noticed the lack of support from the management, as well as immaturity and low investment in knowledge management, information systems, and continuous planning owing to changes in the materials' source location [25]. In addition, some authors reported a lack of specific budgetary allocation for CDW management [60]. Most of the abovementioned barriers are related to the adoption of new approaches and new methods involving more collaboration, communication and holistic and effective strategies For example, as stressed by some authors, the implementation of RL is a challenge for designers [25]. As a result, multidisciplinary teamwork becomes central and requires appropriate management.

3.4.2. Multidisciplinary Teamwork and Management

Many authors have highlighted the need for new methods to improve teamwork when addressing the whole lifecycle of a building [44,46,50,52,54,58,62,66]. Those concerns are related to Cy, De, CL (closed loop), RL, SB, the 3Rs and CDW minimisation. For example, some authors identified the need to change the established design and construction processes to promote the reuse of building components [44]. One paper has mentioned the need for systematic cooperation, while a multidisciplinary approach has been discussed in the case of IFD [52,71,72]. This is supported by authors who studied the decision-making framework used in the steel industry, and for the system thinking used in the construction industry [52,71,72].

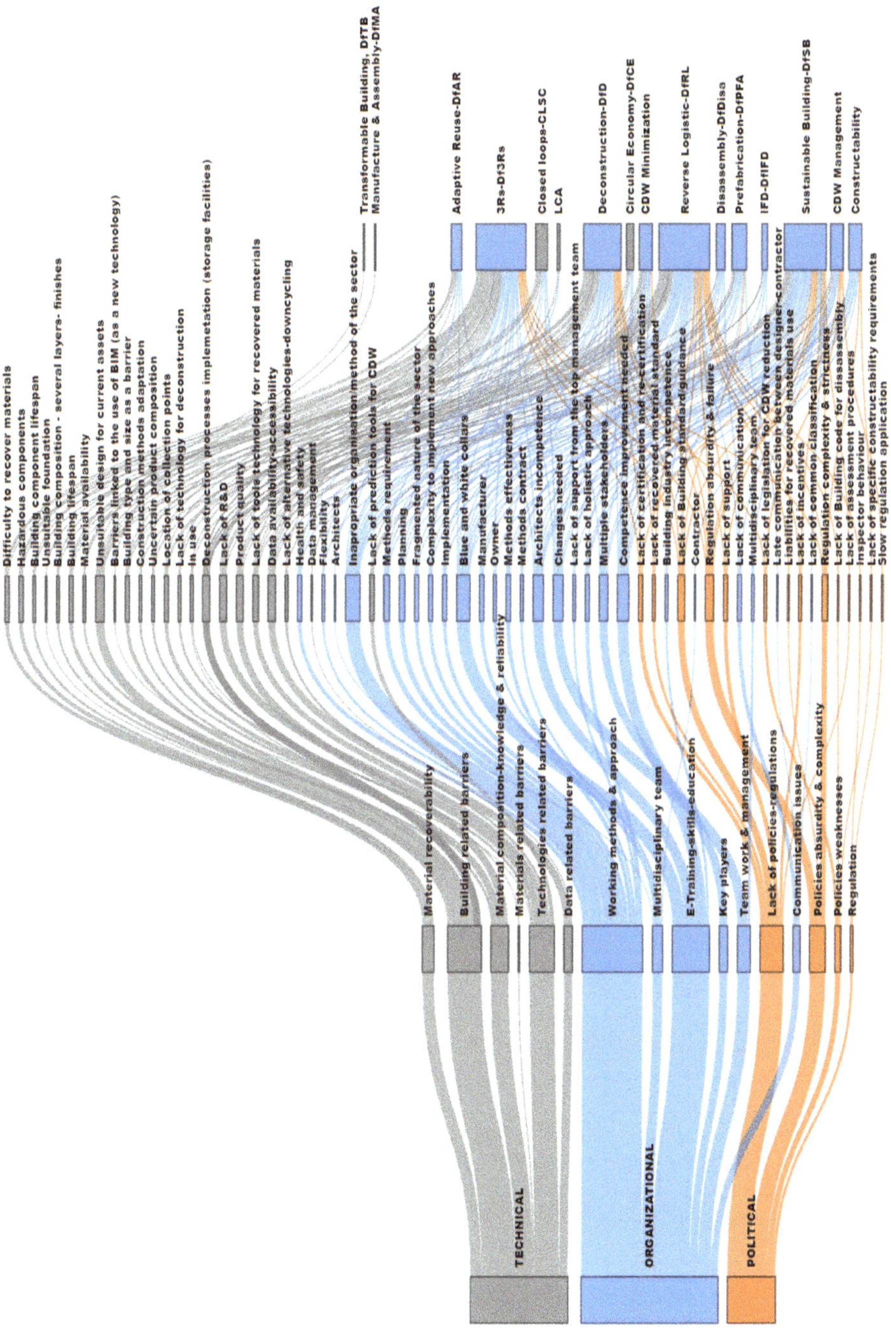

Figure 3. Outline of the technical, organisational and political barriers to the alternative approaches found in the literature.

Some authors have found that there is a lack of communication between the members of the project teams for RL, SB, Disa and PFA [46,51,56]. Specifically, issues related to late communications between the designer and contractor have been identified for RL, SB and PFA, where early collaboration between architects, contractors and manufacturers is required [51].

3.4.3. Key Players

Other barriers are related to the large number of stakeholders, and these have been outlined for many approaches, such as Cy, De, CL, RL, SB, CE and the 3Rs by [46,52,54,61]. The barriers are mainly specific to architects (although they also impact other stakeholders, such as contractors), where the reuse of materials in buildings requires acceptance and change in the design and construction processes [44]. Additionally, other authors have listed a number of the barriers raised by the contractors when implementing Cy. Most of them are linked to communication issues and a lack of skills/knowledge [54]. The manufacturers' lack of involvement and responsibility to minimise waste is stressed by [40]. The supply chain is also a central concern for many authors, including a lack of suppliers for PFA [51] and supply chain complexity in the case of SB [46]. Kifokeris and Xenidis [54] have also listed several barriers specific to the owners in the case of Cy, among others. Regarding the 3Rs approach, the unwillingness of the client to spend money upfront when purchasing materials, at a stage where the contractor is often not appointed yet, is a real issue for several authors [44,54].

3.4.4. Training, Skills and Education Support for a Skilled Workforce

The lack of skills, from an organisational point of view, is different from the skills related to social background already cited in Section 3.2. Logically, obstacles related to competence improvement were also cited with the lack of lessons learned on:

1. The comprehension of SB [46].
2. The application of DfDisa (design for disassembly), which is restrained by uncertainties regarding its global benefits and financial viability [34].
3. How RL remains unexploited or limited in the construction industry [36].

In complement, some authors have spotted the lack of documentation to support competence improvement, referencing:

1. The lack of lessons learned regarding documentation for Cy [54].
2. The lack of empirical evidence to support the widespread use of RL [36].
3. The need for the identification of demonstration projects to illustrate the potential of the different methods [37,61].
4. The lack of IFD studies for high-rise buildings [52].
5. The lack of studies providing clear instructions on how BIM could be used for CDW management [52].
6. The lack of studies that quantitatively demonstrate the effectiveness of the pre-project definition for buildings in the CE context [61].

From this subsection, the managerial implications involved in the shift to CE can be extracted. They consist in mainly revising the whole lifecycle management of an asset to ensure consistency in management from the early design stage to the EOL phase of the asset. These implications affect the role of all the stakeholders, encouraging them to improve the sharing of the information, even after the EOL phase of the asset, with training/education to help them change their way of working.

3.5. Technical Barriers

The technical barriers are split into different scales corresponding to subcategories, from the building scale to the material scale and from data management to the technologies used (Figure 3).

3.5.1. Building-Related Barriers

The long lifecycle of buildings exceeds the lifespan of industrial products and also results in multiple changes of ownership [25,52]. Additionally, the unicity of each building generates a complexity that is difficult to overcome in the modern context [35–37,44]. Firstly, in the design phase, some authors emphasise the barriers related to the designs of buildings that were not made with the SEOL in mind [35]. This includes all of the components, even the foundations, which are most of the time made with concrete [64]. The necessity of adapting the construction methods is emphasised when using reclaimed materials because it adds a whole new level of complexity to the project [44]. One example is given with the use of the in situ connection between precast concrete elements [52]. In the in-use phase, building components are updated or replaced at different intervals during the building's lifetime, adding complexity when updating the data related to the building (e.g., finishes at five-year intervals, lighting at ten-year intervals, HVAC systems at twenty-year intervals, etc.) [66].

Deconstruction processes encounter limitations due to the space available to manage the process and, significantly, to store the materials (see also the section on environmental and economic barriers on this point). The lack of recovery facilities and infrastructure is cited by [25,64]. Deconstruction is more complex than demolition, especially in the case of non-prefabricated components [35], sometimes leading to the impossibility of reusing components [56]. Moreover, demountable connections do not always ensure the possibility of deconstruction, and, in general, the poor connection of these elements is an issue [35,38,52]. Lastly, few demonstration projects have been identified that can help illustrate the potential of the different methods [25,44,49]. As a result, all these issues are increasing the risks associated with the deconstruction process [25,39,44].

3.5.2. Material-Related Barriers (Including Data)

Some barriers are related to the low quality of materials, the poor reliability of the characteristics of recovered materials [25,41,49,53,58,60,66] and the lack of data available for several asset phases. Indeed, in the design phase, the lack of data prevents carrying out an efficient LCA for the EOL phase [32]. In the deconstruction phase, projects and processes are also impacted by a lack of data [35]. At a different level, the behaviour and durability of recycled concrete is difficult to predict accurately without enough data [58]. Importantly, the composition of buildings at the end of their life is essential [47] and the lack of national data on CDW must be overcome [41]. In addition to the lack of data (availability and accessibility), weak data management has also been pointed to by several authors as a concern, especially in the case of national data collection and reporting on CDW [66]. In addition to the general lack of data, there are barriers regarding the limited locations of collection points for recovered materials [35,39,51,52] that generate limitations to material availability [44]. The main source of these issues is the recoverability of construction materials, which is limited by several factors:

1. The use of finishes on building materials reduces the possibility of reusing such materials [31].
2. The use of concrete [37,52].
3. The deconstruction process damages the materials because it is difficult to separate the composites [43,52,56,58,70].
4. Contamination with hazardous materials [25,39,43,56].
5. The deterioration rates are unknown [3,25,35,37].
6. The under-estimation of the resources embedded in the building [61].

3.5.3. Technology-Related Barriers

Regarding technological barriers, most of them are related to the lack of appropriate tools and procedures. Although one barrier concerning the lack of prefabricated building designs with BIM tools was reported by [65], most of the other issues are not related to BIM, but to the lack of several other elements. For example, some authors pointed to the lack of a common framework and automatic calculation procedures for SB [46], whereas other authors stressed the absence of simple processes to reuse a building project [61]. A lack of science-based, user-friendly tools for De, SB and CE was also reported in many studies [37,46,61], as well as the unavailability of proven alternative technologies [46] Some authors reported the lack of tools for designers that would otherwise enable efficient deconstruction [25,37,52,58], help with assessing DW generation [30,55,66], promote the inclusion of new techniques for construction [44,46,61], and help with assessing the costs associated with IFD buildings [48]. Lastly, techniques for reusing reclaimed materials are also missing [25,37,43,48]. Meanwhile, down-cycling cannot be regarded as a closed-loop (CL) approach because of the excessive loss of material value [62].

3.6. Political Barriers

Figure 3 outlines the barriers found in the political category. The main subcategory is the lack of appropriate standards (policies and regulation) and is cited in the cases of Cy, De, CL, RL, SB, the 3Rs, CDW management and PFA [58]. This includes shortcomings in standardised processes, insufficient sharing of the best practices, little clear information and little guidance for designers about the design and procurement procedures that they should adopt when reusing components. The lack of guidance for sustainability in public facilities for facility managers is reported by [46].

The lack of recertification, legal warranties and residual performance analysis of recovered building materials has been noted by Couto and Couto (2010) in the particular cases of De, SB, and the 3Rs, especially when compared to conventional processes. Furthermore, the lack of fiscal incentives or support from governments is cited by [25,46,66] for De, RL and SB. Lastly, the lack of appropriate assessment procedures for architectural competitions, the assessment process being performed late during the design phase and the lack of labelling/measurement standards are shortcomings spotted by [46] for SB.

Some authors argued that the current regulation is too strict to allow innovation, which is a barrier for Cy, De, RL, SB, CDW minimisation and CDW management. Lastly one paper noted that that building inspectors discourage the use of salvaged materials in the case of RL [47], and another author argued that regional governments are slow to apply CDW management plans which have already been approved [60].

4. Conclusions and Perspectives

The debate regarding the implementation of the circular economy (CE) in the architecture, engineering and construction (AEC) sector is relatively new. However, there are already several publications that link the CE with specific alternative approaches that take into account the lifecycle management of assets from design to end-of-life. These alternative approaches (prefabrication, design for change, design for deconstruction, reverse logistics, waste management and closed-loop systems) directly take into account the asset's end-of-life in the design stage, or only consider the manufacture and assembly phase (but may still ease the deconstruction process). Eighteen such approaches have been found that are far from dominant. However, helping them overcome the obstacles that they face may help to remove the future barriers of applying the CE concept to the built environment.

This article has assessed the barriers to those 18 approaches identified in the literature The barriers were classified into six categories (economic, sociological, political, organisational, technological, and environmental). The interrelation between the barriers from different categories is very common due to the holistic nature of the construction lifecycle and this has been illustrated here by two Sankey diagrams. The barriers are also related to the diversity of stakeholders and the phases of the building process.

The organisational barriers are the most cited in this review, and deal with the difficulty with changing the working methods and managing the required teamwork and the multidisciplinary approach. The last series of organisational barriers are linked to the lack of training and support for the skilled workforce and the provision of education. Increased access to data should help to revise the current management of the asset lifecycle.

The socioeconomic barriers are linked to the lack of awareness and demand for alternative approaches. The other socioeconomic barriers are related to people's behaviour (e.g., resistance to change and cultural beliefs), and lastly to the higher cost of alternative approaches (and the seeking of profit first). The political barriers are mainly related to the lack of appropriate standards and policies.

Some of the technical barriers are related to the construction materials (composition knowledge and recoverability) and others to the building scale (e.g., construction methods, project phase adoption and building lifespan). Other barriers are related to data and information management. The last range of technical barriers is linked to a lack of appropriate technology. The barriers linked to the environment are related to the site constraints (limited space), and to the use of non-recoverable materials. Pollution related to the materials (e.g., lead and asbestos) is also cited. Finally, this review gives a detailed map of the interrelations of the barriers that would help stakeholders develop strategies to overcome the current obstacles as they shift to a CE in the AEC sector.

The Sankey diagrams embed information regarding the structuring of the relationship between the barriers of a wide range of approaches. This information may help with the design of new research programmes that question the professionals working in the AEC sector about the barriers to (or the drivers that would foster) the transition to a CE. Moreover, the Sankey diagrams can help policymakers (and therefore standardisation bodies) design appropriate policy (and therefore standards) to foster the shift to a CE in the AEC sector. This information can also be useful to help develop and trigger green finance, which is a means supported by the United Nations to contribute towards delivering several of its sustainable development goals.

The key limitation of this study is that the secondary data analysed cover a wide range of geographic areas, meaning that some of the results may not fit any countries. However, the globalisation of the economy is a factor that would favour the consistency of the study across the world. Another limitation is that the studied approaches may not be consistent with a strict CE framework, knowing that there is no consensus on the definition or implementation of the CE in the AEC sector, especially given the lack of ISO standards on that topic. In this perspective, it should be evidenced also the role of digital technologies (under the umbrella of Industry 4.0 or BIM in the specific context of the AEC sector) in supporting the measurement of performance and data management to sustain the CE.

Author Contributions: Conceptualisation, R.C.; methodology, R.C.; formal analysis, R.C.; data curation, R.C.; writing—original draft preparation, R.C.; writing—review and editing, R.C., J.-C.M. and K.R.; visualisation, R.C. All authors have read and agreed to the published version of the manuscript.

Funding: This research received no external funding.

Institutional Review Board Statement: Not applicable.

Informed Consent Statement: Not applicable.

Conflicts of Interest: The authors declare no conflict of interest.

References

1. Friess, W.A.; Rakhshan, K. A review of passive envelope measures for improved building energy efficiency in the UAE. *Renew. Sustain. Energy Rev.* **2017**, *72*, 485–496. [CrossRef]
2. Villa, S.; Sassanelli, C. The Data-Driven Multi-Step Approach for Dynamic Estimation of Buildings' Interior Temperature. *Energies* **2020**, *13*, 6654. [CrossRef]
3. Rakhshan, K.; Morel, J.-C.; Alaka, H.; Charef, R. Components reuse in the building sector—A systematic review. *Waste Manag. Res.* **2020**, *38*, 347–370. [CrossRef]
4. Kibert, C.J. *Sustainable Construction: Green Building Design and Delivery*; John Wiley & Sons: New York, NY, USA, 2016.
5. Osmani, M. Construction Waste Minimization in the UK: Current Pressures for Change and Approaches. *Procedia-Soc. Behav. Sci.* **2012**, *40*, 37–40. [CrossRef]
6. Charef, R.; Alaka, H.; Emmitt, S. Beyond the Third Dimension of BIM: A Systematic Review of Literature and Assessment of Professional Views. *J. Build. Eng.* **2018**, *19*, 242–257. [CrossRef]

7. Geissdoerfer, M.; Savaget, P.; Bocken, N.; Hultink, E.J. The Circular Economy—A new sustainability paradigm? *J. Clean. Prod.* **2017**, *143*, 757–768. [CrossRef]
8. Ellen MacArthur Foundation. Available online: https://ellenmacarthurfoundation.org/ (accessed on 10 November 2021).
9. Geng, Y.; Doberstein, B. Developing the circular economy in China: Challenges and opportunities for achieving 'leapfrog development'. *Int. J. Sustain. Dev. World Ecol.* **2008**, *15*, 231–239. [CrossRef]
10. Rosa, P.; Sassanelli, C.; Terzi, S. Towards Circular Business Models: A systematic literature review on classification frameworks and archetypes. *J. Clean. Prod.* **2019**, *236*, 117696. [CrossRef]
11. Mhatre, P.; Panchal, R.; Singh, A.; Bibyan, S. A systematic literature review on the circular economy initiatives in the European Union. *Sustain. Prod. Consum.* **2021**, *26*, 187–202. [CrossRef]
12. Potting, J.; Hekkert, M.; Worrell, E.; Hanemaaijer, A. Circular Economy: Measuring Innovation in the Product Chain. Available online: http://www.pbl.nl/sites/default/files/cms/publicaties/pbl-2016-circular-economy-measuring-innovation-in-product-chains-2544.pdf (accessed on 10 November 2021).
13. Friant, M.C.; Vermeulen, W.J.; Salomone, R. Analysing European Union circular economy policies: Words versus actions. *Sustain. Prod. Consum.* **2021**, *27*, 337–353. [CrossRef]
14. Pomponi, F.; Moncaster, A. Circular economy for the built environment: A research framework. *J. Clean. Prod.* **2017**, *143*, 710–718. [CrossRef]
15. Charef, R.; Lu, W. Factor dynamics to facilitate circular economy adoption in construction. *J. Clean. Prod.* **2021**, *319*, 128639. [CrossRef]
16. Iacovidou, E.; Velis, C.A.; Purnell, P.; Zwirner, O.; Brown, A.; Hahladakis, J.; Millward-Hopkins, J.; Williams, P.T. Metrics for optimising the multi-dimensional value of resources recovered from waste in a circular economy: A critical review. *J. Clean. Prod.* **2017**, *166*, 910–938. [CrossRef]
17. Ghisellini, P.; Ji, X.; Liu, G.; Ulgiati, S. Evaluating the transition towards cleaner production in the construction and demolition sector of China: A review. *J. Clean. Prod.* **2018**, *195*, 418–434. [CrossRef]
18. Cai, G.; Waldmann, D. A material and component bank to facilitate material recycling and component reuse for a sustainable construction: Concept and preliminary study. *Clean Technol. Environ. Policy* **2019**, *21*, 2015–2032. [CrossRef]
19. Sassanelli, C.; Urbinati, A.; Rosa, P.; Chiaroni, D.; Terzi, S. Addressing circular economy through design for X approaches: A systematic literature review. *Comput. Ind.* **2020**, *120*, 103245. [CrossRef]
20. Ruiz, L.A.L.; Ramón, X.R.; Domingo, S.G. The circular economy in the construction and demolition waste sector—A review and an integrative model approach. *J. Clean. Prod.* **2020**, *248*, 119238. [CrossRef]
21. Osobajo, O.A.; Oke, A.; Omotayo, T.; Obi, L.I. A systematic review of circular economy research in the construction industry. *Smart Sustain. Built Environ.* **2020**. [CrossRef]
22. Charef, R.; Emmitt, S. Uses of building information modelling for overcoming barriers to a circular economy. *J. Clean. Prod.* **2021**, *285*, 124854. [CrossRef]
23. Dokter, G.; Thuvander, L.; Rahe, U. How circular is current design practice? Investigating perspectives across industrial design and architecture in the transition towards a circular economy. *Sustain. Prod. Consum.* **2021**, *26*, 692–708. [CrossRef]
24. Charef, R.; Ganjian, E.; Emmitt, S. Socio-economic and environmental barriers for a holistic asset lifecycle approach to achieve circular economy: A pattern-matching method. *Technol. Forecast. Soc. Chang.* **2021**, *170*, 120798. [CrossRef]
25. Chileshe, N.; Rameezdeen, R.; Hosseini, M.R.; Lehmann, S. Barriers to implementing reverse logistics in South Australian construction organisations. *Supply Chain Manag. Int. J.* **2015**, *20*, 179–204. [CrossRef]
26. Zou, P.; Hardy, R.; Yang, R. Barriers to building and construction waste reduction, reuse and recycling: A case study of the Australian Capital Region. In *Building Today—Saving Tomorrow, Proceedings of the Sustainability in Construction and Deconstruction Conference, Auckland, New Zealand, 15–17 July 2015*; Panko, M., Kestle, L., Eds.; Unitec Institute of Technology: Auckland, New Zealand, 2015; pp. 27–35.
27. Hossain, U.; Ng, S.T.; Antwi-Afari, P.; Amor, B. Circular economy and the construction industry: Existing trends, challenges and prospective framework for sustainable construction. *Renew. Sustain. Energy Rev.* **2020**, *130*, 109948. [CrossRef]
28. Schlosser, R.W. Appraising the Quality of Systematic Reviews. In *Focus: Technical Briefs*; Held South-West Educational Development Laboratory. Available online: https://ktdrr.org/ktlibrary/articles_pubs/ncddrwork/focus/focus17/ (accessed on 28 May 2021).
29. Grant, M.J.; Booth, A. A typology of reviews: An analysis of 14 review types and associated methodologies. *Health Inf. Libr. J.* **2009**, *26*, 91–108. [CrossRef] [PubMed]
30. Ajayi, S.O.; Oyedele, L.O.; Bilal, M.; Akinade, O.O.; Alaka, H.A.; Owolabi, H.A.; Kadiri, K.O. Waste Effectiveness of the Construction Industry: Understanding the Impediments and Requisites for Improvements. *Resour. Conserv. Recycl.* **2015**, *102*, 101–112. [CrossRef]
31. Akanbi, L.A.; Oyedele, L.O.; Akinade, O.O.; Ajayi, A.O.; Delgado, M.D.; Bilal, M.; Bello, S.A. Salvaging building materials in a circular economy: A BIM-based whole-life performance estimator. *Resour. Conserv. Recycl.* **2018**, *129*, 175–186. [CrossRef]
32. Akinade, O.; Oyedele, L.; Oyedele, A.; Delgado, J.M.D.; Bilal, M.; Akanbi, L.; Ajayi, A.; Owolabi, H. Design for deconstruction using a circular economy approach: Barriers and strategies for improvement. *Prod. Plan. Control.* **2020**, *31*, 829–840. [CrossRef]
33. Bouzon, M.; Spricigo, R.; Rodriguez, C.M.; de Queiroz, A.A.; Miguel, P.A.C. Reverse logistics drivers: Empirical evidence from a case study in an emerging economy. *Prod. Plan. Control.* **2015**, *26*, 1368–1385. [CrossRef]

34. Brancart, S.; Paduart, A.; Vergauwen, A.; Vandervaeren, C.; Laet, L.D.; Temmerman, N.D. Transformable Structures: Materialising Design for Change. *Int. J. Des. Nat. Ecodyn.* **2017**, *12*, 357–366. [CrossRef]
35. Machado, R.C.; de Souza, H.A.; Veríssimo, G.D.S. Analysis of Guidelines and Identification of Characteristics Influencing the Deconstruction Potential of Buildings. *Sustainability* **2018**, *10*, 2604. [CrossRef]
36. Chileshe, N.; Rameezdeen, R.; Hosseini, M.R. Drivers for Adopting Reverse Logistics in the Construction Industry: A Qualitative Study. *Eng. Constr. Archit. Manag.* **2016**, *23*, 134–157. [CrossRef]
37. Couto, J.; Couto, A. Analysis of Barriers and the Potential for Exploration of Deconstruction Techniques in Portuguese Construction Sites—Review. *Sustainability* **2010**, *2*, 428–442. [CrossRef]
38. Crowther, P. Design for Buildability and the Deconstruction Consequences. In Proceedings of the 3rd Annual Meeting of CIB Task Group 39, Rotterdam, The Netherlands, 9 April 2002; Available online: https://www.iip.kit.edu/downloads/CIB_Publication_272.pdf (accessed on 25 October 2021).
39. Crowther, P. Design for Disassembly—Themes and Principles. In *Environment Design Guide*; Royal Australian Institute of Architects: Canberra, Australia, 2005; pp. 1–7.
40. Rios, F.C.; Chong, W.K.; Grau, D. Design for Disassembly and Deconstruction—Challenges and Opportunities. *Procedia Eng.* **2015**, *118*, 1296–1304. [CrossRef]
41. Diyamandoglu, V.; Fortuna, L.M. Deconstruction of Wood-Framed Houses: Material Recovery and Environmental Impact. *Resour. Conserv. Recycl.* **2015**, *100*, 21–30. [CrossRef]
42. Correia, J.M.F.; de Oliveira Neto, G.C.; Leite, R.R.; da Silva, D. Plan to Overcome Barriers to Reverse Logistics in Construction and Demolition Waste: Survey of the Construction Industry. *J. Constr. Eng. Manag.* **2021**, *147*, 78620001966. Available online: https://ascelibrary.org/doi/abs/10.1061/%28ASCE%29CO.1943-7862.0001966 (accessed on 10 November 2021). [CrossRef]
43. Forsythe, P.J. Drivers of Housing Demolition Decision Making and the Impact on Timber Waste Management. *Constr. Econ. Build.* **2011**, *11*, 1–14. [CrossRef]
44. Gorgolewski, M. Designing with reused building components: Some challenges. *Build. Res. Inf.* **2008**, *36*, 175–188. [CrossRef]
45. Leigh, N.G.; Patterson, L.M. Deconstructing to Redevelop: A Sustainable Alternative to Mechanical Demolition: The Economics of Density Development Finance and Pro Formas. *J. Am. Plan. Assoc.* **2006**, *72*, 217–225. [CrossRef]
46. Häkkinen, T.; Belloni, K. Barriers and drivers for sustainable building. *Build. Res. Inf.* **2011**, *39*, 239–255. [CrossRef]
47. Hosseini, M.R.; Chileshe, N.; Rameezdeen, R.; Lehmann, S. Reverse Logistics for the Construction Industry: Lessons from the Manufacturing Context. *Int. J. Constr. Eng. Manag.* **2014**, *3*, 75–90.
48. Hosseini, M.R.; Rameezdeen, M.R.; Chileshe, N.; Lehmann, S. Reverse Logistics in the Construction Indus-try. *Waste Manag. Res.* **2015**, *33*, 499–514. [CrossRef]
49. Huuhka, S.; Hakanen, J.H. Potential and Barriers for Reusing Load-Bearing Building Components in Finland. *Int. J. Hous. Sci.* **2015**, *39*, 215–224.
50. Inglis, M. Construction and Demolition Waste—Best Practice and Cost Saving. *SB07 N. Z.* **2007**, *1*, 57.
51. Jaillon, L.C.; Poon, C.-S. Design issues of using prefabrication in Hong Kong building construction. *Constr. Manag. Econ.* **2010**, *28*, 1025–1042. [CrossRef]
52. Jaillon, L.C.; Poon, C.-S. Life cycle design and prefabrication in buildings: A review and case studies in Hong Kong. *Autom. Constr.* **2014**, *39*, 195–202. [CrossRef]
53. Kibert, C.J. Deconstruction: The Start of a Sustainable Materials Strategy for the Built Environment. *Ind. Environ.* **2003**, *26*, 84–88.
54. Kifokeris, D.; Xenidis, Y. Constructability: Outline of Past, Present, and Future Research. *J. Constr. Eng. Manag.* **2017**, *143*. [CrossRef]
55. Kim, Y.C.; Hong, W.H.; Park, J.W.; Cha, G.W. An Estimation Framework for Building Information Modeling (BIM)-Based Demolition Waste by Type. *Waste Manag. Res.* **2017**, *35*, 1285–1295. [CrossRef] [PubMed]
56. Knecht, B. Designing for Disassembly and Deconstruction. *Archit. Rec.* **2004**, *192*, 181–188.
57. Kohler, N.; Yang, W. Long-term management of building stocks. *Build. Res. Inf.* **2007**, *35*, 351–362. [CrossRef]
58. Nisbet, M.; Venta, G.; Foo, S. Demolition and Deconstruction: Review of the Current Status of Reuse and Recycling of Building Materials. *Air Waste Manag. Assoc.* **2012**, 1–14.
59. Pulaski, M.; Hewitt, C.; Horman, M.; Guy, B. Design for Deconstruction. *Mod. Steel Constr.* **2004**, *44*, 33–37. Available online: https://www.aisc.org/globalassets/modern-steel/archives/2004/06/2004v06_deconstruction.pdf (accessed on 25 October 2021).
60. Merino, M.D.R.; Gracia, P.I.; Azevedo, I.S.-W. Sustainable construction: Construction and demolition waste reconsidered. *Waste Manag. Res.* **2010**, *28*, 118–129. [CrossRef] [PubMed]
61. Sanchez, B.; Haas, C. Capital project planning for a circular economy. *Constr. Manag. Econ.* **2018**, *36*, 303–312. [CrossRef]
62. Sassi, P. Defining closed-loop material cycle construction. *Build. Res. Inf.* **2008**, *36*, 509–519. [CrossRef]
63. Tingley, D.D.; Davison, B. Developing an LCA methodology to account for the environmental benefits of design for deconstruction. *Build. Environ.* **2012**, *57*, 387–395. [CrossRef]
64. Xanthopoulos, A.; Aidonis, D.; Iakovou, E.; Vlachos, D.; Iakovou, E. Reverse Logistics Processes of Multi-Type End-of-Life Buildings/Construction Sites: An Integrated Optimization Framework. *WSEAS Trans. Environ. Dev.* **2009**, *5*, 250–259.
65. Yuan, Z.; Sun, C.; Wang, Y. Design for Manufacture and Assembly-Oriented Para-metric Design of Prefabricated Buildings. *Autom. Constr.* **2018**, *88*, 13–22. [CrossRef]

66. Zaman, A.U.; Arnott, J.; Mclntyre, K.; Hannon, J. Resource Harvesting through a Systematic Deconstruction of the Residential House: A Case Study of the 'Whole House Reuse' Project in Christchurch, New Zealand. *Sustainability* **2018**, *10*, 3430. [CrossRef]
67. Abdulrahman, M.D.; Gunasekaran, A.; Subramanian, N. Critical barriers in implementing reverse logistics in the Chinese manufacturing sectors. *Int. J. Prod. Econ.* **2014**, *147*, 460–471. [CrossRef]
68. Bouzon, M.; Govindan, K.; Rodriguez, C.M.T. Evaluating Barriers for Reverse Logistics Implementation under a Multiple Stakeholders' Perspective Analysis Using Grey Decision Making Approach. *Resour. Conserv. Recycl.* **2018**, *128*, 315–335. [CrossRef]
69. Park, J.; Tucker, R. Overcoming barriers to the reuse of construction waste material in Australia: A review of the literature. *Int. J Constr. Manag.* **2017**, *17*, 228–237. [CrossRef]
70. Tingley, D.D.; Cooper, S.; Cullen, J. Understanding and overcoming the barriers to structural steel reuse, a UK perspective. *J Clean. Prod.* **2017**, *148*, 642–652. [CrossRef]
71. Rose, C.; Stegemann, J. From Waste Management to Component Management in the Construction Industry. *Sustainability* **2018** *10*, 229. [CrossRef]
72. Yeung, J.; Walbridge, S.; Haas, C. The role of geometric characterization in supporting structural steel reuse decisions. *Resour Conserv. Recycl.* **2015**, *104*, 120–130. [CrossRef]

sustainability

MDPI

Article

Beyond Profitable Shifts to Green Energies, Towards Energy Sustainability

Farboud Khatami and Erfan Goharian *

Department of Civil and Environmental Engineering, University of South Carolina, 300 Main St., Columbia, SC 29208, USA; fkhatami@email.sc.edu
* Correspondence: goharian@cec.sc.edu; Tel.: +1-803-777-4625

Abstract: The traditional carbon-based approach towards sustainability has long caused the concepts of green and sustainable energies to be used interchangeably. Recent studies have tried to advance this archaic view by considering more aspects of sustainability. However, almost all major studies have been concerned with only the economic and environmental aspects of electricity generation, whereas the concept of sustainability is beyond these two criteria. In this paper, we seek to provide a methodology for a more comprehensive definition of electricity generation sustainability based on the lessons learned from previous studies and additional metrics suggested by them. The main characteristics of select electricity generation technologies were studied, and their environmental, economic, social, and technical criteria as well as the uncertainties associated with them were selected as the four major factors in our paper. It has also been argued that the utilization of regional resources in addition to the inherent characteristics of electricity generation technologies is vital in providing a realistic view of sustainability. Of the sustainability assessment methods previously introduced, the Relative Aggregate Footprint (RAF) method was used in conjunction with the previously selected criteria as the basis of the study due to its ability to incorporate additional criteria and regional considerations. As such, the framework for sustainability assessment presented in this research accounts for major criteria identified in the literature and takes the available regional resources that affect the feasibility of each electricity technology into account. This study paves the way for the presentation of new guidelines for the creation of more comprehensive electricity generation sustainability measures to distinguish between the concepts of green and profitable vs. sustainable energies to support the development of sustainable energy portfolios.

Keywords: energy sustainability; green energy; carbon emissions; cost of energy; energy portfolio; resource availability

Citation: Khatami, F.; Goharian, E. Beyond Profitable Shifts to Green Energies, TowardsEnergy Sustainability. *Sustainability* **2022**, *14*, 4506. https://doi.org/10.3390/su14084506

Academic Editors: Filippo Vitolla, Nicola Raimo, Ornella Malandrino and Benedetta Esposito

Received: 10 February 2022
Accepted: 4 April 2022
Published: 10 April 2022

1. Introduction

The idea of tracing the footprint of the human impact on the earth is indeed one that dates back a few decades [1,2]. With renewable energies gaining popularity, the concept of how energy sources impact the environment [3] was initially only focused on carbon emissions and how they vary from one technology to the next. Initially, the lower carbon emissions of renewable energies made them synonymous with green energies. However, the concept of energy sustainability goes beyond carbon and embodies other criteria and available natural resources, in addition to their proper utilization. Hence, it is important for the key criteria that play a major role in energy sustainability to be identified and taken into consideration.

The sustainability of energies (and electricity generation in particular) has been discussed numerous times in the past decades. Each study, however, has looked at the issue of energy and its sustainability through the particular lens of the authors' area of expertise. The water–energy nexus works of [4], for example, focus on the allocation of energy to provide water, while [5–7] look at the same issue from the opposite direction: the allocation of water for energy generation. In this regards, ref. [8] emphasizes the need for electricity generation to be reshaped based on water consumption and carbon emissions. In [9–12], the authors look at energy production with the aim of maintaining market sustainability, while [13–15] consider the technological and economic sustainability of the power plants themselves and the need for newer renewable electricity generation methods with lower carbon emissions. Other studies such as [16] look at the sustainability of energy sources in their ability to provide food or even the combined water–food–energy nexus under various scenarios [17–21]. In addition to these, other factors such as the social impact of electricity systems [22] or the risk of providing the required resources for electricity production [23,24] have been proposed as other considerations in the definition of sustainability. As a result of these disjointed studies carried out with vastly different mindsets, the definition of sustainability remains unclear, with the general public and activists still naively associating it solely with carbon emissions [25], vilifying power companies and industries for their selfishness in polluting the environment in the process. The notion of sustainability may be difficult to define as shown by these studies, but it certainly should not be limited to the environmental factors. Other criteria need to be considered to broaden its definition.

Recent studies have addressed these shortcomings by introducing numerous new sustainability indices and indicators as can be seen in the works of [26–29]. While the introduction of newer indicators seems like an improvement at first glance, they are not based on well-known previously introduced metrics in the literature. Likewise, most of these newer indices are based on the authors' definitions of sustainability with the goal of capturing as many criteria as possible as opposed to using the lessons learned from previous studies. Furthermore, these indices are based on different assumptions which make their utilization tricky and their results incomparable to each other. As such, a wise approach would have been to modify an already existing and well-established sustainability metric as opposed to introducing a new one.

Aside from the inherent uncertainties with the definition of energy sustainability, most of the previous studies fail by not considering the region where energy is to be used. As an example, the black and white approach to sustainability by looking only at carbon fails to justify the use of low-tech electricity sources in remote locations without access to other electricity sources or cleaner technologies. All the while, a proper definition of sustainability should be clear enough to consider both the environmental pollution of such older technologies, as well as the added social welfare that accompanies them. In addition, factors such as regional availability of fuel for power plants and the trade-off between using regional resources and purchasing them from neighboring states or countries are among other regional considerations that traditional definitions of sustainability have no answer for aside from economic analysis. As such, the omission of regional resources is another point where most studies fall short of reality.

Despite these shortcomings, some studies propose novel approaches such as looking at sustainability via the lens of reliability and resilience of the system [30], but perhaps the most comprehensive foundation has been laid out by [31] where a sustainability index (Relative Aggregate Footprint—RAF) has been introduced by considering the environmental and economic characteristics of electricity generation technologies. While the definition of RAF may not include a wide range of criteria, it addresses issues such as regional resource availability or uncertainty in data and decision making. As such, their approach has been chosen as the core of the sustainability definition framework presented in this paper. This research aims to look at the concept of electricity generation sustainability more broadly and define it by looking further than the environmental aspects. As such, economic, technical, and social aspects of electricity generation technologies are also taken into account, as well as the characteristics of the region where the technologies are to be used, to provide a clearer picture of available resources and how to best allocate them.

Finally, in defining sustainability, a distinction needs to be made between the use of terms such as clean, green, renewable, and sustainable energies, especially when more than 20 different definitions have been recorded for them in the literature [32]. The carbon-based definition of sustainability has no doubt led to these words being used arbitrarily and interchangeably by many researchers [33]. Generally, clean energies are attributed to carbon-free options, whereas renewable energies are the options with naturally replenishing resources. A notable example of these terms not being equal is the case of nuclear energy and biomass, with the former being a clean energy with small carbon emissions but not renewable, and the latter being a renewable source, despite generating carbon emissions. On the other hand, green energies are a subset of renewable energies with the highest environmental benefits [34]. However, sustainability is often described as the ability to strike a balance between creating a better life for the human race on one hand and dealing with the limitations imposed by nature on the other hand [35], making it a more noble concept that is fundamentally different than the rest of the aforementioned keywords.

In practice, energy users usually define these terms based on their own goals and agendas. Most states in the United States now employ a renewable portfolio standard (RPS) as a means of diversifying electric utilities by including renewable sources. The same standard is sometimes known as the Clean Energy Standard (CES) in other locations. In the US, renewable portfolio standards were first utilized in the 2000s with the aim of allocating a portion of total energy generation to renewable energies. Despite the energy production goals, most states using renewable portfolios either have no preference towards a particular energy type at all, or only have classified select technology tiers rather than a coherent energy portfolio [36]. These issues are exacerbated by knowing that less than half of the American states have developed a renewable energy portfolio. Furthermore, the omission of heterogeneity in the development of renewable energy portfolios has led to similar characterization of energies and often identical portfolios regardless of differences [37]. As such, in practice, the need to study energy suitability based on the region, their available resources, and needs remains in place.

As a result, the goal of this study is to compile a collection of metrics based on the RAF method that states can use to compare the sustainability of electricity generation technologies and develop sustainable electricity portfolios. These metrics are spread out in various groups to account for the environmental, economic, social, and technical footprints of electricity generation technologies and are chosen per suggestions of previous research carried out in each field. Moreover, the regional characteristics and resource availability have been mentioned in addition to the specific energy-related attributes to provide a more realistic view of energy sustainability in each region by distinguishing among green, renewable, and sustainable energy technologies. In most previous studies, only the inherent characteristics of electricity generation technologies have been taken into account, and the specifics of the region are rarely a factor. The metrics mentioned in this study can be used in sustainability analysis of energy technologies for a more comprehensive vision of sustainability.

2. Chosen Energies

Next we provide a short introduction to the energy types chosen to be compared in this study. The choice of energies is important in determining what sustainability factors are to be considered. This especially affects the fuel and supply sources of the energies. Of course, the list of energies mentioned here is not exhaustive, but the majority of the commercially available technologies are considered.

2.1. Biofuels

Biomass resources are one of the oldest forms of energy production, an example of which is the burning wood, which has decreased dramatically throughout the years due to the introduction of coal, oil, and natural gas as alternative sources of energy. However, the use of biofuels comes with the ability to better manage fuel prices in case of fluctuations in the price of other available sources [38]. Currently, the main sources of biofuels are plants, genetically engineered algae, by-products from agricultural activities, methane gases from landfills, or the oil that comes from these sources [39,40]. Another advantage of biofuels is their ability to use residues and by-products of other processes as fuel, thus reducing the overall emissions of said processes [41]. The yield of biofuels is dependent on the crop being used, with corn having the smallest yield of conventional bioenergy and cellulose having the highest. In addition, biomass energy produces carbon, nitrogen, and organic compounds and particulates as by-products that can have adverse effects on air quality and global warming in larger quantities [42]. This is also dependant on the source of energy such that using bioenergy can increase carbon emission up to 20% in the case of corn or decrease it by upwards of 120% in the case of cellulose, making the choice of crop an important consideration [43]. Overall, the sustainability of biofuels is a function of the materials used as fuel, the supply chains in place, management strategies, and policies in place [44].

2.2. Solar

Solar energy started becoming a major source of electricity in the past decade, and its usage is projected to grow slowly but steadily. In reality, however, the energy from the sun has been taken advantage of even in the past century due to the simplicity of the process with its most dominant use being for heating water. Using the energy from the sun to heat water is currently popular in many countries including Germany [45].

For electricity production, solar energy can be used in two forms: Solar thermal and solar photovoltaic (PV). Solar thermal technologies employ the heat from the sun to run electric generators. The fact that the United States is the second largest producer of solar thermal energy in the world [46], shows that the US understands the potential for solar power. In these systems, the sun bounces off parabolic mirrors and is directed to a central tube that is filled with air, vapor, synthetic oil, molten salt, or liquid sodium, heating it in the process [47]. As such, this system is called Concentrated Solar Power (CSP). The hot contents of the central tube later transfer their heat to a source of water to generate steam and drive a turbine. One of the advantages of using molten salt or synthetic oil over water is that they can hold more thermal energy and enable the power plant to keep producing power even when the sun is not shining. For example, the Gemasolar power plant in Spain which uses molten salt, is capable of producing uninterrupted electricity for 24 h. However if these systems remain idle for a long time, heating them back up takes a considerable amount of time and energy. Another issue with solar thermal energy is the effects it can have on wildlife, threatening the habitats of desert creatures and scorching birds that fly in the vicinity. Finally, the output of solar thermal plants relies on sunshine. Therefore, due to cloudy days and bad whether, their production seldom reaches its ideal peak. Lastly for the solar thermal plants to work, large areas of land are required, which increases their land-use footprint.

Solar PV is the other form of solar energy that uses solar cells to convert solar radiation directly to DC electric energy and is the type of solar power with considerable growth in the past decade. The US, South America, North and South Africa, the Middle East and Australia are some of the regions with the highest PV potential [48]. In the US, the Southwest has the highest solar PV potential, and the northern areas have the lowest potential. Currently, the manufacture of panels and batteries are the two items that drive up the cost of solar panels. Some countries such as Germany choose to install solar panels next to highways on land that is otherwise unsuitable for farming or other uses to save on the land footprint. The obvious problem with solar PV is its inability to generate power at night, which makes it an unsuitable choice to provide the baseload. In addition, just like CSP, the actual electricity generation of PV energy is reliant on the sunshine.

2.3. Wind

Wind energy is probably the fastest growing renewable energy source worldwide with its commutative capacity having been multiplied by a factor of 10 since the 1990s [49] Wind is also one of the cheapest renewable energy sources available on the market today especially in the long run. China has the largest installed wind capacity in the world, followed by the US. The fastest growth in wind energy can also be seen in China. In the US, the central regions have the highest onshore wind capacity, and the northeastern and northwestern coasts have the highest offshore wind capacity [50]. About 0.5 percent of solar energy ends up forming winds, so wind ultimately can be considered an alternative form of solar energy. The minimum wind speed required for wind turbines to be feasible is 7 mph at an altitude of 80 m, but in practice, wind farms are constructed in areas with an average wind speed of 12 mph. Because of the slow-moving speed of wind turbines, some of them come with transmissions to adjust the speed of the generator, but the lubricant in these transmission can sometimes be a fire hazard at higher speeds. More modern wind turbines also have the ability to adjust the pitch of their blades towards wind to increase efficiency [51].

In addition to onshore wind, offshore wind is also another alternative that offers a higher potential of power production than its terrestrial counterpart [52]. Sadly, the US is completely missing from the list of major offshore wind farms in the world with most major offshore wind farms being located in the North Sea. Offshore wind turbines are in constant danger of corrosion by seawater that can cause both structural and electrical disruptions. The requirement of large transmission lines is also another disadvantage of offshore wind turbines.

Perhaps the most dominant disadvantage of wind farms is their disruption of the natural landscape. In addition, making room for onshore wind farms sometimes requires the felling of trees which has its own environmental consequences. Furthermore, the general issue with all types of wind turbines is their low capacity factors that often lie in the 25 to 50 percent range [53]. In addition, due to the fluctuations of the wind in different hours of the day and in different seasons, the actual production of wind turbines is much lower than their nominal capacity. This requires huge capacities to be installed, with only a small percentage of which is used to the full potential at any given time.

2.4. Hydropower

While hydropower is a rather clean source of energy, it is not considered a green energy by many sources [54–56]. This is due to the reservoir changing the sediment structure of the bed surface, jeopardizing terrestrial and aquatic species in the vicinity, as well as its high water demand. In particular, the construction of dams has the potential to displace a large number of people and comes with social ramifications. Hydropower is one of the oldest sources of energy with watermills dating back hundreds of years. In 2018, 9 out of the 10 largest power plants in the world were hydropower plants [57], which shows hydropower's high potential for electricity generation. The Three Gorges Dam in China with a capacity of 22,500 MW, for example, is the equivalent of about 32 regular coal power plants. However, US energy generation data in the past decade shows that the share of hydropower in the total electricity production has not grown considerably. At the same time, Chinese hydropower has almost doubled. Globally, the increase in hydropower electricity production has almost been increasing on a linear basis. Overall, hydropower plants have the potential to replace many of the older greenhouse-gas emitting power plants and save considerably on noxious gases and fossil fuels.

In short, hydropower is an almost carbon-emission-free [58] and renewable source of energy with high electricity conversion efficiencies of up to 80% [53] which can be used to supply the baseload of electricity. It is also versatile and can be scaled from as low as 10 kW to 20,000 MW. The operating and maintenance cost of hydropower plants are lower than most other competitors, and they have the advantage of long lifetimes [59]. On the other hand, using hydropower requires a large water footprint [60] and causes the inundation of large portions of land which can have social impacts due to people being displaced. Hydropower's output can vary due to rainfall, and therefore, a level of flooding risk is associated with it. Due to disruptions in natural water flow, hydropower plans can impact aquatic ecology and oxygen depletion in water. Hydropower generation often requires high capital costs and construction can take long amounts of time. The limiting factor of hydropower in many countries, however, is the limited available land and not enough valleys to flood, which hampers its implementation.

2.5. Coal

Coal is a rather old means of electricity production, and only a small percentage of today's largest power plants operate on coal; however, the high use of coal in developing countries is still noticeable [61]. The US is rich in coal reserves and more than half of the states house recoverable reserves [62]. The use of coal is still noticeable in the US, but it is on the decline with the introduction of newer technologies. Less emphasis on coal production in recent years and more automated extraction processes have resulted in the drop of the number of people working in the coal industry [63].

Among the greenhouse gases emitted by the coal power plants are sulfur dioxide, nitrogen oxides, carbon monoxide, carbon dioxide, particulate matters, mercury, lead, selenium and fly ash [64–66]. However, these emissions have seen a considerable decline in the past 30 years due to the improvements of technology. Integrated gasification combined cycle (IGCC) is the newer version of coal power plants that uses a gasifier to turn coal into syngas, a synthetic gas consisting primarily of hydrogen [67]. The carbon in the syngas can be shifted to hydrogen which can result in a carbon free-fuel. Furthermore, the resulting carbon dioxide can be compressed for easier storage. However, the high cost of IGCCs which can drive up the electricity prices is a hindrance to their use.

2.6. Oil

Oil is mostly used directly for the transportation sector rather than for electricity generation. However, the conventional internal combustion engines using oil products lack high efficiencies. Unlike the suggestions made by [68] that the rate of petroleum production follows a bell-shaped curve, the discovery of newer sources of oil and newer extraction technologies has kept the oil industry growing. Aside from the conventional sources of oil, unconventional sources such as oil shales, oil sands, coal-based liquid supplies, biomass-based liquid supplies, and gas to liquid (GTL) have become possibilities [69]. In addition to using oil as fuel, the Fischer–Tropsh synthesis process can be used as shown in Equation (1) to convert carbon monoxide and hydrogen into liquid hydrocarbons as an additional source of fuel.

$$(2n+1)H_2 + nCO_2 \rightarrow C_nH_{(2n+2)} + nH_2O \tag{1}$$

While these new sources provide additional oil, they also impact the environment because of acid drainage, introduction of metals into groundwater, erosion, sulfur emissions, and air pollution, as well as disposal problems, greenhouse gas emissions, and water usage. This adds a layer of uncertainty to the emissions of these new technologies [70].

2.7. Natural Gas

Natural gas can be used for power generation, as well as for transportation, hydrogen production, and the manufacture of fabrics, glass, steel and many other products. The burning of natural gas produces 20 percent less carbon dioxide compared to burning gasoline and 40 percent less compared to coal [71]. In addition, due to the similar nature of natural gas and coal power plants, older coal power plants can be retrofitted to operate on natural gas. The efficiency of natural gas turbines is rather high and can range from 30 percent to 60 percent for the variants with heat recovery. Of course, single cycle power plants are less efficient and therefore more suitable as peaking power plants that operate only a few hours a day. Currently, combined cycle power generation using natural gas is the cleanest source of energy using hydrocarbon fuels in terms of carbon emissions [72]. Another advantage of these power plants is their ability to be turned on and off very quickly which makes them suitable for supplying power over peak demands. The use of natural gas as a replacement for petroleum is also viable as the efficiency of natural gas engines is comparable to that of gasoline engines.

Natural gas can be found in associated (along with oil) or non-associated (isolated) fields. Newer gas resources such as sour gas, tight gas, shale gas, and coalbed gas also exist, but using them can be costly and challenging at the current time. The most common form of natural gas today is probably liquid natural gas (LNG). The energy density of LNG is 2.4 times that of compressed natural gas which makes it an ideal solution for long-distance transport and in situations where pipelines are not available. The liquid form of gas also makes it possible to store the natural gas [73], usually in underground storage facilities; however, leakages in natural gas storage tanks (for example, the Aliso Canyon incident) are still one of the main concerns [74]. Despite natural gas being regarded as one of the greener energies, the losses of up to nine percent of stored methane in storage tanks have a considerable environmental impact, not to mention wasted resources. The issue is not limited to storage facilities and can be witnessed even in urban areas.

2.8. Nuclear

In nuclear power, exothermic nuclear processes create heat which generates electricity. In 2016, nuclear fission powered roughly 10 percent of the world's electricity as well as numerous naval vessels [75]. France and the United States are the leading producers of nuclear power in the world, with the Netherlands and Armenia following as the countries least reliant on nuclear energy [76]. In the southeast US, the state of South Carolina is the main user of nuclear power and the home to four nuclear plants with the Virgil C. Summer Nuclear Station being the most important one. This is partly due to the fact that South Carolina suffers from a lack of gas pipeline infrastructure. Oddly enough, Australia despite being one of the largest producers of uranium in the world, has no nuclear power plants in service.

Most conventional nuclear plants are either of the Boiling Water Reactor (BWR) or Pressurized Water Reactor (PWR) types [77]. In older PWR types, the coolant water also acts as a moderator and goes past the control rods. Therefore, the water becomes contaminated, and in case of leaks, the leaking steam will be dangerous. To remedy this, in newer PWR reactors, boron and control rods are used to maintain system temperature, but the water that goes through the turbine is not directly in contact with nuclear materials and will not be radioactive.

With increased awareness towards probable nuclear catastrophes, Germany and Japan are among countries trying to decommission their nuclear plants [78]. In addition to the direct catastrophes that may be caused by nuclear plants, disposal of nuclear waste is another direct issue of nuclear power as it requires the provision of land and specialized storage grounds and cooling pods.

More optimistically, nuclear energy is rather affordable, with the total fuel costs (including the fuel itself, enrichment, manufacturing, and disposal) being about 5 to 10 cents per kilowatt-hour [79]. Fuel contributes to about 10 percent of this price, so the changes in the price of fuel have minimal impact on the final cost of energy.

2.9. Geothermal

Geothermal energy is the energy stored in the Earth's core in the form of heat. This thermal energy manifests itself in the form of steam and hot water and has been used as a source of electricity for decades. Based on [80], the total geothermal capacity of the Earth has been estimated at 7974 MW_e, but as of year 2000, only 0.3% of this amount was exploited. Extraction of geothermal energy occurs by drilling a hole in the Earth's crust in regions where the geothermal gradient is higher than the average of 30 °C/km in depth. The heat is transferred from the sub-surface regions to the surface via conduction and convection processes through geothermal fluids in underground aquifers. These hot fluids (or steam) are then extracted, and depending on their pressure and temperature, they can be used for electricity generation or heating.

The shortcoming of geothermal energy is the need for three conditions: a thermal anomaly, a productive geothermal source, and a closed reservoir in an accessible depth that is not covered by impenetrable rocks. This greatly limits the viability of geothermal energy despite its potential. On the positive side, when all the aforementioned conditions are met, extraction of geothermal energy can be achieved using conventional technologies. Geothermal energy is generally cleaner than fossil fuels in terms of carbon emissions, but in reality, its use is dependent on the cost of fossil fuels and carbon taxes.

3. Electricity Generation Footprint and Impact

More often than not, studies focus only on one aspect of electricity generation or on many aspects of one technology which does not provide the bigger picture for comparing different electricity generation types [81]. There is more to the concept of sustainability than carbon emissions and costs. Sustainability is looking at the Earth as a whole and using its resources efficiently while considering the limitations. Therefore, not all green energies are necessarily sustainable.

As pictured in Figure 1 based on the data by [82], the current electricity portfolios in the southeast United States are mostly based on hydrocarbon and nuclear sources with renewable sources only contributing a negligible amount to total electricity production. To overcome this shortcoming, some states have incorporated renewable portfolio standards with the goal of improving the sustainability of electricity generation. However, firstly not all states have such standards in place with the states of North and South Carolina being the only examples with renewable portfolio goals in the southeast United States [83]. Secondly, these portfolios are almost exclusively based on increasing the share of renewable energies and whether said energies are actually green is not mentioned. This shows that the states are mostly concerned with lowering their carbon footprints, and other criteria are not a major deciding factor of their definition of sustainability.

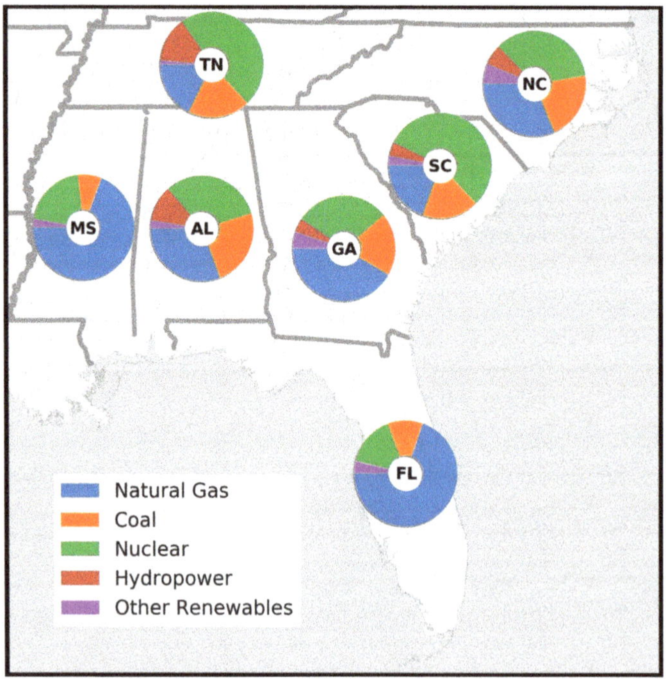

Figure 1. Current energy portfolios in the southeast United States.

As such, the energy impact indexes in the literature are bound by the factors included in them and may not be all-inclusive. These indexes are usually custom-tailored to consider one or a few aspects of electricity production in a very detailed manner. Instead, our paper tries to include more factors to obtain a more coherent look at benefits and drawbacks of electricity generation methods and their sustainability. The considered factors are categorized in four groups of environmental, economic, social, and technical, thus enhancing the traditional way of viewing only the environmental and economic properties of energies. Figure 2 shows the conceptual framework of how sustainability is defined, and the major factors that play a role in defining and measuring sustainability, and forming the energy portfolio. Some of these criteria had been suggested before in the works of [84–86], but a broader range of possible metrics are collected in Figure 2. This enhances the view on sustainability which was based on environmental and economic factors in the past. Since different experts have different priorities in terms of factors they deem important, additional criteria and factors can be added or removed to the list presented here. In addition, different weights can be specified for each criteria to consider the importance of each one based on the experts' opinions.

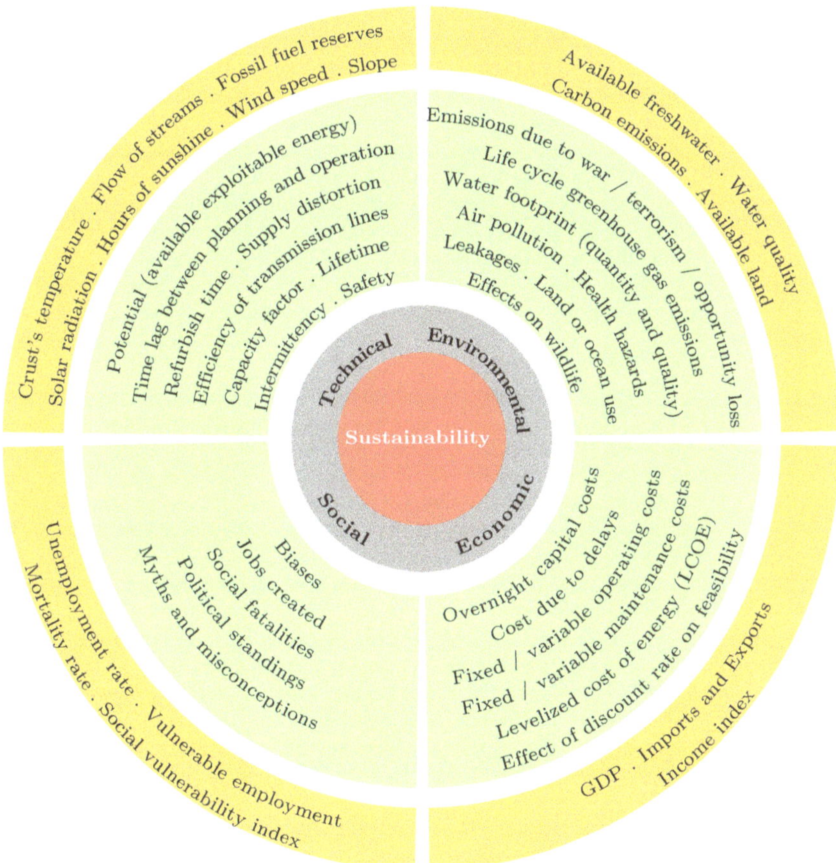

Figure 2. The four main criteria for sustainability: environmental, economic, social, and technical sustainability. The inner layer shows the sustainability factors, while the outer layers depict the corresponding regional and portfolio factors.

To describe the sustainability of each electricity generation technology, the main chosen indicator groups (hereafter referred to as criteria) represent the environmental, economic, social and technical aspects involved in each electricity generation option. Furthermore, special care was taken to make certain each of these criteria are described by using life cycle indicators whenever possible, as these are the most inclusive and therefore capture a better picture of what is involved in electricity production. Of course, as mentioned, all the criteria and indicators presented here can be replaced by other factors or removed entirely based on the discretion of the user.

To complement previous research, in addition to the usual environmental and economic criteria, the social and technological aspects of electricity sources are considered here. For each criteria, an exhaustive list of factors mentioned in the literature was collected, and those that were measurable with available databases were chosen. When studying energy impact, it is important to consider both sustainability and regional factors together. This helps put the numbers in perspective and helps with the decision-making process. Aside from the inherent advantages and disadvantages of each electricity generation technology that dictates their performance, the use of each technology is also dependent on the availability of required resources in a given area. These resources range from the availability of the fuel source (from sunshine to fossil fuels), to the availability of water, land, social acceptance, and many more. For example, even if one generation method is identified as the absolute best choice, as long as the technology for its implementation or the required fuel is not available in a region, that technology will be rendered ineffective.

Some of the most important of these factors are introduced in this paper, each of which is directly related to one of the sustainability factors discussed before. Therefore, the resource availability along with the resource requirements of each electricity generation technology in a location defines the weight of each criteria in that location. As such, the chosen factors are a mix of variables that are directly a result of electricity generation (impact or footprint variables) as well as variables that describe the state of the location where the electricity generation will take place (resource availability or state variables). Special care has been taken to make certain that for each footprint variable, there exists a corresponding variable. These factors are discussed here in more detail.

3.1. *Environmental Impact*

When studying the environmental factors, the most defining of the factors is the level of life cycle greenhouse gas emissions. Aside from the direct and indirect greenhouse gas emissions caused by the operation and construction of power plants, auxiliary emissions are caused due to opportunity losses (i.e., the emissions caused by older technologies that could have been circumvented had a greener energy been used instead) and also because of damages caused to the power plants as a result of accidents or wars. Water and land footprints are also the staples of environmental sustainability analysis. When analyzing the water footprint, aside form the quantity of water needed, the changes in the quality should also be kept in mind. These quality factors include chemical, radioactive, and thermal pollutants. Emitting greenhouse gases also has adverse effects on the air quality and increases the air-pollution-related mortality rates. Another source of contamination is the leakages of fuel from the storage depots or transportation pipelines. Finally, it is important for the effects of each technology on wildlife and the environment to be taken into consideration. Unfortunately, not all these factors can be quantified, and of the ones that can, few have usable databases. As a result, what we describe below is a list of factors that have the potential to be part of the sustainability analysis.

3.1.1. Life Cycle Greenhouse Gas Emissions

Due to the different nature of electricity generation methods, different types of greenhouse gases are emitted. To better quantify the effects of all these greenhouse gases, the CO_2-equivalent (CO_{2e}) values for emissions are used in this study. This is the sum of the amounts of each greenhouse gas emitted in the process multiplied by the global warming potential of each. In addition to the greenhouse gas emissions from the process of electricity generation, emissions due to the planning, construction, and decommissioning phases are also accounted for in the data. The data used in this research are in terms of gCO_{2eq}/kWh for all electricity generation methods. The data for life cycle analysis of greenhouse gas

(GHG) emissions have been presented before in the works of [53,85]. This study suggests the data by [72] as they seem to be the most comprehensive in terms of including most of the processes.

The data show that renewable energies generally emit less greenhouse gases, with the exception of nuclear energy which is on a par with its non- renewable counterparts. The lowest and highest emissions both belong to hydropower which stresses the importance of the scale of the power plant.

3.1.2. Carbon Emissions due to Opportunity Losses

As suggested by [85], some electricity generation methods require a rather short time between their planning and implementation phases, while the same period for others can take much longer. In the meantime, the existing power plants that are assumed to be generating more GHGs are still in operation, and the longer the time before newer technologies are implemented, the longer older technologies generate more GHGs. Therefore, the technologies with lower implementation times are preferable to their counterparts with longer implementation times. The delays in this period are caused by the time it takes for the site permit to be issued, financial sources to be found, permits to be issued, and the construction phase to be completed In addition, retrofitting a plan after its lifetime ends can further add to the delays and emissions resulting from them.

This time is actually not a constant number as plants of the same type are expected to take less time to reach their operational phases. Furthermore, the more cleaner energy sources are in operation, the lower the emissions of GHGs will be over time. This makes the calculation of this metric a complicated matter. The data in this research come from [85], where opportunity losses are defined as the subtraction of the downtime emissions of each technology from the downtime emissions of the technology with the least emissions. Taking into account the emissions in the downtime rather than the actual downtime, is a more reliable method to account for how clean the current systems of electricity production are. Since the available data only give the minimum and maximum values, the median is calculated by taking the arithmetic mean of the two available numbers. In addition, it is implied that the data for coal extend to other fossil fuels as well. Thus, unless otherwise noted by [85], the same numbers are used for oil, natural gas, and coal. It should be noted, however, that some fossil fuels such as natural gas are cleaner than the other ones, and the approach mentioned above can be improved with better data. As shown in the data, the opportunity cost emissions of most renewable energies are either zero or close to it, with the exception once again being hydropower.

3.1.3. Emissions due to Natural Hazards, War, and Terrorism

Due to nuclear energy becoming more common, ref. [85] suggest considering the damages caused in case of accidents, war, terrorist attacks, or even natural disasters involving these facilitates. Depending on the priorities of decision makers, these adverse effects can either be described by the potential casualties or by the carbon emissions due to the destruction of facilities and the long-term warming that they cause, the former of which is recommended by [85]. To estimate the carbon emissions, the countries with nuclear capability and their nuclear facilities in the vicinity of habitable areas are first identified. Then, the damages caused by the potential destruction of each of these facilities (which is proportional to their size) is multiplied by a risk factor. The carbon content of the nearby cities is finally used to estimate the total carbon emissions caused by a nuclear explosion. The final results are reported as the grams of equivalent emitted CO_2 per each kilowatt of released energy.

3.1.4. Storage and Transmission Losses

The emissions caused by storage losses due to leakages has been studied by [85] as well. However, he only considers the leakages associated with carbon capture and sequestration (CCS) on coal plants. As mentioned by [74], leakage in natural gas storage facilities is another important source of emissions that is also considered in the estimations Leakage are defined as the amount of CO_2 that escapes underground formations or storage facilities due to earthquakes. From the given description, it is apparent that accidents storage tank malfunctions, and natural leakages from crevices in the structures are not considered in this estimation (as is normally the case in natural gas storage tanks and pipelines). For the estimation of this factor, ref. [85] first estimated the total amount of CO_2 in a geological formation, and then multiplied it by the leakage rates 1% and 18% as lower and higher estimates. He then compared the amount of leaked carbon with the amount of emissions caused by the same mass of carbon in a power plant with CCS equipment to estimate the mass of released carbon per each kilowatt of energy. While this method is not flawless and fails to take into account other sources of leakages, its results are used in this research mainly because more accurate data are not available.

3.1.5. Adjusted Carbon Footprint

To obtain a better and more generalized metric of the carbon footprint, the emissions due to opportunity losses, war and terrorism, and storage losses were added to the life cycle greenhouse gas emissions to adjust it for the aforementioned cases. The resulting metric is named adjusted carbon footprint and will be the main metric in this research for considering greenhouse gas emissions. Equation (2) shows how this new metric was calculated. The result of Equation (2) offers a slight improvement to the general values of carbon emissions often found in the literature and are given in the Appendix A as part of Table A1.

$$\begin{aligned}\text{Adjusted carbon footprint} =& \text{life cycle carbon footprint} + \\ & \text{Opportunity losses emissions} + \\ & \text{War or terrorism emissions} + \text{Leakage emissions}\end{aligned} \quad (2)$$

3.1.6. Regional Carbon Emissions

The use of energies with a high carbon footprint is particularly undesirable if the regional emission rates are already high. As such, carbon dioxide emissions in each location were also considered as an additional factor to scale the carbon emissions of each technology As a measure of greenhouse gas emissions, the per capita carbon emissions (tons/cap) within the area was selected. Higher values of this number usually indicate that an area is more industrial and can benefit more from the addition of green energies. Other areas with less emissions, on the other hand, are less sensitive to air-quality-related issues. [87] has studied per capita carbon emissions in the US for different sectors. Total carbon dioxide emissions on the state levels have been published by [88]. For carbon emissions to be comparable over a variety of different locations, the emission numbers were divided by the population of each state as given by the US Census Bureau to find the per capita carbon emissions. Naturally, regions with higher carbon emissions find renewable energies more desirable.

3.1.7. Water Footprint

Many parts of the world are currently struggling with water shortages, and with the passing of time, providing safe water sources will become a more prominent issue as a result of global warming. The water footprint of power plants comes from the water used to process the fuels (irrigation of crops, mining operations, etc.), transportation, construction of the facilities, evaporated water in the process of cooling, and, in the case of hydropower, direct water usage.

The water footprint is one of the reliable metrics for quantifying water consumption and is defined in the literature as the total volume of water required for a unit of energy to be generated. Estimates for the water footprint of different electric technologies are given in the works of [60,85,89,90], the latter of which is used in the current study where water consumption is given in units of m^3/GJ. The data for biomass are from [91] as it was missing from the previous study.

3.1.8. Regional Freshwater Withdrawal

To address the state of water availability, the percentage of freshwater withdrawal over the total available fresh water was chosen. This index show how much each area is stretching its available water resources. The lower the value of this index, the less of a concern water supply would be. Regions with higher water availability are less sensitive to energies with a higher water footprint, and it is important for current regional freshwater availability, renewable water sources, freshwater withdrawal, and water demand to be considered. While there are regional documents on water consumption that describe the resources in detail, such as the report by [92], the more comprehensive but lower level data given by [93] were used in this study. Using the available data, annual freshwater withdrawal rates in each region were found. Available annual renewable water sources in each region were also found. Finally, the percentage ratio of freshwater withdrawal to freshwater availability was found for each location. Higher withdrawal numbers show more stress on available water supplies, and such regions prefer technologies with lower water footprints.

3.1.9. Land Use

One of the important factors in the sustainability of energy sources is their land footprint (or sea and ocean usage for offshore technologies). The land used by power plants could have been used more productively by other activities such as farming or building accommodations. In many cases, the potential construction site for power plants used to contain vegetation and various types of wildlife. Any construction work in these areas disrupts the natural habitat. In addition, aside from the land directly occupied by the power plant, the spacing required around it should also be considered in the figures. In the literature, different methods of estimating the spacing for different electricity production methods have been noted. One advantage of aquatic technologies over their terrestrial counterparts is the fact that the turbines are usually spaced more closely together, reducing the land footprint. Aside from the land directly used by the power plant, the additional land that it renders inhabitable or otherwise unusable should be included in the land footprint as well. For example, growing crops to be used as biofuels contributes to the food crisis, and continuous cultivation of a similar crop can have adverse effects on soil health. In the case of hydropower, the large areas behind the reservoir that are flooded become uninhabitable. As another example, nuclear reactors require storage facilities, which add to their land use. In [94], the authors found that although renewable energies may require more direct land, their life cycle land footprint can be lower than their non-renewable counterparts. The authors in [95] have also looked at land usage (in the unit of M^2/GWh) as part of their studies on the effects of climate policy on natural habitats. Since their data have been measured firsthand, and they give the lower and higher bounds, their research has been the basis of the land-use data used here. In addition, due to the unavailability of data, it was assumed that the land usage of onshore and offshore wind turbines is the same. In practice, offshore wind turbines are more efficient but require more elaborate transmission lines. In addition, the median was once again assumed to be the arithmetic mean of the minimum and maximum values. Land footprint is where renewable energies fall short of fossil fuels as their impact on land is usually considerably larger.

3.1.10. Regional Available Land

The availability of land (km^2) is also another environmental factor considered in portfolio analysis. Usually, to find the available land to be utilized by power plants, first, the total available land not occupied by cities and industries should be estimated. Next, the smaller areas and the ones with inappropriate slopes should be eliminated for the total usable land to be found. This is clearly not an easy task and providing the data can prove difficult. For this reason, in this study, the available land is defined as the total available land, whether occupied or not, and land slope has not been accounted for.

The availability of land per region has been documented in many research projects such as [96] where the land use has been reported per each US state. Based on this data, the percentage of available land that can be used for construction of power plants was calculated by subtracting the percentage of urban areas and croplands. This number was then multiplied by the total land area of the region given in [97] to find the total available land.

3.2. Economic Impact

When comparing the newer renewable energies to non-renewables, it should be noted that renewable energies often have higher upfront costs, but their annual costs are cheaper. For this reason, in addition to the capital cost of technologies, their annual operation and maintenance costs should also be taken into account. In addition to these costs, the cost of the power plant's fuel should also be considered. Overall, to calculate the levelized cost of energy, the overnight investment costs, recovery costs, tax rates, depreciation, capacity factor of plants, operation and maintenance costs, discount rates, and operation times should all be taken into consideration. Furthermore, the changes in the discount rates should also be studied as some technologies might not be economically feasible in the case of lower discount rates. Finally, the costs incurred because of delays in construction, interruptions in the operation, and profit losses when the plant is non-operational is another economic factor worthy of consideration. For the purpose of this study, the levelized cost of energy (LCOE) was chosen as the main factor as it incorporates most of the items described above.

3.2.1. Levelized Cost of Energy

In [59], the authors provide one of the most comprehensive publicly available databases on levelized cost of energy (LCOE) in the US and is the main source of prices in this study. However, not all the data in this database include every type of operation and maintenance costs. Furthermore, the major limitation with this price database is the fact that its raw data were obtained from multiple sources, and therefore, not all the assumptions are necessarily consistent. Unless studies on levelized cost of energy become more transparent, these issues will continue to persist in the databases.

In order for the LCOE to be calculated, first, the capital recovery factor (CRF) must be calculated using Equation (3). CRF is the conversion factor of the capital expenditures to annual payments and is dependent on the discount rate (D) and the lifetime of investment (N), which are assumed to be 7% and 30 years, respectively.

$$CRF = \frac{D(1+D)^N}{(1+D)^N - 1} \tag{3}$$

LCOE in \$/kWh is next determined by Equation (4). Aside from the capital costs (\$/kW) and the fixed and variable operation and maintenace costs (\$/kW and \$/MW), LCOE depends on CRF, tax rate ($\tau = 39.2\%$), depreciation value of each technology (D_{PV}), fuel price (\$/MMBtu), and heat rate (\$/MMBtu) if applicable. Heat rate is defined here as the efficiency of the power plant to generate energy from electricity. LCOE is the minimum price that the electricity companies need to charge in order to cover all their expenses.

$$\begin{aligned} \text{LCOE} = & \frac{\text{Capital Cost} \times \text{CRF} \times (1 - \tau D_{PV})}{8760 \times \text{Capacity Factor} \times (1-\tau)} + \frac{\text{Fixed O\&M}}{8760 \times \text{Capacity Factor}} \\ & + \frac{\text{variable O\&M}}{1000 \frac{\text{kWh}}{\text{MWh}}} + \frac{\text{Fuel Price} \times \text{Heat Rate}}{1000000 \frac{\text{BTU}}{\text{mmBTU}}} \end{aligned} \tag{4}$$

3.2.2. Regional GDP

Economic factors are important to show if the location in question is capable of funding the projects or not. GDP is the first and most comprehensive economic index that comes to mind which includes factors such as income, imports, and exports. GDP also helps put the cost of energy production into perspective. The use of GDP gives more insight about how economically affordable each alternative may seem to each region based on their level of economic power. Regions with higher GDPs will have more freedom in choosing energy technologies regardless of the associated costs. The authors is [98] give the annual GDP by state in millions of dollars in 2020 prices. When vastly different locations are compared, the use of GDP alone can be misleading as larger states or countries are likely to have higher GDPs. To eliminate the size, the index of economic power can also be defined as per capita GDP ($/cap).

3.2.3. Average Age of Power Plant Before Retirement

Another important aspect in the economic analysis of a project is its lifespan. In [99], the authors list the average age of power plants in years in the United States based on their fuel type. Naturally, longer lifespans reduce the effects of initial costs over time and are more desirable in terms of economics.

3.2.4. Regional APR Rates

Just like GDP for LCOE, the APR rate in each region makes it easier to decide about the lifespan of each alternative. Regions with higher APRs indicate higher investment risks and normally prefer shorter project time frames as a result. In [100], the authors offer a comprehensive list of APR rates by state in the United States.

3.2.5. Additional Economic Factors

Aside from what was mentioned before, additional factors can also be considered to better convey the stability of markets in a given region. Circularity of goods is one of these factors that shows the percentage use of locally sourced commodities to total commodities used in a certain industry. Higher circularities show higher levels of self-sufficiency and can be important in sourcing fuel for power plants. Resilience of supply chains towards disruption can be another potential factor. A region that relies on multiple supply providers is more resilient, and the industries using those supplies will be more desirable. The Shannon Diversity Index can be used to describe the level of resilience. The importance of a state in the economic market can also be shown through the dependence of other regions on that market. This can be shown by finding what percentage of the resources are being supplied by each source. Likewise, the number of regions that supply a resource is the perfect complement to the dependency. In the electricity market in particular, if a region is the main supplier of electricity to the neighboring areas, the stakeholders are more likely to prioritize production and economic, and technical factors over their environmental or social counterparts. If closer analysis of market prices is needed, factors such as maximum interest rates, average regional electricity prices, land value, minimum wage salary, and the cost of fuel for different types of power plants can also be considered as additional factors. This study however, has omitted these factors entirely for the benefit of an easier data collection process.

3.3. *Social Impact*

When studying the sustainability of electricity sources, the social aspect is often overlooked in favor of more mainstream issues such as emissions and finances. As a result, few articles have considered the social factors. Even the best of technologies have to be accepted in a society before being implemented, hence the importance of considering people's beliefs about certain electricity generation technologies, the urban myths and misconceptions regarding them, and the biases. This level of preparedness in each community in case of a hazardous event is partially determined by the social conditions governing that community, which itself is affected by the industries and technologies present in the region. Politically the use of some technologies might yield other long- or short-term advantages for certain individuals or societies which can be one of the deciding factors in policy making and in determining which electricity generation technologies are to be used. Finally, the operation of power plants creates direct and indirect jobs which are an income source to the society. However, at the same time, the operation of plants is inherently hazardous, causing injuries or fatalities to workers at times. Since the opinion of people towards each electricity generation technology is not easy to scientifically quantify, the chosen social factors in this paper are the ones that were easier to define and measure.

3.3.1. Fatalities

To calculate the fatalities associated with each electricity generation technology per each GW of electricity, ref. [53] divided the yearly fatality rates given in [101] by the net-rated power capacity given by [59] (TCD). Since TCD directly includes the fatality rates per each GigaWatt of energy per year, instead of the method described in the literature, this research uses the TCD data. However, the majority of this data is the average for all OECD countries and using it for the US can be inaccurate. That said, the results are more comprehensive than other similar research. This is the basis for the choice of using them in the current article. The other area where this data fall short is that it does not consider sub-technologies. As such, the results are more or less the average across all sub-technologies. Furthermore, due to the lack of data, as suggested by [53], it was assumed that the fatality rates of CSP are similar to that of geothermal. Just as before, the data for fuel cells are also assumed to be similar to natural gas.

Overall, while the fatality rates of renewable energies are generally lower than the other technologies, the overall fatality rates seem to be mainly directed to the size of the power plant rather than the technology being used. In addition, the minimum and median fatality rates for nuclear power are relatively small, making it look like a safe alternative but in the case of a nuclear catastrophe, the maximum mortality rate of nuclear energy is higher than the other technologies by at least two orders of magnitude.

3.3.2. Regional Social Vulnerability Index (SVI)

The vulnerability of a community towards natural hazards and its ability to prevent or subdue suffering in case of disasters is summarized in the Social Vulnerability Index (SVI). Social vulnerability is also a measure of a community's preparedness to respond to a hazardous event such as outbreaks, droughts, chemical and nuclear spills, and power plant failures. SVI can also be used to determine the amount of emergency supplies that need to be allocated to a region after a disaster or to help plan evacuation routes. One of the most well-known of these indexes, is the SVI index developed jointly by the Centers for Disease Control and Prevention (CDC) and the Agency for Toxic Substances and Disease Registry (ATSDR) based on 15 US Census variables such as income levels, education, disability, housing types, vehicle access, race and age groups among other variables [102]. The data for the SVI index is available online for the US states on county and census tract levels [103]. SVI is measured by ranking the performance of each location in each theme and constructing its percentile rank [104]. To improve the index, ref. [105] expanded SVI by considering more factors, using principal component analysis based on the influence of each of the factors, and proposing the improved SoVI index at county levels. In this study, the weighted average of these numbers as a function of county area was calculated to construct the overall SoVI of a state. The numbers were then scaled to the 0 to 1 range with higher numbers showing more social vulnerability. It must be noted that SoVI values are comparative, and single values cannot be used on their own.

3.3.3. Jobs Created

Recently, the number of jobs created was listed in studies as a tangible social metric for each technology. It is widely agreed that the technologies with a higher number of created jobs are more acceptable both socially and politically. In [106], the authors employ a model proposed by NREL to estimate the ranges of full-time equivalent (FTE) employment rate of each technology. The only exception to this was the employment rate of nuclear technology that was taken from [107]. The employment rates account for direct jobs as the result of construction and operation of facilities, as well as indirect jobs along the supply chain as third-party individuals or companies produce services, goods, and resources beneficial to the power plant. In addition, induced jobs as the result of the money from direct and indirect jobs being spent locally are also considered in the data. The final numbers are reported in terms of the number of jobs created per each GWh of generated electricity. Due to the unavailability of data, the number of created jobs for oil and fuel cells was assumed to be equal to that of natural gas.

3.3.4. Regional Unemployment Rate

The unemployment rate was chosen to put the job creation data in each region into perspective as regions with a higher unemployment rate will be more concerned with the social aspects of energy technologies and the number of new jobs they create rather than the environmental aspects of power generation. The unemployment rate was found based on the data reported by [103].

3.3.5. Regional Politics

While the social repercussions of electricity generation technologies are an important factor in how attractive they are perceived to be, in the end the general policies involving their utilization is set by the political parties in power. As such, the percentages of House and Senate legislators belonging to each of the major political parties in each state as reported by [108] were recorded. This has been included because each of these parties has its own definition of sustainable energies which in turn will be reflected in the policies and regulations they pass. Furthermore, the statewide comprehensive energy plans [109] and the existing capacities of traditional and renewable electricity generation [110] were noted as additional factors of social preference, and as additional evidence of regional electricity generation preferences.

3.3.6. Additional Social Factors

Leakage risk can be considered by compiling a list of oil and gas leakage incidents in the region. Additionally, a list of accidents related to power plants can provide a vision on the overall safety of the technologies and the threat they pose to the society. Finally, the pollution rate of the power plants can optionally be considered. This is important due to the fact that in most studies, only the carbon emissions are considered which results in other forms of pollution such as deterioration of water supplies, air pollution, or disfiguring natural landscapes, being omitted. Of course, in these factors, the number of incidents and their intensity, rather than the amount of pollution, are the main point.

3.4. Technical Impact

Technical factors are mainly concerned with the efficiency and reliability of each technology to deliver electricity. One of the major technical disadvantages of renewable energies as is discussed below is their intermittency and lack of ability to match generation at the time of demand. The capacity factor of plants is another method of evaluating the reliability of each technology. Other factors that can potentially be included in future research when reliable data for them are released include the fluctuations in the fuel price to show the distortions in the supplies, the lifetime of plants, the need for periodic repairs and refurbishments, the time lag between planning and the operation of each plant, and the safety of each technology. The efficiency of transmission lines (in addition to the efficiency of the plant) is also another potential factor that can be considered in future studies.

To address the technical side of energy usage, the available resources in each location were chosen as the most important factor in the analysis. In [111], the authors provide an excellent resource availability study that highlights the most vital factors determining the potential of green energies in a region. Available regional technology is another factor that has been considered through various tech and science indexes. In addition, for widespread energies, the existing infrastructure was considered a resource as it can influence further decision making about energy portfolios. While factors such as intermittency and capacity are important, they are secondary issues if the technology or resources for energy generation are not available in a region.

3.4.1. Resource Availability

When it comes to the resources required for electricity generation, due to inherent differences of energy technologies, various variables need to be considered depending on the electricity type.

- Solar PV and CSP: For solar energy, the availability of regional silicon, gold, copper, and lithium reserves [112] were considered to show how reliant on raw materials a region is in order to locally manufacture solar panels. In terms of generation potential, average temperature [113], daily solar radiation [114] and annual sunshine hours, and mean elevation were chosen. Finally, for the specific case of CSP, the number of regional commercial and private airports [115] was also considered to account for the possible blinding effect of the mirrors in these installations.
- Onshore and Offshore Wind: The potential of wind energy generation is mainly a function of onshore and offshore wind speed [50,116]. The minimum and maximum wind speeds that a wind turbine can capture effectively were also considered, and all locations met these requirements. For the particular case of offshore wind, the length of coastlines [117] were also considered.
- Geothermal: For the generation of geothermal energy, the main resources are the earth's crust's temperature and the availability of a proper aquifer. For this reason, the favorable range of geothermal reservoirs [118], regional freshwater withdrawals, and available renewable water sources [93] were considered.
- Hydropower: Dam elevation, stream flow discharge, and land slope are the main driving factors of hydropower. Aside from the size of the largest regional dams as well as annual, and seasonal stream flows [119], the elevation range of the region was also considered as a measure of the land slope. Elevation range was used since calculation of slope needs to be carried out in a specific direction, whereas elevation is a relative measure. Additionally, the number of notable prior dam failures [120] were recorded as a measure of how safe hydropower can be in the face of natural disasters.
- Nuclear Power: For nuclear power, an inventory of regional uranium and plutonium reserves was compiled [112]. However, none of the southeastern US states house any of these repositories. In addition, the number of nuclear materials storage sites [121] and spent fuel disposal sites [122] were also considered.
- Coal: With the technology of coal power plants being widely available, the main restriction in their use is the availability of coal itself as the fuel. For this reason, the actual recoverable reserves at production mines, estimated recoverable reserves, and demonstrated potential reserve base [62] were taken into account as measures of resources.
- Oil: With the southeast US deprived of oil reserves and near zero current production rates, the undiscovered onshore and offshore technically recoverable oil resources [123] were noted as the oil fuel resources.
- Natural Gas: Similar to oil, with the absence of prominent natural gas fields in the southeast US, the discovered shale gas repositories, as well as existing natural gas processing plants [124] were considered as available resources.
- Biomass: While technically a wide range of biofuels can be used in biomass plants, the sources of energy were limited to wood and dead plants. Thus, the available regional wood stocks as the percentage of forest cover [125], the percentage of land used for plantation crops [96], as well as per capita waste generation and composition [126]

were recorded. For a more in-depth study, Ref. [43] gives a comprehensive list of crops that can be used for bioenergy generation and the level of emission abatement that each one offers. On the other hand, Ref. [127] provides a list of acreage of each crop per US state. As such, a metric of total potential emissions abatement can be found by taking the weighted average of possible abatement with respect to the percentage of each crop available to a specific region.

3.4.2. Science and Technology Index

Aside from the availability of fuel and resources, the availability of each technology should also be considered. This has been achieved by documenting state technology and science indexes and state research and development index, as well as the state technology and science workforce indexes [128]. Additionally, the number of local colleges and universities [129] were also considered as institutions that work towards scientific advancement of energy technologies.

3.4.3. Capacity Factor

In addition to the availability of resources and technologies needed for energy generation, the intermittency of the existing power network can potentially be considered. One of the issues with green energies is that at a single location and at a given time, it is not possible to guarantee their power production. Some green energy sources are more predictable than others. Tidal energy is more predictable than solar for instance. The higher the intermittency of a technology, the more the need for backup generators or alternate sources of energy. However, with all the green energies combined over a larger geographical region, their intermittency drastically decreases as each energy can cover for the other ones. Therefore, intermittency is dependent both on the energies being used and the electricity mix in a power grid. Intermittency can be reduced by geographically dispersing energy sources, load smoothing, energy storage, and reliable weather forecasting.

Since it is complicated to quantify intermittency, reliability or capacity factors of power plants can be used instead. The capacity factor of each power plant is described as the ratio of its actual power production to its nominal capacity. The capacity factor is dependent on the down-time of the power plant due to reliability issues, maintenance, weather conditions, fuel availability, and electricity conversion mechanisms. Since the capacity factor is not a constant number, its annual average is usually reported. The capacity factors of different power plants can range from 18% for solar PV up to 95% for geothermal and nuclear energies. Generally, based on the data provided in [53], renewable energies have lower capacity factors and as such they are not suitable for peak-time electricity generation and need to be used along with non-renewable types.

3.5. Summary of Chosen Factors

Table 1 shows the summary of all the final quantifiable factors and their data sources across different electricity generation technologies. Of all the mentioned criteria, some had to be omitted due to lack of reliable data or subjectivity of the chosen criteria. For the environmental criteria, the adjusted carbon footprint, water footprint, and land footprint are chosen as the deciding factors. The levelized cost of energy, as well as the average age of plant retirement, are candidates from the economics criteria. Additionally, for the social criteria, factors such as fatality rates of each electricity generation technology and jobs created by each energy were chosen. Finally, for the technical criteria, the capacity factor of each technology was considered as the main factor. Example data for these factors have been given in Appendix A.

Table 1. Energy-related sustainability criteria.

Energy Factor	Unit	Criteria	Data Source
	Environmental		
Adjusted carbon footprint	gCO_{2eq}/kWh	L	[72,85]
Water footprint	m^3/GJ	L	[60]
Land footprint	m^2/GWh	L	[95]
	Economic		
Levelized cost of energy	$\$_{2019}/kWh$	L	[59]
Plant retirement age	Yr	H	[99]
	Social		
Fatality rate	$\#/GW_{eq}Yr$	L	[59]
Job creation	#/GWh	H	[106]
	Technical		
Capacity factor	%	H	[53]

L: Lower is better, H: Higher is better.

Although the intrinsic properties of energy technologies are important, it is also vital to consider the regional characteristics as well. Of all the criteria mentioned before, those that can be expressed either in quantitative or qualitative ways are summarized in Table 2. Since some of the qualitative data are essential in understanding the state of an energy's usage in a certain area, it is suggested that the regional factors be graded by the experts in the field, rather than via mathematical means. Among the environmental factors, carbon emissions, freshwater withdrawal, and available land were chosen. Regional GDP and APR rates are the indicators for the economic criteria. In terms of social aspects, the social vulnerability index and unemployment rate are the indicators of choice. In addition, the political standing and the previous policies have been considered as well. Finally the regional technical characteristics are captured by looking at the available regional resources, as well as various science and technology indexes. Resource availability ranges from the availability of raw materials and fuels, to the existing infrastructure.

Table 2. Location-Related Sustainability Criteria.

Energy Factor	Unit	Criteria	Data Source
	Environmental		
Carbon emissions	ton/cap	L	[88]
Freshwater withdrawal	%	L	[93]
Available land	km^2	H	[97]
	Economic		
Regional GDP	M$	H	[98]
Regional APR	Yr	L	[100]
	Social		
Social Vulnerability	-	L	[105]
Unemployment rate	%	L	[103]
	Technical		
Regional resource avalability	Varies	Varies	See Section 3.4
Tech and science index	%	H	[128]
R&D index	%	H	[128]
Technology workforce index	%	H	[128]
Number of universities	#	H	[129]

L: Lower is better, H: Higher is better.

4. Conclusions

In this research, a collection of criteria was compiled to study the sustainability of energy technologies. Previous research studies have only looked at a few of these criteria. Thus, the concept of sustainability can still be improved by looking at energy technologies more comprehensively. Still, the use of all the facets of the aforementioned criteria is not possible in practice due to gaps in data availability, uncertainties in the quantification of data, regional characteristics not being included in sustainability studies, and sustainability indexes having been designed with only a few of the aforementioned criteria in mind. This research aims to provide a list of the criteria that needs to be included in the definition of sustainability to elevate it from the traditional carbon-based view. It also serves to identify RAF as a solid basis for a sustainability index for definitions of sustainability.

4.1. Data Availability and Gaps

Upon studying the literature on the sustainability of energy sources, the first visible issue with creating a more comprehensive sustainability index is the databases not sharing the same assumptions. As such, in many occasions, further assumptions were made or part of the data was estimated based on other technologies. Next, some of the parameters mentioned in the studies are extremely difficult, if not impossible, to quantify. Variables such as energy mix intermittency, biases, myths and misconceptions, and wildlife diversity are examples of these that cannot be expressed in numbers due to their nature or the ambiguity in their definition. Of course, the inclusion of such variables in sustainability studies can be greatly beneficial.

Finally, the lack of a comprehensive index or framework for studying sustainability is noticeable. Most sustainability indexes are concerned only with environmental and economic aspects of electricity generation, and almost no framework considers other criteria such as social and technical aspects. Even then, to date, only the RAF index [31] includes regional resource availability. The omission of such important aspect from sustainability studies is puzzling.

4.2. Future of Sustainability

Based on the review of sustainability factors presented in this paper, the need for change in the future of sustainability studies is evident. The previous studies defined sustainability using only a limited number of criteria that were chosen for their specific case study. We argue that the definition of sustainability needs to be broader than confining it to typical criteria such as emissions or cost. In this research, a review of the previous methods was given, and the criteria presented in them was summarized under the main four categories of environmental, economic, social, and technical criteria, improving the previous viewpoint on sustainability as a result. In doing so, only the aspects that could be quantified were kept. In addition, it was argued that most studies miss the regional characteristics of the location in which electricity is to be generated. This step is important as regional characteristics indicate the feasibility of each technology for the specific location in which they are to be used. Thus, proper location-based factors were also given to complement each of the electricity generation criteria. Finally, instead of creating a new index, this study suggests using the RAF index as a basis to build a more comprehensive definition of sustainability to keep the results relevant and comparable. To improve the definition of energy sustainability, the following actions are suggested:

- Studies need to consider additional criteria aside from environmental and economic factors. Even then, the definition of environmental factors can be enhanced by looking beyond carbon emissions and considering other environmental factors such as wildlife. Almost no global standards exist for the inclusion of additional criteria and the variables that need to be considered in each. Thus, it is hoped that the information provided in this paper is useful for other researchers.

- Some of the variables can be placed in different categories depending on the point of view. For example, damages caused due to natural hazards to the power plants can be looked at from the point of view of environmental emissions, economic costs that arise from the repairs or outages, or even the social consequences of them. It is inevitable to include similar variables in each of the criteria; however, caution should be exercised to keep these variables from skewing the final direction of the sustainability analysis. Furthermore, it is suggested that a similar number of variables be used in each criteria to make certain no aspect of electricity generation receives more attention that the other parts.
- The same can be said about the technical resource availability of different energy technologies. Some technologies share the same resources, thus duplicate items will be present. Additionally, the researchers need to make certain that either the life-cycle variables are used, or if not, the variables are not biased towards a certain technology or a certain resource.
- To date, the most comprehensive sustainability indexes are based only on environmental and economic aspects. Ideally, a more comprehensive metric needs to be developed that considers other aspects of electricity generation as well. The outdated view of sustainability being analogous to carbon emissions is not the best approach, and a better index can help change this mentality.
- Except for the RAF index, almost no other study takes into account the regional characteristics of where the electricity technologies are to be used. Naturally, the sustainability of energy sources are not static and can change based on the regional and environmental status. As such, studying the concept of sustainability without considering the regional aspects will not be realistic. The authors suggest treating the regional values as weights that change the importance of each sustainability criteria. However, other approaches can also be taken.
- Finally, with the plethora of sustainability indexes that are constantly being introduced via the literature, it is vital for the new works to be based on previous studies. When possible, they should augment and complete the current indexes, and when not possible, they should be built based on one of the existing indexes. This way, recording the assumptions of each method will become much easier, and the findings of each method can be taken advantage of, while the shortcomings of a method are addressed.

Throughout this paper, a summary of different approaches towards capturing the essence of energy sustainability was given. Still, no single study manages to present an all-inclusive approach towards sustainability. It is evident that despite all the recent studies, a better framework for defining sustainability needs to be developed. The change in approach towards sustainability can help enhance the understanding of sustainability, which in turn leads to the design of a better energy portfolio. The current sustainability studies are limited in their scopes, and the authors hope the provided materials will be beneficial in future policy- making and sustainability analysis endeavors. However, the most important outcome of this comprehensive review of sustainability is perhaps to inform the energy managers, planners, and stakeholders that a renewable energy is not always green; a green energy is not always sustainable; one energy that may be sustainable in one location may not necessarily be so in another location.

Author Contributions: Conceptualization, F.K. and E.G.; methodology, F.K.; validation, F.K. and E.G.; formal analysis, F.K.; investigation, F.K. and E.G.; resources, F.K.; data curation, F.K.; writing—original draft preparation, F.K.; writing—review and editing, F.K. and E.G.; visualization, F.K.; supervision, E.G.; project administration, E.G. All authors have read and agreed to the published version of the manuscript.

Funding: This research received no external funding.

Institutional Review Board Statement: Not applicable.

Informed Consent Statement: Not applicable.

Data Availability Statement: All the data used in this research has been referenced in Tables 1 and 2.

Conflicts of Interest: The authors declare no conflict of interest.

Abbreviations

The following abbreviations are used in this manuscript:

ATSDR	Agency for Toxic Substances and Disease Registry
CCS	Carbon Capture and Storage\Sequestration
CDC	Center for Disease Control
CRF	Capital Recovery Factor
FTE	Full-Time Equivalent
GDP	Gross Domestic Product
LCOE	Levelized Cost of Energy
RAF	Relative Aggregate Footprint
SVI	Social Vulnerability Index
TCD	Transparent Cost Dataset

Appendix A. Sustainability Criteria Values for Different Electricity Generation Technologies

Tables A1 and A2 list the numeric values of environmental, economic, social, and technical sustainability values as expressed in Section 3.5. These numbers describe the inherent characteristics of electricity generation technologies under the aforementioned four criteria. Of these, the environmental criteria has been broken down into three sub-criteria. The values presented in Tables A1 and A2 can be used to calculate the RAF index. It should be noted that in some criteria, smaller numbers indicate a better value, while for others, larger numbers are preferable. RAF has the ability of using both of these data types without further need for standardization. Of course, for a real case study, these tables should be matched with a corresponding table of regional characteristics (criteria presented in Table 2) for the weight and importance of each criteria to be determined. As such, no general advisement can be given just by considering the inherent characteristics of electricity generation technologies alone, and the results of regional RAF metrics should be the main source of comparison.

Table A1. Environmental sustainability criteria values.

Technology	Carbon Footprint gCO_{2eq}/kWh Lower Is Better			Land Footprint m^2/GWh Lower Is Better			Water Footprint m^3/GJ Lower Is Better		
	Min	Median	Max	Min	Median	Max	Min	Median	Max
Biomass	130	230	420	4433	13,116.5	21,800	20	42	64
Conc. Sol. P.	8.8	27	63	340	510	680	0.118	1.149	2.18
Solar PV	18	48	180	704	1232	1760	0.00064	0.1518	0.303
Wind: Onshore	7	11	56	2168	2404	2640	0.0002	0.0007	0.0012
Wind: Offshore	8	12	35	2168	2404	2640	0.0002	0.0007	0.0012
Hydropower	32	39.5	2200	538	1803	3068	0.3	425.15	850
Coal	792.8	910.9	1039	83	325	567	0.079	1.0895	2.1
Oil	708	831	953	1490	1490	1490	0.214	0.702	1.19
Natural Gas	461	559	737	623	623	623	0.076	0.658	1.24
Nuclear	62.7	96.55	220.1	63	78	93	0.018	0.734	1.45
Geothermal	7	41.5	85	33	248	463	0.0073	0.3832	0.759

Table A2. Economic, Social, and Technical sustainability criteria values.

Technology	Economic		Social		Technical
	LCOE $/kWh L	Avg. Lifespan yr H	Job Creation Jobs/GWh H	Fatality Fatality/GW_{eq}yr L	Capacity Factor % H
Biomass	0.08	41	2.06	0.0149	84
Conc. Sol. P.	0.18	40	2.02	0.0017	50
Solar PV	0.27	10	1.47	0.0002	22
Wind: Onshore	0.07	15	0.33	0.0019	41
Wind: Offshore	0.13	15	0.62	0.0064	43
Hydropower	0.07	70	0.96	0.0027	87
Coal	0.07	54	0.66	0.1200	85
Oil	0.09	38	0.48	0.0932	85
Natural Gas	0.05	38	0.48	0.0721	85
Nuclear	0.08	35	0.47	0.0069	90
Geothermal	0.09	40	0.91	0.0017	90

H: Higher is better, L: Lower is better.

References

1. Wackernagel, M. The ecological footprint of Santiago de Chile. *Local Environ.* **1998**, *3*, 7–25. [CrossRef]
2. Wackernagel, M.; Onisto, L.; Bello, P.; Linares, A.C.; Falfán, I.S.L.; Garcıa, J.M.; Guerrero, A.I.S.; Guerrero, M.G.S. National natural capital accounting with the ecological footprint concept. *Ecol. Econ.* **1999**, *29*, 375–390. [CrossRef]
3. Elliott, D. Renewable energy and sustainable futures. *Futures* **2000**, *32*, 261–274. [CrossRef]
4. Chini, C.M.; Stillwell, A.S. The state of US urban water: Data and the energy-water nexus. *Water Resour. Res.* **2018**, *54*, 1796–1811 [CrossRef]
5. Torcellini, P.; Long, N.; Judkoff, R. *Consumptive Water Use for US Power Production*; Technical Report; National Renewable Energy Lab.: Golden, CO, USA, 2003.
6. Macknick, J.; Newmark, R.; Heath, G.; Hallett, K. *Review of Operational Water Consumption and Withdrawal Factors for Electricity Generating Technologies*; Technical Report; National Renewable Energy Lab.: Golden, CO, USA, 2011.
7. Macknick, J.; Newmark, R.; Heath, G.; Hallett, K.C. Operational water consumption and withdrawal factors for electricity generating technologies: A review of existing literature. *Environ. Res. Lett.* **2012**, *7*, 045802. [CrossRef]
8. Ackerman, F.; Fisher, J. Is there a water–energy nexus in electricity generation? Long-term scenarios for the western United States *Energy Policy* **2013**, *59*, 235–241. [CrossRef]
9. Steinhurst, W.; White, D.; Roschelle, A.; Napoleon, A.; Hornby, R.; Biewald, B. *Energy Portfolio Management: Tools & Resources for State Public Utility Commissions*; Synapse Energy Economics Inc.: Cambridge, MA, USA, 2006.
10. Cappers, P.; Goldman, C.; Kathan, D. Demand response in US electricity markets: Empirical evidence. *Energy* **2010**, *35*, 1526–1535 [CrossRef]
11. Ervural, B.C.; Evren, R.; Delen, D. A multi-objective decision-making approach for sustainable energy investment planning *Renew. Energy* **2018**, *126*, 387–402. [CrossRef]
12. Espey, S. Renewables portfolio standard: A means for trade with electricity from renewable energy sources? *Energy Policy* **2001** *29*, 557–566. [CrossRef]
13. Chatzimouratidis, A.I.; Pilavachi, P.A. Technological, economic and sustainability evaluation of power plants using the Analytic Hierarchy Process. *Energy Policy* **2009**, *37*, 778–787. [CrossRef]
14. Jacobsson, S.; Johnson, A. The diffusion of renewable energy technology: An analytical framework and key issues for research *Energy Policy* **2000**, *28*, 625–640. [CrossRef]
15. Mai, T.; Mulcahy, D.; Hand, M.M.; Baldwin, S.F. Envisioning a renewable electricity future for the United States. *Energy* **2014**, *65*, 374–386. [CrossRef]
16. Khan, S.; Hanjra, M.A. Footprints of water and energy inputs in food production–Global perspectives. *Food Policy* **2009**, *34*, 130–140. [CrossRef]
17. Biggs, E.M.; Bruce, E.; Boruff, B.; Duncan, J.M.; Horsley, J.; Pauli, N.; McNeill, K.; Neef, A.; Van Ogtrop, F.; Curnow, J.; et al Sustainable development and the water–energy–food nexus: A perspective on livelihoods. *Environ. Sci. Policy* **2015**, *54*, 389–397 [CrossRef]
18. Bazilian, M.; Rogner, H.; Howells, M.; Hermann, S.; Arent, D.; Gielen, D.; Steduto, P.; Mueller, A.; Komor, P.; Tol, R.S.; et al Considering the energy, water and food nexus: Towards an integrated modelling approach. *Energy Policy* **2011**, *39*, 7896–7906 [CrossRef]
19. Memarzadeh, M.; Moura, S.; Horvath, A. Optimizing dynamics of integrated food–energy–water systems under the risk of climate change. *Environ. Res. Lett.* **2019**, *14*, 074010. [CrossRef]

20. Payet-Burin, R.; Kromann, M.; Pereira-Cardenal, S.; Strzepek, K.M.; Bauer-Gottwein, P. WHAT-IF: An open-source decision support tool for water infrastructure investment planning within the water–energy–food–climate nexus. *Hydrol. Earth Syst. Sci.* **2019**, *23*, 4129–4152. [CrossRef]
21. Ringler, C.; Bhaduri, A.; Lawford, R. The nexus across water, energy, land and food (WELF): Potential for improved resource use efficiency? *Curr. Opin. Environ. Sustain.* **2013**, *5*, 617–624. [CrossRef]
22. Akella, A.; Saini, R.; Sharma, M.P. Social, economical and environmental impacts of renewable energy systems. *Renew. Energy* **2009**, *34*, 390–396. [CrossRef]
23. Angelis-Dimakis, A.; Biberacher, M.; Dominguez, J.; Fiorese, G.; Gadocha, S.; Gnansounou, E.; Guariso, G.; Kartalidis, A.; Panichelli, L.; Pinedo, I.; et al. Methods and tools to evaluate the availability of renewable energy sources. *Renew. Sustain. Energy Rev.* **2011**, *15*, 1182–1200. [CrossRef]
24. Arent, D.; Pless, J.; Mai, T.; Wiser, R.; Hand, M.; Baldwin, S.; Heath, G.; Macknick, J.; Bazilian, M.; Schlosser, A.; et al. Implications of high renewable electricity penetration in the US for water use, greenhouse gas emissions, land-use, and materials supply. *Appl. Energy* **2014**, *123*, 368–377. [CrossRef]
25. Thunberg, G. *No One Is too Small to Make a Difference*; Penguin: New York, NY, USA, 2019.
26. Elavarasan, R.M.; Pugazhendhi, R.; Irfan, M.; Mihet-Popa, L.; Campana, P.E.; Khan, I.A. A novel Sustainable Development Goal 7 composite index as the paradigm for energy sustainability assessment: A case study from Europe. *Appl. Energy* **2022**, *307*, 118173. [CrossRef]
27. Silva, R.M.; Silva, A.R.; Lima, T.M.; Charrua-Santos, F.; Osório, G.J. Energy Sustainability Universal Index (ESUI): A proposed framework applied to the decision-making evaluation in power system generation. *J. Clean. Prod.* **2020**, *275*, 124167. [CrossRef]
28. Marquez-Ballesteros, M.J.; Mora-López, L.; Lloret-Gallego, P.; Sumper, A.; Sidrach-de-Cardona, M. Measuring urban energy sustainability and its application to two Spanish cities: Malaga and Barcelona. *Sustain. Cities Soc.* **2019**, *45*, 335–347. [CrossRef]
29. Shah, S.; Zhou, P.; Walasai, G.; Mohsin, M. Energy security and environmental sustainability index of South Asian countries: A composite index approach. *Ecol. Indic.* **2019**, *106*, 105507. [CrossRef]
30. McLellan, B.; Zhang, Q.; Farzaneh, H.; Utama, N.A.; Ishihara, K.N. Resilience, sustainability and risk management: A focus on energy. *Challenges* **2012**, *3*, 153–182. [CrossRef]
31. Ristic, B.; Mahlooji, M.; Gaudard, L.; Madani, K. The relative aggregate footprint of electricity generation technologies in the European Union (EU): A system of systems approach. *Resour. Conserv. Recycl.* **2019**, *143*, 282–290. [CrossRef]
32. Ahi, P.; Searcy, C. A comparative literature analysis of definitions for green and sustainable supply chain management. *J. Clean. Prod.* **2013**, *52*, 329–341. [CrossRef]
33. Hosseini, S.E.; Wahid, M.A. Hydrogen production from renewable and sustainable energy resources: Promising green energy carrier for clean development. *Renew. Sustain. Energy Rev.* **2016**, *57*, 850–866. [CrossRef]
34. Environmental Protection Agency. What is Green power? 2021. Available online: https://www.epa.gov/greenpower/what-green-power (accessed on 15 August 2021).
35. Kuhlman, T.; Farrington, J. What is sustainability? *Sustainability* **2010**, *2*, 3436–3448. [CrossRef]
36. Wiser, R.; Namovicz, C.; Gielecki, M.; Smith, R. The experience with renewable portfolio standards in the United States. *Electr. J.* **2007**, *20*, 8–20. [CrossRef]
37. Yin, H.; Powers, N. Do state renewable portfolio standards promote in-state renewable generation? *Energy Policy* **2010**, *38*, 1140–1149. [CrossRef]
38. Gupta, V.K.; Tuohy, M.G. *Biofuel Technologies. Recent Developments*; Springer: Berlin/Heidelberg, Germany, 2013.
39. Liew, W.H.; Hassim, M.H.; Ng, D.K. Review of evolution, technology and sustainability assessments of biofuel production. *J. Clean. Prod.* **2014**, *71*, 11–29. [CrossRef]
40. Yuan, P.S.; Du, M.A.; Te Huang, I.; Liu, I.H.; Chang, E. Strategies on Implementation of Waste-to-Energy (WTE) Supply Chain for Circular Economy System: A Review. *J. Clean. Prod.* **2015**, *108*, 409–421.
41. Habagil, M.; Keucken, A.; Sárvári Horváth, I. Biogas production from food residues—The role of trace metals and co-digestion with primary sludge. *Environments* **2020**, *7*, 42. [CrossRef]
42. Searchinger, T.D. Biofuels and the need for additional carbon. *Environ. Res. Lett.* **2010**, *5*, 024007. [CrossRef]
43. Sims, R.E.; Mabee, W.; Saddler, J.N.; Taylor, M. An overview of second generation biofuel technologies. *Bioresour. Technol.* **2010**, *101*, 1570–1580. [CrossRef]
44. Efroymson, R.A.; Dale, V.H.; Kline, K.L.; McBride, A.C.; Bielicki, J.M.; Smith, R.L.; Parish, E.S.; Schweizer, P.E.; Shaw, D.M. Environmental indicators of biofuel sustainability: What about context? *Environ. Manag.* **2013**, *51*, 291–306. [CrossRef]
45. Schmidt, T.; Mangold, D.; Müller-Steinhagen, H. Central solar heating plants with seasonal storage in Germany. *Sol. Energy* **2004**, *76*, 165–174. [CrossRef]
46. Sawin, J.L.; Sverrisson, F. *Renewables 2014 Global Status Report*; Renewable Energy Policy Network: Paris, France, 2015.
47. Tian, Y.; Zhao, C.Y. A review of solar collectors and thermal energy storage in solar thermal applications. *Appl. Energy* **2013**, *104*, 538–553. [CrossRef]
48. Hoogwijk, M.M. On the Global and Regional Potential of Renewable Energy Sources. Ph.D. Thesis, Utrecht University, Utrecht, The Netherlands, 2004.
49. Mabel, M.C.; Fernandez, E. Growth and future trends of wind energy in India. *Renew. Sustain. Energy Rev.* **2008**, *12*, 1745–1757. [CrossRef]

50. National Renewable Energy Laboratory. Eastern Wind Data Set. 2010. Available online: https://www.nrel.gov/grid/eastern-wind-data.html (accessed on 3 April 2022).
51. Muljadi, E.; Butterfield, C.P. Pitch-controlled variable-speed wind turbine generation. *IEEE Trans. Ind. Appl.* **2001**, *37*, 240–246. [CrossRef]
52. Esteban, M.D.; Diez, J.J.; López, J.S.; Negro, V. Why offshore wind energy? *Renew. Energy* **2011**, *36*, 444–450. [CrossRef]
53. Klein, S.J.; Whalley, S. Comparing the sustainability of US electricity options through multi-criteria decision analysis. *Energy Policy* **2015**, *79*, 127–149. [CrossRef]
54. Gunkel, G. Hydropower–A green energy? Tropical reservoirs and greenhouse gas emissions. *CLEAN Soil Air Water* **2009**, *37*, 726–734. [CrossRef]
55. Gibson, L.; Wilman, E.N.; Laurance, W.F. How green is 'green' energy? *Trends Ecol. Evol.* **2017**, *32*, 922–935. [CrossRef] [PubMed]
56. Penghao, C.; Pingkuo, L.; Hua, P. Prospects of hydropower industry in the Yangtze River Basin: China's green energy choice. *Renew. Energy* **2019**, *131*, 1168–1185. [CrossRef]
57. Energy Information Administration. The World'S Nine Largest Operating Power Plants are Hydroelectric Facilities. 2016. Available online: https://www.eia.gov/todayinenergy/detail.php?id=28392 (accessed on 3 April 2022).
58. Steinhurst, W.; Knight, P.; Schultz, M. *Hydropower Greenhouse Gas Emissions*; Synapse Energy Economics Inc.: Cambride, MA, USA, 2012.
59. National Renewable Energy Laboratory. Transparent Cost Database. 2019. Available online: https://openei.org/wiki/Transparent_Cost_Database (accessed on 3 April 2022).
60. Mekonnen, M.M.; Gerbens-Leenes, P.; Hoekstra, A.Y. The consumptive water footprint of electricity and heat: A global assessment. *Environ. Sci. Water Res. Technol.* **2015**, *1*, 285–297. [CrossRef]
61. Jin-ke, L.; Feng-hua, W.; Hua-ling, S. Differences in coal consumption patterns and economic growth between developed and developing countries. *Procedia Earth Planet. Sci.* **2009**, *1*, 1744–1750. [CrossRef]
62. Energy Information Administration. Recoverable Coal Reserves at Producing mines, Estimated Recoverable Reserves, and Demonstrated Reserve Base by Mining Method. 2019. Available online: https://www.eia.gov/coal/annual/pdf/table15.pdf (accessed on 3 April 2022).
63. Kolstad, C.D. *What Is Killing the US Coal Industry*? Stanford Institute for Economic Policy: Stanford, CA, USA, 2017; pp. 1–6.
64. Stanger, R.; Wall, T. Sulphur impacts during pulverised coal combustion in oxy-fuel technology for carbon capture and storage. *Prog. Energy Combust. Sci.* **2011**, *37*, 69–88. [CrossRef]
65. Saarnio, K.; Frey, A.; Niemi, J.V.; Timonen, H.; Rönkkö, T.; Karjalainen, P.; Vestenius, M.; Teinilä, K.; Pirjola, L.; Niemelä, V.; et al. Chemical composition and size of particles in emissions of a coal-fired power plant with flue gas desulfurization. *J. Aerosol Sci.* **2014**, *73*, 14–26. [CrossRef]
66. Zhu, H.; Chang, M.; Wang, H. Study on primal CO gas generation and emission of coal seam. *Int. J. Min. Sci. Technol.* **2017**, *27*, 973–979. [CrossRef]
67. Descamps, C.; Bouallou, C.; Kanniche, M. Efficiency of an Integrated Gasification Combined Cycle (IGCC) power plant including CO_2 removal. *Energy* **2008**, *33*, 874–881. [CrossRef]
68. Hubbert, M.K. Hubbert estimates from 1956 to 1974 of US oil and gas. In *Methods and Models for Assessing Energy Resources*; Elsevier: Amsterdam, The Netherlands, 1979; pp. 370–383.
69. Grushevenko, E.; Grushevenko, D. Unconventional oil potential tends to change the world oil market. *Energy Sci. Technol.* **2012**, *4*, 68–74.
70. McKellar, J.M.; Charpentier, A.D.; Bergerson, J.A.; MacLean, H.L. A life cycle greenhouse gas emissions perspective on liquid fuels from unconventional Canadian and US fossil sources. *Int. J. Glob. Warm.* **2009**, *1*, 160–178. [CrossRef]
71. Burnham, A.; Han, J.; Clark, C.E.; Wang, M.; Dunn, J.B.; Palou-Rivera, I. Life-cycle greenhouse gas emissions of shale gas, natural gas, coal, and petroleum. *Environ. Sci. Technol.* **2012**, *46*, 619–627. [CrossRef]
72. Schlomer, S.; Bruckner, T.; Fulton, L.; Hertwich, E.; McKinnon, A.; Perczyk, D.; Roy, J.; Schaeffer, R.; Sims, R.; Smith, P.; et al. Annex III: Technology-specific cost and performance parameters, climate change 2014: Mitigation of climate change. In *Climate Change 2014: Mitigation of Climate Change: Contribution of Working Group III to the Fifth Assessment Report of the Intergovernmental Panel on Climate Change*; Cambridge University Press: Cambridge, UK, 2014.
73. Peng, X.; She, X.; Li, C.; Luo, Y.; Zhang, T.; Li, Y.; Ding, Y. Liquid air energy storage flexibly coupled with LNG regasification for improving air liquefaction. *Appl. Energy* **2019**, *250*, 1190–1201. [CrossRef]
74. Weller, Z.D.; Roscioli, J.R.; Daube, W.C.; Lamb, B.K.; Ferrara, T.W.; Brewer, P.E.; von Fischer, J.C. Vehicle-based methane surveys for finding natural gas leaks and estimating their size: Validation and uncertainty. *Environ. Sci. Technol.* **2018**, *52*, 11922–11930. [CrossRef]
75. Michalski, M.Ł. Development of nuclear power as an alternative to fossil fuels. *Acta Innov.* **2019**, *30*, 38–47. [CrossRef]
76. Sadekin, S.; Zaman, S.; Mahfuz, M.; Sarkar, R. Nuclear power as foundation of a clean energy future: A review. *Energy Procedia* **2019**, *160*, 513–518. [CrossRef]
77. Aksan, N. An overview on thermal-hydraulic phenomena for water cooled nuclear reactors; part I: SETs, and ITFs of PWRs, BWRs, VVERs. *Nucl. Eng. Des.* **2019**, *354*, 110212. [CrossRef]
78. Cherp, A.; Vinichenko, V.; Jewell, J.; Suzuki, M.; Antal, M. Comparing electricity transitions: A historical analysis of nuclear, wind and solar power in Germany and Japan. *Energy Policy* **2017**, *101*, 612–628. [CrossRef]

79. Aminov, R.; Shkret, A.; Garievskii, M. Thermal and nuclear power plants: Competitiveness in the new economic conditions. *Therm. Eng.* **2017**, *64*, 319–328. [CrossRef]
80. Barbier, E. Geothermal energy technology and current status: An overview. *Renew. Sustain. Energy Rev.* **2002**, *6*, 3–65. [CrossRef]
81. Singh, R.K.; Murty, H.R.; Gupta, S.K.; Dikshit, A.K. An overview of sustainability assessment methodologies. *Ecol. Indic.* **2009**, *9*, 189–212. [CrossRef]
82. Energy Information Administration. U.S. States—State Profiles and Energy Estimates. 2019. Available online: https://www.eia.gov/state/ (accessed on 3 April 2022).
83. National Conference of State Legislatures. State Renewable Portfolio Standards and Goals. 2021. Available online: https://www.ncsl.org/research/energy/renewable-portfolio-standards.aspx (accessed on 3 April 2022).
84. Wang, J.J.; Jing, Y.Y.; Zhang, C.F.; Zhao, J.H. Review on multi-criteria decision analysis aid in sustainable energy decision-making. *Renew. Sustain. Energy Rev.* **2009**, *13*, 2263–2278. [CrossRef]
85. Jacobson, M.Z. Review of solutions to global warming, air pollution, and energy security. *Energy Environ. Sci.* **2009**, *2*, 148–173. [CrossRef]
86. Jacobson, M.Z.; Delucchi, M.A.; Bazouin, G.; Bauer, Z.A.; Heavey, C.C.; Fisher, E.; Morris, S.B.; Piekutowski, D.J.; Vencill, T.A.; Yeskoo, T.W. 100% clean and renewable wind, water, and sunlight (WWS) all-sector energy roadmaps for the 50 United States. *Energy Environ. Sci.* **2015**, *8*, 2093–2117. [CrossRef]
87. Apergis, N.; Payne, J.E. Per capita carbon dioxide emissions across US states by sector and fossil fuel source: Evidence from club convergence tests. *Energy Econ.* **2017**, *63*, 365–372. [CrossRef]
88. Energy Information Administration. Rankings: Total Carbon Dioxide Emissions. 2018. Available online: https://www.eia.gov/state/rankings/?sid=SC#/series/226 (accessed on 3 April 2022).
89. Hadian, S.; Madani, K. The water demand of energy: Implications for sustainable energy policy development. *Sustainability* **2013**, *5*, 4674–4687. [CrossRef]
90. Gleick, P.H. Water and energy. *Annu. Rev. Energy Environ.* **1994**, *19*, 267–299. [CrossRef]
91. Gerbens-Leenes, P.; Hoekstra, A.Y.; Van der Meer, T. The water footprint of energy from biomass: A quantitative assessment and consequences of an increasing share of bio-energy in energy supply. *Ecol. Econ.* **2009**, *68*, 1052–1060. [CrossRef]
92. Wachob, A.; Park, A.; Newcome, R., Jr. *South Carolina State Water Assessment*; South Carolina Department of Natural Resources: Columbia, SC, USA, 2009.
93. Dieter, C.A.; Maupin, M.A.; Caldwell, R.R.; Harris, M.A.; Ivahnenko, T.I.; Lovelace, J.K.; Barber, N.L.; Linsey, K.S. *Estimated Use of Water in the United States in 2015*; Number 1441; US Geological Survey: Reston, VA, USA, 2018.
94. Fthenakis, V.; Kim, H.C. Land use and electricity generation: A life-cycle analysis. *Renew. Sustain. Energy Rev.* **2009**, *13*, 1465–1474. [CrossRef]
95. McDonald, R.I.; Fargione, J.; Kiesecker, J.; Miller, W.M.; Powell, J. Energy sprawl or energy efficiency: Climate policy impacts on natural habitat for the United States of America. *PLoS ONE* **2009**, *4*, e6802. [CrossRef] [PubMed]
96. Milesi, C.; Elvidge, C.D.; Nemani, R.R.; Running, S.W. Assessing the impact of urban land development on net primary productivity in the southeastern United States. *Remote Sens. Environ.* **2003**, *86*, 401–410. [CrossRef]
97. United States Census Bureau. State Area Measurements and Internal Point Coordinates. 2010. Available online: https://www.census.gov/geographies/reference-files/2010/geo/state-area.html (accessed on 3 April 2022).
98. Bureau of Economic Analysis. GDP by State. 2020. Available online: https://www.bea.gov/data/gdp/gdp-state (accessed on 3 April 2022).
99. Sonnichsen, N. Average Age of Retired Power Plants in the United States between 2000 and 2015, by Fuel Type. 2018. Available online: https://www.statista.com/statistics/828521/average-age-of-retired-power-plants-by-fuel-type-in-the-us/ (accessed on 3 April 2022).
100. Findlaw. State Interest Rate Laws. 2016. Available online: https://statelaws.findlaw.com/consumer-laws/interest-rates.html (accessed on 3 April 2022).
101. Edenhofer, O.; Pichs-Madruga, R.; Sokona, Y.; Seyboth, K.; Matschoss, P.; Kadner, S.; Zwickel, T.; Eickemeier, P.; Hansen, G.; Schlömer, S.; et al. *IPCC Special Report on Renewable Energy Sources and Climate Change Mitigation*; Prepared By Working Group III of the Intergovernmental Panel on Climate Change; Cambridge University Press: Cambridge, UK, 2011.
102. Lehnert, E.A.; Wilt, G.; Flanagan, B.; Hallisey, E. Spatial exploration of the CDC's Social Vulnerability Index and heat-related health outcomes in Georgia. *Int. J. Disaster Risk Reduct.* **2020**, *46*, 101517. [CrossRef]
103. Center for Disease Control and Prevention. CDC's Social Vulnerability Index (SVI): A Tool to Identify Socially Vulnerable Communities. 2018. Available online: http://svi.cdc.gov/ (accessed on 3 April 2022).
104. Flanagan, B.E.; Gregory, E.W.; Hallisey, E.J.; Heitgerd, J.L.; Lewis, B. A social vulnerability index for disaster management. *J. Homel. Secur. Emerg. Manag.* **2011**, *8*, 1–22. [CrossRef]
105. Cutter, S.L.; Boruff, B.J.; Shirley, W.L. *Social Vulnerability to Environmental Hazards*; Routledge: London, UK, 2012.
106. National Renewable Energy Laboratory. JEDI: Jobs & Economic Development Impact Models. 2014. Available online: https://www.nrel.gov/analysis/jedi/ (accessed on 3 April 2022).
107. Kenley, C.; Klingler, R.; Plowman, C.; Soto, R.; Turk, R.; Baker, R.; Close, S.; McDonnell, V.; Paul, S.; Rabideau, L.; et al. Job creation due to nuclear power resurgence in the United States. *Energy Policy* **2009**, *37*, 4894–4900. [CrossRef]

108. The New York Times. Senate Election Results: GOP Keeps Control. 2017. Available online: https://www.nytimes.com/elections/2016/results/senate (accessed on 3 April 2022).
109. Missouri Energy Initiative. Implications of State Energy Plans. 2014. Available online: https://www.eenews.net/assets/2014/09/29/document_ew_02.pdf (accessed on 3 April 2022).
110. US Energy Information Administration. Total Energy Consumption, Price, and Expenditure Estimates, 2016. 2019. Available online: http://www.eia.gov (accessed on 3 April 2022).
111. Lopez, A.; Roberts, B.; Heimiller, D.; Blair, N.; Porro, G. *US Renewable Energy Technical Potentials: A GIS-Based Analysis*; Technical Report; National Renewable Energy Lab (NREL): Golden, CO, USA, 2012.
112. United States Geological Survey. Minerals. 2020. Available online: https://www.usgs.gov/products/maps/map-topics/minerals (accessed on 15 June 2021).
113. National Oceanic and Atmospheric Administration. Data Snapshot Details: Average Monthly Temperature. 2021. Available online https://www.climate.gov/maps-data/data-snapshots/data-source-average-monthly-temperature (accessed on 7 April 2022).
114. National Renewable Energy Laboratory. Solar Resource Data, Tools, and Maps. 2018. Available online: https://www.nrel.gov/gis/solar.html (accessed on 3 April 2022).
115. Global Aviation Navigator. United States Airports. 2020. Available online: https://www.globalair.com/airport/state.aspx (accessed on 3 April 2022).
116. USA.com. U.S. Average Wind Speed State Rank. 2020. Available online: http://www.usa.com/rank/us--average-wind-speed--state-rank.htm (accessed on 3 April 2022).
117. National Oceanic and Atmospheric Administration. *The Coastline of the United States*; Number v.55 in the Coastline of the United States; U.S. Department of Commerce, National Oceanic and Atmospheric Administration: Silver Spring, MA, USA, 1975.
118. National Renewable Energy Laboratory. Geothermal Resource Data, Tools, and Maps. 2018. Available online: https://www.nrel.gov/gis/geothermal.html (accessed on 3 April 2022).
119. United States Geological Survey. Streamflow–Water Year. 2018. Available online: https://waterwatch.usgs.gov/publications/wysummary/2018/ (accessed on 3 April 2022).
120. Association of State Dam Safety Officials. Dam Failures and Incidents. 2019. Available online: https://damsafety.org/dam-failures (accessed on 3 April 2022).
121. Congressional Research Service. Nuclear Waste Storage Sites in the United States. 2020. Available online: https://fas.org/sgp/crs/nuke/IF11201.pdf (accessed on 3 April 2022).
122. McMahon, J. New Map Shows Expanse of U.S. Nuclear Waste Sites. *Forbes*, 31 May 2019. Available online: https://www.forbes.com/sites/jeffmcmahon/2019/05/31/new-map-shows-expanse-of-u-s-nuclear-waste-sites/?sh=aa5cb97c2cf7 (accessed on 3 April 2022).
123. Post, P.; Klazynski, R.; Klocek, E.; DeCort, T.; Riches, T.; Li, K.; Elliott, E.; Poling, R. Assessment of undiscovered technically recoverable oil and gas resources of the Atlantic Outer Continental Shelf 2011 as of January 1, 2009. *Bur. Ocean. Energy Manag BOEM* **2012**, *16*, 39.
124. Energy Information Administration. Natural Gas. 2021. Available online: https://www.eia.gov/naturalgas/ (accessed on 3 April 2022).
125. Vogt, J.T.; Smith, W.B. *Forest Inventory and Analysis Fiscal Year 2016 Business Report*; USDA: Washington, DC, USA, 2017.
126. Center for Sustainable Systems. Municipal Solid Waste Factsheet. 2020. Available online: http://css.umich.edu/factsheets/municipal-solid-waste-factsheet (accessed on 3 April 2022).
127. United States Department of Agriculture. *Crop Production 2020 Summary*; Technical Report; USDA: Washington, DC, USA, 2021
128. Milken Institute. State Tech and Science Index. 2014. Available online: http://statetechandscience.org/state-ranking.html (accessed on 3 April 2022).
129. US College and University Directory. Colleges and Universities in the United States of America (USA) by State/ Possession 2013. Available online: http://univsearch.com/state.php (accessed on 3 April 2022).

sustainability

Article

Agricultural Biogas Production—Climate and Environmental Impacts

Henrik B. Møller [1,*], Peter Sørensen [2], Jørgen E. Olesen [2], Søren O. Petersen [2], Tavs Nyord [3] and Sven G. Sommer [1]

[1] Department of Biological and Chemical Engineering, Aarhus University, Blichers Allé 20, 8830 Tjele, Denmark; sgs@bce.au.dk
[2] Department of Agroecology, Aarhus University, Blichers Allé 20, 8830 Tjele, Denmark; ps@agro.au.dk (P.S.); jeo@agro.au.dk (J.E.O.); sop@agro.au.dk (S.O.P.)
[3] Concito, Læderstræde 20, 1201 København, Denmark; tn@concito.dk
* Correspondence: henrikb.moller@bce.au.dk

Abstract: Livestock manure is a major source of the greenhouse gases (GHGs) methane (CH_4) and nitrous oxide (N_2O). The emissions can be mitigated by production of biogas through anaerobic digestion (AD) of manure, mostly together with other biowastes, which can substitute fossil energy and thereby reduce CO_2 emissions and postdigestion GHG emissions. This paper presents GHG balances for manure and biowaste management as affected by AD for five Danish biogas scenarios in which pig and cattle slurry were codigested with one or more of the following biomasses: deep litter, straw, energy crops, slaughterhouse waste, grass–clover green manure, and household waste. The calculated effects of AD on the GHG balance of each scenario included fossil fuel substitution, energy use for transport, leakage of CH_4 from biogas production plants, CH_4 emissions during storage of animal manure and biowaste, N_2O emissions from stored and field applied biomass, N_2O emissions related to nitrate (NO_3^-) leaching and ammonia (NH_3) losses, N_2O emissions from cultivation of energy crops, and soil C sequestration. All scenarios caused significant reductions in GHG emissions. Most of the reductions resulted from fossil fuel substitution and reduced emissions of CH_4 during storage of codigestates. The total reductions in GHG emissions ranged from 65 to 105 kg CO_2-eq ton^{-1} biomass. This wide range showed the importance of biomass composition. Reductions were highest when straw and grass–clover were used as codigestates, whereas reductions per unit energy produced were highest when deep litter or deep litter plus energy crops were used. Potential effects of iLUC were ignored but may have a negative impact on the GHG balance when using energy crops, and this may potentially exceed the calculated positive climate impacts of biogas production. The ammonia emission potential of digestate applied in the field is higher than that from cattle slurry and pig slurry because of the higher pH of the digestate. This effect, and the higher content of TAN in digestate, resulted in increasing ammonia emissions at 0.14 to 0.3 kg NH_3-N ton^{-1} biomass. Nitrate leaching was reduced in all scenarios and ranged from 0.04 to 0.45 kg NO_3-N ton^{-1} biomass. In the scenario in which maize silage was introduced, the maize production increased leaching and almost negated the effect of AD. Methane leakage caused a 7% reduction in the positive climate impact for each percentage point of leakage in a manure-based biogas scenario.

Keywords: biogas; anaerobic digestion; manure; greenhouse gases; methane; nitrous oxide; environmental impacts

Citation: Møller, H.B.; Sørensen, P.; Olesen, J.E.; Petersen, S.O.; Nyord, T.; Sommer, S.G. Agricultural Biogas Production—Climate and Environmental Impacts. *Sustainability* **2022**, *14*, 1849. https://doi.org/10.3390/su14031849

Academic Editors: Ornella Malandrino, Nicola Raimo, Filippo Vitolla and Benedetta Esposito

Received: 5 January 2022
Accepted: 3 February 2022
Published: 6 February 2022

Publisher's Note: MDPI stays neutral with regard to jurisdictional claims in published maps and institutional affiliations.

1. Introduction

Emissions of nitrous oxide (N_2O) and methane (CH_4) from livestock manure management contribute around 10% of the total non-CO_2 greenhouse gas (GHG) emissions globally calculated as CO_2 equivalents [1]. N_2O and CH_4 have global warming potentials for time horizons of 100 years (GWP100) of 298 and 25 times, respectively, higher than GWP100 per kg of carbon dioxide (CO_2) [2]. Ammonia (NH_3) emission and leaching losses of nitrate (NO_3^-) are important indirect sources of N_2O [3,4].

Globally, manure management contributes about 10% of agricultural CH_4 emissions [5], but in confined livestock production systems (e.g., dairies and piggeries) with liquid manure management, this proportion can exceed 50% depending on climate [1]. Animal manure applied to soil contributes to maintenance of soil carbon (C) insofar as a fraction of the manure C is sequestered. The Danish Energy Agency has calculated that of the Danish GHG emissions from livestock production, manure management contributes 22%, manure and mineral fertilizers applied to soil contribute 39%, and enteric fermentation contributes 39% to total agricultural GHG emissions [6].

Emissions of CH_4 and N_2O are regulated and accounted under the UNFCCC as part of the Paris Agreement. The reduction target for the EU on GHG is 55% by 2030 with reference to the year 1990 [7]. The Danish parliament has decided that GHG emissions from Denmark must be reduced by 70% with reference to 1990 by 2030 [6], and agriculture must contribute to this reduction. This calls for the implementation of technologies that cost-effectively reduce GHG emissions from the livestock slurry management chain. Slurry is in focus because 80% of Danish livestock manures are managed in the form of slurries [8].

GHG emissions from slurry management systems can be reduced by AD treatment, frequent export of slurry from livestock buildings to colder outside stores, acidification, or separation of slurry combined with incineration [9–13].

In Denmark, more than 25% (weight basis) of animal manure is today anaerobically codigested on centralized biogas plants with organic wastes from the food industry, slaughterhouses, dairies, and the fish industry with the aim to produce CH_4 for bioenergy. The residues from AD must be recycled as fertilizer and meet the requirements for content of pathogens, heavy metals, and environmentally harmful substances. Biogas plants use almost all industrial residues available in Denmark, and increasing amounts of straw, grass, deep litter, etc. are used in the codigestion of slurry.

The economy is critical when making decisions about the introduction of technologies in farming aiming to reduce GHG emissions, and socioeconomic impacts of different types of biomass for AD have been calculated [14,15]. In scenarios with codigestion of slurry with fibre fraction from slurry separation, maize silage, grass, and sugar beet, NO_3^- leaching was assessed, but the effects of the AD treatment on GHG emissions and NO_3^- leaching were not well documented [14]. More recently, a refined model of CH_4 from manure management that accounted for different storage conditions was used to calculate the effects of biogas and frequent export of slurry from livestock housing to an outside storage tank [16]. However, no biomasses other than slurry were accounted for, nor were any other effects, such as energy and environmental impacts.

When assessing the effect of AD as a potential mitigation measure, a whole-farm approach is needed to estimate GHG emissions from the pig or dairy farm, and calculations must include evaluation of side effects in the form of increased NH_3 emissions, reduced C sequestration, leakage of CH_4 from the biogas plant, etc. [17]. It is relevant to improve estimates of the potential of AD to reduce the negative GHG balance of livestock farming. Therefore, in the present study, the effect of AD on the GHG gas balance was calculated using a "system analysis approach", which included substituting CO_2 emission from power and heat production using fossil fuel; leakage of CH_4 from biogas production plants; CH_4 emissions during storage of animal manure and organic waste; N_2O emissions from stored and field applied manure, organic waste, and digestate; N_2O emissions related to NO_3^- leaching and NH_3 emission; N_2O emission from cultivating energy crops; and effects on soil C sequestration (Figure 1).

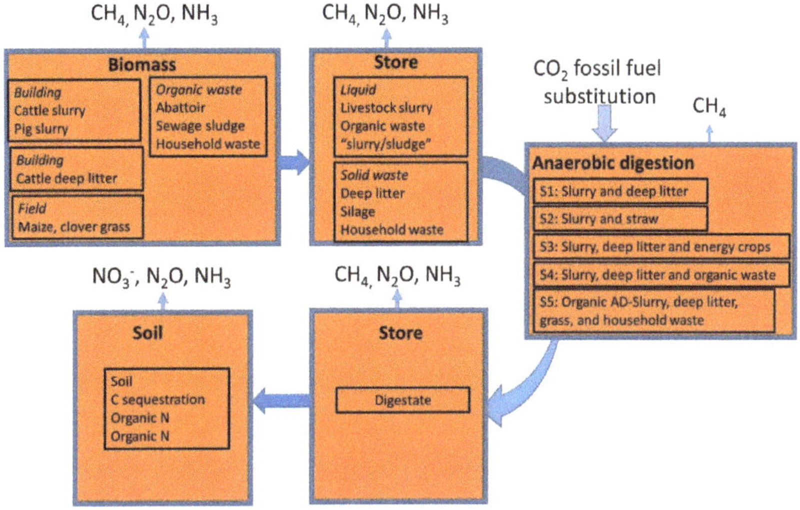

Figure 1. Flow diagram depicting the transfer of biomass among the five compartments of the biomass transfer continuum or system.

This quantification of the climate and environmental effects of biogas production constitutes an important basis for designing and targeting future biogas subsidies to optimize the climate and environmental benefits of production. In this study, assessments are reported for five biomass scenarios in a Danish biogas context with different retention times and biomass compositions. The assessment was based on the best technologies currently used by the Danish biogas sector.

2. Materials and Methods

The calculated climate and environmental effects of introducing centralized codigestion AD in the livestock sector were compared with typical reference management of slurry and waste biomasses. Environmental impacts of introducing AD in five manure management systems (Figure 1; Table 1) were calculated with a whole-system calculation approach for each scenario using one ton of dry matter of biomass as the functional unit. The calculations included CH_4 and N_2O emissions from each scenario (AD and reference farm without AD), the effect on soil C storage, and AD effects on N_2O emissions related to NO_3^- leaching and NH_3 emission. GHG and environmental effects were calculated using the models applied in the Danish national inventory. Global warming potentials for a 100-year time horizon (GWP100) of CH_4 and N_2O at 25 and 298 CO_2-eq kg^{-1}, respectively, were used [18].

2.1. Slurry and Biowaste Management

In this study, the starting point was the collection of excreta and urine in slurry channels. The retention time was set at 20 days for cattle and 19 days for pigs [16], and the average slurry temperature was set at 13.8 °C for cattle and 18.6 °C for pigs. From the channels, slurry was transferred to an outside store and subsequently applied to grassland or arable land between March and August. In the outdoor storage tank, the temperature of untreated slurry was calculated based on monthly mean temperatures [16]. Untreated slurry and digestate must be stored according to national regulations to minimize ammonia emissions [19]. About 80% of untreated slurry was covered by a floating layer of natural crust, straw, or clay pebbles, and 20% was covered by a tent cover. Solid manure was stored for an average of 5 months in heaps covered by a PVC sheet. The deep litter in the reference system was covered with PVC and stored for an average of 5 months.

Table 1. Model plants used in the five scenarios. Distribution of biomass input is given in weight percentage. The slurry input was a mixture of 50% cattle slurry and 50% pig slurry, except in S5 where only cattle slurry was used.

Scenario	Input	Input g DM kg^{-1} Biomass	Reactor g DM kg^{-1} Biomass	Reference Scenario
1	Slurry (80%) + deep litter (20%)	112	95	Animal slurry is stored in a slurry tank and then applied by injection or trailing hose. Deep litter is stored in covered stacks/manure piles for five months and applied before sowing spring cereals.
2	Slurry (80%) + straw (20%)	220	95	Straw is cut and incorporated.
3	Slurry (80%) + deep litter (8%) + energy crops (12%)	94	51	The land farmed with energy crops is used for cereal crops.
4	Slurry (70%) + deep litter (10%) + organic waste (20%)	141	53	The organic waste is stored as slurry and then spread directly on the field (slaughterhouse waste), incinerated (glycerine), or composted and then applied (biowaste).
5	Organic grass–clover (25%) + cattle slurry (50%) + deep litter (20%) + biowaste (5%)	97	95	At an organic farm without a biogas plant, the grass–clover is managed as green manure.

Slurries were applied by injection to bare soil and grassland and by trail hoses to autumn-sown crops such as winter cereals, whereas deep litter had to be incorporated within 4 h after application to soil.

In the reference system, slaughterhouse waste was stored together with livestock slurry and then spread on fields in the growing season. Glycerine was used for energy production by incineration, and source-separated household and industrial organic wastes were composted and used as fertilizer. The maize grown for codigestion in S3 was assumed to substitute a cereal crop in the reference system, and the grass–clover grown for codigestion in the organic farming scenario (S5) was assumed to be cut and mulched in the reference system.

In the AD concepts, liquid manure was transferred from livestock buildings to stores on the farm, as in the reference system (Figure 1). Every 3–30 d, the slurry in the outside store and slurry channels within the building were assumed to be emptied and transferred to stores on the AD plant and covered according to regulations. Deep litter was assumed to be transferred from the farms to biogas plants, where it was stored in covered heaps until used for biogas production. Straw was used for biogas production, and in the organic farm scenario, grass–clover crops were fed to the digester as silage. Organic waste in the form of abattoir waste was fed to the biogas reactors after storage in concrete stores similar to those used for animal slurry, and glycerine was assumed to be stored in containers until use. Source-separated household waste and industrial organic waste were assumed to be stored in covered heaps.

In the five scenarios examined in this study (Table 1), biomass was fed to centralized biogas plants with 45 d hydraulic retention times (HRT) in primary thermophilic reactors (53 °C) and an average retention time of 45 d. In newly built biogas plants in Denmark, HRT tends to be longer, and therefore, the calculations included scenarios with retention times of 60 and 90 days. Digestate was defined to be cooled by heat exchange to 25 °C and then stored at the plant for 20 d in a storage tank with CH_4 gas collection. In all model plants, digestion took place by serial operation in two reactors with the same retention time in each unit, and the biogas was assumed to be upgraded and transferred to the natural gas distribution network by removal of CO_2 in the biogas.

The digestate was transferred from the AD plant to farmer concrete stores, where 50% of the digestate was assumed to be covered with a tent and the rest by a floating cover, straw, or natural surface crust. The digestate was applied by injection to bare soil and grass crops (mandatory in Denmark) and by trail hoses to other growing crops.

2.2. Calculations

Greenhouse gases, NH_3 emissions, and leaching of NO_3^- from different farm compartments in the reference and AD biomass management continuum are depicted in Figure 1. An overview of the calculation of biogas production, emissions, leaching, and soil C storage is given in the following. Short reviews about the algorithms used to calculate emissions, transformation, and leaching losses are presented in Supplementary Materials, which also contain tables with parameters for the algorithms and emission factors.

2.2.1. Biogas Production

The CH_4 production from the different biomasses was estimated from our own experiments and other studies (Supplementary Materials, Table S1). The ultimate CH_4 yield in terms of volatile solids (VS) is the yield achieved at a retention time of more than 90 days, and to determine the yield at shorter retention times, the CH_4 produced in the biogas reactor was determined for each biomass through modelling using the Gompertz equation [20]. The gas potential at a given time was calculated as follows:

$$M(t) = B_0 \cdot \left(1 - e^{-k \cdot t}\right) \quad (1)$$

where M is the cumulative CH_4 yield (mL g^{-1} (VS)), B_0 is the theoretical CH_4 yield (mL g^{-1} (VS)), k is a first-order kinetic rate constant representing the hydrolysis constant, and t is retention time (days). The biogas production was calculated from this equation and thus determined by the amount and quality of biomasses and the retention time in the biogas reactor.

2.2.2. Energy Production and Consumption

The calculations of CO_2 substitution from energy production were based on the displacement of natural gas, in which CH_4 substituted CO_2 corresponding to 0.057 kg CO_2-eq MJ^{-1} [21]. The electricity demand for agitators, pumps, etc. at the plant was assumed to be covered by a mix of the Danish electricity production, which in 2019 was estimated at 0.150 g CO_2 kWh^{-1}, as calculated using data from [22]. The volume of biomass transported in one truckload to and from the biogas plant varied among the different biomass types, and diesel consumption was given per kilometre driven. The distances were calculated as the average additional transport of biomass compared to the reference scenario without AD and varied from 0.8 kg CO_2-eq ton^{-1} for grass and maize silage to 11.6 kg CO_2-eq ton^{-1} for glycerol. The effect on GHGs of replacing mineral N fertilizer due to higher N availability in the digestates was estimated by assuming that the long-term N availability equivalent of total N was 5% higher in the treated than in untreated manure [23]. The potential reduction in N fertilizer use was assumed to give a reduction in GHGs of 5.6 kg CO_2 kg^{-1} N [24], equivalent to 0.28 kg CO_2 kg^{-1} treated N in the biogas plant.

2.2.3. Methane Emission from Slurry and Digestate

Daily methane emissions from slurry and digestate during storage were estimated from the volumes of readily degradable (VS_d, kg kg^{-1}) and slowly degradable organic matter (VS_{nd}, kg kg^{-1}) as proposed by Sommer et al. [9]:

$$F_t = (VS_d + 0.01VS_{nd})e^{(lnA - \frac{E_\alpha}{RT})} \quad (2)$$

where F_t is the methane production rate (g CH_4 kg^{-1} VS h^{-1}), E_α is the process activation energy (J mol^{-1}), lnA (g CH_4 kg^{-1} VS h^{-1}) represents the methanogenic potential of the substrate, R is the universal gas constant (J K^{-1} mol^{-1}), and T is the temperature (K). Equation (1) assumes that the amount and degradability of biomass organic matter, and storage temperature, are main controlling variables.

E_α was set to 81,000 J mol^{-1} [25]. The parameter lnA was highly variable and depended on slurry origin, treatment, storage conditions, and age [4]. Petersen et al. [26] estimated *lnA* for slurry collected in pig and cattle barns by measuring CH_4 production rates at the storage temperature and calculating *lnA* from Equation (1) after rearrangement:

$$lnA = ln\left[\frac{F_t}{(VS_d + 0.01VS_{nd})}\right] + \frac{E_\alpha}{RT} \quad (3)$$

The degradability of VS in slurry changes during storage, and no data were available from outside storage tanks. Instead, a different approach was used in which *lnA* estimates were related to total VS:

$$lnA' = ln\left[\frac{F_t}{VS_{total}}\right] + \frac{E_\alpha}{RT} \quad (4)$$

Note that the parameter value derived from total VS is referred to as lnA' to distinguish from the original calculation of *lnA* with reference to degradable VS. A limited number of studies were identified for which information about storage temperature and VS content were available to allow estimation of lnA' in pig and cattle slurry, as well as digestate (Supplementary Materials, Table S4). For the present study, the *lnA* of pig and cattle slurry in barns reported by Petersen et al. [26] were recalculated to lnA'; Table 2 summarizes the values used.

Table 2. Values of the parameter lnA' used in scenario analyses to represent methanogenic potential; for derivation, see text. In the table, $\overline{x} \pm$ s.e. refers to average and standard error.

Category	Storage Period	lnA' g CH_4 kg^{-1} VS h^{-1}	Reference
Cattle slurry	Barn	30.1	[20]
	Outside store	29.2 ± 0.1	[25,27,28]
Pig slurry	Barn	30.6	[20]
	Outside store	30.3 ± 0.4	[25,27,28]
Digestate	Outside store	27.9 ± 0.4	[28,29]

Methane emissions were calculated separately for cattle and pig slurry in barns and for untreated slurry, digestate, and other biomasses assumed to be stored in outside storage tanks. Assumptions regarding retention time, storage temperatures, etc. are given in Supplementary Materials.

After field application, manure environments are predominantly at a redox level at which little, if any, CH_4 is produced. Transient emissions have sometimes been reported; these are probably due to release of dissolved methane produced during storage [30,31].

2.2.4. Methane Emission from Solid Manure

Methane may be emitted from solid manure (deep litter, fibre fraction from separated slurry or digestate) during storage. The emission level is determined by VS degradability, air- and water-filled porosity, and coverage, which in turn determine biological oxygen demand, gas exchange rates, temperature, and anaerobic volume developing during storage. The Danish emission inventory estimates this source with a model proposed in [4]:

$$EF = BMP * MCF * 0.67 \quad (5)$$

where *EF* (kg CH_4 kg^{-1}(VS)) is the CH_4 emission factor, *BMP* (m^3 CH_4 kg^{-1}(VS)) is the biochemical methane production potential, and *MCF* (%) is a country-specific CH_4 conversion factor. For deep litter, an MCF of 3% is assumed if manure is exported at 1-month intervals or less, and an MCF of 17% is assumed if manure is removed at longer intervals. With a BMP of 0.24 m^3 kg^{-1}(VS), the overall CH_4 emission from barns and during outside storage were as shown in Table 3.

Table 3. Emission factors for CH_4 and N_2O from deep litter in housing and storage facilities (IPCC 2006) as well as emission factors used for the storage period in this report. BMP was 0.240 m^3 (CH_4) kg^{-1} (VS).

Categories	Methane		Nitrous Oxide
	IPCC (housing and outside storage)		
	MCF (% of BMP)	kg CH_4 kg^{-1} (VS)	N_2O-N % of total N
<1 month in housing	3	0.005	1
>1 month in housing	17	0.027	1
	This study (outside storage)		
	kg CH_4 kg^{-1} (C)	kg CH_4 kg^{-1} (VS)	% of total-N
Storage in covered heaps	0.015	0.0075	0.5
Composting	0.03	0.015	2.2

We assumed that half of these emissions would come from outside stores, corresponding to 0.005 and 0.027 kg CH_4 kg^{-1}(VS) for short- and long-term storage periods, respectively (Table 3). Recent studies have shown that emission of CH_4 from uncovered and uncompacted manure heaps are at a level of 0.027 kg CH_4 kg^{-1}(VS) [32–34], and we therefore assumed that the emission factor was 0.03 kg CH_4 kg^{-1}(VS) for uncovered manure heaps and half as much for heaps with PVC tent covers (i.e., 0.015 kg CH_4 kg^{-1}(VS)) (Table 3) [34].

In organic farming, animal manure is often actively composted by turning the heap. This reduces the development of anaerobic volumes in the heap and contributes to lower CH_4 emission compared to undisturbed heaps. Based on a CH_4 emission from actively composted organic waste corresponding to 3% C [32], and assuming the same C/VS ratio in organic waste and deep litter, CH_4 emissions from deep litter were estimated (Table 3).

2.2.5. Nitrous Oxide Emission

Nitrous oxide may be emitted from slurry and digestate, as well as solid manure, during storage and after field application. Nitrous oxide emissions are associated with nitrification and denitrification, two interdependent processes occurring under aerobic and anaerobic conditions, respectively. Oxic–anoxic gradients occur in slurry storages with surface crusts and in the outer layers of manure heaps.

Nitrous oxide emissions during storage of slurry or digestate depend on the development of a floating crust where populations of nitrifying and denitrifying bacteria live. The IPCC guidelines give a default emission factor for storage tanks with a surface crust of 0.5%, i.e., 0.5% of total N entering the storage tank is converted to N_2O [4]. Danish pilot-scale measurements indicated lower emissions, 0.2–0.4% [35], but the level of emissions is influenced by climatic conditions, especially the water balance (rain and evaporation). The emission factor of 0.5% of N in the total flow of slurry was used here even though only the surface can be a source of N_2O. Without a surface crust, the emission factor for N_2O was set to 0 for both untreated slurry and digestate.

For cattle deep litter, IPCC [4] recommends a N_2O emission factor for barn and storage of 1% regardless of retention time in the barn. Assuming half of these emissions occur during outdoor storage, this effectively corresponded to 0.5% of total N. In a review by Pardo et al. [32], N_2O-N emissions from compost heaps with organic waste corresponded to 2.2% of total N (Table 3).

The default emission factor for nitrous oxide emissions from N in field-applied liquid and solid manure is 1% [4]. In soil with organic amendments, be they manure, digestate, or crop residues, the balance between oxygen (O_2) demand and O_2 supply is an important control of denitrification and N_2O emissions [36]. While anaerobic digestion reduces the availability of degradable VS, and hence O_2 demand, the net effect on N_2O emissions depends on the interaction with specific soil conditions. A review of field studies [37] reported mostly reductions in N_2O emissions with AD, but increases have also been reported. In the present study, no effect of anaerobic digestion on N_2O emissions was assumed. Nitrous oxide emissions related to NH_3 emissions and nitrate (NO_3^-) leaching were accounted for with emission factors of 1% and 0.75%, respectively, in accordance with the national inventory of Denmark [18].

2.2.6. Ammonia Emission

The emissions of NH_3 from slurry storage tanks and solid manure heaps were calculated using emission factors estimated by [38]. These were within the ranges given in the recent review by Kupper et al. [39]. The NH_3 emission factor for composting of source separated organic waste was taken from the review by Pardo et al. [32]. When deposited, the emitted NH_3 contributes to N_2O emissions [4], and this indirect source of N_2O was included in the calculations. Ammonia emission factors for applied livestock liquid manure given by [38] were used in this study, while emission factors for each month were calculated with the ALFAM model [40] using average monthly weather conditions and average slurry compositions for Denmark. Based on a review of recent studies, we assumed that NH_3 emission from digestate would be higher than emission from untreated slurry (Supplementary Materials, Table S6).

Data from studies on emissions of NH_3 from deep litter applied to soil were limited in 2008 [38], and only one emission factor for application during different months was given. The evidence about the effects of climatic conditions on emission of NH_3 from solid manure applied to soil is still limited, and emission factors cannot be assessed at a monthly scale [41]. We therefore used the same emission factor for applied deep litter regardless of whether the application took place in the spring or autumn. Deep litter must, in Denmark be incorporated into the soil within 4 h of application, and it was assumed that within this timespan, 25% of total ammoniacal N (TAN) was emitted as NH_3.

2.2.7. Crop Production and Nitrate Leaching

AD processing of animal slurry and biowaste affects NO_3^- leaching from crop production in both direct and indirect ways. The direct effects are through the effects of applied N in both the first and following year after application of fertilizer or manure; here, AD affects the quality and quantity of N applied. The indirect effects of AD on NO_3^- leaching occur through AD effects on NH_3 volatilization, which affects the N available for plants through both the N loss and the deposited N.

Nitrate leaching during the first year after fertilizer or manure application was assumed to be proportional to the amount of total N (mineral N + organic N) applied [23,42,43]. Total N application was assumed to be similar before and after AD, in accordance with Danish fertilizer regulations, despite more N being plant available after AD. This means that more N is taken up in the first crop and less organic N is left in the soil from digestates. In the scenario with an energy crop (S3), total N application increased by AD, as the energy crop contributed with extra organic N to the system, and the N derived from the energy crop was assumed to replace only mineral N fertilizers with an efficiency of 40% as in the Danish legislation. In the other scenarios, total N application remained the same before and after digestion.

The effect of AD on NO_3^- leaching over a 10-year period was calculated using a model based on the principles described by Sørensen et al. [23]. It was assumed that about 40% of the organic N input would be mineralized during years 2–10 after application, and that 34% of the mineralized N would be leached as nitrate [23]. Because of the lower N mineralization after AD, NO_3^- leaching was also slightly reduced after digestion. The model uses information on soil and climatic conditions, and it was assumed that 80% of the manures were used on sandy soils with precipitation above average Danish levels [44].

The direct and indirect effects of increased NH_3 volatilization after AD on NO_3^- leaching were calculated separately. It was assumed that increased NH_3 emission caused a net reduction in NO_3^- leaching [8] as estimated with the following assumptions. The empirical NLES5 model estimates an average marginal leaching of 17% during the first three years after mineral N application in spring under Danish conditions and at N rates near the economic optimum [44]. In Denmark, 80% of livestock manure is applied to sandy soils [45] with higher-than-average marginal leaching, and therefore, 20% of NH_4^+-N in applied manure/digestate was assumed to be lost by leaching over a 3-year time period. Furthermore, NH_4^+-N applied to soil contributes to organic N from plant residues that may give an extra NO_3^- N leaching equivalent to 2% of the N input in years 3–10 after application [8]. The total reduction in NO_3^- N leaching from NH_4^+-N was therefore 22% of the increase in NH_4^+-N volatilization loss over a 10-year period. However, part of the lost NH_3 would be deposited on agricultural land where part of that pool is leached. It was estimated that 10% the NH_4^+-N lost by volatilization was land deposited and leached as nitrate [8]. Thus, the net effect of increased NH_3 loss on reduction in NO_3^- N leaching was set to 22% − 10% =12% of the increase in NH_3-N loss over a 10-year period. This factor was used to calculate the leaching reduction due to increased ammonia volatilization.

2.2.8. Soil Carbon Storage

The effect of biogas treatment of slurry and other livestock manure on soil C storage is still relatively poorly quantified, but a study based on laboratory incubations measured slightly smaller soil C storage in connection with AD treatment [46]. Based on Thomsen et al. [46], the amount of C digested in the biogas plant was assumed to have contributed to C storage by 25% of the effect achieved when adding C in fresh plant material and straw, i.e., 0.25 × 15% = 3.75% of the C transformed to CO_2 and CH_4 in biogas during the digestion process would alternatively have been stored after a 20-year period, assuming retention in the soil of 15% of the C added in plant material over 20 years [46].

3. Results

3.1. Biogas Production

The largest biogas production was achieved in the scenarios in which slurry was codigested with straw or grass, which were the scenarios with largest energy output. In the energy balances showing reduced CO_2-eq emissions, it was assumed that the CH_4 produced substituted natural gas (Figure 2). Power is used in biogas plants for heating, pumping, and mixing, which reduces the net energy production, but most Danish biogas plants limit the energy consumption using heat exchangers, reducing the digestate temperature to 25 °C. Without heat exchange, the digestate would be stored at higher temperatures and be a significant source of CH_4 emissions. In the calculations, the use of heat exchangers reduced the need for process heat and thereby enhanced the CO_2 balance by around 10% (Figure 2).

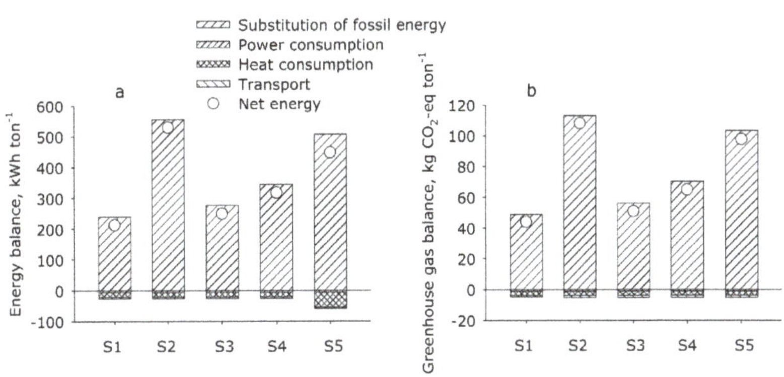

Figure 2. The energy (**a**) and GHG (**b**) balances of the energy production at model plants with heat exchangers installed. Scenarios were (S1) slurry and deep litter; (S2) slurry and straw; (S3) slurry, deep litter, and maize silage; (S4) slurry, deep litter, and organic waste; and (S5) slurry, deep litter, organic waste, and organic grass–clover.

The use of diesel for transport of biomass had little influence on GHG balances (Figures 2 and 3) in the biogas plants and constituted less than 3% of the energy produced.

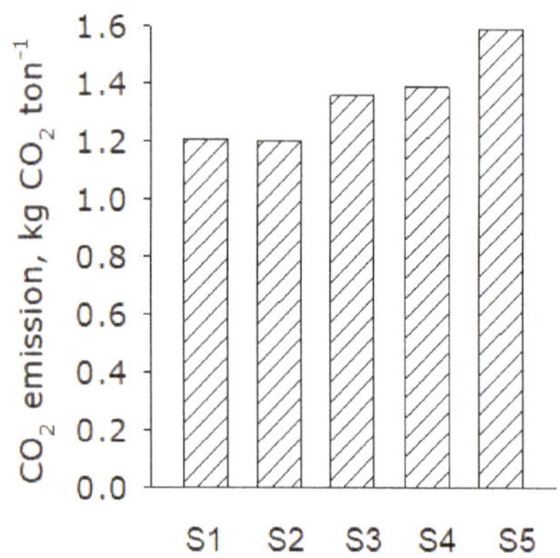

Figure 3. CO_2 emissions from biomass transport in model plants in which slurry was codigested with the following biomasses: (S1) deep litter, (S2) straw, (S3) deep litter and maize silage, (S4) deep litter and organic waste, and (S5) deep litter, organic waste, and organic grass–clover.

3.2. Methane Emissions

Whereas experimental data were available for the estimation of daily CH_4 emissions from cattle and pig slurry in barns, parameters representing the methanogenic potential during storage had to be extracted from published storage experiments. The *lnA′* values (Table 2) indicated that the methanogenic potential in stored digestate was around 70% lower than that in cattle slurry and 90% lower than that in pig slurry. This was in accordance with the assumption that anaerobic digestion would remove 90% of the degradable VS. With these differences in methanogenic potential, the five biogas scenarios showed reductions in CH_4 emissions from barns and subsequent outside storage that varied between 41 and 56% (Figure 4).

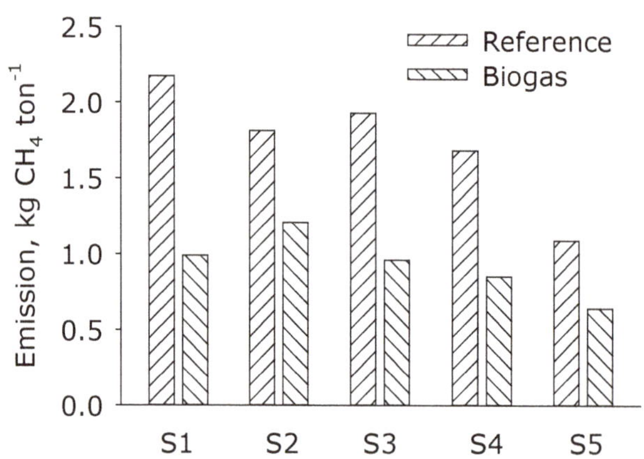

Figure 4. Calculated CH_4 emissions from barn and slurry storage tanks for the five reference and biogas scenarios. The model plants included slurry codigestates as follows (S1): deep litter, (S2) straw, (S3): deep litter and maize silage, (S4) deep litter and organic waste, and (S5): deep litter, organic waste, and organic grass–clover.

In the model calculations, the relationship between VS degradation and CH_4 production depended on assumptions regarding the proportions of C in VS converted to CH_4 and

CO_2, a ratio that is subject to large uncertainty (cf. Supplementary Materials). In the scenario calculations, the CH_4/CO_2 ratio was set at 25:75 for untreated slurry and codigestates stored anaerobically and 10:90 for digestates. The importance of these ratios was evaluated in a sensitivity analysis calculating the effects of reducing by half, or doubling, the share of CH_4 produced in pig and cattle slurry (Table 4). Untreated pig slurry was sensitive to the assumption regarding CH_4 share, with 21% lower CH_4 emissions if the assumed CH_4 share was reduced by half and 17% higher emissions if the assumed CH_4 share was doubled. All other relative changes were negligible. Experimental data on CH_4 emissions from slurry and digestate were the reference for model calculations, and as a consequence, a lower or higher CH_4/CO_2 ratio would lead to less or more residual VS, respectively, being exported and available as substrate for CH_4 emissions from the outside storage. The limited effect of changing the CH_4/CO_2 ratio for digested slurry was due to the fact that pretreatment emissions in barns were identical, and the CH_4 emissions following biogas treatment were greatly reduced.

Table 4. Sensitivity analysis of the importance of the CH_4/CO_2 ratio in the gas produced from VS degradation of untreated and digested slurry for the cumulated CH_4 emissions from barn and outside slurry storage (relative differences with scenario results as basis).

Untreated Slurry			Digested Slurry		
CH_4/CO_2	**Cattle**	**Pig**	**CH_4/CO_2**	**Cattle**	**Pig**
12.5:87.5	0.99	0.79	5:95	0.98	0.99
25:75	1.00	1.00	10:90	1.00	1.00
50:50	1.01	1.17	20:80	1.01	1.00

For each scenario, the CH_4 emissions were calculated using 45, 60, and 90 days of HRT in the reactor, which may affect the predicted CH_4 emissions during subsequent storage of the digestate. This is exemplified in Figure 5b, which shows the sources of CH_4 for scenario S5 (organic biogas) with cattle slurry, deep litter, grass–clover silage, and biowaste, as well as CH_4 emissions without treatment (HRT 0 d) for reference. It was mainly the emission of CH_4 from cattle slurry that was affected by increasing HRT, the emission being 16% less at 90 than at 45 days HRT. The reduction for deep litter was 5–6%, and the changes for grass silage and biowaste were <1%.

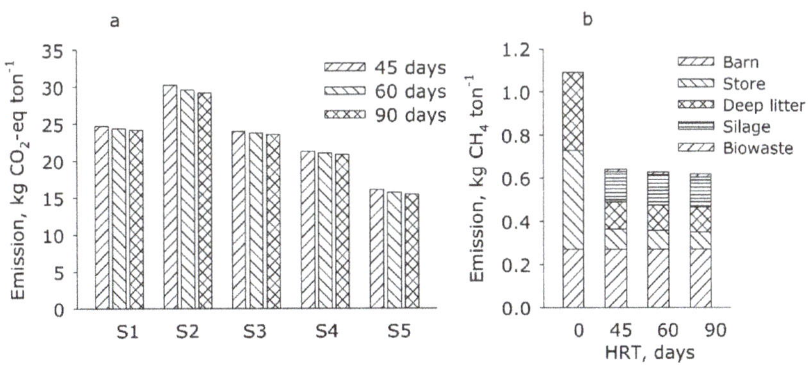

Figure 5. (**a**): Total GHG emissions in each of the five scenarios with 45, 60, and 90 d HRT calculated as CO_2-eq. (**b**): Methane emissions from stored biomasses without biogas treatment (HRT 0 d) or (S5) with biogas treatment at increasing hydraulic retention time (HRT). In the biogas scenarios, livestock slurry was codigested with (S1) deep litter, (S2) straw, (S3) deep litter and maize silage, (S4) deep litter and organic waste, and (S5) deep litter, organic waste, and grass–clover.

The total CH_4 emissions for scenarios with HRT at 45, 60, or 90 d, expressed as CO_2 equivalents, are shown in Figure 5a. All scenarios showed the same trend, with 2–3% lower GHG emission at 90 than at 45 d HRT.

Overall GHG balances were calculated for biogenic CH_4 and N_2O emissions from barns and outside storage, and after field application (Figure 6). While biogas treatment gave a substantial reduction in CH_4 emissions, ranging from 41% to 56%, as described above, no effect on N_2O emissions was assumed, and N_2O emissions are therefore directly proportional to the N content of the biomasses used in each scenario, which were identical in reference and biogas scenarios. As a result, the overall GHG balances corresponded to reductions through biogas treatment ranging between 21% and 40%.

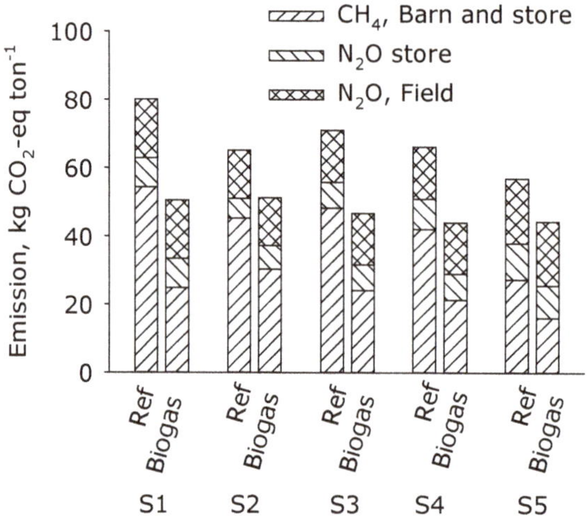

Figure 6. Total biogenic CH_4 and N_2O emissions, calculated as CO_2-eq, from stored slurry and other biomasses in the reference (Ref) and biogas scenarios (Biogas). The scenarios were (S1) slurry and deep litter; (S2) slurry and straw; (S3) slurry, deep litter, and maize silage; (S4) slurry, deep litter, and organic waste; and (S5) slurry, deep litter, organic waste, and organic grass–clover.

3.3. Ammonia Emission

The digestion of animal slurry, crops, and biowaste increases emission of NH_3 because of the higher emission potential of digestate due to its higher pH and TAN to N ratio Consequently, AD increased NH_3 emission in all five scenarios. The highest increases were in scenarios S3 and S5, in which slurry was codigested with maize silage and grass–clover which increased the N content in the digestate. This was, in the scenarios with deep litter, to some extent counteracted by reduced emissions from the solid manure management, as emissions were lower from the fraction of digested solid manure than from the solid manure managed in the reference scenarios (Table 5).

Table 5. Changes in NO_3^- leaching, NH_3 emission, and NO_x emission by introducing AD in manure management. In the model plants, pig and cattle slurry was codigested with (S1) deep litter, (S2) straw, (S3) deep litter and maize silage, (S4) deep litter and organic waste, and (S5) deep litter, organic waste, and organic grass–clover.

Source	Scenarios				
	S1	S2	S3	S4	S5
NO_3^- (kg NO_3^--N ton^{-1} (biomass))	−0.19	−0.13	−0.04	−0.18	−0.45
NH_3 (kg NH_3-N ton^{-1} (biomass))	0.19	0.18	0.21	0.14	0.30
NO_x (g NOx ton^{-1} (biomass))	2.49	2.48	2.30	3.97	2.13

3.4. Crop Production and Nitrate Leaching

The change in NO_3^- leaching by implementing AD was estimated for a mixture of the most common crops in Denmark and calculated with and without the inclusion of NH_3 emission after digestion. The increase in NH_3 emission reduced NO_3^- leaching by only 0.02–0.04 kg NO_3^--N ton^{-1} (biomass) (data not shown). In the reference scenario, slaughterhouse waste was assumed to be applied to crops as untreated biofertilizer hygienized and mixed with slurry. The source-separated organic household waste would in the reference system be composted and applied to a spring-sown crop as an alternative to digestion. The reduction in NO_3^- leaching by AD was 0.04–0.45 kg NO_3^--N ton^{-1} (biomass), with the lowest reduction in the scenario with an energy crop (S3) and the largest in the organic system with digestion of a grass–clover green manure crop that would alternatively be cut and mulched in the field (S5). The reduction in NO_3^- leaching was lowest in S3 because the overall input of N increased in this scenario. We did not include the effect of a crop change to maize in the leaching effect because of the uncertainty of which crop was being replaced by maize, although it was expected to be mainly cereals. Nitrate leaching is higher from maize than from most other crops [44], and if this effect had been accounted for, or if a higher proportion of maize had been applied in S3, then an increase in NO_3^- leaching would be expected by AD [47].

3.5. Soil Carbon Storage

For the scenario including straw as a cosubstrate, the alternative was assumed to be incorporation of straw into the soil. For one ton of straw with a dry matter content of 85%, and 45% C in the dry matter, this corresponded to reduced soil C storage of 7.7 kg C, equal to 28.1 kg CO_2-eq. It was assumed that substituting cereals for silage maize would not affect soil C storage [48] and that changes from mulching grass–clover to a cutting regime with return of the digestate would also have little effect [49].

3.6. GHG Emissions

Introducing AD in the five different scenarios reduced the net GHG emissions of livestock farming by 67–111 kg CO_2-eq ton^{-1} of biomass at 60 d HRT (Table 6). The corresponding effect in terms of DM was 479–613 kg CO_2-eq ton^{-1} DM. Most of this reduction was ascribed to substitution of fossil energy in energy production and avoided CH_4 emission during storage of biomass. The difference in biogas production in the five scenarios was due to the different amounts of dry matter and biogas potentials in the substrates supplied to the plants.

The biogas plants in scenarios S2 and S5 were supplied with biomasses with high dry matter concentrations, which contributed to a high amount of energy produced per ton of biomass and thus a considerable GHG reduction. If the comparison were to be made independently of these differences in dry matter content, the assessment should be based on the GHG effect per GJ or dry matter input. For example, scenario S2 with straw had a significantly higher GHG effect per ton of biomass than S1 with deep litter. The reason is that considerably more dry matter was supplied in the straw scenario. If the same amount of dry matter had been supplied in the straw scenario as in the deep litter scenario, then the straw scenario would have had a lower GHG effect. This was reflected in the lower GHG effect per GJ of straw than of deep litter. The model plant with the greatest GHG reduction per ton of biomass was the scenario in which 20% straw was added (S2). However, this was also the scenario with the lowest GHG effect in terms of dry matter.

Table 6. Calculated GHG emissions per ton of biomass and per kg of dry matter (DM) (values in brackets) for five model plants at 60 d HRT and 1% methane leakage from the biogas plant. The model plants were (S1) slurry and deep litter; (S2) slurry and straw; (S3) slurry, deep litter, and maize silage; (S4) slurry, deep litter, and organic waste; and (S5): slurry, deep litter, organic waste, and organic grass–clover. Positive values indicate lower emissions, and negative values indicate higher emissions, from biogas.

Source	Unit	S1	S2	S3	S4	S5
Energy	kg CO_2-eq ton^{-1} biomass or kg CO_2-eq ton^{-1} DM	50.44 (450.4)	117.05 (532.1)	57.84 (507.4)	71.83 (509.4)	105.30 (534.5)
Glycerol for heating					−13.80 (−97.9)	
Process energy		−4.08 (−36.4)	−4.08 (−18.5)	−4.08 (−35.8)	−4.08 (−28.9)	−4.08 (−20.9)
Transport		−1.21 (−10.8)	−1.20 (−5.5)	−1.15 (−10.1)	−1.62 (−11.5)	−1.20 (−6.1)
Fertilizer production, N		1.61 (14.4)	1.32 (6.0)	1.43 (12.5)	1.43 (10.1)	1.77 (9.0)
Methane leakage from biogas plant		−4.50 (−40.1)	−10.3 (−46.9)	−5.42 (−47.5)	−6.34 (−44.9)	−9.29 (−47.2)
Methane from storage *		29.91 (267.1)	15.75 (71.6)	24.50 (214.9)	21.04 (149.2)	11.54 (58.6)
Nitrous oxide from storage *		0.00	−1.26 (−5.7)	0.00	1.32 (9.3)	1.32 (6.7)
Nitrous oxide after application		0.00	0.00	0.00	0.00	0.00
Nitrous oxide from nitrogen leaching		0.40 (3.6)	0.27 (1.3)	0.04 (0.8)	0.40 (2.8)	1.01 (4.9)
Nitrous oxide from ammonia emission		−0.69 (−6.2)	−0.66 (−3.0)	−0.76 (−6.6)	−0.51 (−3.6)	−1.11 (5.63)
Nitrous oxide from maize cropping		0.00	0.00	−0.74 (−12.1)	0.00	0.00
Soil C storage (digested biomass)		−3.14 (−28.0)	−6.16 (−28.0)	−2.12 (−18.6)	−2.11 (−15.0)	−2.64 (13.4)
Total impact		68.8 (613)	110.7 (503)	69.6 (604)	67.6 (479)	102.6 (520)
Energy production	GJ gross energy ton^{-1} biomass	0.90	2.07	1.02	1.27	1.86
Total impact	kg CO_2-eq GJ^{-1} gross energy	76.47	53.61	68.19	53.29	55.21
Nitrate leaching	kg NO_3-N ton^{-1} biomass	0.19 (1.7)	0.13 (0.6)	0.04 (0.4)	0.18 (1.3)	0.45 (2.3)
NH_3	kg NH_3-N ton^{-1} biomass	−0.19 (−1.7)	−0.18 (−0.8)	−0.21 (−1.8)	−0.14 (1.0)	−0.30 (1.5)
NO_x	g NO_x ton^{-1} biomass	−2.49 (22.2)	−2.48 (11.3)	−2.30 (20.2)	−3.97 (28.1)	−2.13 (10.8)

* Methane and N_2O from storage relate to emissions from storage of biomasses, especially slurry, deep litter, and slaughterhouse waste.

4. Discussion

The analysis presented here showed that anaerobic codigestion of slurry with biowaste, crop residues, and crops primarily reduced GHG emissions by substituting fossil fuel for power and heat production and reducing CH_4 emission during postdigestion storage. The main environmental benefits from biogas energy systems compared to fossil fuel energy systems occurred in terms of reduced GHG emissions and reduced resource consumption [50]. The impact of utilizing animal manure for biogas production is important in this respect, since avoided emissions from the reference system of conventional manure management could be credited to the biogas system [50], which was not the case for energy crops and straw that would not be sources of GHGs in a non-AD scenario.

In our study, AD was assumed not to affect N_2O emissions, in contrast to calculations presented in a previous study by Sommer et al. [9], in which the removal of degradable organic matter during AD was assumed to reduce N_2O emissions from digestate applied to soil by reducing the potential for denitrification. However, experimental results on this aspect have conflicted [37], and recent studies have made it clear that the amount and composition of denitrification products depend on complex interactions between digestate and soil properties. Thus, accounting for the effect of AD on N_2O emissions would probably require considering the composition of residual VS in digestates, as modified by codigestates and soil gas exchange controlling the exchange of oxygen and denitrification products [51]. The consideration of specific site and weather conditions was beyond the scope of this study, but this should be investigated further.

In our study, the total impact on GHG was 67–111 kg CO_2-eq ton^{-1} of biomass at 60 d HRT, which demonstrates that the biomass mix played an important role. In a study by Poeschl et al. [52] the GHG impact was 75 kg CO_2-eq ton^{-1} for a small-scale plant and 120 kg CO_2-eq ton^{-1} for a large-scale plant. In this study, liquid manure accounted for 55% in the small-scale plant, while in the large-scale plant, only wastes from industry and household were included [53]. Including corn silage had a high positive GHG impact, while grass silage had a negative GHG impact. This was in contrast to our study, in which the scenario with maize had the lowest GHG impact while the scenario with glass–clover had significantly higher impact. However, in a study by Hijazi et al. [50], nonleguminous perennial grass was used. In organic farming, leguminous perennial grass is used as a source for N fertilizer because of its ability to fix atmospheric N [53]. This type of grass might be better than nonleguminous perennial grass in terms of the savings of direct N_2O emission from N input in the form of mineral or organic fertilizers [53].

4.1. Biomass Sources for Biogas

The GHG impacts calculated in our study were based on a Danish territorial perspective; in general, only impacts on Danish national GHG emissions were included. However, the production of commercial N fertilizer was taken into account with a minor climate impact corresponding to 0.28 kg CO_2-eq per kg biomass N, or about 1.5 kg CO_2-eq per ton of biomass. The fertilizer replacement value of biomass is increased after digestion. However, Danish legislation with quotas on mineral N fertilizer application does not account for this, and it is uncertain whether farmers take the higher N availability into account. Therefore, this effect is uncertain in practice but has potential to be utilized. Since there is no fertilizer production in Denmark, such emissions are not included in Denmark's national GHG inventory and therefore were not considered. Neither were the possible effects of changed land use elsewhere on the planet (iLUC) considered. This was relevant only for scenario S2, in which the cultivation of maize as an energy crop was set to replace the production of cereals and could, thus, potentially have iLUC impacts. The iLUC impacts are uncertain, but they may potentially exceed the calculated positive climate impacts of biogas production [53]. In the inventory of biofuels under the EU's Renewable Energy Directive (RED), iLUC impacts are included, although they are not included in the EU requirements for compliance with the RED II [54]. For maize for biogas production, the iLUC impact in a RED context was most recently calculated as 21 kg CO_2-eq MJ^{-1} [55].

Besides livestock manure, the model plants used different types of biomasses from the agricultural sector. Addition of 20% straw to the slurry (S2) gave the largest GHG reduction per ton, but the smallest reduction in terms of DM. The Danish biogas sector is considered a cornerstone in Danish green energy production, and there is increasing demand for more biomass for codigestion with slurry. While there is plenty of unused straw available as cosubstrate, the amount of straw added in scenario S2 cannot be managed with existing biogas technology because of problems related to pumping and agitating the biomass. However, the Danish biogas industry has projected that future technologies in form of pretreatment, pumps, and agitators will be able to manage this amount of dry matter. The calculations in this report assumed that the alternative to the use of straw for biogas was incorporation in the field. If the alternative had been incineration for combined heat and power, the climate impact would have been considerably lower. On the other hand, biogas from straw has the advantage that plant nutrients and part of the slowly degradable C in the straw is returned to the field.

Substituting some of the deep litter (S1) with maize silage (S3) improved the GHG emission reduction per ton slightly. The high degradability of the organic matter in maize contributes to high biogas production per ton of biomass, and energy production was therefore higher for S3 than for S1. In a study from 2013 [14], the GHG emission by codigestion of slurry and 10% maize was reduced by 72 kg CO_2-eq ton^{-1} (wet basis), which was about 10 kg CO_2-eq ton^{-1} (wet basis) more than in our study. In the mentioned study [14], the fibre fraction from separation of slurry was included in the calculation, and it was assumed that it had a higher CH_4 emission potential during storage than deep litter.

Glycerol can be used for energy production in power plants and Otto engines [56], and it is also a useful raw material for biogas production. When glycerol is used in conventional heat and power (CHP) production, the CO_2 reduction effect is 690 kg CO_2 ton^{-1} when it substitutes natural gas, and when producing biogas substituting natural gas, it reduces GHG emissions with an equivalent of 558 kg CO_2 per ton. When compared to incineration in power plants, biogas is considered a high-value energy carrier that can be stored and converted to electricity.

Among the biogas plants, the GHG effect was lowest in the plant with deep litter (S1) since CH_4 emissions during storage increased because of the higher CH_4 emission from digestate than from heaps of deep litter. In S4, with added glycerol, it was assumed that the glycerol would otherwise have been used efficiently for heat and power production. This assumption has not previously been used and is one of the reasons why the GHG reduction potential of an "industrial waste plant" was lower than in the study by Nielsen et al. [57], in which the total GHG effect was a reduction at 90 kg CO_2-eq ton^{-1}.

The organic farming biogas plant (S5) had, next to the deep litter and straw model plant (S2), the largest GHG reduction potential per ton (102 CO_2-eq ton^{-1}), which may be attributed to the high biogas yield as a result of the high proportion of grass, deep litter and biowaste. In a previous analysis from 2013, the GHG reduction calculated for the introduction of AD in an organic farming system was 83 kg CO_2-eq ton^{-1} [15], but lower effects of both energy production and CH_4 emission during storage was assumed. In our study, the effect of substituting fossil fuels contributed more to the overall GHG impact than in previous studies, in which the importance of reducing CH_4 emissions was greater and a reduction in N_2O emissions was included in the calculations [57].

Energy crops, such as maize, are still a significant source of substrates used by Danish biogas plants, but the amount that can be used is constrained by restrictions under subsidy schemes. Compared with cereal crops such as wheat, there are only limited negative effects of growing maize and other energy crops. However, grass and sugar beets have better environmental and GHG profiles for biogas production than maize [58], and in the future, cover crops are also expected to be used for biogas production. This may reduce the N_2O emissions currently seen after the incorporation of cover crops [59] while at the same time maintaining and possibly improving soil C storage potential. Cover crops and straw together provide a promising source of biogas while at the same time increasing the recycling of nutrients in crop production [60]. Utilization of the expanding area with cover crops as a source of biomass for biogas production would not have the negative iLUC effects that are associated with the cultivation of energy crops for biogas.

4.2. Biogas Plant Configuration

Increasing the reactor size and hydraulic retention time (HRT) reduced net GHG emissions via an increase in the production of biogas and a reduction in the amount of degradable VS in digestate transferred for downstream storage, which will reduced CH_4 emissions from the digestate (Figure 7). The effect of increasing HRT was related to the degradability of the organic matter in the biomass used, and the highest effect was calculated when adding straw or deep litter, which feature high concentrations of slowly degradable biomass. There were increases in the GHG reduction potential by increasing HRT from 45 to 60 days in all scenarios, whereas the effect of extending HRT further was positive only for scenarios S1 to S4, because the positive effect in S5 was outweighed by greater consumption of process energy.

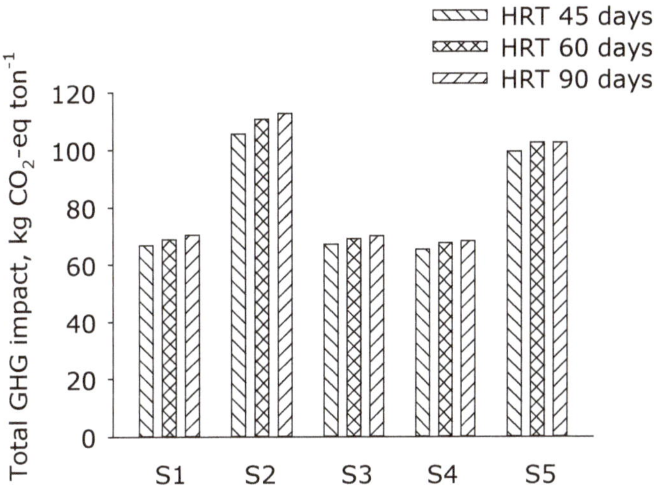

Figure 7. Effect of hydraulic retention time in AD reactor on total GHG impact.

Methane leakage may have a significant effect on total climate impact, which would mainly be due to the global warming potential of CH_4 and, to a lesser extent, the unrealized energy production (Figure 8). The total climate impact was almost linearly reduced with increasing CH_4 leakage (Figure 8). The total impact was reduced by about 5 kg CO_2-eq per ton of biomass from scenario S1 at a methane leakage of 1% to about 10 kg CO_2-eq ton^{-1} of biomass at a methane leakage of 2%. This means that about 7% of the positive climate impact of the plant was lost for each percentage point of leakage. For a leakage of 15%, biogas no longer had a positive effect for scenario S1. Release and leakage of CH_4 from small, unheated digesters may, in a scenario in which biogas energy substitutes coal, negate the GHG-reducing effect of biogas production if 40% of the biogas produced is emitted to the atmosphere [61].

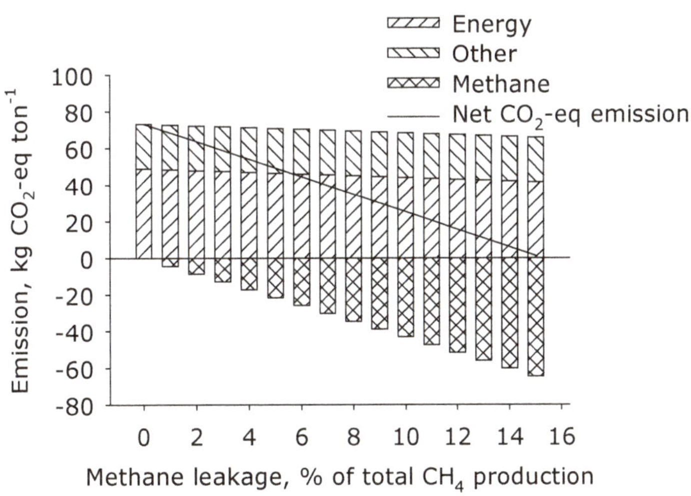

Figure 8. Effect of methane leakage on the climate impact for model plant (S1) with 45 days HRT. Methane and nitrous oxide emissions shown are from storage of livestock manure and biogas slurry.

Nielsen et al. [57] assumed that the organic waste used as codigestate would be stored under anaerobic conditions in the reference scenario and therefore be a significant source of CH_4 emissions. Compared with [57], the biogas scenarios in the present study used a greater diversity of biomasses as codigestates. The relevant alternative management of these biomasses was incorporation (straw), ensiling (maize), storage in a heap (deep litter), or composting (biowaste), and only the slaughterhouse waste included in scenario S4 was assumed to be stored under anaerobic conditions if not digested. To the extent that codigestates were not stored anaerobically in the reference scenarios, this reduced the CH_4 mitigation potential of biogas treatment compared with that in the previous analyses.

4.3. Nitrogen Losses

The reduction in NO_3^- leaching from the AD scenarios was lowest in the system with energy crops (S3), because the overall N input increased in this system with the introduction of more organic N from plant material. If 12% maize silage was used, the effect of digestion on leaching was close to zero, in accordance with previous estimates by Sørensen and Børgesen [23]. Nitrate leaching is typically greater for maize cultivation than for most other crops [44], and if this effect could be included, or if a higher proportion of energy crops were applied in S3, then an increase in NO_3^- leaching would be expected with AD compared with the corresponding reference scenario. However, since the effect is much affected by the assumption about which crop would be replaced by maize, we did not include the effect of a crop change to maize. On organic farms, AD of plant biomass from green manure crops, as in S5, increased the average plant availability of N [60]. This meant that less organic N was left in the soil to contribute to leaching by mineralization in the following years, and the effect of AD in such a system was estimated to reduce NO_3^- leaching by 0.45 kg N ton^{-1} biomass.

The NH_3 emission potential of digestate applied in the field was higher than that of cattle slurry and pig slurry because of the higher pH of the digestate. This effect, and the higher content of TAN in digestate, contributed to higher emission from the AD systems. The effect of higher TAN concentration contributed 60–70% of the increase in NH_3 emission from AD systems, and most of this increase was due to higher emission from digestate applied to soil. The relatively small increase in NH_3 emission in scenario S3 was due to the emissions from storage and application of deep litter and biowaste being large in the reference system. Ammonia emission in the organic scenario, S5, was high because of the increase in the TAN content of digestate originating from N in the codigested grass–clover and the low NH_3 emission from the reference system. This increase in NH_3 can be avoided if organic farms inject slurry into the soil, which is feasible because of the high share of spring crops in organic crop rotations for which slurry can be injected prior to seeding or on grassland during the growing season.

In Denmark, slurry acidification is an alternative to incorporation and injection, but this technology is not suitable for digestates, because the high pH buffer capacity of the digestate results in high demand for acid and thus high cost for this treatment; furthermore, acidification with sulfuric acid is not allowed in organic farming.

4.4. Uncertainties

The calculations were based on the current knowledge about energy production in the form of biogas from different types of biomasses and their related environmental and GHG impacts. Biogas technology is constantly developing, and so is the alternative use of biomasses. This leads to uncertainty with respect to the representativeness of the biogas plants defined and their composition of biomasses. However, the model plants analysed in this study represent the types of biomasses currently used for biogas production in Denmark.

Other uncertainties are associated with the way in which impacts were quantified. The study was based on the models and data used in Denmark's inventories of environmental and GHG impacts. These models are constantly being developed, particularly to better account for variation in environmental controls of the biological processes that determine the impacts and in the properties of biomasses and how they are managed in practice. With the current knowledge, it is not possible to quantify those uncertainties, but a qualitative discussion follows.

The gas potentials of the different biomasses, and the rate at which the gas was produced, are significant sources of uncertainty, especially in the assessment of the effect of retention time in the biogas reactor. The degradation profiles used were, therefore, essential for the estimates of gas yield with different retention times and for the assessment of residual VS in the digestates. The degradation profiles further influenced the CH_4 emissions during the subsequent storage. There is a considerable need for documentation of the rate at which biogas is produced and identification of any interactions between biomasses that may give rise to synergies and/or antagonism. Moreover, it is necessary to provide better documentation of the correlation of CH_4 production between batch tests and continuous systems. These sources of uncertainty may have affected the reported differences among the effects of 45, 60, and 90 d retention time.

Our study assumed that electricity for the process energy used in biogas production was covered by a mix of Danish electricity production, which was estimated to be 0.150 g CO_2 kWh^{-1} in 2019. It may be argued that such emissions could have been 0 g kWh^{-1} if only renewable energy had been used. If it were assumed that no CO_2 was emitted from the production of process energy, the total positive climate impact would increase by about 0.97 kg CO_2 ton^{-1} of biomass at 45 d retention time, increasing the total climate effect by a maximum of 1.7%. Hence, the emission factor assumed for electricity for process energy was less important.

The estimation of CH_4 emissions and the related degradation of organic matter during storage of biomasses was based on simplified input data and assumptions. Firstly, CH_4 emissions were assumed to be a product of VS and temperature alone, but this does not always well explain temporal dynamics [62], and the composition and growth of methanogenic communities would probably be part of an improved model [63]. Furthermore, CH_4 emissions were calculated from total VS and not degradable VS because of a lack of relevant experimental data regarding VS composition. In the analysis presented here, increasing the retention time in biogas reactors from 45 to 90 days showed only a limited effect on CH_4 emissions during subsequent storage, with a 15% reduction in posttreatment emissions from cattle slurry being the most significant effect. A possible reason for the limited sensitivity to HRT could be that CH_4 emissions were estimated on the based on total VS and using the same lnA' value for digestate regardless of retention time. This parameter also represents the slowly degradable parts of VS for which anaerobic degradation was small whether HRT was 45, 60, or 90 days. It is likely that a model in which the VS degradability of each biomass was defined would better capture differences among digestates from the five scenarios with very different feedstocks.

Biogas treatment of pig and cattle slurry (and codigestates) reduces the availability of easily degradable organic matter. All other factors being equal, the resulting decline in demand for oxygen to degrade residual organic matter after field application should reduce the extent and lifetime of anoxic conditions with a potential for N_2O production in well-drained soil. There are, however, confounding effects. For example, soil compaction and periods with rainfall reduce soil oxygen status and change the conditions for N_2O production in untreated manure and digestates applied to soil [36]. Furthermore, fibre-rich codigestates such as maize silage or deep litter may change the distribution of degradable VS in the soil of digestates compared with that of untreated manure [64]. A systematic investigation is needed to elucidate interactions among manure and digestate properties, soil conditions, and N_2O emissions before the effect of anaerobic digestion on N_2O emissions can be included in biogas assessments.

The calculations included an assumption that digestion of slurry and biomasses reduced soil C storage compared with direct field application of those biomasses. Only very limited documentation exists of this effect [46], which is very difficult to determine experimentally. Therefore, the effect is also uncertain, and further studies are needed.

In our calculations, the higher content of TAN in the digestate caused a significant increase in NH_3 emissions in the scenario calculations, corresponding to 60–70% of the higher NH_3 emission. However, there is great uncertainty in these estimates due to a lack of knowledge about how the different combinations of substrates and slurry in the biogas scenarios would affect digestate physical characteristics and infiltration in the soil. Ammonia emissions from slurry or digestates are reduced with faster infiltration, which is a function of viscosity, dry matter content, and adhesiveness. It is poorly understood how infiltration of digestate is affected by the substrates used for AD.

Regardless the types of biomasses used for biogas production, the following measures are important to achieve the potential environmental and climate benefits:

- CH_4 leaks from the biogas installation should be minimized.
- Digestate storage should be covered, and low-NH_3-emission technology should be used for field application.
- Heat exchangers should be employed to cool down the digestate to ambient temperature before storage to improve the energy balance and reduce GHG and NH_3 emissions.

5. Conclusions

Environmental and climate assessments were conducted for different biogas scenarios to evaluate the sustainability of this treatment technology and identify potential improvements of environmental and climate impacts. The scenarios were analysed considering (i) biomass composition; (ii) process temperatures; (iii) hydraulic retention time; (iv) methane leakage from biogas installations; and (v) digestate storage and field application. With respect to energy production, only upgrading for the natural gas grid and substitution of natural gas with biogas were considered.

On the basis of this study, the following conclusions were drawn:

(1) The scenarios investigated resulted in GHG mitigation ranging from 65 to 105 kg CO_2-eq ton^{-1} biomass. Reductions per ton of biomass were greatest when straw or grass–clover was used for codigestion, whereas reductions per unit energy produced were highest with deep litter and deep litter plus maize silage.

(2) The ammonia emission potential of digestate applied in the field was higher than that from untreated cattle and pig slurry because of digestates' higher pH, resulting in an increase in ammonia emission of 0.14 to 0.3 kg NH_3-N ton $^{-1}$ biomass. The use of low-emissions application technology for a larger share of the digestate should limit these higher emissions.

(3) All scenarios reduced nitrate leaching (0.04 to 0.45 kg NO_3-N ton^{-1} biomass). However, introducing maize silage almost eliminated this reduction.

(4) Increasing the hydraulic retention times led to higher climate impact via increased energy production and lower amounts of volatile solids available for degradation and subsequent CH_4 emission during digestate storage.

(5) Methane leakages can have a significant effect on the total climate impact, with about 7% of the positive climate impact being lost for each percentage point of leakage in a manure-based biogas scenario.

(6) The methodology used predicted significant reductions in CH_4 emissions but assumed there was no reduction in direct emissions of N_2O from digestates, which is not always true. Furthermore, iLUC, which was ignored which for bioenergy use, may have a negative impact on the GHG balance.

These and other examples given above show the importance of the assumptions chosen for this type of analysis. Still, it was concluded that biogas treatment of livestock slurry and biowastes has the potential to reduce GHG emissions, improve N use efficiency and reduce nitrate leaching losses. However, the risks of higher ammonia emission and CH_4 leakage during AD need to be managed.

Supplementary Materials: The following supporting information can be downloaded at: https://www.mdpi.com/article/10.3390/su14031849/su14031849/s1, Table S1: Assumptions on dry matter content and gas potential of different biomasses used in biogas production. The ultimate CH_4 yield is the yield achieved at a retention time of more than 90 days. The CH_4 yields after 45 and 60 days and ultimate gas yield are based on data from tests at the Foulum biogas plant Aarhus University. Sources; (1) Average of 50 analyses of slurry supplied to two biogas plants, (2) Olesen et al. (2018), (3) Data from Foulum biogas plant, Aarhus University and (4) Data from tests at Foulum biogas plant, Aarhus University; Table S2: Data used for the calculation of energy consumption on a standard Danish biogas plant; Table S3: Energy use and CO_2 emission due to transportation of biomass. CO_2 emissions of 2.7 kg per litre of diesel is assumed; Table S4: The methane production potential values, lnA', for digestate, cattle slurry and pig slurry were calculated based on information extracted from published studies about methane production rate, total VS and temperature. In the table, $\bar{x} \pm$ s.e. refers to average and standard errors, Table S5: Ammonia emission factors for stored liquid and solid manure (Hansen et al. 2008) and organic food waste (Pardo et al. 2015); Table S6: Ammonia emission factors for cattle and pig slurry applied to soil (Hansen et al. 2008), and the novel estimates emission from digestate; Table S7: Assumptions about plant available N (NH_4^+-N) in biomasses before and after biogas treatment during the first crop growing season after application of manure, required N use efficiency for manures and organic wastes (by Danish legislation), and calculated reduction in NO_3^- leaching due to AD of manure—not accounting for changed NH_3 loss. The share of NH_4^+-N in manure is based on Sørensen and Børgesen (2015). Organic N expected to be transformed to NH_4^+ within the first season is included in the NH_4^+-N share of total N; Figure S1: Average cumulative net N mineralisation from organic N applied in livestock manure over a 10-year period after application.

Author Contributions: Funding acquisition, H.B.M. and J.E.O.; methodology, H.B.M., P.S., J.E.O., S.O.P., T.N. and S.G.S.; project administration, J.E.O.; writing—original draft, J.E.O.; writing—review and editing, H.B.M., P.S., S.O.P., T.N. and S.G.S. All authors have read and agreed to the published version of the manuscript.

Funding: This report was prepared as part of the agreement between the Danish Energy Agency and Aarhus University.

Institutional Review Board Statement: Not applicable.

Informed Consent Statement: Not Applicable.

Conflicts of Interest: The authors declare no conflict of interest.

References

1. Owen, J.J.; Silver, W.L. Greenhouse gas emissions from dairy manure management: A review of field-based studies. *Glob. Chang. Biol.* **2014**, *21*, 550–565. [CrossRef] [PubMed]
2. Olesen, J.E.; Møller, H.B.; Petersen, S.O.; Sørensen, P.; Nyord, T.; Sommer, S.G. *Sustainable Biogas—Climate and Environmental Impacts of Biogas Production*; DCA report nr. 181; Aarhus Universitet, DCA—Danish Centre for Food and Agriculture: Aarhus, Denmark, 2020; p. 85. Available online: https://dcapub.au.dk/djfpdf/DCArapport181.pdf (accessed on 1 January 2021).
3. Venterea, R.T.; Halvorson, A.D.; Kitchen, N.; Liebig, M.A.; Cavigelli, M.A.; Del Grosso, S.J.; Motavalli, P.P.; Nelson, K.A.; Spokas, K.; Singh, B.P.; et al. Challenges and opportunities for mitigating nitrous oxide emissions from fertilized cropping systems. *Front. Ecol. Environ.* **2012**, *10*, 562–570. [CrossRef]
4. IPCC. *2006 IPCC Guidelines for National Greenhouse Gas In-ventories, Prepared by the National Greenhouse Gas Inventories Programme*; Eggleston, H.S., Buendia, L., Miwa, K., Ngara, T., Tanabe, K., Eds.; IGES: Yokohama, Japan, 2006; Available online: www.ipcc-nggip.iges.or.jp/support/Primer_2006GLs.pdf (accessed on 1 January 2010).
5. Yusuf, R.O.; Noor, Z.Z.; Abba, A.H.; Hassan, M.A.A.; Din, M.F.M. Methane emission by sectors: A comprehensive review of emission sources and mitigation methods. *Renew. Sustain. Energy Rev.* **2012**, *16*, 5059–5070. [CrossRef]
6. Anonymous. *Basisfremskrivning 2020—Danmarks Klima- og Energifremskrivning*; Energistyrelsen: København, Denmark, 2020. Available online: https://ens.dk/sites/ens.dk/files/Basisfremskrivning/basisfremskrivning_2020-webtilg.pdf (accessed on 14 March 2021).
7. EU Commission. 2030 Climate Target Plan. EU climate action and the European Green Deal. 2020. Available online: https://ec.europa.eu/clima/policies/eu-climate-action/2030_ctp (accessed on 14 March 2021).
8. Eriksen, J.; Thomsen, I.K.; Hoffmann, C.C.; Hasler, B.; Jacobsen, B.H. *Virkemidler Til Reduktion af Kvælstofbelastningen af Bandmiljøet*; DCA rapport nr 174; Aarhus Universitet: Aarhus, Denmark, 2020; p. 452.
9. Sommer, S.; Petersen, S.; Møller, H. Algorithms for calculating methane and nitrous oxide emissions from manure management. *Nutr. Cycl. Agroecosyst.* **2004**, *69*, 143–154. [CrossRef]

10. Olesen, J.E.; Schelde, K.; Weiske, A.; Weisbjerg, M.; Asman, W.; Djurhuus, J. Modelling greenhouse gas emissions from European conventional and organic dairy farms. *Agric. Ecosyst. Environ.* **2006**, *112*, 207–220. [CrossRef]
11. Sommer, S.G.; Olesen, J.E.; Petersen, S.O.; Weisbjerg, M.R.; Valli, L.; Rodhe, L.; Béline, F. Region-specific assessment of greenhouse gas mitigation with different manure management strategies in four agroecological zones. *Glob. Chang. Biol.* **2009**, *15*, 2825–2837. [CrossRef]
12. Petersen, S.; Højberg, O.; Poulsen, M.; Schwab, C.; Eriksen, J. Methanogenic community changes, and emissions of methane and other gases, during storage of acidified and untreated pig slurry. *J. Appl. Microbiol.* **2014**, *117*, 160–172. [CrossRef]
13. Baral, K.R.; Arthur, E.; Olesen, J.E.; Petersen, S.O. Predicting nitrous oxide emissions from manure properties and soil moisture: An incubation experiment. *Soil Biol. Biochem.* **2016**, *97*, 112–120. [CrossRef]
14. Jacobsen, B.H.; Laugesen, F.M.; Dubgaard, A.; Bojesen, M. *Biogas Production in Denmark—Assessments of Operation and Socioeconomics (Biogasproduktion i Danmark—Vurderinger af Drifts og Samfundsøkonomi)*; IFRO Report 220; University of Copenhagen: Copenhagen, Denmark, 2013.
15. Møller, F.; Martinsen, L. *Socioeconomic Evaluation of Selected Biogas Technologies*; Scientific Report from DCE no. 62; University of Aarhus: Aarhus, Denmark, 2013.
16. Mikkelsen, M.H.; Albrektsen, R.; Gyldenkærne, S. *Consequences of Biogas Production for Greenhouse Gas Emissions in the Agricultural Sector (Biogasproduktions Konsekvenser for Drivhusgasudledning i Landbruget)*; Scientific Report from DCE—Danish Centre for Environment and Energy no. 197; Aarhus University: Aarhus, Denmark, 2016; p. 41.
17. Chadwick, D.; Sommer, S.G.; Thorman, R.; Fangueiro, D.; Cardenas, L.; Amon, B.; Misselbrook, T. Manure management: Implications for greenhouse gas emissions. *Anim. Feed Sci. Technol.* **2011**, *166–167*, 514–531. [CrossRef]
18. Nielsen, O.K.; Plejdrup, M.S.; Winther, M.; Nielsen, M.; Gyldenkærne, S.; Mikkelsen, M.H.; Albrektsen, R.; Thomsen, M.; Hjelgaard, K.; Fauser, P.; et al. *Denmark's National Inventory Report 2019. Emission Inventories 1990–2017—Submitted under the United Nations Framework Convention on Climate Change and the Kyoto Protocol*; Scientific Report No. 318; Aarhus University, DCE—Danish Centre for Environment and Energy: Aarhus, Denmark, 2019; p. 886.
19. Sommer, S.G.; Knudsen, L. Impact of Danish livestock and manure management regulations on nitrogen pollution, crop production and economy. *Front. Sustain.* **2021**, *2*, 658231. [CrossRef]
20. Li, P.; Li, W.; Sun, M.; Xu, X.; Zhang, B.; Sun, Y. Evaluation of Biochemical Methane Potential and Kinetics on the Anaerobic Digestion of Vegetable Crop Residues. *Energies* **2018**, *12*, 26. [CrossRef]
21. Møller, H.B.; Nielsen, A.; Murto, M.; Christensson, K. *Manure and Energy Crops for Biogas Production*; Nordic Council of Ministers TemaNord: Copenhagen, Denmark, 2008.
22. Energinet.dk. 2019. Available online: https://energinet.dk/Om-nyheder/Nyheder/2020/01/16/Rekord-lav-CO2udledning-fradanskernes-elforbrug-i-2019 (accessed on 1 January 2021).
23. Sørensen, P.; Børgesen, C.D. *Nitrogen Leaching and Fertilising Effect from Digestate (Kvælstofudvaskning og GødningsVirkning ved Anvendelse af Afgasset Biomasse)*; DCA report no. 65; Aarhus University, DCA—Danish Centre for Food and Agriculture: Aarhus, Denmark, 2015; p. 46. Available online: https://pure.au.dk/portal/files/93094219/DCArapport065.pdf (accessed on 3 January 2022).
24. Chojnacka, K.; Kowalski, Z.; Kulczycka, J.; Dmytryk, A.; Górecki, H.; Ligas, B.; Gramza, M. Carbon footprint of fertilizer technologies. *J. Environ. Manag.* **2018**, *231*, 962–967. [CrossRef] [PubMed]
25. Elsgaard, L.; Olsen, A.B.; Petersen, S.O. Temperature response of methane production in liquid manures and co-digestates. *Sci. Total Environ.* **2016**, *539*, 78–84. [CrossRef]
26. Petersen, S.O.; Olsen, A.B.; Elsgaard, L.; Triolo, J.M.; Sommer, S.G. Estimation of Methane Emissions from Slurry Pits below Pig and Cattle Confinements. *PLoS ONE* **2016**, *11*, e0160968. [CrossRef] [PubMed]
27. Maldaner, L.; Wagner-Riddle, C.; VanderZaag, A.C.; Gordon, R.; Duke, C. Methane emissions from storage of digestate at a dairy manure biogas facility. *Agric. For. Meteorol.* **2018**, *258*, 96–107. [CrossRef]
28. Husted, S. Seasonal Variation in Methane Emission from Stored Slurry and Solid Manures. *J. Environ. Qual.* **1994**, *23*, 585–592. [CrossRef]
29. Sharpe, R.; Harper, L.; Byers, F. Methane emissions from swine lagoons in Southeastern US. *Agric. Ecosyst. Environ.* **2002**, *90*, 17–24. [CrossRef]
30. Clemens, J.; Trimborn, M.; Weiland, P.; Amon, B. Mitigation of greenhouse gas emissions by anaerobic digestion of cattle slurry. *Agric. Ecosyst. Environ.* **2006**, *112*, 171–177. [CrossRef]
31. Sherlock, R.R.; Sommer, S.G.; Khan, R.Z.; Wood, C.W.; Guertal, E.A.; Freney, J.R.; Dawson, C.O.; Cameron, K.C. Ammonia, Methane, and Nitrous Oxide Emission from Pig Slurry Applied to a Pasture in New Zealand. *J. Environ. Qual.* **2002**, *31*, 1491–1501. [CrossRef]
32. Pardo, G.; Moral, R.; Aguilera, E.; del Prado, A. Gaseous emissions from management of solid waste: A systematic review. *Glob. Chang. Biol.* **2014**, *21*, 1313–1327. [CrossRef]
33. Chadwick, D. Emissions of ammonia, nitrous oxide and methane from cattle manure heaps: Effect of compaction and covering. *Atmos. Environ.* **2005**, *39*, 787–799. [CrossRef]
34. Hansen, M.N.; Henriksen, K.; Sommer, S.G. Observations of production and emission of greenhouse gases and ammonia during storage of solids separated from pig slurry: Effects of covering. *Atmos. Environ.* **2006**, *40*, 4172–4181. [CrossRef]

35. Petersen, S.O.; Dorno, N.; Lindholst, S.; Feilberg, A.; Eriksen, J. Emissions of CH4, N2O, NH3 and odorants from pig slurry during winter and summer storage. *Nutr. Cycl. Agroecosyst.* **2013**, *95*, 103–113. [CrossRef]
36. Thomsen, I.K.; Pedersen, A.R.; Nyord, T.; Petersen, S.O. Effects of slurry pre-treatment and application technique on short-term N2O emissions as determined by a new non-linear approach. *Agric. Ecosyst. Environ.* **2010**, *136*, 227–235. [CrossRef]
37. Möller, K. Effects of anaerobic digestion on soil carbon and nitrogen turnover, N emissions, and soil biological activity. A review. *Agron. Sustain. Dev.* **2015**, *35*, 1021–1041. [CrossRef]
38. Hansen, M.N.; Sommer, S.G.; Hutchings, N.J.; Sørensen, P. *Emission Factors for Calculating Ammonia Evaporation from Storage and Application of Livestock Manure (Emissionsfaktorer til Beregning af Ammoniakfordampning ved Lagring og Udbringning af Husdyrgødning)*; DJF Husdyrbrug no. 84; Ministry of Agriculture: Aarhus, Denmark, 2008; p. 43. Available online: https://pure.au.dk/portal/files/2424282/djfhus84.pdf (accessed on 1 June 2019).
39. Kupper, T.; Häni, C.; Neftel, A.; Kincaid, C.; Bühler, M.; Amon, B.; VanderZaag, A. Ammonia and greenhouse gas emissions from slurry storage—A review. *Agric. Ecosyst. Environ.* **2020**, *300*, 106963. [CrossRef]
40. Søgaard, H.; Sommer, S.; Hutchings, N.; Huijsmans, J.; Bussink, D.; Nicholson, F. Ammonia volatilization from field-applied animal slurry—The ALFAM model. *Atmos. Environ.* **2002**, *36*, 3309–3319. [CrossRef]
41. Sommer, S.G.; Webb, J.; Hutchings, N. New Emission Factors for Calculation of Ammonia Volatilization From European Livestock Manure Management Systems. *Front. Sustain. Food Syst.* **2019**, *3*, 101. [CrossRef]
42. De Notaris, C.; Rasmussen, J.; Sørensen, P.; Olesen, J.E. Nitrogen leaching: A crop rotation perspective on the effect of N surplus, field management and use of catch crops. *Agric. Ecosyst. Environ.* **2018**, *255*, 1–18. [CrossRef]
43. Pedersen, B.N.; Eriksen, J.; Christensen, B.T.; Sørensen, P. Fertilizer replacement value and leaching of nitrogen applied to spring barley in cattle deep litter: A 3-year lysimeter study. *Soil Tillage Res.* **2021**, *209*, 104954. [CrossRef]
44. Børgesen, C.D.; Sørensen, P.; Blicher-Mathiesen, G.; Kristensen, K.; Pullens, J.W.M.; Zhao, J.; Olesen, J.E. *NLES5: An Empirical Model for Predicting Nitrate Leaching from the Root Zone of Agricultural Land in Denmark*; DCA report No. 163; Aarhus University, DCA—Danish Centre for Food and Agriculture: Aarhus, Denmark, 2019; p. 116. Available online: http://web.agrsci.dk/djfpublikation/index.asp?action=show&id=1313 (accessed on 1 June 2020).
45. Blicher-Mathiesen, G.; Olesen, J.E.; Jung-Madsen, S. *Update of Baseline 2021 (Opdatering af Baseline 2021)*; Technical report from DCE no. 162; Aarhus University, DCA—Danish Centre for Food and Agriculture: Aarhus, Denmark, 2020; p. 135.
46. Thomsen, I.K.; Olesen, J.E.; Møller, H.; Sørensen, P.; Christensen, B.T. Carbon dynamics and retention in soil after anaerobic digestion of dairy cattle feed and faeces. *Soil Biol. Biochem.* **2013**, *58*, 82–87. [CrossRef]
47. Christensen, B.T. Kulstoflagring ved nedmuldning af halm og efterafgrøder. In *Jordbrug og klimaændringer—Samspil til Vandmiljøplaner*; DJF rapport Markbrug nr. 109. s; Olesen, J.E., Petersen, S.O., Gyldenkærne, S., Mikkelsen, M.H., Jacobsen, B.H., Vesterdal, L., Jørgensen, A.M.K., Christensen, B.T., Abildtrup, J., Heidmann, T., et al., Eds.; Ministry of Agriculture: Copenhagen, Denmark, 2004; pp. 157–166.
48. Taghizadeh-Toosi, A.; Olesen, J.E. Modelling soil organic carbon in Danish agricultural soils suggests low potential for future carbon sequestration. *Agric. Syst.* **2016**, *145*, 83–89. [CrossRef]
49. Hu, T.; Sørensen, P.; Olesen, J.E. Soil carbon varies between different organic and conventional management schemes in arable agriculture. *Eur. J. Agron.* **2018**, *94*, 79–88. [CrossRef]
50. Hijazi, O.; Munro, S.; Zerhusen, B.; Effenberger, M. Review of life cycle assessment for biogas production in Europe. *Renew. Sustain. Energy Rev.* **2016**, *54*, 1291–1300. [CrossRef]
51. Petersen, S.O.; Sommer, S.G. Ammonia and nitrous oxide interactions: Roles of manure organic matter management. *Anim. Feed Sci. Technol.* **2011**, *166–167*, 503–513. [CrossRef]
52. Poeschl, M.; Ward, S.; Owende, P. Environmental impacts of biogas deployment e Part II: Life cycle assessment of multiple production and utilization pathways. *J. Clean. Prod.* **2012**, *24*, 184–201. [CrossRef]
53. Searchinger, T.D.; Wirsenius, S.; Beringer, T.; Dumas, P. Assessing the efficiency of changes in land use for mitigating climate change. *Nature* **2018**, *564*, 249–253. [CrossRef]
54. European Commission. *Directive (EU) 2015/1513 of the European Parliament and of the Council of 9 September 2015 Amending Directive 98/70/EC Relating to the Quality of Petrol and Diesel Fuels and Amending Directive 2009/28/EC on the Promotion of the Use of Energy from Renewable Sources*; Law, European Commission: Brussels, Belgium, 15 September 2015.
55. Valin, H.; Peters, D.; van den Berg, M.; Frank, S.; Havlik, P.; Forsell, N.; Hamelinck, C.; Pirker, J.; Mosnier, A.; Balkovic, J.; et al. *The Land Use Changes the Impact of Biofuels Consumed in the EU. Quantification of Area and Greenhouse Gas Impacts*; European Commission Project: BIENL13120; Ecofys: Utrecht, The Netherlands, 2015; p. 242.
56. Bohon, M.D.; Metzger, B.A.; Linak, W.P.; King, C.J.; Roberts, W.L. Glycerol combustion and emissions. *Proc. Combust. Inst.* **2010**, *33*, 2717–2724. [CrossRef]
57. Nielsen, L.H.; Hjort-Gregersen, K.; Thygesen, P.; Christensen, J. *Socioeconomic Analyses of Central Biogas Plants—with Technical and Financial Background Analyses (Samfundsøkonomiske Analyser af Biogasfællesanlæg—Med Tekniske og Selskabsøkonomiske Baggrundsanalyser)*; Report No. 136; Institute of Food and Resource Economics: Copenhagen, Denmark, 2002.
58. Olesen, J.E.; Petersen, S.O.; Lund, P.; Jørgensen, U.; Kristensen, T.; Elsgaard, L.; Sørensen, P.; Lassen, J. *Measures to Reduce Greenhouse Gasses in the Agricultural Sector (Virkemidler til Reduktion af Klimagasser i Landbruget)*; DCA Report No. 130; DCA—Danish Centre for Food and Agriculture: Aarhus, Denmark, 2018; p. 119.

59. Li, X.; Petersen, S.O.; Sørensen, P.; Olesen, J.E. Effects of contrasting catch crops on nitrogen availability and nitrous oxide emissions in an organic cropping system. *Agric. Ecosyst. Environ.* **2015**, *199*, 382–393. [CrossRef]
60. Fontaine, D.; Feng, L.; Labouriau, R.; Møller, H.B.; Eriksen, J.; Sørensen, P. Nitrogen and Sulfur Availability in Digestates from Anaerobic Co-digestion of Cover Crops, Straw and Cattle Manure. *J. Soil Sci. Plant Nutr.* **2019**, *20*, 621–636. [CrossRef]
61. Bruun, S.; Jensen, L.S.; Vu, V.T.K.; Sommer, S.G. Small-scale household biogas digesters: An option for global warming mitigation or a potential climate bomb? *Renew. Sustain. Energy Rev.* **2014**, *33*, 736–741. [CrossRef]
62. Baral, K.R.; Jégo, G.; Amon, B.; Bol, R.; Chantigny, M.H.; Olesen, J.E.; Petersen, S.O. Greenhouse gas emissions during storage of manure and digestates: Key role of methane for prediction and mitigation. *Agric. Syst.* **2018**, *166*, 26–35. [CrossRef]
63. Dalby, F.R.; Hafner, S.D.; Petersen, S.O.; VanderZaag, A.C.; Habtewold, J.; Dunfield, K.; Chantigny, M.H.; Sommer, S.G. Understanding methane emission from stored animal manure: A review to guide model development. *J. Environ. Qual.* **2021** [CrossRef] [PubMed]
64. Petersen, S.O.; Nissen, H.H.; Lund, I.; Ambus, P. Redistribution of slurry in soil as influenced by organic matter content and injection method. *J. Environ. Qual.* **2003**, *32*, 2399–2409. [CrossRef] [PubMed]

sustainability

MDPI

Article

Can Renewable Energy and Export Help in Reducing Ecological Footprint of India? Empirical Evidence from Augmented ARDL Co-Integration and Dynamic ARDL Simulations

Soumen Rej [1,2,*], Barnali Nag [1] and Md. Emran Hossain [3,*]

1 Vinod Gupta School of Management, Indian Institute of Technology, Kharagpur 721302, India
2 School of Business, University of Petroleum and Energy Studies, Dehradun 248007, India
3 Department of Agricultural Finance and Banking, Bangladesh Agricultural University, Mymensingh 2202, Bangladesh
* Correspondence: soumen.rej@iitkgp.ac.in (S.R.); emranaerd@gmail.com (M.E.H.)

Abstract: The objective of this study is to investigate the impact of exports, renewable energy, and industrialization on the ecological footprint (EF) of India over the period spanning from 1970–2017 by employing the newly developed augmented ARDL (A-ARDL) co-integration approach and the novel dynamic ARDL (D-ARDL) technique. The empirical results demonstrate that exports and renewable energy consumption reduce the EF, while industrialization intensifies the EF. More precisely, a 1% increase in export (renewable energy consumption) reduces the EF by 0.05% (0.09%). In addition, the short-run elasticity of the GDP is found to be larger than the long-run elasticity indicating the possibility of the existence of the Environmental Kuznets Curve (EKC) of the EF for India. The study indicates that the income effect and increased policy focus on renewable energy usage can be expected to reduce India's per capita EF in the long run. Moreover, India's export sector has been traditionally less energy intensive, which reflects in our findings of export growth leading to a reduction in EF. Based on the empirical findings, this study recommends some policy insights that may assist India to effectively reduce its ecological footprint.

Keywords: ecological footprint; renewable energy; augmented ARDL; dynamic ARDL; frequency domain causality; export; India

Citation: Rej, S.; Nag, B.; Hossain, M.E. Can Renewable Energy and Export Help in Reducing Ecological Footprint of India? Empirical Evidence from Augmented ARDL Co-Integration and Dynamic ARDL Simulations. *Sustainability* **2022**, *14*, 15494. https://doi.org/10.3390/su142315494

Academic Editors: Nicola Raimo, Filippo Vitolla, Ornella Malandrino and Benedetta Esposito

Received: 20 October 2022
Accepted: 18 November 2022
Published: 22 November 2022

1. Introduction

In the contemporary period, controlling greenhouse gas emissions and mitigating climate change is the pivotal policy agenda around the globe and one of the pioneering goals under the United Nations Sustainable Development Goals (SDG-13) [1–3]. While global commitment is to reduce the world's average increase of temperature below 1.5 °C, India is not an exception, and is obligated toward net zero emissions by 2070. Moreover, India, the second most populated country of more than 1.2 billion has emerged as one of the most important markets in the world, with the third largest GDP in terms of purchasing power parity. According to World Bank estimates, India is also expected to be one of the fastest-growing economies, with a growth percentage of 6–7% in the coming decade. This enormous amount of growth, however, comes with immense natural resource costs. The natural resources available to a country take time to regenerate but can be used up in no time, and hence, a country's growth is unlikely to be sustainable for future generations if the rate of its resource usage is much higher than the rate of the regeneration of those resources. Thus, the balance between resource usage and the ecological capacity to replenish them, i.e., biocapacity, is imperative for the sustainable development of any economy. The Global Footprint Network pioneered the process of systematic measurement and accounting of an ecological footprint in 2003, in order to trace anthropogenic usage of ecosystems in comparison to a renewal of those ecosystems. This is done every year to keep track of the economy's demand on the environment and to assess its sustainability.

The total demand of an economy from nature can be in the form of forests and fisheries and cropping and grazing land that is either converted into various kinds of infrastructure for people or can be used to absorb waste generated from various economic activities. The total of these two usages represents an economy's ecological footprint. If the economy's footprint exceeds its biocapacity, there is an "ecological overshoot", which indicates an unsustainable economic growth path [4]. As shown in Figure 1, the ecological deficit of India has been widening since 1980, when the deficit was around 56% of the biocapacity of the country. The gap increased to around 100% at the beginning of the new millennium and was more than 170% in 2018. It needs to be noted here that the consistent overshoot of ecological consumption in the 1990s coincided with economic liberalization and the opening up of the economy, and the distinct widening of the gap in the 2000s, as evident from Figure 1, coincided with the newfound economic prosperity of the country with the worldwide boom in the IT sector.

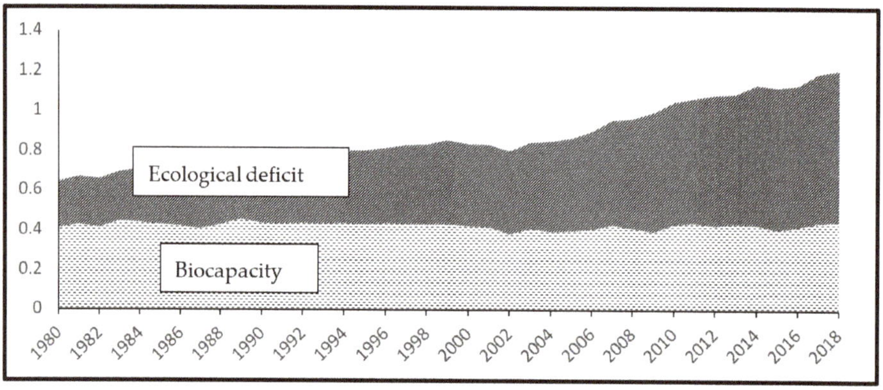

Figure 1. Per capita biocapacity and ecological deficit of the Indian economy from 1980–2018. Data source: Global Footprint Network, https://www.footprintnetwork.org/ (accessed on 10 September 2022).

Most of the climate literature in social sciences has, until date, focused primarily on the nexus between economic growth, in the form of GDP, and carbon emissions, caused primarily by the anthropogenic consumption of energy [5,6]. However, while carbon emissions are an important GHG, the issue of climate change is too complex to just account for anthropogenic carbon emissions in isolation. The dynamic change in the depletion of the natural resource space in totality vis a vis the economic growth that it is contributing to, the environment needs to be examined together to understand if an economy is moving towards sustainability or away from it. Economic growth in turn is also not uniform, and the various components of the GDP contribute in different ways to the various economic activities that impose their demand on natural resources. One such important component is exports, and this study attempts to find the impact of exports on the change in the ecological footprint (EF) of India. India's rapid economic growth, especially in the post-liberalization era has been accompanied by substantial growth in exports, with the share of exports in total GDP going up to 20.8% in 2021. The share of merchandise export in total GDP increased from 11.4% in 2017 to 13.3% in 2021–2022 [7].

It is well established in the literature that exports not only contribute to a direct increase in the GDP of a country but also contribute to the growth and development of a country through specialization, productivity increase, improved technology and R&D, economies of scale, the expansion of associated industries and service sections, and the employment and income generation that leads to a better standard of living [8,9]. However, this expansion in economic activity through production and consumption rise in a country cannot be fulfilled without drawing from the environment and using up natural resources, land area, and water bodies nor without the emission of pollutants that need land cover for adequate absorption. Moreover, the extent of the export sector's demand on ecology would be dependent on the nature of the export goods. Hence, in the context of the sustainability of export-led growth—or even growth leading to higher exports—it is imperative to understand its impact on the ecological footprint.

While there are some studies on the impact of trade and export diversification on the environment, as discussed in detail in the next section, there are not many studies that have explored the link between exports as part of GDP and the environment. The export sector, being the comparative advantage of an economy, can be assumed to be more efficient than most other sectors of the economy and is likely to attract more investments and eventually expand more. Hence, it is important to link exports, per se, with the EF of a country without considering imports like in variables like trade openness, which may not have any direct impact on the ecological balance of an economy.

It is also known that the industrialization of an economy raises the demand on the ecology through increased energy and land, water, and natural resource consumption as well as the higher production of pollutants. The higher penetration of renewable energy usage in both production and consumption sectors is likely to impact biodiversity usage also. India's commitment to integrating the UN Sustainable Development Goals into its economic growth path has led to a rapid expansion of its renewable energy sector. The installed capacity of renewable energy grew by almost 286% over the past 7.5 years, and today, India ranks fourth in overall renewable energy capacity in the world [10].

Consequently, a vast number of studies in the existing body of climate literature provide immense attention to studying the influence of export and industrialization in the pathway of environmental correction. While the sustainable growth of any economy has been highly reliant on the expansion of export and industrial development, the extant literature provides mixed outcomes while examining their influence on ecological corrosion. While various studies in the existing body of knowledge articulate the beneficial impact of exports in attaining ecological correction [11,12], Rahman et al. [13] and Gozgor and Can [14] refuted the former argument and provided the detrimental association of exports in attaining ecological fortification. In addition, while various studies investigate the influence of industrialization in the carbon neutralization of India [15,16], such investigation on an ecological footprint solely for India has been missing in the existing body of knowledge.

Against this milieu, this research has been expected to add novel contributions to the existing body of knowledge in the following ways by utilizing 48 years of annual time series data from 1970–2017: (i) in contrast to the earlier empirical investigations, this study utilizes ecological footprint as a proxy of an environmental quality indicator for India and investigates the dynamic association among the ecological footprint, exports, renewable energy consumption, industrialization, and GDP. (ii) To the best of our knowledge, this is the first study that investigates the impact of exports on the ecological footprint, specifically for India. (iii) While the existing body of the Environmental Kuznets Curve (EKC) literature available for India fails to provide any specific shape of the EKC for India, this study finds evidence of the EKC of ecological footprints for India in the existing framework. (iv) Employing an augmented ARDL (A-ARDL) co-integration technique to overcome the degeneracy problems and shortcomings of the classical ARDL cointegration model in the existing framework is a novel contribution. (v) Finally, the application of a dynamic ARDL simulation and Breitung and Candelon's (2006) spectral causality test is expected to provide a robust policy-oriented outcome in the existing framework.

The paper is organized as follows—Section 2 provides a literature review on earlier studies; Section 3 describes the data and methodology used in the study; Section 4 is on results and their discussions; and Section 5 concludes with the policy implications of inferences made from the empirical analyses.

2. Literature Review

2.1. Renewable Energy and Environment

Most nations have been focusing on cleaner energy supplies in order to achieve sustainable economic development while not harming the environment, as the substantial usage of non-renewable energy is discovered to be more carbon-intensive [17]. It cannot be refuted that using REC might worsen environmental degradation in this case [18]. However, the influence of REC on climate change is less destructive and more affordable when compared to non-renewable energy [19,20]. Sharma et al. [21], Inglesi-Lotz and Dogan [22], and Dogan and Seker [23] also found that using REC is substantially more ecologically friendly over time.

The surveillance of air, water, and land conditions has also evolved, in addition to CO_2 emissions, since it is impossible to ignore how negatively economic activity affects these natural parameters [24]. In this sense, the EF has come into action and plays an important role in terms of environmental quality measurement among academics and researchers. Similar to the reasons that cause CO_2 emissions, a number of economic, societal, and political factors affect the EF. Researchers believe that cleaner technology and renewable power sources cause environmental degradation to decline [25]. The impact of using fossil fuels and renewable energy on the ecosystem has been examined by Sharif et al. [26] for 74 economies between 1990 and 2015 and demonstrates that the use of green energy does have a negative coefficient. Bello et al. [27] looked at the environmental impacts of Malaysia's hydroelectric energy consumption from 1971 to 2016. They discovered that the usage of hydroelectric electricity decreased environmental damage. According to Ben Jebli et al. [28], REC improves the environment in 22 nations in Central and South America over the long term. In order to revisit the effects of REC and non-renewable energy usage on Turkey's EF, Sharif et al. [29] used the QARDL approach for the years 1965Q1–2017Q4. The results revealed that REC has a long-term negative impact on each quantile of the EF. Christoforidis and Katrakilidis [30] for OECD countries follow a similar line. Xue et al. [31] have also concluded the favorable influence of adopting renewable energy penetration on the EF of four South-Asian nations over the data period 1990–2014. Similar findings has also been evidenced by Dogan and Shah [32] for GCC economies, and by Kim [33] for South Korea.

A few studies have also been done in the instance of India regarding REC and environmental quality; however, most of them use CO_2 emissions as a stand-in for environmental quality rather than EF. For instance, Akadiri and Adebayo [34] examined these correlations utilizing annual data spanning from 1970 to 2018 using a variety of econometric techniques. The findings showed that carbon emissions in India are decreased by favorable changes in the REC. It is now even more crucial to look at the function of the REC for India, as Rej and Nag [35] discovered evidence of an N-shaped EKC for CO_2 emissions for India with a potential policy divergence between capital generation and REC penetration in the economy. As a result, there is a glaring research vacuum when it comes to evaluating the relationship between REC and the environment, while using EF as a stand-in for environmental conditions in India.

2.2. Export and Environmental Quality

Trade openness, which is determined by the proportion of total exports to imports in the GDP, is a typical way to depict international trade in previous empirical investigations [36]. However, in addition to trade openness, the international trade basket's makeup can have a significant impact on environmental quality [37]. As a result, subsequent studies have evaluated the environmental effects of international trade by using the value of export items (or concentration) as a percentage of the GDP to reflect the makeup of the global trade portfolio. Conflicting views on how diversifying the export mix would affect the environment have been supported by prior empirical studies. Export diversification has been linked to increased environmental degradation in certain studies [38,39], but it has also been shown to improve the quality of the environment [40,41]. In a similar vein, trade openness has also shown either positive or negative effects on the environmental quality, depending on the industrialization and development level of a country [42]. For instance, to examine the link between EF and trade openness in Qatar, Charfeddine [43] discovered a two-way causal relationship between EF and total foreign trade. Alola et al. [44] found that trade openness had a negative influence on the EF in the European region; as a result, it had a positive effect on environmental quality.

In another view, Hu et al. [45] found that exports and imports have a positive effect on the EF. According to Dogan et al. [46], urbanization promotes environmental pollution at the upper quantiles, and so does export quality, based on the nations' wealth levels. Rahman [47] came to a similar conclusion on the negative linkage between exports and the states of the environment. Contrarily, in their analysis of the relationship between export product concentration and CO_2 emissions in 19 industrialized economies, Apergis et al. [11] came to the conclusion that greater levels of export product concentration result in reduced CO_2 emissions. There has not been consensus on this relationship, though, which gives scholars more space to delve deeper into this area of environmental economics.

Scholars have underlined that the export content is just as pertinent as the export volume [48]. Export attributes are influenced by structural change, particularly in nations that are in the middle of their economic development [49]. The development of competitive advantages through the production of higher-quality versions of existing items boosts productivity and export earnings, which in turn encourages the export of environmentally friendly goods. The diversification of new goods is a must for making significant progress in quality improvement for developing nations like India. These show that increasing exports is a crucial policy concern, and they are backed up by instances from developed nations that have been successful in accomplishing structural improvements in terms of both the economy and the environment through the improvement of export quality. However, since its big export volume necessitates considerable resource exploitation, India, a major exporter, has so far received less attention in the analysis of the environmental effect of its export volume. As a result, there is still a significant knowledge gap regarding the nexus between exports and the environment in India. This gap would be filled by this research, as well as contribute to the body of existing literature.

2.3. Industrialization and Environmental Quality

The causality between industrialization and environmental quality has been intensively studied in this era of industrialization development. Alam [50] looked at how selected South Asian nations' GDP value addition in the service, industrial, and agricultural sectors affected their CO_2 emissions. The results of this study revealed that whereas industrial and service value-added have favorable significant impacts on CO_2 emissions, agricultural value-added has a negative influence on CO_2 emissions. In the Chinese economy, Huan et al. [51] look at the connection between industrial productivity and environmental quality. The findings supported the existence of a long- and short-term N-shaped relationship between these variables in China. The results also provide two industrial output thresholds for environmental quality. The highest threshold point of industrial output is 29.21%, at which the quality of the environment is getting worse off, and the lower threshold point is 30.71%, at which the quality of the environment turns out to be good in the long run in China. Using ARDL and VECM, Liu and Bae [52] examined the impact of IVA on CO_2 emissions in China. The study discovered that IVA increases CO_2 emissions while causality findings revealed that long-term feedback occurs between them. In 69 nations, Liu and Hao [53] used the FMOLS and DOLS techniques to confirm the existence of a positive association between IVA and CO_2 emissions. Using the dynamic ARDL simulation model, Hossain et al. [15] examined the relationship between IVA and CO_2 emission in India and found that when income and IVA grow, environmental quality declines as a result of rising CO_2 emissions.

In addition, a threshold regression model was used by Dong et al. [54] to assess the effects of urbanization and industrialization on carbon emissions in 14 developed nations. Industrialization increases carbon emissions from the standpoint of income levels. Industrialization progressively has a positive impact on carbon emissions, especially at low- and middle-income levels. However, at the high-income level, this promotional impact starts to wane. Using a nonlinear ARDL model, Ullah et al. [55] tried to establish the link between industrialization and CO_2 emissions in Pakistan over the years 1980–2018. The findings show that industrialization raises emissions, whereas deindustrialization lowers emissions over the long and short terms and subsequently improves the environmental quality. In a sample of 44 Sub-Saharan African nations over the years from 2000–2015, Mentel et al. [56] sought to investigate the connection between industrial value added, renewable energy, and CO_2 emissions. They discovered that the proportion of industry in GDP has a considerable beneficial influence on CO_2 emissions using a two-step system GMM estimator. However, Appiah et al. [57] showed that over the long run, fossil fuel use, industrialization, and urbanization had a non-significant beneficial effect on CO_2 emissions for SSA nations.

Interestingly, according to Yang and Khan's [58] research, capital creation and industrial value-added (IVA) increase environmental sustainability in IEA member nations. Similarly, IVA massively increases CO_2 emissions over the short term while boosting the environmental quality in the long term, according to research by Abbasi et al. [59]. Khan [60] asserts that despite China's primary industries' decreased CO_2 emissions, the secondary and tertiary sectors continue to have a beneficial relationship with one another. The discussion above indicates that there is disagreement over the impact of IVA on the environment, as there are diverse findings on this nexus reported by the researchers. However, the impact of IVA on EF in the context of India is concerning. Thus, the focus of this study was on the influence of IVA on the EF in the setting of India, where it is grave and requires in-depth investigation.

2.4. Economic Development and Environmental Quality

Numerous studies have looked at the nexus between environmental quality and economic growth (GDP) in light of the Environmental Kuznets Curve (EKC) hypothesis. As the economy reaches a specific level of per-capita income and reaches economic maturity, it is projected that the connection between the GDP and CO_2 emissions would take the shape of an inverted U. The EKC and curve shape utilizing CO_2 emissions have been the subject of extensive investigation and found diluted results. For instance, in their study, Shahbaz et al. [61] examined the impact of the GDP on carbonization in Tunisia from 1971 to 2010, when EKC was still operational in that country. They found that the connection between carbonization and GDP was inverted U-shaped. Furuoka [62] investigated the relationship between the GDP and CO_2 emissions using the EKC model; however, there was no proof of an EKC between the two. The association between environmental destruction and GDP in EU countries was examined in the study of [63]. A one-way association between the GDP and CO_2 was discovered by the study. Gyamfi et al. [64] used data from 1995 to 2018 to create their N-shaped EKC research for the E7 countries.

In 35 OECD countries, between 2000 and 2014, Ozcan et al. [65] discovered that economic development and energy consumption patterns improve countries' environmental performance levels. The EKC only exists in the long term, according to Shahbaz et al. [66], who used a sample of Vietnam's yearly data from 1974 to 2016 to make this discovery. The link between long-term income and pollution is, however, best captured by the N-shape. This suggests that, during a certain stage of economic expansion, Vietnam might anticipate a brief decrease in CO_2 emissions. However, after reaching another income tipping point, this will be followed by an additional increase in CO_2 emissions. In the long run, economic growth and environmental deterioration are linked in the BRICS nations, according to Naseem et al. [67]. On the other hand, Rahman et al. [68] discovered a high correlation between environmental degradation and economic development, but were unable to detect EKC in the BRICS region.

Other researchers have also used the EF as a proxy of environmental pollution and reported the shape of the relation between EF and GDP. According to Danish et al. [69], economic expansion results in a larger ecological footprint, which exacerbates environmental deterioration. The connections between natural resources, technical advancements, GDP, and the consequent EF in rising nations were examined by Ahmad et al. [70]. Long-term growth and the expansion of the EF are driven by natural resources and GDP, while technology advancements assist in slowing down environmental deterioration. Furthermore, in the context of the EKC hypothesis, the quadric term of GDP had a negative influence on the EF. Similar findings for other nations and the economic bloc have also been reported in other research [71]. Additionally, some study on this connection was done in India [72]. In India, Ahmed and Wang [73] discovered an inverted U-shaped relationship between the GDP and EF. According to Murshed et al. [74], it is essential to foster intra-regional trade and hasten economic expansion if South Asian countries are to significantly reduce their EFs. Additionally, for the South Asian panel, as well as for Bangladesh, Sri Lanka, Nepal, and Bhutan, the results confirm the EKC hypothesis, but not for India and Pakistan. Economic growth has a favorable long- and short-term impact on EF, according to Adebayo et al. [75], who also supported the EKC theory in India.

3. Data Definition and Empirical Approach

This section entails the detailed data description and empirical strategy adopted for obtaining robust policy-oriented outcomes.

3.1. Data Description

This research exploits 48 years of annual time series data from 1970–2017 to investigate the influence of GDP, export, industrialization, and aggregate renewable energy deployment on the EF of India. For this multivariate time series analysis, we have included the following variables: EF in global hectares/capita as the proxy of environmental decay, real GDP per capita in constant 2015 USD as a proxy of economic prosperity, industry value added (% of GDP) as a proxy of industrialization, export of goods and services (% of GDP), and REC in tonnes of oil equivalent/capita. The data for GDP, exports, and industrialization have been outsourced from World Development Indicator (WDI). EF data has been outsourced from the Global Footprint Network. REC data has been taken from the BP Statistical Review of World Energy. Further, the data sources with a detailed explanation have been outlined in Table 1. The descriptive measures of the variables, as shown in Table 2, say that all the variables except IVA are right-skewed. All the variables seem to follow a normal distribution. EF seems to have a very strong positive association with GDP and exports, a moderate association with IVA, and a very strong negative linkage with REC.

Table 1. Variables Specification.

Variable	Definition	Source
EF	Ecological Footprint (global hectares/capita)	Global Footprint Network
GDP	Economic growth (real GDP per capita constant 2015 US$)	WDI
EXP	Export of goods and services (% of GDP)	WDI
IVA	Industry (including construction), value added (% of GDP)	WDI
REC	Renewable energy consumption (tonnes of oil equivalent/capita)	BP Statistical Review of World Energy

Table 2. Descriptive statistics.

	lnEF	lnGDP	lnEXP	lnIVA	lnREC
Mean	−0.22	6.56	2.29	3.27	−3.97
Median	−0.23	6.44	2.28	3.29	−4.05
Maximum	0.18	7.59	3.23	3.44	−3.24
Minimum	−0.47	5.94	1.29	3.07	−4.39
Std. Dev.	0.19	0.51	0.59	0.09	0.31
Skewness	0.53	0.49	0.13	−0.59	0.85
Kurtosis	2.16	1.97	1.68	3.26	2.69
Jarque-Bera	3.67	4.05	3.61	3.01	2.78
Probability	0.16	0.13	0.17	0.22	0.48
Observations	48	48	48	48	48
			Correlation		
	lnEF	**lnGDP**	**lnEXP**	**lnIVA**	**lnREC**
lnEF	1.00				
lnGDP	0.99	1.00			
lnEXP	0.93	0.95	1.00		
lnIVA	0.69	0.71	0.81	1.00	
lnREC	−0.91	0.89	0.84	0.62	1.00

3.2. Empirical Strategy

Before beginning an econometric study, it is essential to look at the stationary qualities of the time series variables to prevent unneeded regression problems. We have used the traditional unit root tests, i.e., Augmented Dickey–Fuller (ADF) and the Phillips–Perron (PP) tests to serve that purpose. However, traditional unit root tests have been criticized for having low power to reject the null hypothesis in the presence of a structural break The second generation unit root test technique, i.e., Zivot and Andrews (1992) structural break unit root test, has been done additionally to determine the break date. Traditional co-integration techniques, i.e., Engle–Granger [76], and Johansen–Juselius [77] require all the variables to be integrated into I(1). The frequently used ARDL bounds test of the co-integration approach, developed by Pesaran et al. [78], requires the independent variables to be integrated at I(1) or fractionally integrated to I(0)/I(1) under the assumptions that the dependent variable needs to be stationary at first differenced form. McNown et al. [79] and Sam et al. [80] counter-argued that in many of the research articles, one of the pre-conditions of applying the ARDL bounds testing approach, i.e., the dependent variable needs to be integrated at I(1) has not been followed. The augmented ARDL (A-ARDL) bounds test co-integration approach, proposed by McNown et al. [79], overcomes the critics of the ARDL bounds test co-integration approach, developed by Pesaran et al. [78]. This model allows the variables to be integrated in any order other than integrated to I(2). The augmented ARDL model under log-linear specification can be presented as follows:

$$\begin{aligned}\Delta lnEF_t = \phi_0 \; & + \gamma_1\, DU_t + \phi_1\, lnEF_{t-1} + \phi_2\, lnGDP_{t-1} + \phi_3\, lnEXP_{t-1} + \phi_4\, lnIVA_{t-1} + \phi_5\, lnREC_{t-1} \\ & + \textstyle\sum_{j=1}^{k_1} \beta_{1j}\Delta lnEF_{t-j} + \sum_{j=0}^{k_2} \beta_{2j}\,\Delta lnGDP_{t-j} + \sum_{j=0}^{k_3} \beta_{3j}\Delta lnEXP_{t-j} + \sum_{j=0}^{k_4} \beta_{4j}\Delta lnIVA_{t-j} \\ & + \textstyle\sum_{j=0}^{k_5} \beta_{5j}\Delta lnREC_{t-j} + \varepsilon_{1t}\end{aligned} \tag{1}$$

In Equation (1), the first difference operator has been denoted by Δ, γ_1 is the coefficient of the break dummy variable, the lag accompanied with each of the variables have been represented by k_1–k_5, the nomenclature of the variables is the same as given in Table 1, ln represents the logarithmic transformation of the variables, the summation term captures the short run coefficients, and ε_{1t} represents the white noise error term.

The bound F test has been performed by considering the optimal lag structure through Akaike Information Criteria (AIC) statistics and setting the null hypothesis of the lagged term of variables to zero (i.e., H_0: $\phi_1 = \phi_2 = \phi_3 = \phi_4 = \phi_5 = 0$). The cointegrating situation has arrived if the estimated $F_{overall}$ surpasses the upper bound critical values designed by

Narayana [81]. This model is also capable of eliminating the degeneracy problem that arises in the case of the ARDL bounds testing approach. In the ARDL bounds testing approach, two degeneracy cases may arise: (i) lagged explanatory variables are not statistically significant, and (ii) lagged level of the outcome variable is not statistically significant. In order to address these degeneracy issues, McNown et al. [79] and Sam et al. [80] have proposed two additional tests in the existing ARDL bounds testing framework: (i) examining the significance level of the lagged dependent variable through a t-test (t_{DV}) by setting the null hypothesis, (H_0: $\phi_1 = 0$) and a cointegration situation is feasible if the estimated t_{DV} is greater than the upper bound critical values computed by Pesaran et al. (2001); (ii) investigating the significance level of the lagged independent variables through the F test ($F_{IDV)}$ by setting null hypothesis H_0: $\phi_2 = \phi_3 = \phi_4 = \phi_5 = 0$ and cointegration situation can be reached if the estimated F_{IDV} surpasses the upper bounds critical values designed by Sam et al. [81].

After having the confirmation of long-run co-integration among the core variables, the dynamic ARDL technique (D-ARDL), proposed by Jordan and Philips [68], has been employed to compute the long and short-run coefficient estimates. The D-ARDL technique can be considered as a superior technique over the classical ARDL model in case of low convergence speed, and the variables can be integrated into any order except I(2) [82]. The DARDL can also construct graphs showing positive and negative counterfactual adjustments in independent variables and their influence on dependent variables, which is not facilitated in the classical ARDL approach [82–84]. The D-ARDL model in our present study can be expressed as follows:

$$
\begin{aligned}
(lnEF)_t = \lambda_0 + \theta_0 lnEF_{t-1} + \beta_1 \Delta lnGDP_t + \theta_1 lnGDP_{t-1} + \quad \beta_2 \Delta lnEXP_t + \theta_2 lnEXP_{t-1} + \\
\beta_3 \Delta lnIVA_t + \theta_3 lnIVA_{t-1} + \beta_4 \Delta lnREC_t + \theta_4 lnREC_{t-1} + \xi ECT_{t-1} + u_t
\end{aligned} \tag{2}
$$

The classical D-ARDL technique helps in the computation of the short and long-run influence of the determinants on EF in the context of India, but in order to understand the consequences for policy, it is also required to analyze the causal relationship between the two variables. The causality between variables at different frequencies, i.e., short-term (ωi = 2.50), medium-term (ωi = 1.50), and long-term (ωi = 0.05) can be explored by augmenting the frequency-domain causality test, proposed by Breitung and Candelon [85]. The null hypothesis of Breitung and Candelon [85] in a bivariate framework can be demonstrated as follows:

The null hypothesis of $M_{y \to x}(\omega) = 0$ is represented as

$$H_0 : R(\omega)\beta = 0 \tag{3}$$

where $\beta = (\beta_1, \ldots, \beta_p)$ and $R(\omega) = \begin{pmatrix} cos(\omega)cos(2\omega)\ldots.cos(p\omega) \\ sin(\omega)sin(2\omega)\ldots.sin(p\omega) \end{pmatrix}$.

In this approach, the long-run (short-run) causality has been explored through low (high) frequency.

4. Results and Discussions

The outcomes of the entire econometric analysis have been portrayed through the following Sections 4.1–4.6.

4.1. Unit Root Analysis Results

The empirical journey has been initiated by conducting the classical unit root test, i.e., Augmented Dickey–Fuller (ADF) and Phillips–Perron (PP) unit root test for the following two reasons: (i) to test the stationary behavior (synonymous to not having unit root complications) of the time series variables to get rid of the spurious outcomes, and (ii) to determine the order of the integration of the variables to employ the A-ARDL bounds test of co-integration and D-ARDL technique. The unit root test findings, as reported in Table 3, say that all the core variables of our study are stationary in first difference form, which is suitable to apply both the A-ARDL bounds test of co-integration and the D-ARDL technique.

Table 3. Unit Root test.

Variable	Form	ADF (t-Statistics)		PP (t-Statistics)		Order of Integration
		Intercept	Trend + Intercept	Intercept	Trend + Intercept	
lnEF	Level	1.33 (0.998)	−2.13 (0.519)	1.71 (0.999)	−1.95 (0.612)	I(1)
	First Difference	−8.43 *** (0.000)	−9.22 *** (0.000)	−8.28 *** (0.000)	−9.17 *** (0.000)	
lnGDP	Level	4.19 (1.000)	−1.64 (0.761)	5.16 (1.000)	−1.87 (0.652)	I(1)
	First Difference	−5.69 *** (0.000)	−8.25 *** (0.000)	−5.75 *** (0.000)	−11.94 *** (0.000)	
lnEXP	Level	−1.25 (0.643)	−2.26 (0.448)	−1.23 (0.654)	−1.69 (0.738)	I(1)
	First Difference	−3.19 ** (0.027)	−3.27 ** (0.044)	−6.48 *** (0.001)	−6.57 *** (0.000)	
lnIVA	Level	−2.25 (0.194)	−1.59 (0.779)	−2.38 (0.152)	−1.17 (0.905)	I(1)
	First Difference	−3.23 ** (0.025)	−3.66 ** (0.036)	−6.93 *** (0.000)	−7.39 *** (0.000)	
lnREC	Level	−0.04 (0.949)	−1.45 (0.832)	−0.04 (0.949)	−1.56 (0.792)	I(1)
	First Difference	−6.27 *** (0.000)	−6.34 *** (0.000)	−6.27 *** (0.000)	−6.34 *** (0.000)	

Note: In parentheses, the probability value is given. ***, ** and * denotes a 1%, 5% and 10% level of significance.

In this study, we have also employed the second generation unit root test, i.e., Zivot and Andrews's (1992) structural break unit root test to detect the structural break and the order of integration of the variables under the presence of structural break. Table 4 outlines the outcomes of Zivot and Andrews's (1992) structural break unit root test, and essentially indicates that all the variables are stationary at levels, except the GDP, exports, and renewable energy consumption. However, all the variables are found to be stationary at the first difference form. These results ensure there are no stationarity issues in the series of variables.

Table 4. Zivot and Andrews structural break test.

Variable	Level		First Difference	
	t-Statistics	Time Break	t-Statistics	Time Break
lnEF	−4.31 ***	2000	−9.86 **	2006
lnGDP	−1.31	2003	−5.11 **	1991
lnEXP	−2.96	2002	−4.87 ***	1987
lnLIVA	−3.04 **	2006	−4.93 ***	2002
lnREC	−3.65	2000	−4.37 **	2004

Note: *** denotes a 1% level of significance, ** denotes a 5% level of significance, and * denotes a 10% level of significance.

4.2. Augmented ARDL Co-Integration Analysis Results

After evidencing all the variables are I(1), the long-run co-integration among the core variables of our study has been examined over the data period 1970–2017 by employing the augmented ARDL bounds test of the co-integration technique, proposed by McNown et al. [79]. The empirical outcome of the A-ARDL co-integration technique, as shown in Table 5 suggests that the overall F statistics exceed the upper bounds critical value designed by Narayan (2005) at a 1% level of significance. Additionally, two new sets of tests, proposed by McNown et al. [79] and Sam et al. [80], have been done in parallel to avoid the degeneracy complications which may happen in the classical ARDL co-integration model. Table 4 exhibits that t_{DV} of the lagged dependent variable is negative and exceeds the upper bounds critical value designed by Pesaran et al. (2001) at a 5% significance level. Moreover, the F_{IDV} of the lagged independent variables seems to surpass the upper bounds critical value, designed by Sam et al. [81], at a 1% significance level. In all three cases, the null hypothesis has been rejected and hereby confirmed the long-run co-integrating relationship among the variables over the stipulated time period, as mentioned in this study.

Table 5. Augmented-ARDL Cointegration Results.

DV I LDV			Lag-Length		Test Statistics		Results
LEF_t I LGDP, LEXP, LIVA, LREC, Dummy			4, 1, 4, 1, 3, 2		$F_{Overall}$: 6.95 ***		Cointegrated
					t_{DV}: −4.11 *		
					F_{IDV}: 8.30 ***		
Table CVs	1%		5%		10%		For k = 5
Tests	I(0)	I(1)	I(0)	I(1)	I(0)	I(1)	Sources
$F_{Overall}$	3.41	4.68	2.62	3.79	2.26	3.35	Narayan (2005)
t_{DV}	−3.43	−4.79	−2.86	−4.19	−2.57	−3.86	Pesaran et al. [79]
F_{LDV}	3.05	5.02	2.24	3.90	1.86	3.39	Sam et al. [81]

Note: *** denotes a 1% level of significance, ** denotes a 5% level of significance, and * denotes a 10% level of significance.

It is worthwhile mentioning that the dummy variable associated with structural break was found to be insignificant and the model's coefficients were found to be stable in the absence of a dummy variable. That is why we are not including the dummy variable in the dynamic ARDL estimations. Adebayo et al. [75] have also not considered the break dummy in the dynamic ARDL estimations while examining the impact of structural change, and hydro and coal energy consumption on the ecological footprint of India; however, the structural break persists in the ecological footprint data.

4.3. Long Run and Short Run Coefficient Estimates from D-ARDL Analysis

After establishing the co-integrating relationship among the variables studied in this research, the accompanied short- and long-run coefficient of the EF determinants of India can be computed through the D-ARDL technique (see Table 6). The results from the baseline D-ARDL regression simulations say that both the long- and short-run coefficient of the GDP is positive and significant at a 5% significance level. In the long run, a 1% upsurge in the GDP enhances the EF by 0.14%, and similarly, a 1% upswing in the GDP intensifies the EF by 0.33% in the short run. These results provide empirical evidence of holding the EKC hypothesis in India through the lens of Narayan and Narayan [86], as the short-run co-efficient of GDP is larger than the long-run co-efficient. Studies by Rej et al. [87], Kanjilal and Ghosh [88], Hossain et al. [15], and Jayanthakumaran et al. [89] also conclude similar findings for India while investigating the existence of the EKC hypothesis. This finding does not support the empirical findings by Rej and Nag [35], Bandyopadhyay and Rej [90], and Rej et al. [91], who fail to prove the existence of the EKC hypothesis in India. This finding suggests that the present economic growth pattern acts as a stumbling block to achieving environmental sustainability in India. One of the possible reasons for the higher detrimental impact of the GDP in the short run, with respect to the long run, lies in the absence of substitution possibility and reserved policy action.

Table 6. Long- and short-run coefficients from the dynamic ARDL model.

Variables	Co-Efficient	t-Stat	Prob.
lnGDP	0.14 **	2.18	0.036
ΔlnGDP	0.33 **	2.17	0.037
lnEXP	−0.05 **	−1.62	0.048
ΔlnEXP	0.06	1.21	0.234
lnIVA	0.26 ***	3.71	0.001
ΔlnIVA	0.16	0.98	0.331
lnREC	−0.09 **	−1.9	0.045
ΔlnREC	−0.06 **	−2.08	0.041
Cons.	−1.51 ***	−3.09	0.004
ECT (-1)	−0.37 ***	−5.56	0.000
R^2	0.99	Adjusted R^2	0.98
F- Statistics [Prob.]	364.37 [0.000]	Simulation	5000

Note: ***, **, and * denotes a 1%, 5%, and 10% level of significance.

The long-run coefficient of exports seems to be negative and significant in the long run but not significant in the short run. Precisely, a 1% upsurge in exports corroborates to a corresponding 0.05% decrease in the EF in the long run. This finding supports the empirical findings by Apergis et al. [11] for 19 developed economies, Haug and Ucal [12] for Turkey, but refutes the findings of previous studies by Rahman et al. [13] for Newly Industrialized Countries (NIC), and Gozgor and Can [14] for China. Our results illustrate that India's present export basket consists of relatively fewer carbon-intensive goods that stimulate the lower EF of India. In other words, Indian firms may produce relatively fewer carbon-intensive goods to manage the environmental quality but may engage in importing the carbon-intensive goods.

Industry value added also corresponds to an increase in the ecological footprint of India. As found, a 1% increase in the IVA appears to stimulate the EF by 0.26% in the long run but carries insignificant influence in the short run. Our finding is consistent with the previous empirical findings by Li and Lin [16] for 73 nations, Liu and Bae [52] for China, Shahbaz et al. [61] for Bangladesh, and Hossain et al. [15] for India. Our findings suggest that India's current policy focuses on accelerating carbon-intensive home-grown manufacturing as an integral part of "Make in India" initiatives which stimulates the ecological footprint of India.

The coefficient of REC illustrates placate the EF of India in both the short and long run. Precisely, a 1% increase in REC corresponds to a shrinking 0.09% (0.06%) of the EF of India in the long run (short run). Our finding is in line with the previous findings by Rej and Nag [35], Bandyopadhyay et al. [92], and Rej et al. [93], who have studied the same in the pathway of environmental sustainability for India. This finding further stimulates the fact of accelerating the acceptance of renewable energy usage in the Indian economy as a source of prime fuel that not only can safeguard the environment but also can secure the 24 × 7 electricity availability of the Indian economy.

4.4. Visualization of Counterfactual Changes

Figure 2 illustrates the impulse response plot between the export and the ecological footprint. The graph shows how changes in exports impact the ecological footprint in India. The anticipated value is shown by the dots in the figure, while the deep blue line represents the 75 percent confidence interval, and the light blue to lightest blue lines reflect the 90 percent and 95 percent confidence intervals, respectively. Figure 2 unveils that every 10% increase in exports is associated with a corresponding decrease in the ecological footprint of India. In contrast, a 10% decrease in exports corroborates to the corresponding increase in the ecological footprint of India. Figure 2 also uncovers that both the positive and negative shock of exports have almost the same impact on the ecological footprint of India. Figure 3 depicts the impulse response plot between the industry value added and the ecological footprint. This figure tells that every 10% increase in IVA deepens the ecological footprint of India, while every 10% decrease in IVA plummets the environmental dilapidation. Furthermore, the impulse response plot between the GDP and the ecological footprint, as shown in Figure 4, illustrates that every 10% increase in the GDP soars the environmental decrepitude, while every 10% decrease in the GDP contributes towards the environmental correction. Moreover, the impulse response plot between renewable energy consumption and the ecological footprint, as depicted in Figure 5, says that every 10% increase in renewable energy consumption contributes to the pathway of ecological footprint neutralization, while every 10% decrease in renewable energy consumption intensifies the ecological footprint of India.

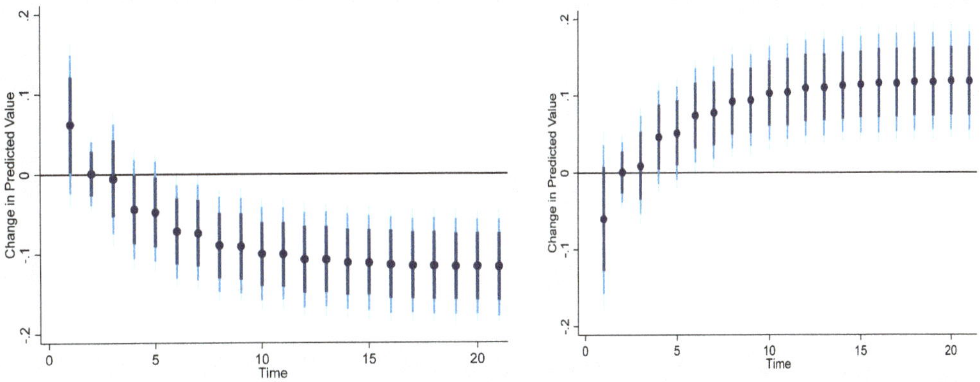

Figure 2. The above figure signifies a 10% increase or decrease in export and their effect on the ecological footprint in India. Note: Years are used as the unit of time in the figure's horizontal axis.

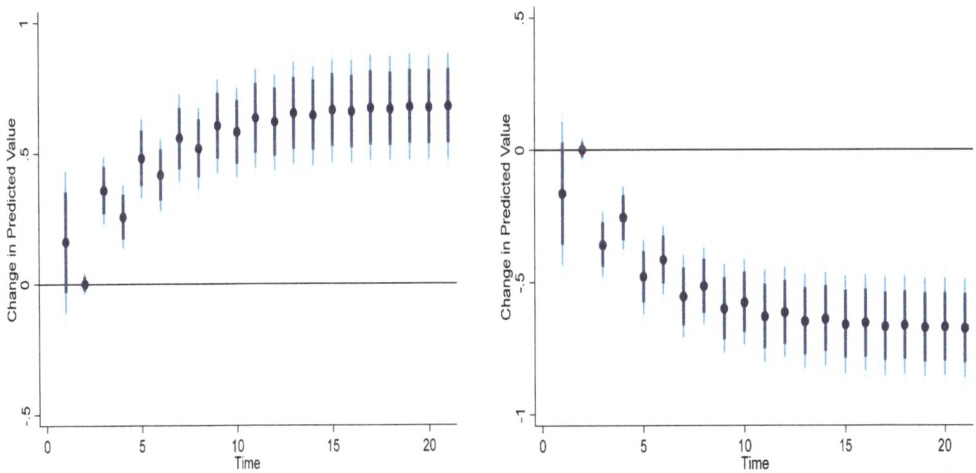

Figure 3. The above figure signifies a 10% increase or decrease in industry value added and their effect on the ecological footprint in India. Note: Years are used as the unit of time in the figure's horizontal axis.

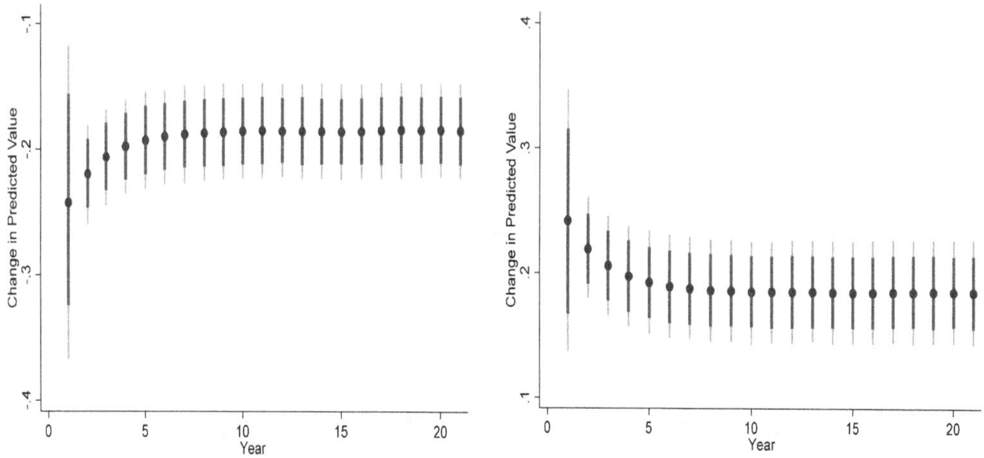

Figure 4. The above figure signifies a 10% increase or decrease in GDP and their effect on the ecological footprint in India.

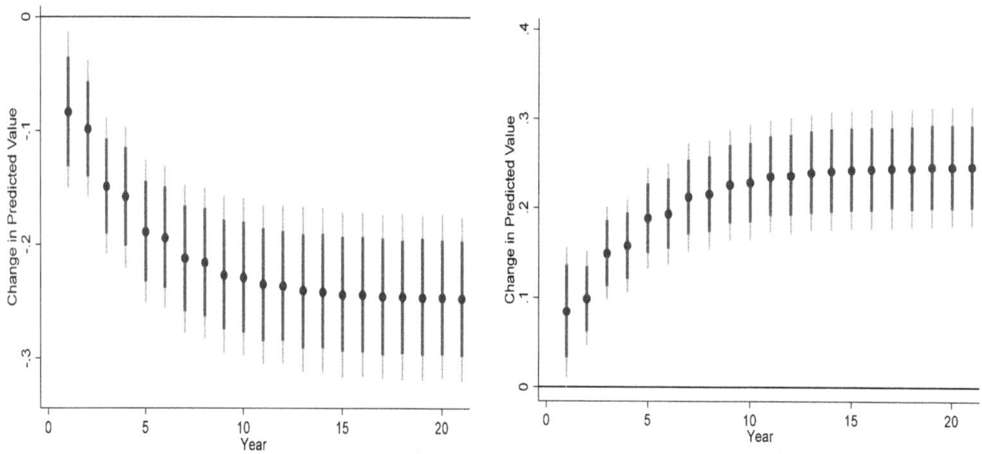

Figure 5. The above figure signifies a 10% increase or decrease in renewable energy consumption and their effect on the ecological footprint in India.

4.5. Model Diagnostics and Robustness Test

The credibility of our model and outcomes have been further verified through some post-diagnostic tests and by applying ARDL, FMOLS, and DOLS, respectively. The outcomes of the post-diagnostics tests, as shown in Table 7, unveils that our estimated model has passed the Jarque–Bera test of normality, Breusch–Godfrey serial correlation LM test, Breusch–Pagan–Godfrey test of heteroscedasticity, and the specification test. The outcomes of the stability test, as depicted in Figure 6, certify that the estimated D-ARDL model has passed the stability test, as the plots of both CUSUM and CUSUMQ are well within the 95% critical limit.

Table 7. Diagnostic tests for the D-ARDL model.

Diagnostic Test	Statistics	Decision
Breusch–Godfrey serial correlation LM test	F-stat: 0.13 Prob: 0.879	No serial correlation
Jarque–Bera test	χ^2: 1.13 Prob: 0.569	Error terms are normally distributed
Breusch–Pagan–Godfrey test	F-stat: 1.07 Prob: 0.414	No heteroskedasticity
Ramsey RESET test	F-stat: 2.59 Prob: 0.117	Model is correctly specified

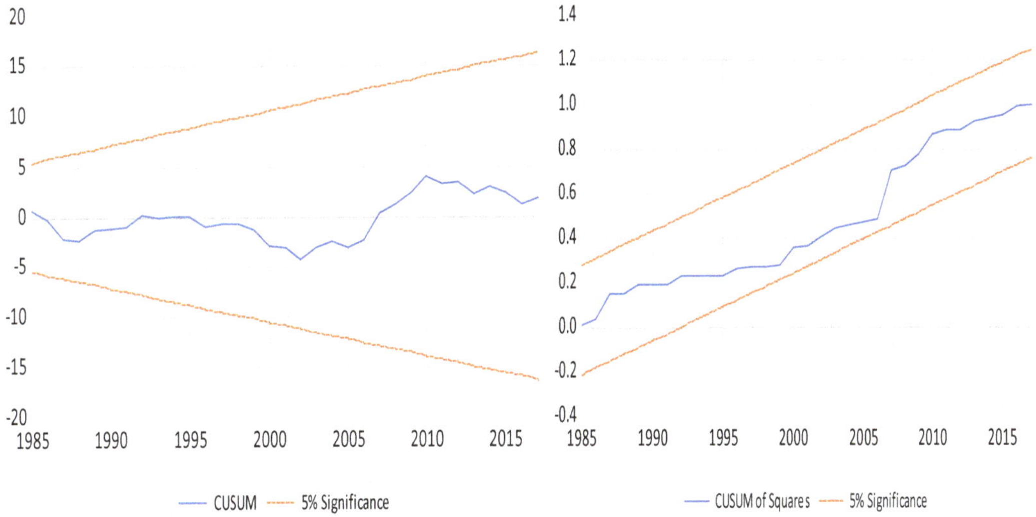

Figure 6. The plot of CUSUM and CUSUM of square test.

The essence of augmenting the ARDL, FMOLS, and CCR approaches is to cross-validate the outcomes of the D-ARDL estimations. The outcomes of the long-run coefficient estimates, as given in Table 8, seem to be consistent with the D-ARDL estimations, as given in Table 6. This further certifies the usability of the D-ARDL outcomes for further policy inferences.

Table 8. Results of ARDL, FMOLS, and CCR.

Variable	ARDL		FMOLS		CCR	
	Coefficient	t-Stat.	Coefficient	t-Stat.	Coefficient	t-Stat.
lnGDP	0.29 ***	4.05	0.37 ***	9.87	0.38 ***	10.31
lnEXP	−0.11 **	−1.56	−0.06 **	−1.87	−0.06 **	−2.05
lnIVA	0.42 **	1.94	0.12 *	1.25	0.09 *	1.12
lnREC	−0.19 **	−2.61	−0.11 ***	−2.66	−0.09 **	−2.62

Note: *** denotes a 1% level of significance, ** denotes a 5% level of significance, and * denotes a 10% level of significance.

4.6. Analysis of Causality

The pairwise causality analysis has also been done by augmenting the Breitung and Candelon [85] frequency domain causality test (Table 9). The results of this test have provided empirical evidence of (i) unidirectional causality from the GDP to the ecological footprint in long run, (ii) unidirectional causality from the industry value added to the ecological footprint in long run, and (iii) a short-run causality from renewable energy consumption to the ecological footprint. This could be because of the comparatively larger land requirement for setting up renewable energy plants, which possibly get compensated in long run through clean energy consumption in the economy.

Table 9. Spectral causality test analysis.

	Long Term $\omega i = 0.05$	**Medium Term** $\omega i = 1.50$	**Short Term** $\omega i = 2.50$
GDP→EF	6.672 ** (0.036)	1.148 (0.563)	2.581 (0.275)
EXP→EF	2.594 (0.273)	2.821 (0.244)	0.079 (0.961)
IVA→EF	2.799 * (0.087)	0.143 (0.931)	0.659 (0.719)
REC→EF	0.081 (0.961)	1.655 (0.437)	3.899 ** (0.042)

Note: The probability value has been given in parentheses. *** denotes a 1% level of significance, ** denotes a 5% level of significance, and * denotes a 10% level of significance.

5. Conclusions and Policy Recommendations

5.1. Conclusions

The main theme of this study is to empirically examine the influence of exports, the GDP, industry value added, and renewable energy consumption on the EF of India during the period spanning from 1970–2017. For this purpose, the recently developed augmented ARDL (A-ARDL) bounds test, by McNown et al. [79], has been employed to explore the long-run co-integration among the variables. The dynamic ARDL (D-ARDL) simulations model has also been adopted to estimate the short- and long-run coefficient estimates of the EF determinants for India. The outcomes of the D-ARDL approach provide some fresh insights. This study not only confirms the existence of long-run dynamics among the core variables of this study but also confirms the EKC hypothesis for India While export and renewable energy consumption seem to improve the environmental quality, industry value added seems to be accompanied by the corresponding obliteration of environmental well-being. Further, the association of industrialization with negative environmental externalities heightens the conflict between two seemingly ambitious goals of India, i.e., "Make in India" by achieving higher economic growth through enriching indigenous manufacturing, and "INDC" by reducing the targeted carbon emissions. In addition, the spectral causality test has also been used to predict the pairwise causality among the variables. The findings of this approach reveal unidirectional long-run causality from both the GDP and the industry value added to EF. However, the short-run causality from renewable energy consumption to the ecological footprint shows that renewable energy penetration may impact the ecological footprint in the short run.

5.2. Policy Recommendations

The results of this study can have implications for some significant policy decisions. As we found that renewable energy has a significant beneficial impact on the environmental conditions in India, policy emphasis is needed to prompt greater renewable energy use, from household consumption to the industry level. The use of renewable energy at the household level may be encouraged by raising environmental awareness among people, offering subsidies, and securing tax exemptions from the state and federal governments. However, at the industry level, the necessary funds, credit facilities, and tax incentives are needed to promote the use of clean energy sources at the industrial level. Conversely, the legislation of the Pigouvian tax can be applied to industries that utilize non-renewable energy more frequently. These policy measures can be beneficial in reducing the ecological footprint of industrialization as well.

According to our empirical results, India may be able to reduce its ecological footprint and achieve sustainable growth through the expansion of the export sector. The Indian government must develop long-term plans to increase the country's trading basket in order to reap the benefits of exports. The government must consider the cost-effectiveness of producing exportable goods, the use of cutting-edge technology in the manufacturing process, the adoption of energy-efficient equipment, the reliable supply of raw materials, and the political stability of the nation when developing a long-term strategy for the green export sector. In this context, it is essential to increase the export mix of the nation's knowledge-based products, which are typically less harmful to environmental quality. In 2021, India witnessed an 18.7% growth in exports from the services sector as compared to a 4.82% increase in merchandise exports in 2020. Although this could be an outcome of a two-year pandemic, India still needs to plan for similar growth trends in exports. The results of this study also indicate that India's export basket possibly represents its comparative advantage in terms of its ecological footprint as well. In addition, more sustainable growth through exports can be possible if there is policy emphasis on improving the efficiency of production in this sector, reducing energy intensity, and reducing resource intensity through technology and innovation. Close monitoring, data collection, and political will are, however, needed to channel trade instruments, such as subsidies, to boost the export sector and ensure that the resource intensity of exports is not just maintained but reduced through strong efficiency norms and environmental standards of production that can additionally make Indian exports more attractive to the world market.

Given that the strategies proposed in this study only took a few aggregated economic elements into account, further studies can consider demographic factors to increase the robustness of the policy formation. However, the policies outlined in this study can still serve as a baseline report that can be developed further. Future research on this topic can take into account the geographical aspects of demographic disparities, as this can provide a number of fresh perspectives on the pattern of changes in the ecological footprint. Therefore, more research may be done by considering a larger economic area or group of nations, as well as other crucial factors such as export diversification, R&D spending, green innovation, etc.

Author Contributions: Conceptualization, S.R. and B.N.; methodology, S.R.; software, S.R.; validation, S.R., B.N. and M.E.H.; formal analysis, B.N.; investigation, M.E.H.; resources, S.R.; data curation, B.N. and M.E.H.; writing—original draft preparation, S.R., B.N. and M.E.H.; writing—review and editing, S.R., B.N. and M.E.H.; visualization, M.E.H.; supervision, B.N.; project administration, B.N.; funding acquisition, S.R. and B.N. All authors have read and agreed to the published version of the manuscript.

Funding: This research received no external funding.

Institutional Review Board Statement: Not applicable.

Informed Consent Statement: Not applicable.

Data Availability Statement: Data can be freely available from https://www.footprintnetwork.org/ and https://data.worldbank.org/ (accessed on 10 September 2022) and https://www.bp.com/en/global/corporate/energy-economics/statistical-review-of-world-energy.html (accessed on 10 September 2022).

Conflicts of Interest: The authors declare no conflict of interest.

References

1. Caglar, A.E. Can Nuclear Energy Technology Budgets Pave the Way for a Transition toward Low-Carbon Economy: Insights from the United Kingdom. *Sustain. Dev.* **2022**. [CrossRef]
2. Caglar, A.E.; Ulug, M. The Role of Government Spending on Energy Efficiency R&D Budgets in the Green Transformation Process: Insight from the Top-Five Countries. *Environ. Sci. Pollut. Res.* **2022**, *29*, 76472–76484.

3. Caglar, A.E.; Mert, M. Carbon Hysteresis Hypothesis as a New Approach to Emission Behavior: A Case of Top Five Emitters. *Gondwana Res.* **2022**, *109*, 171–182. [CrossRef]
4. Global Footprint Network Google Scholar. Available online: https://www.footprintnetwork.org/content/uploads/2022/09/BAK_Economics_Sustainability_22092022.pdf (accessed on 19 October 2022).
5. Mitić, P.; Munitlak Ivanović, O.; Zdravković, A. A Cointegration Analysis of Real GDP and CO_2 Emissions in Transitional Countries. *Sustainability* **2017**, *9*, 568. [CrossRef]
6. Gong, B.; Zheng, X.; Guo, Q.; Ordieres-Meré, J. Discovering the Patterns of Energy Consumption, GDP, and CO2 Emissions in China Using the Cluster Method. *Energy* **2019**, *166*, 1149–1167. [CrossRef]
7. MoC Google Scholar. Available online: https://commerce.gov.in/wp-content/uploads/2020/02/MOC_636434269763910839_international_tpp_DFTP.pdf (accessed on 19 October 2022).
8. Ram, R. Exports and Economic Growth in Developing Countries: Evidence from Time-Series and Cross-Section Data. *Econ. Dev. Cult. Chang.* **1987**, *36*, 51–72. [CrossRef]
9. Ekanayake, E.M. Exports and Economic Growth in Asian Developing Countries: Cointegration and Error-Correction Models. *J. Econ. Dev.* **1999**, *24*, 43–56.
10. MoRE Google Scholar. Available online: https://mnre.gov.in/img/documents/uploads/file_f-1618564141288.pdf (accessed on 19 October 2022).
11. Apergis, N.; Can, M.; Gozgor, G.; Lau, C.K.M. Effects of Export Concentration on CO2 Emissions in Developed Countries: An Empirical Analysis. *Environ. Sci. Pollut. Res.* **2018**, *25*, 14106–14116. [CrossRef]
12. Haug, A.A.; Ucal, M. The Role of Trade and FDI for CO2 Emissions in Turkey: Nonlinear Relationships. *Energy Econ.* **2019**, *81*, 297–307. [CrossRef]
13. Rahman, M.M.; Nepal, R.; Alam, K. Impacts of Human Capital, Exports, Economic Growth and Energy Consumption on CO2 Emissions of a Cross-Sectionally Dependent Panel: Evidence from the Newly Industrialized Countries (NICs). *Environ. Sci. Policy* **2021**, *121*, 24–36. [CrossRef]
14. Gozgor, G.; Can, M. Does Export Product Quality Matter for CO2 Emissions? Evidence from China. *Environ. Sci. Pollut. Res.* **2017**, *24*, 2866–2875. [CrossRef] [PubMed]
15. Hossain, M.E.; Rej, S.; Saha, S.M.; Onwe, J.C.; Nwulu, N.; Bekun, F.V.; Taha, A. Can Energy Efficiency Help in Achieving Carbon-Neutrality Pledges? A Developing Country Perspective Using Dynamic ARDL Simulations. *Sustainability* **2022**, *14*, 7537. [CrossRef]
16. Li, K.; Lin, B. Impacts of Urbanization and Industrialization on Energy Consumption/CO2 Emissions: Does the Level of Development Matter? *Renew. Sustain. Energy Rev.* **2015**, *52*, 1107–1122. [CrossRef]
17. Zaidi, S.A.H.; Hou, F.; Mirza, F.M. The Role of Renewable and Non-Renewable Energy Consumption in CO2 Emissions: A Disaggregate Analysis of Pakistan. *Environ. Sci. Pollut. Res.* **2018**, *25*, 31616–31629. [CrossRef] [PubMed]
18. Bulut, U. The Impacts of Non-Renewable and Renewable Energy on CO2 Emissions in Turkey. *Environ. Sci. Pollut. Res.* **2017**, *24*, 15416–15426. [CrossRef]
19. Chen, Y.; Wang, Z.; Zhong, Z. CO_2 Emissions, Economic Growth, Renewable and Non-Renewable Energy Production and Foreign Trade in China. *Renew. Energy* **2019**, *131*, 208–216. [CrossRef]
20. Long, X.; Naminse, E.Y.; Du, J.; Zhuang, J. Nonrenewable Energy, Renewable Energy, Carbon Dioxide Emissions and Economic Growth in China from 1952 to 2012. *Renew. Sustain. Energy Rev.* **2015**, *52*, 680–688. [CrossRef]
21. Sharma, R.; Sinha, A.; Kautish, P. Does Renewable Energy Consumption Reduce Ecological Footprint? Evidence from Eight Developing Countries of Asia. *J. Clean. Prod.* **2021**, *285*, 124867. [CrossRef]
22. Inglesi-Lotz, R.; Dogan, E. The Role of Renewable versus Non-Renewable Energy to the Level of CO2 Emissions a Panel Analysis of Sub-Saharan Africa's Big 10 Electricity Generators. *Renew. Energy* **2018**, *123*, 36–43. [CrossRef]
23. Dogan, E.; Seker, F. Determinants of CO2 Emissions in the European Union: The Role of Renewable and Non-Renewable Energy. *Renew. Energy* **2016**, *94*, 429–439. [CrossRef]
24. Wackernagel, M.; Rees, W. *Our Ecological Footprint: Reducing Human Impact on the Earth*; New Society Publishers: Gabriola, Canada, 1998; Volume 9.
25. Destek, M.A.; Sinha, A. Renewable, Non-Renewable Energy Consumption, Economic Growth, Trade Openness and Ecological Footprint: Evidence from Organisation for Economic Co-Operation and Development Countries. *J. Clean. Prod.* **2020**, *242*, 118537. [CrossRef]
26. Sharif, A.; Raza, S.A.; Ozturk, I.; Afshan, S. The Dynamic Relationship of Renewable and Nonrenewable Energy Consumption with Carbon Emission: A Global Study with the Application of Heterogeneous Panel Estimations. *Renew. Energy* **2019**, *133*, 685–691. [CrossRef]
27. Bello, M.O.; Solarin, S.A.; Yen, Y.Y. The Impact of Electricity Consumption on CO2 Emission, Carbon Footprint, Water Footprint and Ecological Footprint: The Role of Hydropower in an Emerging Economy. *J. Environ. Manag.* **2018**, *219*, 218–230. [CrossRef] [PubMed]
28. Ben Jebli, M.; Ben Youssef, S.; Apergis, N. The Dynamic Linkage between Renewable Energy, Tourism, CO_2 Emissions, Economic Growth, Foreign Direct Investment, and Trade. *Lat. Am. Econ. Rev.* **2019**, *28*, 1–19. [CrossRef]

29. Sharif, A.; Baris-Tuzemen, O.; Uzuner, G.; Ozturk, I.; Sinha, A. Revisiting the Role of Renewable and Non-Renewable Energy Consumption on Turkey's Ecological Footprint: Evidence from Quantile ARDL Approach. *Sustain. Cities Soc.* **2020**, *57*, 102138. [CrossRef]
30. Christoforidis, T.; Katrakilidis, C. Does Foreign Direct Investment Matter for Environmental Degradation? Empirical Evidence from Central–Eastern European Countries. *J. Knowl. Econ.* **2021**, 1–30. [CrossRef]
31. Xue, L.; Haseeb, M.; Mahmood, H.; Alkhateeb, T.T.Y.; Murshed, M. Renewable Energy Use and Ecological Footprints Mitigation: Evidence from Selected South Asian Economies. *Sustainability* **2021**, *13*, 1613. [CrossRef]
32. Dogan, E.; Shah, S.F. Analyzing the Role of Renewable Energy and Energy Intensity in the Ecological Footprint of the United Arab Emirates. *Sustainability* **2022**, *14*, 227. [CrossRef]
33. Kim, S. The Effects of Foreign Direct Investment, Economic Growth, Industrial Structure, Renewable and Nuclear Energy, and Urbanization on Korean Greenhouse Gas Emissions. *Sustainability* **2020**, *12*, 1625. [CrossRef]
34. Akadiri, S.S.; Adebayo, T.S. Asymmetric Nexus among Financial Globalization, Non-Renewable Energy, Renewable Energy Use, Economic Growth, and Carbon Emissions: Impact on Environmental Sustainability Targets in India. *Environ. Sci. Pollut. Res.* **2021**, *29*, 16311–16323. [CrossRef]
35. Rej, S.; Nag, B. Investigating the Role of Capital Formation to Achieve Carbon Neutrality in India. *Environ. Sci. Pollut. Res.* **2022**, *29*, 60472–60490. [CrossRef] [PubMed]
36. Sadorsky, P. Energy Consumption, Output and Trade in South America. *Energy Econ.* **2012**, *34*, 476–488. [CrossRef]
37. Liu, H.; Kim, H.; Choe, J. Export Diversification, CO2 Emissions and EKC: Panel Data Analysis of 125 Countries. *Asia-Pac. J. Reg. Sci.* **2019**, *3*, 361–393. [CrossRef]
38. Gozgor, G.; Can, M. Export Product Diversification and the Environmental Kuznets Curve: Evidence from Turkey. *Environ. Sci. Pollut. Res.* **2016**, *23*, 21594–21603. [CrossRef]
39. Mania, E. Export Diversification and CO_2 Emissions: An Augmented Environmental Kuznets Curve. *J. Int. Dev.* **2020**, *32*, 168–185. [CrossRef]
40. Sharma, R.; Sinha, A.; Kautish, P. Examining the Nexus between Export Diversification and Environmental Pollution: Evidence from BRICS Nations. *Environ. Sci. Pollut. Res.* **2021**, *28*, 61732–61747. [CrossRef] [PubMed]
41. Zafar, M.W.; Saleem, M.M.; Destek, M.A.; Caglar, A.E. The Dynamic Linkage between Remittances, Export Diversification, Education, Renewable Energy Consumption, Economic Growth, and CO_2 Emissions in Top Remittance-Receiving Countries. *Sustain. Dev.* **2022**, *30*, 165–175. [CrossRef]
42. Islam, M.; Hossain, M.; Khan, M.; Rana, M.; Ema, N.S.; Bekun, F.V. Heading towards Sustainable Environment: Exploring the Dynamic Linkage among Selected Macroeconomic Variables and Ecological Footprint Using a Novel Dynamic ARDL Simulations Approach. *Environ. Sci. Pollut. Res.* **2022**, *29*, 22260–22279. [CrossRef]
43. Charfeddine, L. The Impact of Energy Consumption and Economic Development on Ecological Footprint and CO_2 Emissions: Evidence from a Markov Switching Equilibrium Correction Model. *Energy Econ.* **2017**, *65*, 355–374. [CrossRef]
44. Alola, A.A.; Bekun, F.V.; Sarkodie, S.A. Dynamic Impact of Trade Policy, Economic Growth, Fertility Rate, Renewable and Non-Renewable Energy Consumption on Ecological Footprint in Europe. *Sci. Total Environ.* **2019**, *685*, 702–709. [CrossRef]
45. Hu, H.; Xie, N.; Fang, D.; Zhang, X. The Role of Renewable Energy Consumption and Commercial Services Trade in Carbon Dioxide Reduction: Evidence from 25 Developing Countries. *Appl. Energy* **2018**, *211*, 1229–1244. [CrossRef]
46. Dogan, B.; Madaleno, M.; Tiwari, A.K.; Hammoudeh, S. Impacts of Export Quality on Environmental Degradation: Does Income Matter? *Environ. Sci. Pollut. Res.* **2020**, *27*, 13735–13772. [CrossRef]
47. Rahman, M.M. Do Population Density, Economic Growth, Energy Use and Exports Adversely Affect Environmental Quality in Asian Populous Countries? *Renew. Sustain. Energy Rev.* **2017**, *77*, 506–514. [CrossRef]
48. Hausmann, R.; Hwang, J.; Rodrik, D. What You Export Matters. *J. Econ. Growth* **2007**, *12*, 1–25. [CrossRef]
49. Harding, T.; Javorcik, B.S. Foreign Direct Investment and Export Upgrading. *Rev. Econ. Stat.* **2012**, *94*, 964–980. [CrossRef]
50. Alam, J. Impact of Agriculture, Industry and Service Sector's Value Added in the GDP on CO_2 Emissions of Selected South Asian Countries. *World Rev. Bus. Res.* **2015**, *5*, 39–59.
51. Huan, Y.; Hassan, M.S.; Tahir, M.N.; Mahmood, H.; Al-Darwesh, H.R.I. The Role of Energy Use in Testing N–Shaped Relation between Industrial Development and Environmental Quality for Chinese Economy. *Energy Strategy Rev.* **2022**, *43*, 100905. [CrossRef]
52. Liu, X.; Bae, J. Urbanization and Industrialization Impact of CO_2 Emissions in China. *J. Clean. Prod.* **2018**, *172*, 178–186. [CrossRef]
53. Liu, Y.; Hao, Y. The Dynamic Links between CO_2 Emissions, Energy Consumption and Economic Development in the Countries along "the Belt and Road". *Sci. Total Environ.* **2018**, *645*, 674–683. [CrossRef]
54. Dong, F.; Wang, Y.; Su, B.; Hua, Y.; Zhang, Y. The Process of Peak CO_2 Emissions in Developed Economies: A Perspective of Industrialization and Urbanization. *Resour. Conserv. Recycl.* **2019**, *141*, 61–75. [CrossRef]
55. Ullah, S.; Ozturk, I.; Usman, A.; Majeed, M.T.; Akhtar, P. On the Asymmetric Effects of Premature Deindustrialization on CO_2 Emissions: Evidence from Pakistan. *Environ. Sci. Pollut. Res.* **2020**, *27*, 13692–13702. [CrossRef]
56. Mentel, U.; Wolanin, E.; Eshov, M.; Salahodjaev, R. Industrialization and CO_2 Emissions in Sub-Saharan Africa: The Mitigating Role of Renewable Electricity. *Energies* **2022**, *15*, 946. [CrossRef]
57. Appiah, M.; Li, F.; Korankye, B. Modeling the Linkages among CO_2 Emission, Energy Consumption, and Industrialization in Sub-Saharan African (SSA) Countries. *Environ. Sci. Pollut. Res.* **2021**, *28*, 38506–38521. [CrossRef]

58. Yang, X.; Khan, I. Dynamics among Economic Growth, Urbanization, and Environmental Sustainability in IEA Countries: The Role of Industry Value-Added. *Environ. Sci. Pollut. Res.* **2022**, *29*, 4116–4127. [CrossRef]
59. Abbasi, K.R.; Hussain, K.; Redulescu, M.; Ozturk, I. Does Natural Resources Depletion and Economic Growth Achieve the Carbon Neutrality Target of the UK? A Way Forward towards Sustainable Development. *Resour. Policy* **2021**, *74*, 102341. [CrossRef]
60. Khan, R. Beta Decoupling Relationship between CO_2 Emissions by GDP, Energy Consumption, Electricity Production, Value Added Industries, and Population in China. *PLoS ONE* **2021**, *16*, e0249444. [CrossRef]
61. Shahbaz, M.; Khraief, N.; Uddin, G.S.; Ozturk, I. Environmental Kuznets Curve in an Open Economy: A Bounds Testing and Causality Analysis for Tunisia. *Renew. Sustain. Energy Rev.* **2014**, *34*, 325–336. [CrossRef]
62. Furuoka, F. The CO_2 Emissions–Development Nexus Revisited. *Renew. Sustain. Energy Rev.* **2015**, *51*, 1256–1275. [CrossRef]
63. Magazzino, C.; Porrini, D.; Fusco, G.; Schneider, N. Investigating the Link among ICT, Electricity Consumption, Air Pollution and Economic Growth in EU Countries. *Energy Sour. Part B Econ. Plan. Policy* **2021**, *16*, 976–998.
64. Gyamfi, B.A.; Adedoyin, F.F.; Bein, M.A.; Bekun, F.V. Environmental Implications of N-Shaped Environmental Kuznets Curve for E7 Countries. *Environ. Sci. Pollut. Res.* **2021**, *28*, 33072–33082. [CrossRef]
65. Ozcan, B.; Tzeremes, P.G.; Tzeremes, N.G. Energy Consumption, Economic Growth and Environmental Degradation in OECD Countries. *Econ. Model.* **2020**, *84*, 203–213. [CrossRef]
66. Shahbaz, M.; Haouas, I.; Van Hoang, T.H. Economic Growth and Environmental Degradation in Vietnam: Is the Environmental Kuznets Curve a Complete Picture? *Emerg. Mark. Rev.* **2019**, *38*, 197–218. [CrossRef]
67. Naseem, S.; Mohsin, M.; Zia-UR-Rehman, M.; Baig, S.A.; Sarfraz, M. The Influence of Energy Consumption and Economic Growth on Environmental Degradation in BRICS Countries: An Application of the ARDL Model and Decoupling Index. *Environ. Sci Pollut. Res.* **2022**, *29*, 13042–13055. [CrossRef]
68. Rahman, H.U.; Zaman, U.; Górecki, J. The Role of Energy Consumption, Economic Growth and Globalization in Environmental Degradation: Empirical Evidence from the Brics Region. *Sustainability* **2021**, *13*, 1924. [CrossRef]
69. Danish, H.S.; Baloch, M.A.; Mahmood, N.; Zhang, J.W. Linking Economic Growth and Ecological Footprint through Human Capital and Biocapacity. *Sustain. Cities Soc.* **2019**, *47*, 101516. [CrossRef]
70. Ahmad, M.; Jiang, P.; Majeed, A.; Umar, M.; Khan, Z.; Muhammad, S. The Dynamic Impact of Natural Resources, Technological Innovations and Economic Growth on Ecological Footprint: An Advanced Panel Data Estimation. *Resour. Policy* **2020**, *69*, 101817 [CrossRef]
71. Ahmed, Z.; Asghar, M.M.; Malik, M.N.; Nawaz, K. Moving towards a Sustainable Environment: The Dynamic Linkage between Natural Resources, Human Capital, Urbanization, Economic Growth, and Ecological Footprint in China. *Resour. Policy* **2020**, *67*, 101677. [CrossRef]
72. Galli, A.; Kitzes, J.; Niccolucci, V.; Wackernagel, M.; Wada, Y.; Marchettini, N. Assessing the Global Environmental Consequences of Economic Growth through the Ecological Footprint: A Focus on China and India. *Ecol. Indic.* **2012**, *17*, 99–107. [CrossRef]
73. Ahmed, Z.; Wang, Z. Investigating the Impact of Human Capital on the Ecological Footprint in India: An Empirical Analysis *Environ. Sci. Pollut. Res.* **2019**, *26*, 26782–26796. [CrossRef]
74. Murshed, M.; Nurmakhanova, M.; Al-Tal, R.; Mahmood, H.; Elheddad, M.; Ahmed, R. Can Intra-Regional Trade, Renewable Energy Use, Foreign Direct Investments, and Economic Growth Mitigate Ecological Footprints in South Asia? *Energy Sour. Part B Econ. Plan. Policy* **2022**, 1–26. [CrossRef]
75. Adebayo, T.S.; Akadiri, S.S.; Altuntaş, M.; Awosusi, A.A. Environmental Effects of Structural Change, Hydro and Coal Energy Consumption on Ecological Footprint in India: Insights from the Novel Dynamic ARDL Simulation. *Environ. Dev. Sustain.* **2022** 1–24. [CrossRef]
76. Engle, R.F.; Granger, C.W. Co-Integration and Error Correction: Representation, Estimation, and Testing. *Econom. J. Econom. Soc* **1987**, *55*, 251–276. [CrossRef]
77. Johansen, S.; Juselius, K. Maximum Likelihood Estimation and Inference on Cointegration—With Appucations to the Demand for Money. *Oxf. Bull. Econ. Stat.* **1990**, *52*, 169–210. [CrossRef]
78. Pesaran, M.H.; Shin, Y.; Smith, R.J. Bounds Testing Approaches to the Analysis of Level Relationships. *J. Appl. Econom.* **2001**, *16*, 289–326. [CrossRef]
79. McNown, R.; Sam, C.Y.; Goh, S.K. Bootstrapping the Autoregressive Distributed Lag Test for Cointegration. *Appl. Econ.* **2018**, *50*, 1509–1521. [CrossRef]
80. Sam, C.Y.; McNown, R.; Goh, S.K. An Augmented Autoregressive Distributed Lag Bounds Test for Cointegration. *Econ. Model.* **2019**, *80*, 130–141. [CrossRef]
81. Narayan, P.K. The Saving and Investment Nexus for China: Evidence from Cointegration Tests. *Appl. Econ.* **2005**, *37*, 1979–1990 [CrossRef]
82. Jordan, S.; Philips, A.Q. Cointegration Testing and Dynamic Simulations of Autoregressive Distributed Lag Models. *Stata J.* **2018**, *18*, 902–923. [CrossRef]
83. Hossain, M.E.; Islam, M.S.; Bandyopadhyay, A.; Awan, A.; Hossain, M.R.; Rej, S. Mexico at the Crossroads of Natural Resource Dependence and COP26 Pledge: Does Technological Innovation Help? *Resour. Policy* **2022**, *77*, 102710. [CrossRef]
84. Hossain, M.; Islam, M.; Sujan, M.; Khan, H.; Tuhin, M.; Bekun, F.V. Towards a Clean Production by Exploring the Nexus between Agricultural Ecosystem and Environmental Degradation Using Novel Dynamic ARDL Simulations Approach. *Environ. Sci. Pollut. Res.* **2022**, *29*, 53768–53784. [CrossRef]

85. Breitung, J.; Candelon, B. Testing for Short-and Long-Run Causality: A Frequency-Domain Approach. *J. Econom.* **2006**, *132*, 363–378. [CrossRef]
86. Narayan, P.K.; Narayan, S. Carbon Dioxide Emissions and Economic Growth: Panel Data Evidence from Developing Countries. *Energy Policy* **2010**, *38*, 661–666. [CrossRef]
87. Rej, S.; Bandyopadhyay, A.; Das, N.; Hossain, M.; Islam, M.; Bera, P.; Yeedibali, T. The Asymmetric Influence of Environmental-Related Technological Innovation on Climate Change Mitigation: What Role Do FDI and Renewable Energy Play? *Environ. Sci. Pollut. Res.* **2022**, 1–16. [CrossRef]
88. Kanjilal, K.; Ghosh, S. Environmental Kuznet's Curve for India: Evidence from Tests for Cointegration with Unknown Structural-breaks. *Energy Policy* **2013**, *56*, 509–515. [CrossRef]
89. Jayanthakumaran, K.; Verma, R.; Liu, Y. CO_2 Emissions, Energy Consumption, Trade and Income: A Comparative Analysis of China and India. *Energy Policy* **2012**, *42*, 450–460. [CrossRef]
90. Bandyopadhyay, A.; Rej, S. Can Nuclear Energy Fuel an Environmentally Sustainable Economic Growth? Revisiting the EKC Hypothesis for India. *Environ. Sci. Pollut. Res.* **2021**, *28*, 63065–63086. [CrossRef]
91. Rej, S.; Bandyopadhyay, A.; Mahmood, H.; Murshed, M.; Mahmud, S. The Role of Liquefied Petroleum Gas in Decarbonizing India: Fresh Evidence from Wavelet–Partial Wavelet Coherence Approach. *Environ. Sci. Pollut. Res.* **2022**, *29*, 35862–35883. [CrossRef]
92. Bandyopadhyay, A.; Rej, S.; Abbasi, K.R.; Awan, A. Nexus between Tourism, Hydropower, and CO_2 Emissions in India: Fresh Insights from ARDL and Cumulative Fourier Frequency Domain Causality. *Environ. Dev. Sustain.* **2022**, 1–25. [CrossRef]
93. Rej, S.; Bandyopadhyay, A.; Murshed, M.; Mahmood, H.; Razzaq, A. Pathways to Decarbonization in India: The Role of Environmentally Friendly Tourism Development. *Environ. Sci. Pollut. Res.* **2022**, *29*, 50281–50302. [CrossRef]

sustainability

MDPI

Article

Key Determinants of Municipal Waste Sorting in Slovakia

Michal Stričík * and Monika Čonková

Faculty of Business Economics with Seat in Košice, University of Economics in Bratislava, 04130 Košice, Slovakia; monika.conkova@euba.sk
* Correspondence: michal.stricik@euba.sk

Abstract: The Slovak Republic does not meet the targets of the waste economy in the long run. In order to meet these objectives, it is necessary to make changes to the current system of municipal waste management. Building on an empirical analysis, this paper focuses on the evaluation of the production of municipal waste and the factors that influence the level of municipal waste sorting as a prerequisite for the maximal re-use, recovery, or recycling of municipal waste. The type of fee for municipal waste was confirmed as the most significant factor for the higher rate of municipal waste sorting, and pertinent recommendations were suggested according to the needs of Slovak municipalities.

Keywords: municipal waste; waste sorting; waste management; waste recovery; municipal waste charging; waste disposal

Citation: Stričík, M.; Čonková, M. Key Determinants of Municipal Waste Sorting in Slovakia. *Sustainability* **2021**, *13*, 13723. https://doi.org/10.3390/su132413723

Academic Editors: Nicola Raimo, Filippo Vitolla, Ornella Malandrino and Benedetta Esposito

Received: 3 November 2021
Accepted: 4 December 2021
Published: 13 December 2021

1. Introduction

The highest priority of the sophisticated waste economy is to prevent waste generation [1]. The binding waste management hierarchy is the cornerstone of European policies and waste legislation. It is key for the transition to the circular economy. The basic goal of the waste management hierarchy is to minimize the adverse effects on the environment and increase, as well as optimize, the efficiency of waste management resources [2,3].

The European Commission (EC) adopted a European Union (EU) action plan for the circular economy in 2015. The circular economy (CE) considers waste as a usable resource. The banning prohibition increases the waste recovery rate. Countries with limited municipal landfills, as well as recyclable and reusable waste, have a higher rate of utilization and packaging of municipal waste [4]. The circular economy does not arise as an isolated European project because, in particular, Asian countries have applied this concept for several years. The importance of these changes also recognizes North America and some South American countries. Adoption of a circuit concept in most countries, or optimally worldwide, is a basic condition for the success of its application [5].

Analysis of waste separating for recycling in households is crucial, particularly at a global scale [6]. According to Eurostat data, 221,610 thousand tons of municipal waste (MW) were produced in 2018 in 27 EU countries [7]—i.e., 496 kg per capita. Compared to other EU countries, the production of MW in Slovakia is relatively low. In fact, the production of MW per capita in Slovakia is lower than the EU's average.

According to the Report on the State of the Environment of the Slovak Republic, over 13,478 million tons of waste was produced in Slovakia in 2018. Almost 2325 million tons of this amount was made up of municipal waste, which means 427 kg of MW per capita. The total amount of MW, as well as the MW per capita, has mainly, in the last several years, had the tendency to annually grow [8,9]—especially during the period from 2013–2018 where this amount increased at an annual average of 5.24%.

The Slovak Republic has set objectives in the area of waste management based on the objectives adopted by the EU [2]. One of these goals was to achieve a 50% recycling rate of municipal waste by 2020 [10]. In 2018, the EC sent a warning note, wherein the

EC perceived a risk that in Slovakia in 2020 the re-use/recycling rate of MW would not increase to 50% [11]. However, we can already know that the given objective was not fulfilled since the rate of recycling in 2020 was 43.67% [9].

The actual objective of the Slovak Republic by 2025 is to increase the rate of separated MW to 60% and the rate of preparation for reuse and recycling of MW to 55%. To achieve this goal, it is necessary to increase the level of separation of recyclable parts of MW. Since collected and sorted MW constituents are not fully recyclable, targets for sorted MW collections exceed the recycling objective itself. The targets for the sorted collection rate were increasing gradually from 20% in 2016 to 65% in 2020, with 60% as the target for the recycling rate in 2025 [10,12].

The EU considers the area of MW management very important. The European Parliament approved ambitious objectives in April 2018, namely that by 2025 the rate of municipal waste recycling would increase to 55% and then to 65% by 2035. Another ambitious goal is to achieve that, by the year 2035, more than 10% of waste does not end up in landfills [13].

In order to meet these objectives, it is necessary to make changes to the current system of MW management in individual cities, boroughs, and villages in Slovakia (as there has, for a long time, been a high proportion of landfilled total waste). It is striking that despite this fact, in 2017 more than 61% of MW in Slovakia was landfilled (5% year-to-year decrease). The EU average amount of landfilled waste is below 25%, and is gradually decreasing. The aim of the Slovak Republic is therefore to reduce the amount of waste disposal in such a way, and thus achieve a gradual transition to more appropriate forms of MW management. In 2017, 29% of the total processed MW, including composting, was recycled in Slovakia [14,15].

Waste recycling is largely affected by policy, recycling program and social norm variables as they influence incentives and intentions for recycling.

The level of waste recycling is largely influenced by the policy and objectives of recycling and social standards because these initiate incentives and recycling efforts [16,17]. Some of the important success factors in implementing preventive measures for municipal waste production and increasing municipal waste recycling rates are the attitudes of the residents on the issues of municipal waste management, the perception of the consequences of environmental contamination, and the legislative measures [18].

Waste segregation at the source (i.e., at the household level) plays a very important role in effectively managing municipal waste [19]. There are various research findings about the influence of incentives to increase the ratio of separated parts of MW. For example, the positive impact of motivation factors were documented by studies in the Czech Republic [20,21], as well as other countries [22]. Research in Italy has shown that the motivation of individuals does not correlate with an increase in behavior that affects the proportion of a recyclable part of MW, while the behavior responsible for minimizing the quantity of the MW depends only on the internal conviction [23,24]. The more knowledge citizens have about the impact of MW on the environment and, at the same time, believing in the meaning of the measures in increasing recycling rates, the more they are willing to cooperate and sort waste [25,26]. Of course, ultimately, for recycling and overall waste management, a multidimensional approach is decisive when in addition to the motivation of citizens, as well as when legislative, financial, and logistical aspects are considered [27,28]. The use of waste gases and waste within the circular economy can also contribute to the matter [29]. Several studies were concerned at the influence of the mean, and the size of the individual expenses, related to the MW production and its management (for example [17,30,31]). The study by [32] analyzed the choice of the method for calculating charges for waste disposal and discussed the consequences of the introduction of the different options—for example, according to the water consumption. The focus on finance, legislation, and information stems from traditional economic models based on rational choices, which assume that consumers make choices by calculating the costs and benefits to them in each situation, optimizing their own personal gains [33]. Study [34] proves the role of environmental awareness in waste sorting.

The linear economy must be changed to the functional CE. Successful initiatives in the field of CE reduce societal dependence on natural resources, and, at the same time, create a value for society. The portal of World Business Council for Sustainable Development estimates that worldwide there is a yet exploited potential for CE of $4.5 trillion [35].

In recent years, the EC adopted from member states several legislative proposals to support the transition from the linear model of production and consumption ("take-make-consume-throw away") to a new model of CE—leading to the conclusion of the imaginary material flow. The aim is to prevent waste generation, emphasize eco-design, and reuse, amongst other similar measures. According to estimates, it could bring net savings of €600 billion, while reducing total year-old greenhouse gas emissions by 2–4%. The CE also has a positive impact in reducing unemployment in Europe—it is estimated to create 580,000 new waste processing jobs [36].

The EU Action Plan for CE was created by the EC as a tool to achieve the objectives of the Sustainable Development Agenda by 2020, and, in particular, focus on step no. 12 for "sustainable consumption and production". It dealt with topics such as production, consumption, waste management (i.e., prevention, preparation for reuse, recycling, energy recovery, and disposal), market support with secondary raw materials, and others [37].

The above-mentioned aim was not fulfilled at last, but although recycling level in Slovakia is continuously low, and despite the growing volume of communal waste, a positive development was recorded in waste recycling [38]. Since the aims were known and measures for both the public and private sector, as well as the efforts of respective organizations for producer responsibility, were not fully successful, it is necessary to continuously try to improve the Slovak MW management system. In such, situation countries could strive to replicate "success stories" of high-income countries, but it must not be conducted without paying particular attention to the respective costs, required skills, education, and technical expertise [39], as well as without appropriate analyses and tests.

2. The Aim, Materials, and Methods

2.1. The Aim

Slovakia and its municipalities entering the route of purposefulness, efficiency, effectiveness, and sustainability in solid MW management are still relatively at the beginning, therefore different adequate analyses based on a sufficient amount of relevant data are necessary for the decision-making and determining of the assumptions and parameters for the MW management system. Scientific projects conducted by a university team aimed at the aspects of MW management in the Slovak Republic are the means for gaining useful systematic analytical information.

The aim of this article, as a partial output of the project started in 2018 in Slovakia, is to assess key determinants of MW management and to draft measures that would contribute to improvement in the area.

2.2. Specification of Research Areas and Research Methodology

Concerning the purposeful managing of waste management issues at rural and urban authorities, we expect that we obtained relevant information from the relevant respondents. In the analyses, we did not evaluate the data applicable in the population of Slovak citizens, only that within the population of municipal employees and municipal authorities competent in municipal waste, from which we obtained a sufficiently large representative sample of data. The representation of other citizens participating in our project is statistically unrepresentative, but within the project we were interested in more-or-less relatively significant differences in the compared groups. The results of the statistical tests, as well as conclusions of our experiences and the personal opinions of the professionals participating in the survey, were finally summarized as the recommendations in connection with the improvement of the situation in Slovak waste management.

Our main research hypothesis results from our experiences (for example [40–44]) in the studied area and from literature review (noted not only above in the introduction). We suppose that there exists one or more fac9tors influencing the ratio of the sorted MW in municipalities. The main research hypothesis was examined through the partial statistical hypotheses concerning particular examined areas using appropriate statistical tests. Statistically-confirmed results were fundamental for relevant suggestions for the legislative changes.

In the analysis of data from the questionnaire described below, we focused on a simplified illustrative classification of factors of the degree of sorting of municipal waste and its prediction of the level of sorting. For automatic determination of factors influencing the level of municipal waste sorting, or classification of data according to whether they belong to the particular level of MW classification, we used the method of classification (decision) tree. Then we used cluster analysis to assess the classification tendencies of the selected variables.

These targeted data analyses were preceded by a statistical verification of the properties of the questionnaire.

2.3. Source of Data

In order to assess key aspects of municipal waste management in the Slovak Republic, we conducted a systemic analysis from 2018–2019. For the analysis, we used the data collected through the questionnaire survey of citizens' views for whom communal waste management issues are part of their responsibilities and labor powers. That is to say, we have systematically asked self-government staff or representatives of business waste management companies. These citizens were considered proper respondents given their knowledge of the relevant information from the area in question. At the same time, they should actively pay attention and consider the time of their workload in this matter. Their awareness should be largely based on knowledge, and not only on feelings, intuition, or random experiences. In the case of the creation decisions by these citizens, such decisions are verified in practice or at least tested, and the need for changes in the problem of these issues should be accompanied by previous experience, solutions, considerations, discussions, and team evaluations, and should be carried out by more responsible self-government officers.

Questions in the survey were focused on personal opinions, attitudes, the behavior of citizens, and, if the citizen was a representative of the municipal office (i.e., village/town/local in the city), the questionnaire was also extended with a part in which the respondent, no longer as a citizen but on behalf of the self-government, was asked for answers specifically regarding quantification issues on the state of municipal waste management.

In the first section, personal part respondents were asked to answer questions related to their place of residence (I.e., permanent or transient), where they pay municipal waste charges, and where they live during most of the year. If the respondent worked in a municipal office that was not the same municipality in which he lived, we gained information from one respondent about two different locations. If the respondent had an identical location of residence and workplace, we have gained a view of the citizen and officer, and we estimated that such citizens would show more frequent characteristics supporting waste management optimization as conventional citizens due to their conventional behavior with labor knowledge, responsibility, and purposefulness.

The general part of the survey form about the respondents and their residence data are used for stratifying responses and finding relevant relationships, and subsequently for finding improvement solutions—at least in the field of municipal waste sorting in Slovakia. In the questionnaire, above-mentioned factors of the condition and development of ecology and waste management were captured directly or indirectly. In the questionnaire, the questions made up several planes representing the environmental and waste management status, and whose responses could be evaluated alone, but they were also used as stratification variables—or as variables that assigned relevant weight to other variables—and they should have the roles of causality, effect, or bonds.

Due to the breadth of topics, we could not devote attention to each field extensively enough, so the accuracy and reliability of the questionnaire was measured particularly by a group of four questions—no. 9, 10, 11 and 12—adopted from a similar foreign survey [45]. These issues also confirmed the validity of the survey. Areas covered within the questions in the questionnaire are summarized in Table 1. In the article, we retain the questions with the numbering used in their original version. Not all questions from the survey have been used in this article.

Table 1. Areas of the survey.

	Survey Areas
A.	Perception and attitude towards environmental issues—expecting that pro-environmental orientation is a prerequisite for the spontaneous purposeful approach of citizens to minimize waste production. (Question 1)
B.	Involvement in environmental protection—like attitudes, actions in favor of environmental sustainability should be an even stronger argument in favor of the optimization of waste production. (Questions 2, 3)
C.	Perceptions, knowledge, their application, and declared behavior on municipal waste issues—are the basis for a purposeful conduct towards improving waste management. (Questions 4–12)
D.	Levels in the sorting of municipal waste components—as an objective indicator of the status and performance indicator of MW management targets. (Questions 13–21)
E.	Motivation, internal or external, in sorting or generally in municipal waste management—assuming that personal motivation of citizens is in addition to the legislative measures and influential factors on the level of MW management (Questions 22–27)
F.	Improvement proposals—suggestions for dealing with the MW management. (Question 28)
G.	Basic categorization of respondents to self-government workers and other citizens. (Question 29–33)
H.	For municipal representatives, questions related to MW management in their municipality, including specifications of quantity. (Questions 34–66)
I.	General demographic data designed for categorization of responses and voluntary insights, feedback. (Questions 67–80)
J.	Significant relationships and dependencies related to the perception of the environment, with the production, prevention, sorting, motivation, and other aspects of municipal waste management that will be determined by using statistical methods.

Source: primary data.

2.4. Data File Characteristics

The questionnaire was distributed electronically using a form tool from Google. All towns/cities (in the case of Bratislava and Košice all urban parts) and villages in Slovakia (hereinafter referred to as the municipalities) with the available email addresses were addressed during several weeks—the total selection method was used. We requested municipalities to cooperate in acquiring information on the ways of MW management and sorting, and the further distribution of questionnaires.

Overall, in the first stage, following the pilot testing of the questionnaire, the 2927 unique e-mails on the mostly publicly available addresses of Slovak municipalities, cities, and urban parts (a few tens of addresses came from private sources of project team members were used. The email addresses of two municipalities were not traceable.

We also distributed messages to the alternative email addresses of employees of the self-government who were called by phone when we found that the email was undeliverable to any known address, or when the email was deleted without reading and submitting responses about the municipality in the Google form table. In the second stage, we again sent emails to the addresses of the offices from which we did not receive the answers in an adequate time.

Thus, in the first May–June stage, we addressed all 2927 [46] offices (total number of municipalities without urban parts, including Bratislava and Košice as a whole: 2890, total number of urban parts: Bratislava (17), Košice (22), Total number of towns/cities: 140, total number of rural municipalities: 2750 (including 3 military circuits)) (Figure 1).

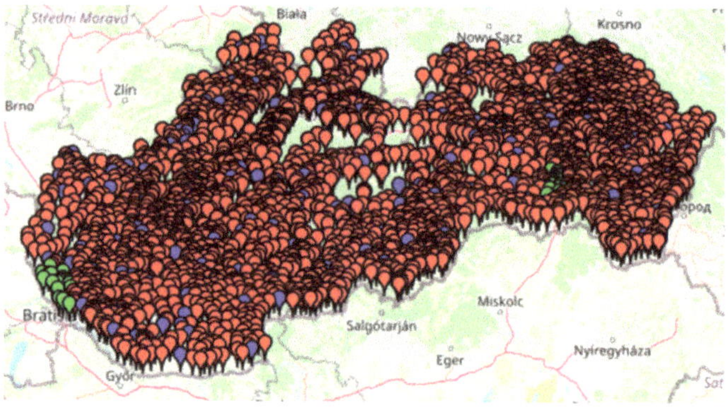

Figure 1. Dot map of the municipalities of the Slovak Republic (red-villages, blue-towns, green-urban parts). They were all addressed in the survey. Source: http://www.sodbtn.sk/obce/obce_body.php (accessed on 15 September 2020). Reprinted with permission from [46]. © Vladimír Bačík.

In the stage conducted during July, in which return rate was higher, we sent again 2438 requests to municipalities with functioning e-mail addresses (44 offices were not addressed due to non-working or unfindable e-mail address.

The specific Slovak population sample was a group of citizens and employees of the self-government, relevant by its quantity, who represented not only themselves but the office agenda of other citizens in the municipality, as well as executives of the municipality. This is why we can consider them as the indicator and the catalyzer for the municipal waste management, and particularly, the locality's ecological state and development (Figure 2).

Another group of survey respondents consisted of randomly addressed citizens, although they were partly included from chain-addressing by other respondents. The randomly addressed citizens were used as a contrast to the target group since they were not professionally included in the MW issues. In this group, citizens were not working in the relevant positions, and given this, the group represented ordinary citizens without the responsibility for meeting the objectives and did not have the expected knowledge in the area of MW management. The intention of collecting data from this group of citizens was to obtain a control sample as well as awareness and feedback on the existing reality in the area of municipal waste management from the ordinary citizens' circuit, given that they were not involved in the creation, securing, controlling, or recording of the municipal waste measures. In the analysis of this supplementary group of respondents, unlike the target group of municipal respondents, the conclusions must be presented prudently with regard to the possible shift of results due to the absence in randomness of selection of the respondents coming from the surroundings of the self-government staff, as well as readers of selected periodicals, general newspapers, and so on (Figure 3).

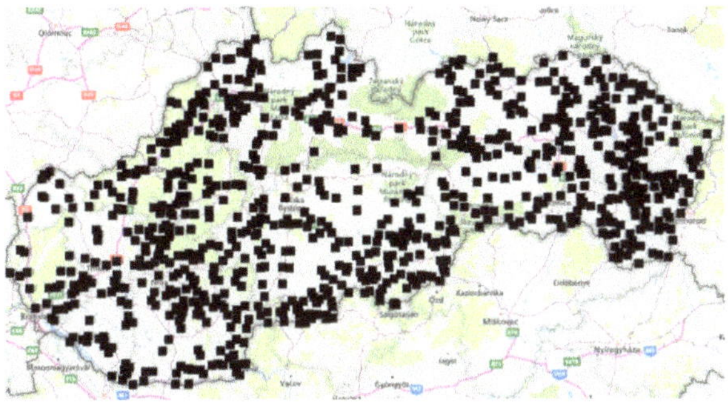

Figure 2. Municipalities involved in the survey with the response of one or more employees of the self-government office. Source: Map of the MS Excel, own data processing.

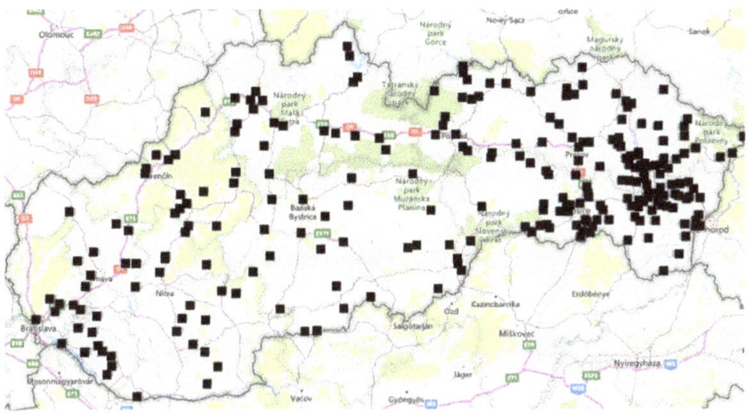

Figure 3. Residence places of the respondents that were not employed by the self-government that were involved in the survey, based on their response to the questionnaire. Source: Maps MS Excel, own data processing.

During the whole survey period, we had gained 1439 relevant responses from all respondents, but replays from the non-repeated localities of municipal representatives with the knowledge of municipal waste management were from 947 villages, towns/cities, which is a 32.23% proportion of all self-government offices of the municipalities. Two respondents from the self-government offices did not indicate the particular location of their office, along with some other demographic data.

Two cases were deleted from the database due to the identity of values with another two observations (except the observation time track that varied within several seconds). These dual observations could have been sent repeatedly and been double recorded in the result table due to an unstable internet connection of the respondents. From among the citizens, another response was further excluded since the respondent's answers on open issues were empty or without meaning.

Respondents from the municipalities included in the survey multiple times were corrected by weights so in the analysis they were counted only once. If various answers for a respective municipality entered the analysis in different categories with the quantity expressed as a decimal number, this was rounded.

2.5. Characteristics of Data Set of Respondents Working at the Self-Government Office

Since we addressed all municipalities, we are not conducting a random sampling. Our municipalities samples are all authorities that expressed their willingness to participate in the questionnaire survey. Potential distortion we will verify by comparing some of the acquired information with known published statistical facts. By addressing all municipalities, we expected the representation of respondents to not be shifted with respect to the demographic data, which was confirmed by the comparison of the ratios of the counts of municipalities that participated in the survey with the actual distribution of the municipalities in the individual areas, regions, districts, and urban/rural locations (–4). The actual proportion of women versus men working in the relevant positions of self-government offices, as well as representations of other social characteristics, are not known, so we did not verify this detail.

In the case of an unbiased data sample, conclusions are estimated with the statistical error for estimating the proportion of the examined characteristics for the entire data file, which is maximally equal to 3.2% (Equation (2) according to the formula for determining the minimum size of the sample of respondents [47] (Equation (1), which we used when conducting the survey based on the random selection:

$$n = Z^2 \cdot \frac{p \cdot (1-p)}{C^2} \tag{1}$$

$$C = \sqrt{\frac{Z^2 \cdot p \cdot (1-p)}{n}} = \sqrt{\frac{1.96^2 \cdot 0.5 \cdot (1-0.5)}{n}} \doteq 0.03185 \tag{2}$$

where:

Z—statistical table value. For statistical significance equal to 95%, value Z equals to 1.96, for significance 99%, $Z = 2.58$;

n—sample size (population size is not taken into account);

p—proportion of the examined characteristic (if the proportion is not known then p is thus replaced with the value of the sample size without considering the population size 0.5, which leads to the highest sample size estimation;

C—acceptable error interval.

To ensure the basic representativeness of our sample, we verified the relative representation of the participating municipalities/districts/regions/counties/country, corresponding approximately to the actual proportional representation of all the municipalities in the relevant territorial units. In the case of comparing the relative proportions of municipalities in Western, Central and Eastern Slovakia, in reality and in our survey, the differences were +0.1, −1.9% and +1.8%, respectively (Table 2).

Table 2. Proportional representation of municipalities participating in the survey in comparison with the actual proportional representation of municipalities in individual areas of Slovakia.

Counties in Slovakia	Municipalities	Municipalities Proportion	Municipalities in the Survey	Municipalities Proportion in the Survey	Difference between the Proportions
Western Slovakia	694	23.7%	223	23.6%	0.1%
Central Slovakia	1107	37.8%	375	39.7%	−1.9%
Eastern Slovakia	1126	38.5%	347	36.7%	1.8%
Unspecified location			2	not considered	
Total	2927	100%	945(+2)	100%	0%

Source: www.sodbtn.sk (accessed on 15 September 2020) and own data processing. Reprinted with permission from [46]. © Vladimír Bačík.

The representations of the municipalities in individual regions, in reality and in the survey, and the differences in representations are summarized in Table 3. The largest differences are lower in representation in the survey when compared to reality by −5.6% in the Prešov region and higher by 3.9% in the Košice region, which is negligible.

Table 3. Proportional representation of municipalities in individual regions of Slovakia.

Region	Number of Municipalities in Regions	Proportion of Municipalities in Regions	Number of Individual Municipalities in Regions Participating in the Survey	Proportion of Individual Municipalities in Regions Participating in the Survey	Difference between Proportions of Individual Municipalities in Regions in Reality and in the Survey
Bratislava	89	3.0%	27	2.9%	−0.1%
Trnava	251	8.6%	53	5.6%	−3.0%
Trenčín	276	9.4%	90	9.5%	0.1%
Nitria	354	12.1%	143	15.1%	3.0%
Žilina	315	10.8%	108	11.4%	0.6%
Banská Bystrica	516	17.6%	177	18.7%	1.1%
Prešov	665	22.7%	162	17.1%	**−5.6%**
Košice	461	15.7%	185	19.6%	**3.9%**
Unspecified location	-	-	2	not considered	
Total	2927	100.0%	945(+2)	100.0%	0%

Source: www.sodbtn.sk (accessed on 19 September 2020) and own data processing. Reprinted with permission from [46]. © Vladimír Bačík.

Verification of the representation of the urban or rural municipalities is summarized in Table 4. The differences in the representation of the urban and rural municipalities are maximally ±4.6%, whereby in the survey there is a slightly higher ratio of urban municipalities and a slightly lower representation of rural municipalities.

Table 4. Proportional representation of Slovak municipalities and their representation in the survey.

Type of Municipality	Number of Municipalities	Proportion of Municipalities in Slovakia	Number of Individual Municipalities in the Survey	Proportion of Individual Municipalities in the Survey	Difference between Proportions
Towns/Cities	140	4.8%	89	9.4%	4.6%
Villages	2750	95.2%	858	90.6%	−4.6%
Total	2890	100.0%	947	100.0%	

Source: www.sodbtn.sk (accessed on 15 September 2020) and own data processing. Reprinted with permission from [46]. © Vladimír Bačík.

The representativeness of the sample of respondents among local self-government employees by gender, age, education, number of household members, and other characteristics has not been verified due to the unknown distribution in the population of local self-government employees, but we do not anticipate bias due to the availability of a sufficiently large research sample.

2.6. Survey Items (Questions)

The questionnaire consisted of a relatively large number of open, closed, and most often semi-closed questions, with the possibility of commenting or further specification of the answer. The questionnaire was branched and not all questions were displayed to each respondent. The answers to the questions on the questionnaire, and their analysis, should help in specifying the situation in the field of municipal waste management, the perception of environmental issues, waste sorting, and the proposals for optimizing waste management.

Although the sample of respondents among citizens not working at self-government offices is not representative, and we cannot apply all the conclusions of the survey automatically to all citizens, the sample size of the self-government staff and its representativeness, at least with respect to all self-government staff, allows us to apply the found results and statistically significant relationships, but also the individual opinions of the respondents to expanding the knowledge base on which it will be possible to design a more efficient waste management system.

The items of the questionnaire cover 10 areas of interest, and studies and analyses are listed in Section 2.3. Most of the questions are for determining the attitudes, perceptions and activities, which were conceived using a 6-point Likert scale. This allows us, if the sample size is sufficient, to more accurately determine the intensity of the perception interest, severity, and relationships of the question. Questions with a Likert scale were also used to analyze the properties of the questionnaire. In case of an insufficient representation of answers at the individual levels of the scale, we can combine these into a smaller number of levels. In order to clarify the content of more complex questions, we left open the possibility of specifying the answer. The six-point scale remained in use. As such additionally specified answers were attached to the basic scale with the possibility for additional consideration in the discussion on the analysis of the questionnaire.

2.7. Statistical Tools Used in Data Analyses

In the analyses of the questionnaire data, the statistical tools summarized below are used using the software MS Excel [48], IBM SPSS Statistics [49], or Gretl [50]. The analyses' quantitative conclusions are based on statistically verified results using methods briefly characterized below.

2.8. Significance Level

The significance level is an estimated likelihood for rejection of a null hypothesis assuming it was right (probability of error of the first type), and to determine it each time before calculating the test criterion (i.e., before the test). The p-value is the lowest possible level of significance, designed based on the test criterion, where we can also reject the null hypothesis. It is one of the options for which we decide whether the result is statistically significant. In our analyses, we consider it a statistically significant result when the p-value ≤ 0.05.

2.9. Factor Analysis in Assessing the Validity of the Questionnaire

Validity is an important indicator of the questionnaire's quality as a measurement tool. The degree of validity indicates how the test/questionnaire measures the concept (i.e., phenomenon or construct) on which the questionnaire is focused. There are more types of validity and more methods of its destination. The statistical test of the questionnaire is a factor analysis that identifies those that are related to each other between multiple variables. Analyzing those that are closely linked to the so-called factors thereby reduces the number of variables to a smaller number of factors and confirms the eligibility of the measuring tool.

2.10. Questionnaire Reliability Analysis

The reliability index evaluates the accuracy of such measurements. If measuring tools are not valid and reliable, there is an incorrect interpretation of the phenomena and the application of incorrectly-related decisions. A reliable measuring tool provides the same results by repeating the evaluation.

2.11. Decision Trees

Decision trees are used for classification or prediction of categorical or continuous values. They are most commonly used as a non-parametric and non-linear alternative to a linear model. They are not affected by extreme or missing values. For analysis of relations between target and input variables, we used a method of a full chi-square automatic interactive detection, or so-called Exhaustive CHAID (Chi-Square Automatic Interaction Detector) that identifies the file's explanatory variables to classify/predict the target variable.

A CHAID Classification tree divides a tree node only if the statistical significance criterion is met. CHAID tries to prevent this from the beginning of the so-called overfitting.

In our work, we will use the decision tree to specify (i.e., predict) the level of sorting of the municipal waste based on the relevant input variables.

2.12. Cluster Analysis

Cluster Analysis was used for searching in the empirical data and grouping similar objects–types [51]. We used it to find characters for generating object classification assumptions. In the paper, a two-step cluster was used by IBM SPSS software that automates and solves some problems of standard cluster techniques.

By means of the cluster method, we specify the characteristic clusters for the municipal waste sorting levels corresponding to the individual levels of the considered input variables.

3. Results

3.1. Verification of the Properties of the Questionnaire

One of the fundamental conditions for the questionnaire's research is the use of valid and reliable research instruments. By determining the questionnaire's validity, we verify whether the questionnaire measures what we intended to detect. The instrument validation is important when the respondents provide, in essence, inaccurate, subjective, or opinion-based characteristics. In addition to the confirmation of the construct validity (i.e., measurement of a certain feature of a man) questionnaire in its further processing, it is then possible to replace multiple variables by one factor representing the construct examined.

Several variables are substitutable by one latent variable, a factor, and arise if we determine the characteristics of the respondents, their attitudes, tendencies, and other features. It is not sufficient to ask the relevant specificity directly with a single question because the respondent may either unconsciously or consciously distort the reality of the assessment for that feature under investigation. For example, a respondent may state the subjectively or objectively desired state of affairs, while even to himself/herself may not admit the fact. Several more specifically focused questions related to the examined characteristics forming the so-called manifest variables can more reliably characterize the respondent, concerning the examined feature, than only his own answer could. The more the questions are aimed at identifying the investigated feature, the more objective the resulting findings are in terms of the concomitant treatment of the counterproductive redundancy of the number of questions. Statistical methods, and other ways of assessing the properties of the questionnaire as a measuring tool in its preliminary verification or its subsequent evaluation, help in deciding on the appropriateness of including the question in the structure of the measuring tool and form a picture of the extent to which the examined property manifests itself in the respondent.

In our research, the area for determining the respondents' environmental orientation was more closely represented by the questions, which was also the subject of the assessment of validity and reliability, and for which we singled out four questions taken from similarly focused foreign research [45]. Our questionnaire was based on the objectives of a large project covering a wide area and mapping the issues of municipal waste management from the perspective of citizens and municipalities (see Section 2.3). Given this, it was not possible to include enough questions for each area to create a comprehensive, statistically valid, and reliable questionnaire without negatively influencing perceptions and some of the respondents not completing the questionnaire. Therefore, other areas were not confirmed by multiple inquiries, which could lead to incorrect categorization/evaluation of the respondent in individual cases. However, due to the purpose of the questionnaire, which was not to evaluate individual respondents or individual municipalities, the lower overall variability is sufficient. As a priority, we focused on obtaining answers to a greater number of questions, which enabled us to cover and analyze more areas with regard to the state of municipal waste management in the Slovak Republic.

3.1.1. Validity of the Research Instrument

In this article, we will not use the FA analysis results with any other examination or modeling, and only for the purpose of confirming the questionnaire's validity.

The validity of the questionnaire and its reliability were analyzed in a pre-survey among students with the expected results being similar to the survey in the target group of respondents. There was agreement of a significant result in defining the factors for the questionnaire, which focused on the environmental self-perception of the respondent.

The application of a factor analysis (FA) is indicated by the Kaiser-Meyer-Olkin (KMO) test statistics, which express the extent to which each variable can be predicted without errors through other variables, the definition of the unambiguous and reliable factors in our case that represent the investigated areas (i.e., constructs and factors) of the environmental profile of respondents, and with a recommended minimum test statistic value of 0.6. The higher the value of the KMO statistics, the more reliable the individual factors that are defined. The significance of the Barlett sphericity test confirms the existence of interrelationships between variables, which also justifies the use of a factor analysis [52] (Table 5).

Table 5. KMO a Bartlettov Test.

Kaiser-Meyer-Olkin Measure of Sampling Adequacy		**0.854**
Bartlett's Test of Sphericity	Approx. Chi-Square	4173.901
	df	105
	Sig.	0

Source: primary data, IBM SPSS Statistics analysis.

For the implementation of an FA, the minimum number of observations is 10 per variable, which in our survey of respondents included in the factor analysis is 1260. This far exceeds the minimum number, even after subtracting those who, in one of the analyzed variables in the open part of the answer, stated "I do not know"; "?"; or left an empty value or answer that could not be included in any of the offered categories.

For the FA, the principal components method was used to define the factors (Principal Component Analysis—PCA), and for better interpretability of factors Varimax orthogonal rotation for uncorrelated factors was used.

Initial communalities (i.e., the number of extracted factors) was defined by the number of eigenvalues from the analysis—the so-called eigenvalues were greater than 1. These values represent the extracted factors involved in explaining the variability of the original variables in order from the factor with the largest proportion of variability to the extracted factor with the lowest increment of the proportion of the explained variability (Table 6).

Table 6. Communalities of factor analysis of questionnaire items.

Communalities (Questionnaire Issues)	**Initial**	**Extracted**
26. Can the municipality/city impose a fine for non-compliance with the basic principles of municipal waste management?	1.000	0.897
11. I consider myself a consumer who cares about the protection of natural resources.	1.000	0.773
12. Protecting the environment is part of my lifestyle.	1.000	0.761
2. Do you buy energy-efficient electrical appliances?	1.000	0.743
4. How do you see the quality of the environment in your city/town?	1.000	0.708
13. Is separate municipal waste collection organized in your city/municipality?	1.000	0.668
9. I organize my daily life so that I use as few natural resources as possible (I save water, heat, energy).	1.000	0.659
21. According to you, the fee for municipal waste is ...	1.000	0.602

Table 6. *Cont.*

Communalities (Questionnaire Issues)	Initial	Extracted
3. In case of product failure (electrical appliances/clothing/toys/ means of transport), if the product is repairable ...	1.000	0.584
18. Do you know what fee you pay for municipal waste per person in your town/village?	1.000	0.552
1. Are you interested in the current state and future of the environment?	1.000	0.549
10. I try to use as few natural resources as possible, even if it requires additional costs and effort (instead of a car I use a bus or bicycle or walk; I buy more expensive organic food and just enough to use everything without waste, I use reusable packaging).	1.000	0.538
7. To what extent do you care to minimize the amount of produced municipal waste?	1.000	0.504
14. Are you involved in the separate collection of municipal waste?	1.000	0.471
22. Are you motivated enough to sort waste?	1.000	0.449
Extraction Method: Principal Component Analysis.		

Source: primary data, IBM SPSS Statistics analysis.

The cumulative proportion of the variability of the 15 questions examined by the part of the questionnaire (Table 6), which can be explained by the defined factors, adds up to only 50% due to the above-mentioned lack of room for the inclusion of additional variables representing individual factors and many original unrelated variables (Table 7).

Table 7. Total explained variability in factor analysis.

Total Variance Explained									
Component	Initial Eigenvalues			Extraction Sums of Squared Loadings			Rotation Sums of Squared Loadings		
	Total	% of Variance	Cumulative %	Total	% of Variance	Cumulative %	Total	% of Variance	Cumulative %
1	3.955	26.367	26.367	3.955	26.367	26.367	3.372	22.478	22.478
2	1.560	10.403	36.770	1.560	10.403	36.770	1.555	10.364	32.842
3	1.062	7.080	43.850	1.062	7.080	43.850	1.464	9.758	42.600
4	1.043	6.952	50.803	1.043	6.952	50.803	1.230	8.202	50.803
5	0.949	6.324	57.126						
6	0.888	5.923	63.049						
7	0.854	5.693	68.742						
8	0.849	5.658	74.400						
9	0.774	5.163	79.563						
10	0.729	4.862	84.425						
11	0.645	4.298	88.722						
12	0.609	4.062	92.784						
13	0.470	3.130	95.914						
14	0.414	2.762	98.676						
15	0.199	1.324	100.000						
Extraction Method: Principal Component Analysis.									

Source: primary data, IBM SPSS Statistics analysis.

The number of extracted factors was defined by the number of eigenvalues from the analysis, which were greater than 1. These values represent the extracted factors involved in explaining the variability of the original variables in order from the factor with the largest proportion of variability to the extracted factor with the lowest increment of the proportion of variability.

In (Table 7), there are the squares of multiple correlations for individual variables with all other variables. In the case of the PCA method, all are equal to 1, as the variables are standardized with a variance of 1. The extracted communities are parts of the variability (i.e., variance estimates) of the variables explained by all extracted factors. Small commu nality values mean that variables are not sufficiently explained by the extracted factors and should be excluded from the analysis if possible. The extracted communities in our case are acceptable, although values lower than 0.5 for two questions indicate that they do not correspond to the extracted factors as well as the other variables.

The cumulative proportion of the variability of the 15 items of the questionnaire which can be explained by defined factors, adds up to only 50%, which is due to the above mentioned insufficient space for inclusion of additional variables representing individual factors and many original unrelated (non-correlated) variables (Table 7). Graphically, this is the variance that is associated with each factor shown by a scree plot in Figure 4. Typically the plot shows a distinct break between the steep slope of the large factors and the gradual trailing of the rest (the scree).

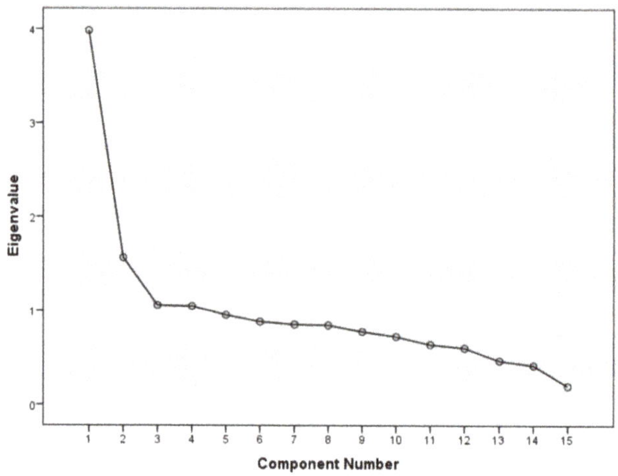

Figure 4. Scree Plot. Source: primary data, IBM SPSS Statistics analysis.

The extracted factors created dimensions/constructs, to which the individual variables after rotation were bound by the highest load, i.e., the so-called loadings, for the selected method of rotation. The values are both regression coefficients for the linear combination of the relevant factors explaining the manifest original variable, as well as correlations of the variables with the relevant factors. The factors form the highlighted row groups in the table with the highest numeric values in both the rows and the columns.

As can be seen, the variables from questions no. 9, 10, 11 and 12, representing the environmental self-perception of respondents included in the questionnaire based on other similar research to verify the validity and reliability of our questionnaire, as well as for comparison with the relevant research, [45] are bound by a high number indicating a high proportion of variability in the observed variable, which is explained by the extracted factor.

The best-covered latent variable (extracted factor) is a factor called "self-perception", which correlates with questions 9 to 12 (the above-mentioned questions taken from the foreign survey [45]) and for which the proportion of variability explained by the extracted factors is calculated to be 0.682 to 0.857. In the rotated matrix of components (Table 8), these variables are correlated with a defined factor and a significant force of 0.721 to 0.857.

In the table of the rotated matrix of components, in the column of the factor of the environmental self-perception, we can see that the factor was linked to the variable in question number 7 from our questionnaire to determine the extent to which the respondent cares to minimize the amount of municipal waste produced. It is not related to any other factor and it is indeed a variable from the category of self-perception, and not from the category of proven behavior as originally defined. Given this, we left the question for analysis in a new group and can replace the five original variables from the questionnaire with a single new latent variable, called the respondent's environmental "self-perception".

Table 8. Rotated [a] factor analysis matrix.

Manifest Variables (Observations) \ Components (Factors, Dimensions, Constructs)	Self-Perception	Prerequisites for Sorting	Confidence in the Future	Financial Burden
11. I consider myself a consumer who cares about the protection of natural resources.	**0.857**			
12. Protecting the environment is part of my lifestyle.	**0.839**			
9. I organize my daily life so that I use as few natural resources as possible (I save water, heat, energy).	**0.784**			
10. I try to use as few natural resources as possible, even if it requires additional costs and effort (instead of a car I use a bus or bicycle or walk; I buy more expensive organic food and just enough to use everything without waste, I use reusable packaging).	**0.721**			
7. To what extent do you care to minimize the amount of produced municipal waste?	**0.682**			
13. Is separate municipal waste collection organized in your city/municipality?		**0.650**		
4. How do you see the quality of the environment in your city/town?		**0.577**		
22. You are motivated enough to sort waste?	0.202	**0.538**		
14. Are you involved in separate municipal waste collection?	0.228	**0.512**	0.311	
21. According to you, the fee for municipal waste is ...			**−0.669**	0.350
1. Are you interested in the current state and future of the environment?	0.389		**0.575**	0.236
18. Do you know what fee you pay for municipal waste per person in your town/village?		0.320	**0.515**	
2. Do you buy energy-efficient electrical appliances?		−0.221	**0.478**	0.385
3. In case of product failure (electrical appliances/clothing/toys/means of transport), if the product is repairable ...				**0.698**
26. Can the municipality/city impose a fine for non-compliance with the basic principles of municipal waste management?				**0.523**

Extraction Method: Principal Component Analysis.
Rotation Method: Varimax with Kaiser Normalization.

Source: primary data, IBM SPSS Statistics analysis. [a] Rotation converged in 6 iterations.

The other second most important extracted factor also has a relatively high load with the four original variables (questions no. 4, 13, 14, 22). Three of the questions are from category F on the classification, and specifically, question 4 asks about the quality of the environment in the city/municipality concerned. By identifying the common features of the variables, we decided to name the factor "The preconditions for sorting", because most of the variables are dedicated to the sorting of municipal waste, namely: "Is separate collection organized in your municipality?", "Are you motivated enough to sort waste?", and "Are you involved in the separate collection of municipal waste?". The correlation of the perception of the quality of the environment with the conditions for sorting is caused by the lack of other relevant questions about the quality of the environment, but there is also a clear connection, or vice versa, if the quality of the environment is taken into account in the municipality—of which it is assumed that measures will be introduced to support sorting.

The other two questions in the group "The preconditions for separate collection" are also weakly linked to another factor for known reasons. At the same time, however, the connection with another factor is justified, because many respondents stated that they do not need external motivation and that they sort out their own belief that it is right. Even the question, "whether people are involved in the separate collection of waste?", is justified in the group of preconditions for sorting since knowing the answers of the respondents that are not involved is important. This is because they may only have a mixed waste container at the house, or only a certain type of sorted waste is collected, or waste collection takes longer time intervals, between which the containers are filled in such a way that they no longer fit into them, and sorting is thus prevented.

The third factor to be extracted will be the matters of interest for the future, from Group A. Although the second question on purchasing energy-efficient appliances was theoretically categorized in group B (real demonstration of interest in environmental sustainability) when designing the questionnaire, the FA revealed that it has to do with a focus on the future when we know that input costs of more energy-intensive equipment are higher, and therefore, such products are uninteresting for people who do not calculate with the future in mind.

Question no. 21 in the group about the perception of the fee includes a variant of the answer that the fee could be higher in order to solve as many waste problems as possible, resp. that it is insufficient. These insufficient fee responses are on the opposite side of the scale, i.e., dissatisfaction with the high fee was in positions 1 and 2, and complete dissatisfaction with the amount, in the sense that it is insufficient, is at 5 and 6. The question, therefore, follows the factor with a negative coefficient, i.e., those who are interested in the future also advocate the need for a higher fee and at the same time have an overview of the fee (question no. 18). In this group, too, the questions are linked to other factors that are more or less interpretable.

The last extracted factor is related to the financial burden of the respondent, which can be avoided. The strong correlation of questions 3 and 26 with the financial burden factor explains the connection between the answers, and that when the respondent is aware that it is possible to obtain a fine for acting in violation of the municipality's regulation, he prefers to avoid it, and similarly, repairs a repairable product not to save the environment, but because the FA suggests that it is to save money—since a repair, especially with one's own hands, is more financially advantageous than buying a new product.

The last three factors, mainly due to the weak explanation of the percentage variability of the original variables and the multiple links of the questions to several factors, will not be applied to the analyses but will remain with the original assumption and proof that only the environmental self-perception factor is sufficiently valid and usable.

Next we also verify the reliability of the defined factor/construct/new latent variable.

3.1.2. Reliability of a Research Tool

A good research tool is not a mixture of different items; on the contrary, the items are designed to detect the same construct/property. In this case, the research tool then has a good internal consistency. Cronbach's Alpha is used to detect it in polytomy items with a larger scoring range (in our case 1 to 6). It is this coefficient that is used in questionnaires where the items are scaled (all items in the questionnaire have scales of the same value and length) [53]. We used IBM SPSS Statistics software to calculate it.

We evaluated the reliability of the questionnaire with variables identical to those we applied in the factor analysis (see the list in Table 8). The expected result with lower reliability in the inclusion of all questions due to a larger number of inconsistent constructs is shown in Table 9. The value of Cronbach's Alpha is greater than 0.7, which is in line with the recommendation for the value of Cronbach's Alpha evaluated in the social and economic sciences [54]. In the pre-survey, the same questionnaire answered by students similarly showed the reliability of the 15 components of the several inconsistent constructs at the level of 0.779.

Due to further analysis, we can use a valid dimension named "Environmental self-perception", which will also evaluate its internal consistency separately (Table 9). The five-component construct provides a quantified reliability greater than 0.831 (the student pre-survey had a higher reliability of 0.858). 119 respondents participated in the student survey, of which 96 fully completed answers were valid for the 15-component construct and 97 for the 5-component construct.

In conclusion, the analysis of the data representativeness, and the structure and reliability of the questionnaire confirms more than 30% representation and geographical representativeness of the target group of respondents—employees of the local authorities, the applicability of data from citizens outside the target group of respondents, and environmental self-perception verifying the non-randomness of the answers and, in turn, the entire questionnaire.

Table 9. Analysis of the internal consistency of the questionnaire (reliability) with all 15 polyatomic questions and only with five questions representing the construct of environmental self-perception.

Reliability Statistics		
Cronbach's Alpha	**Items**	**Number of Respondents with Non-Empty Answers**
0.707	15	1260
0.831	5	1426

Source: primary data, IBM SPSS Statistics analysis.

3.2. Data Analysis

The questionnaire represented a wide area of the municipal waste management issues related to the opinions, perceptions, declared behaviors, justifications, or recommendations that are accountable to all the respondents. The aim of the survey and its analysis was to evaluate the situation in Slovakia's waste management through anonymous data from local government documents, but also through the attitudes and perceptions of the environmental and municipal waste issues by citizens with a focus on competent local government representatives as an influencing factor of the level of municipal waste management. Based on the part of the questionnaire survey conducted within local governments in Slovakia, using two classification methods, we present partial results and the recommendations that could contribute to the gradual reduction of waste disposal in landfills, allow more intensive sorting of waste, and their material recovery.

3.2.1. Classification of Explanatory Variables and Prediction of Classification Level

Based on the data of the questionnaire (with the verified validity and reliability on a selected part of the questions), the target dependent variable for the decision tree technique was set as the *"Sorted municipal waste ratio to the total quantity per inhabitant"*. For this variable, we would need to obtain the greatest possible value—the largest proportion of sorted waste. As the independent variables were used, all possible variables in the dataset were used as well. Missing values in this analysis were included in the tree-growing process as a floating category that was allowed to merge with other categories in the tree nodes. Together with the missing values, we processed 811 observations, which were randomly divided into test subgroups by the method of cross-validation [55] of the decision tree. The result of the cross-validation is an estimate of how well the tree generalizes the entire data population. The risk of our tree is, according to the cross-validation based on five subsets, calculating the average risk and the risk of resubstitution of 0.016 and 0.017, respectively, with a standard error of 0.001. For multiple comparisons, significance values for the merging and splitting of criteria are adjusted using the Bonferroni method.

The generated decision tree consists of seven nodes and five final leaves of the tree. It consists of three automatically created levels (Figure 5).

The root node consists of 100 % of all the values with the current value of the target variable at the level of 0.173 (17.3% level of sorting of municipal waste). The value should roughly correspond to the average proportion of the recyclable waste in Slovakia in 2017. The difference of 5.7% is caused by the aiming of the survey while conducting on and determining the factors for the higher rate of sorting that is not estimating the representative population's parameter as it is stated at the end of the introduction part of the paper.

Several main classification/explanatory variables classifying/predicting the level of the ratio of the sorted waste in the total amount of municipal waste were identified and included in the decision tree model. In order to simplify and present more potential factors of the rate of sorting, we defined three intervals for maximally branching the continuous variables with a minimum node size of 70 observations, and on the level of the final nodes—the tree leaves—at least 35 observations.

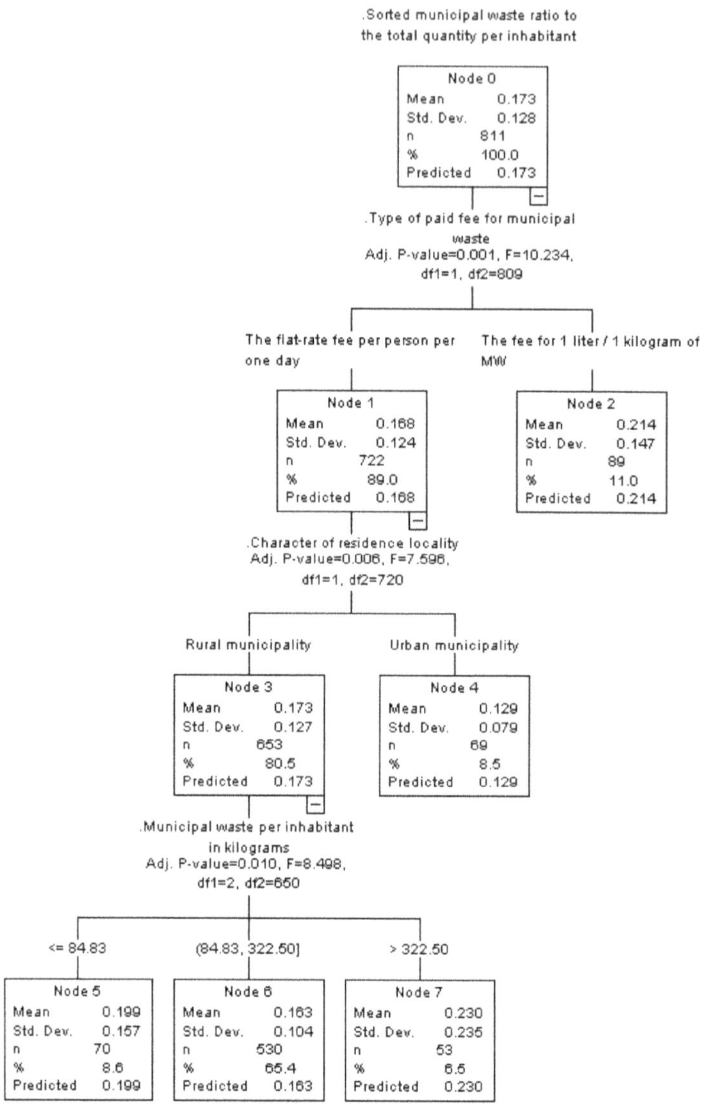

Figure 5. Decision tree for the rate of sorted waste in the total amount of MW produced in the municipality. Source: primary data, IBM SPSS Statistics analysis.

The most important statistically significant factor of the MW sorting that was identified by the decision tree is the categorical variable: *Type of paid fee for municipal waste* The decision tree technique thus split all the data into two branches, unbalanced by abundance (89% versus 11 %) but informatively relevant. This criterion distinguishes the data according to the type of paid fee for the data of the flat-rate fee and the data of the fee paid according to the MW amount. On the basis of the available data, the nodes are statistically significantly (p-value = 0.001; F = 10.234; df_1 = 1, df_2 = 809) characterizing or predicting the municipal waste sorting level of 16.8% in the case of a node with a flat-rate charging, and for the sorting joined with a fee for 1 L or 1 kg of waste at the higher sorting degree of 21.4%

The first branch's node in the current model is next broken down according to another binomial categorical variable, *Character of residence locality,* and again into another two branches (p-value = 0.006). The tree identified that the urban character of a municipality has less probability of a higher level of sorting than the smaller municipalities. Cities with the flat-rate fee, according to the model, are capable of sorting waste at the average level of 12.9% and rural villages at the average level of 17.3%.

The next significant factor that split node no. 3 into three automatically (set up to 10 intervals) created intervals. The most numerous group (530 respondents) for the production of MW per inhabitants was from 84.8 to 322.5 kg, who are not possible to be split to the next level, even with the manual settings. No other statistically significant subgroups can be created using selected factors. That big group remains similarly in the Cluster analysis the average group with the level of sorting 16.3%, which is slightly under the whole average level of sorting. However, there is a group (Node 5) with an annual municipal waste per inhabitant on a level under 84.8 kg, which consisted of rural municipalities with a population under 1000 where they can manage a higher rate of sorting compared with the average rate. At the opposite end, with the largest annual municipal waste per inhabitant (over 322.5 kg), we can conclude according to the additional data analysis that 81.7% of all respondents live in the economically strongest district of Slovakia, with the highest average charging for MW corresponding with the highest sorting level.

Other variables have not been specified as relevant by the decision tree with the given data set.

3.2.2. Clustering of Variables into Relatively Homogeneous Groups

The method by which we can identify the groups that are the most similar to each other within a cluster with the relevant properties, and at the same time differ from each other as much as possible, is through cluster analysis. The clusters thus identified (i.e., the municipalities with the relevant characteristics defined within them) can then be the object of the appropriately selected tools and procedures, in order to contribute to the optimization in the decision-making aimed at increasing the level of municipal waste sorting or minimizing the total amount of solid non-recyclable waste.

We used the IBM SPSS Statistics' two-step clustering method to identify such groups, which combines the benefits of other clustering methods with both the categorical and continuous variables, and applies a hierarchical clustering algorithm to the larger amounts of data. We did not treat the missing data in the variables; respondents who did not provide data on the amounts of waste components produced were excluded from this analysis. About 5% of the extreme data were also excluded from the analysis so as not to affect the results in the clusters formed, or to form separate clusters. Thus, a total of 735 respondents from the local government representatives were included in the analysis who then provided the necessary data.

The results of the analysis are shown in Figure 6 with a good degree of consistency within the clusters versus a degree of difference between the clusters of 0.6. The result table in the fields graphically shows the relative distribution of data. The clusters are arranged according to the waste sorting ratio from the highest sorting ratio to the lowest. Clusters are defined by three selected categorical variables and by three continuous variables of concern. Categorical variables with the highest degree of predictive importance in the classification of respondents into clusters from 77% to 100% are the *Type of the fee, Character of residence locality,* and *Perception of the fee.* The three continuous variables, and thus possess a lower predictor's importance, are the *Sorted municipal waste ratio to the total quantity per inhabitant, Annual municipal waste per inhabitant in kilograms,* and *the Fee per inhabitant per year* for MW, which is paid by the respondents. Continuous variables were standardized as part of the application of the method.

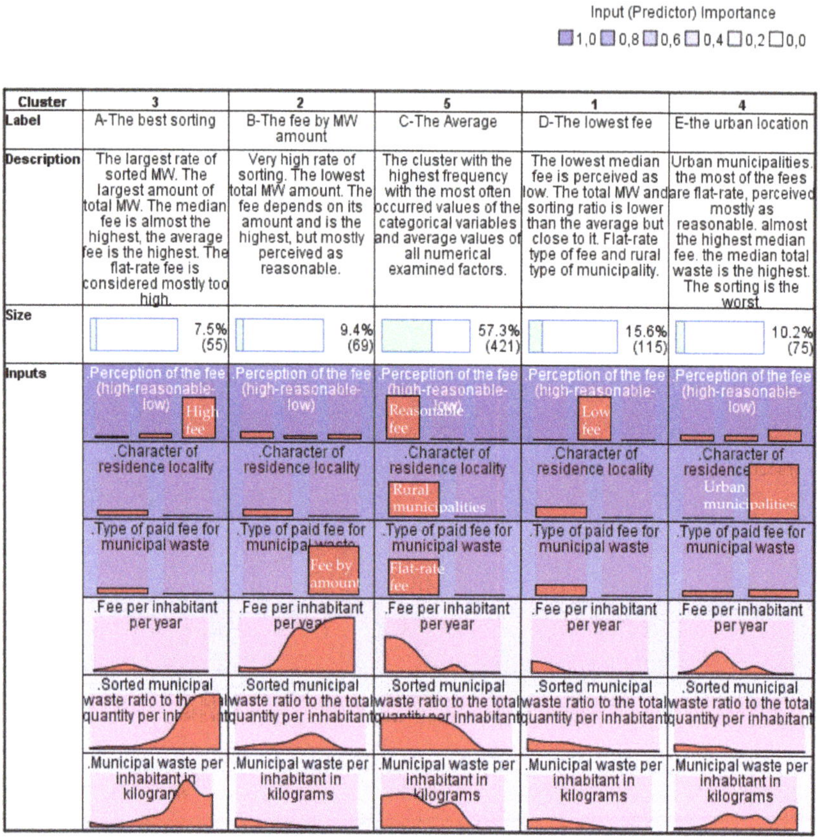

Figure 6. Clusters of respondents with externally different, internally similar properties for the purpose of defining the determinants of the ratio of sorted waste and the total amount of municipal waste (cells show relative distributions). Source: primary data, IBM SPSS Statistics analysis.

The characteristics of the clusters are shown directly in the figure. We can state that the largest group is formed by cluster no. 5, called C. It is the third in the order according to the ratio of the sorted waste per citizen of the municipality. It groups 421 respondents from rural municipalities with the MW flat-rate fee. Compared to other respondents in other clusters, it has a low-to-average level of waste production and level of sorting, and a low-to-average fee (median of the annual fee for the MW per year per capita is €13.05), which is perceived by 100% of cluster's respondents as appropriate. This is the most widespread example of smaller municipalities in particular, which are essentially minimally active. The minimum size of the fee (at the level of neighboring municipalities in the form of a flat-rate fee) does not irritate or motivate any to change either in the production of mixed municipal waste or a higher level of sorting. These are mainly municipalities with a lower standard of living, which is reflected in the lower waste production (median of total MW per capita—183.2 kg) as well as the lower level of waste sorting (median of sorted waste rate—15%).

The second-largest group of rural municipalities (115 respondents) with a flat-rate fee has a lower level of waste sorting, but also a lower level of MW production. This cluster is the one with the lowest fee. It is the cluster no. 1, called D. Even the respondents themselves (100% of the cluster) state that the fee is low. Objectively, the median of the MW fee is the lowest one—€11.94. The annual amount of the total MW per capita and the rate of sorted MW is lower than the average, but not the lowest. Precisely because of the perception of the low fee, the cluster is set aside separately, but to solve the optimization of waste management it is appropriate to assess and manage the situation in the same way as in cluster C.

Other clusters are relatively small (10.2%; 9.4%; 7.5% of the dataset). At the unfavorable end of the spectrum of the sorting level are cities (100% in cluster 5, named E) where the perception of the level of the flat fee is irrelevant, or in other words, there are all options represented, but numerically the fee is mostly perceived as reasonable (69.3% of cluster's respondents). Cities are characterized by the lowest level of waste sorting (11 %) and the highest median level of total annual waste production per capita (276.54 kg/person/year), and by almost the highest median annual fee (€18.9) (arithmetical average €20.94, which is the highest among the clusters with the flat-rate fee).

A little smaller cluster, already with at least a higher proportion of sorted waste identified in cluster no. 4, is called B (17% of all respondents). Its characteristic feature is the type of fee being estimated by the amount of the produced unsorted MW applied in the rural municipalities. The median of the fee size is the highest one, as well as the unweighted informative arithmetic average. The high sorting rate is probably supported by that high fee. Controversially, most of the respondents don't consider the fees as too high. About 84.1% of them marked their fee as reasonable. Finally, the highest fees led to the lowest amount in the median annual total waste per inhabitant (143.07 kg/person/year), which aimed at reducing the final paid fee through the higher degree of sorting. Sorted municipal waste (median of data is 17 % of the total MW) financially does not burden the waste producers, but rather the companies responsible for the collection, recovery, and recycling of municipal waste.

The highest proportion of the sorted MW out of all, even with a flat type fee (100%), has the smallest cluster, named A (Cluster 2—7.5% of all respondents). The annual flat-rate fee (€16.85) is higher than the overall median annual fee (€14.1), but not the highest one. More than half of respondents of the cluster (54.5%) deem that the fee is too high. Representatives of these rural municipalities in this cluster might constitute some of the richest municipalities because they possess the largest amount of total MW (median = 272.8 kg/person/year versus 191.4 kg/person/year of the whole dataset—similar to the amount in the urban municipalities). Fortunately, relatively tons of waste can be collected for the recycling—33% (still a low number, but the highest ratio among clusters). Thus, not the fee size alone can be responsible for the sorted MW ratio, but also the socio-economic situation of the inhabitants. This cluster is the only one with a mixed type of municipalities—given that 29.7% of them are from urban municipalities.

4. Discussion

As we can simplistically conclude, our decision tree analysis confirmed that the strongest determinant of the sorted MW rate is the financial factor represented by the variable, *Type of paid fee for municipal waste (Type of fee)*. The type of paid fee determining its size, mainly the higher fee for MW paid for the weighted amount of the unsorted MW, is the way that can lead to minimizing the total waste and maximizing the ratio of the sorted recyclable or recoverable waste. Our results are in accordance with the statements of other studies, for example [56], where authors state that separation of solid waste is mostly done for a financial motive among households. Similarly to [57], we can also conclude that the urban and wealthier households, headed by older and more literate individuals, are more likely to use municipal waste collection arrangements.

We expected also that a motivation and positive attitude towards nature and the environment could increase the MW sorting level, but since we have not tested the individual measure of sorting, but rather the sorting level of a whole village/town, the pro-environmental status of the municipality representative was not recognized as relevant. Other studies, like [24,45], confirmed that, although the individuals have a positive attitude towards recycling, they have not been able to practice such positive behavior that could improve the quality of their natural environment and the MW sorting rate.

From 1 January 2010, Slovak towns and municipalities were obliged to introduce a sorted waste collection of four components of municipal waste: paper, plastics, glass, and metals [58]. Despite this measure regulated by the older Waste Act, the results of our survey point to the fact that 1.7% of respondents stated that separate municipal waste collection was not introduced in their municipality, even in 2018. This situation is also confirmed by the findings of the Slovak Statistical Office, according to which, in 2017, not all but only 99.48% of municipalities were participating in municipal waste recovering [15].

It is important to introduce the right motivation system in individual municipalities. Also, experience from the Czech Republic presents a large difference in the waste sorting between households and municipalities for which the PAYT (Pay-As-You-Throw) system is in place, as well as among those for which this system is not in place. In the municipalities and households that have the PAYT system in place, citizens sort more waste and produce less residual waste. Conversely, in municipalities and households where they do not have this system, citizens sort less, which confirms our survey analysis, as well as that of other publications—for example [17,30,31]. In Slovakia, by the Waste Act, collection of municipal waste by its amount has been introduced in many municipalities, as was allowed by § 81(10) of the Waste Act as amended [59,60]. They are also aware of the seriousness of municipal waste management in the Czech Republic, where the current Waste Act stipulates that from 2024 on, it is prohibited to landfill mixed municipal waste and recyclable and usable waste provided by implementing legislation, although the EU plans to ban landfilling in 2030.

It is important to respond to this promptly and to provide an efficient and capacity-friendly infrastructure for the treatment of landfill waste, which will attract potential investors and should be acceptable to both professionals and the public [4]. At the end of 2015, the Ministry of the Environment of the Slovak Republic warned that Slovakia would not meet the valid recycling target for 2020 i.e., 50% of municipal waste. In 2015, despite the Ministry's efforts, only 14.9% of municipal waste was recycled. In the following years, we can observe only a slight improvement. On the basis of the assessment underlying the early warning, the Commission concludes that:

- Separate collection of recyclable materials, including bio-waste, is not yet efficient;
- There is a lack of economic incentives for households to sort waste;
- Extended producer responsibility schemes in Slovakia do not fully cover the costs of separate collection, and;
- More investment is needed in higher level waste hierarchy projects (such as recycling) that go beyond the treatment of residual waste.

Summary Recommendations Resulting from the Analysis of the Questionnaire Survey

Our analyses of the conclusions are based on the analyses performed, as well as the more-or-less subjective opinions of the local government representatives acquired during the the conducting of the survey. Based on the questionnaire survey and the results obtained, we propose the following recommendations, which, if applied, will help to increase interest in the sorting of municipal waste and reduce the production of mixed municipal waste:

1. To prefer the introduction of a collection by the amount of municipal waste and sorted components of municipal waste in municipalities and cities, or their parts;
2. Adjust the fee for the export of MW so that those who sort the municipal waste pay less or have other benefits from the sorting;
3. To improve the organization of waste management so that citizens have enough bags for the sorted components of municipal waste in family houses and enough containers next to the residential blocks. As well, ensure the regular collection of these sorted components to avoid overfilling of the collection containers;
4. To introduce municipal waste sorting in those cities and municipalities where this collection has not yet been introduced or has not been implemented effectively;

5. To apply restrictive measures following the valid regulations and legislation, in case of their violation by citizens not sorting their municipal waste (i.e., a warning, non-export of unsorted mixed municipal waste or contaminated sorted waste, as well as the application of fines if necessary);
6. To increase awareness and promotion among citizens about the importance of municipal waste sorting.

5. Conclusions

Based on the performed quantitative analyses, the following variables appear as significant factors of the level of municipal waste sorting in Slovak municipalities: 1. *Type of paid fee* (flat-rate charging versus charging by the amount of municipal waste), 2. *Character of residence locality* (urban or rural character of the seat of residence), and 3. *Municipal waste per inhabitant*. Together with the other analyzed variables (4. *Perception of the fee*, 5. *Sorted municipal waste ratio, and 6. Fee per inhabitant*), they confirm the municipality positions in MW sorting, and indicate the direction in which the municipalities should go. Since some of the factors are not influenceable by human decisions and actions, the others can be regulated directly or indirectly by governmental/municipal measures, as well as by motivational factors for human activities and their consequences.

Our research focused on the analysis and solutions supporting the fulfillment of the objectives of the EU and the Slovak Republic, regarding the reduction of landfill use for municipal waste and increasing the level of sorting and material recovery of municipal waste, which represents a variation of the current and still partial solution to global waste problems. However, for the sustainability of the state of the environment, it is necessary to ensure at the same time a reliable and stable solution for the sale and processing of sorted components.

The need to increase the ratio of sorting and recovery of municipal waste is indisputable. This can be achieved mainly by increased activity in the area of separate collection directly at the producers of the individual waste components. Despite clear environmental, economic, and social benefits, high-quality and efficient separate collection systems are still not widespread in Slovakia. For this reason, it is necessary to make changes that are in line with the objectives and legislation of the Slovak Republic and the European Union. The need for these changes, especially in the area of motivating citizens, also emerged from our survey.

Author Contributions: Both authors (M.S., M.Č.) has equally contributed to this publication. All authors have read and agreed to the published version of the manuscript.

Funding: The APC was funded by Lungo Mare, n. o., Slovakia.

Institutional Review Board Statement: Not applicable.

Informed Consent Statement: Not applicable. The research described in the article involved anonymous, confidential and protected records and data sets that exist in the public domains.

Data Availability Statement: Details of the data and how to request access are available from Michal Stričík at Department of Economics, Faculty of Business Economics with seat in Košice, University of Economics in Bratislava.

Acknowledgments: Authors would like to thank all survey participants that honestly contributed with their time, knowledge and willingness and to editors and reviewers for their valuable comments and suggestions, which significantly helped us to improve the paper.

Conflicts of Interest: The authors declare no conflict of interest.

References

1. Hutner, P.; Thorenz, A.; Tuma, A. Waste prevention in communities: A comprehensive survey analyzing status quo, potentials, barriers and measures. *J. Clean. Prod.* **2017**, *141*, 837. [CrossRef]
2. Iqbal, S.; Naz, T.; Abdul Sattar, M. Challenges and opportunities linked with waste management under global perspective: A mini review. *J. Qual. Assur. Agric. Sci.* **2021**, *1*, 9–13. Available online: https://doi.org/10.52862/jqaas.2021.1.1.2 (accessed on 15 November 2021).
3. Circular Economy Package: Questions & Answers. 2015. Available online: https://ec.europa.eu/commission/presscorner/detail/en/MEMO_15_6204 (accessed on 2 September 2020).
4. Kulhánková, P. Aktuální Otázky Oběhového Hospodářství se Zaměřením na Plasty. TVIP. 2018. Available online: http://www.odpadoveforum.cz/TVIP2018/prispevky/308.pdf (accessed on 9 December 2021).
5. Horsák, Z. Oběhové Hospodářství v Praxi. TVIP. 2018. Available online: http://www.odpadoveforum.cz/TVIP2018/prispevky/301.pdf (accessed on 9 December 2021).
6. Singh, N.; Lavina, A. Waste separation at household level. *Indian J. Appl. Res.* **2015**, *1*, 558.
7. Municipal Waste by Waste Management Operations. Eurostat. Available online: https://appsso.eurostat.ec.europa.eu/nui/show.do?dataset=env_wasmun&lang=en (accessed on 2 December 2018).
8. Ministerstvo životného prostredia Slovenskej republiky, Slovenská agentúra životného prostredia. *Správa o Stave Životného Prostredia Slovenskej Republiky v Roku 2016*; Ministerstvo životného prostredia Slovenskej republiky, Slovenská agentúra životného prostredia: Bratislava, Slovakia, 2018.
9. Relative Indicators from the Area of Treatment with Municipal Waste. Available online: http://datacube.statistics.sk/#!/view/en/VBD_SK_WIN/zp3002rr/v_zp3002rr_00_00_00_en (accessed on 9 December 2021).
10. Program Odpadového Hospodárstva SR na Roky 2016–2020. 2015. Available online: https://www.minzp.sk/files/sekcia-enviromentalneho-hodnotenia-riadenia/odpady-a-obaly/registre-a-zoznamy/poh-sr-2016-2020_vestnik.pdf (accessed on 9 December 2021).
11. Európska Komisia. Správa Komisie o Vykonávaní Právnych Predpisov EÚ o Odpadoch. 2018. Available online: https://eur-lex.europa.eu/legal-content/SK/TXT/HTML/?uri=CELEX:52018DC0656&from=EN (accessed on 9 December 2021).
12. Program Odpadového Hospodárstva SR na Roky 2021–2025. 2020. Available online: https://www.enviroportal.sk/sk/eia/detail/program-odpadoveho-hospodarstva-slovenskej-republiky-na-roky-2021-2025 (accessed on 9 December 2021).
13. News European Parliament: Circular Economy: MEPs Back Plans to Boost Recycling and Cut Landfilling. 2015. Available online: https://www.europarl.europa.eu/news/en/press-room/20180227IPR98710/circular-economy-meps-back-plans-to-boost-recycling-and-cut-landfilling (accessed on 9 December 2021).
14. Pobožná, M. *Odpady v Slovenskej Republike, 2016*; ŠÚ SR: Bratislava, Slovakia, 2017; ISBN 978-80-8121-581-0.
15. Pobožná, M. *Odpady v Slovenskej Republike, 2017*; ŠÚ SR: Bratislava, Slovakia, 2018; ISBN 978-80-8121-648-0.
16. Kannangara, M.; Dua, R.; Ahmadi, L.; Bensebaa, F. Modeling and prediction of regional municipal solid waste generation and diversion in Canada using machine learning approaches. *Waste Manag.* **2018**, *74*, 3. [CrossRef] [PubMed]
17. Sidique, S.F.; Satish, V.J.; Lupi, F. Factors influencing the rate of recycling: An analysis of Minnesota counties. *Resour. Conserv. Recycl.* **2010**, *54*, 242–249. [CrossRef]
18. Corsini, F.; Gusmerotti, N.M.; Tesla, F.; Iraldo, F. Exploring waste prevention behaviour through empirical research. *Waste Manag.* **2018**, *79*, 132. [CrossRef] [PubMed]
19. Maskey, B. Determinants of Household Waste Segregation in Gorkha Municipality, Nepal. *J. Sustain. Dev.* **2018**, *11*. [CrossRef]
20. Struk, M. Distance and incentives matter: The separation of recyclable municipal waste. *Resour. Conserv. Recycl.* **2017**, *122*, 155. [CrossRef]
21. Slavik, J.; Pavel, J. Do the variable charges really increase the effectiveness and economy of waste management? A case study of the Czech Republic. *Resour. Conserv. Recycl* **2013**, *70*, 68. [CrossRef]
22. Xiao, L.; Zhang, G.; Zhu, Y.; Lin, T. Promoting public participation in household waste management: A survey based method and case study in Xiamen city, China. *J. Clean. Prod.* **2017**, *144*, 313. [CrossRef]
23. Gilli, M.; Nicolli, F.; Farinelli, P. Behavioural attitudes towards waste prevention and recycling. *Ecol. Econ.* **2018**, *154*, 294. [CrossRef]
24. Cheng, K.W.; Osman, S.; Jusoh, Z.M.; Leby, J. The Determinants of Intention to Practise Solid Waste Segregation-at-Source among Selangor Households. *Malays. J. Consum. Fam. Econ.* **2020**, *25*, 67–90. Available online: https://www.researchgate.net/publication/346659098_The_Determinants_of_Intention_to_Practise_Solid_Waste_Segregation-at-Source_among_Selangor_Households (accessed on 9 December 2021).
25. Junquera, B.; del Brío, J.Á.; Muñiz, M. Citizens' attitude to reuse of municipal solid waste: A practical application. *Resour. Conserv. Recycl.* **2001**, *33*, 51–60. [CrossRef]
26. Vidová, J. Households as an important part of the circular economy. *Waste Forum* **2019**, *3*, 178.
27. Chang, N.B.; Pires, A.; Martinho, G. Empowering Systems Analysis for Solid Waste Management: Challenges, Trends, and Perspectives. *Crit. Rev. Environ. Sci. Technol.* **2011**, *16*, 1449. [CrossRef]
28. Bing, X.; Bloemhof, J.M.; Ramos TR, P.; Barbosa-Povoa, A.P.; Wong, C.Y.; van der Vorst, J.G. Research challenges in municipal solid waste logistics management. *Waste Manag.* **2016**, *48*, 584. [CrossRef] [PubMed]

29. El-Emam, R.S.; Ozcan, H.; Zamfirescu, C. Updates on promising thermochemical cycles for clean hydrogen production using nuclear energy. *J. Clean. Prod.* **2020**, *262*, 121424. [CrossRef]
30. Šauer, P.; Pařízková, L.; Hadrabová, A. Charging systems for municipal solid waste: Experience from the Czech Republic. *Waste Manag.* **2008**, *28*, 2772–2777. Available online: https://doi.org/10.1016/j.wasman.2008.03.030 (accessed on 9 December 2021).
31. Puig-Ventosa, I. Charging systems and PAYT experiences for waste management in Spain. *Waste Manag.* **2008**, *28*, 2767. [CrossRef]
32. Ulfik, A.; Nowak, S. Determinants of Municipal Waste Management in Sustainable Development of Regions in Poland. *Pol. J. Environ. Stud.* **2014**, *23*, 1039–1044. Available online: http://www.pjoes.com/pdf-89280-23138?filename=Determinants%20of%20Municipal.pdf (accessed on 9 December 2021).
33. Dai, Y.C.; Lin, Z.Y.; Li, C.J.; Xu, D.Y.; Huang, W.F.; Harder, M.K. Information strategy failure: Personal interaction success, in urban residential food waste segregation. *J. Clean. Prod.* **2016**, *134*, 298–309. Available online: https://doi.org/10.1016/j.jclepro.2015.12.104 (accessed on 9 December 2021).
34. Cheng, K.W.; Osman, S.; Jusoh, Z.M.; Lau, J.L. Does Environmental Knowledge Moderate the Relationship between Household Determinants' Intention to Practice Solid Waste Segregation-at-Source? A Conceptual Paper. In Proceedings of the International e-Conference on Green and Renewable Energy, Bintulu, Malaysia, 8–19 August 2020; pp. 93–104. Available online: http://psasir.upm.edu.my/id/eprint/83496/ (accessed on 9 December 2021).
35. World Business Council for Sustainable Development. *Circular Economy-Factor10*; World Business Council for Sustainable Development: Geneva, Switzerland, 2020; Available online: https://www.wbcsd.org/Clusters/Circular-Economy-Factor10 (accessed on 15 May 2020).
36. Dupáková, M. *Obehové Hospodárstvo: Viac ako Recyklácia*; Euractiv: Bratislava, Slovakia, 2016; Available online: https://euractiv.sk/section/veda-a-inovacie/linksdossier/obehove-hospodarstvo-viac-ako-recyklacia-000342/ (accessed on 9 December 2021).
37. European Commision. Kruh sa Uzatvára–Akčný Plán EÚ pre Obehové Hospodárstvo. 2015. Available online: https://ec.europa.eu/transparency/regdoc/rep/1/2015/SK/1-2015-614-SKF1-1.PDF (accessed on 9 December 2021).
38. Taušová, M.; Mihaliková, E.; Čulková, K.; Stehlíková, B.; Tauš, P.; Kudelas, D.; Štrba, Ľ. Recycling of Communal Waste: Current State and Future Potential for Sustainable Development in the EU. *Sustainability* **2019**, *11*, 2904. Available online: https://doi.org/10.3390/su11102904 (accessed on 9 December 2021).
39. Zurbrügg, C.; Gfrerer, M.; Ashadi, H.; Brenner, W.; Küper, D. Determinants of sustainability in solid waste management–The Gianyar Waste Recovery Project in Indonesia. *Waste Manag.* **2012**, 32, 2126. Available online: https://doi.org/10.1016/j.wasman.2012.01.011 (accessed on 9 December 2021).
40. Čonková, M.; Stričík, M.; Bačová, M. Pohľad samospráv na vybrané otázky nakladania s komunálnym odpadom [Views of municipalities on selected municipal waste management issues]. *Waste Forum* **2018**, *4*, 544–559. Available online: http://www.wasteforum.cz/cisla/WF_4_2018.pdf#page=138 (accessed on 9 December 2021).
41. Stričík, M.; Bačová, M.; Čonková, M. Large-volume collection of municipal waste. In *Production Management and Business Development*; CRC Press: London, UK, 2018; pp. 211–216. Available online: https://doi.org/10.1201/9780429468667 (accessed on 9 December 2021).
42. Stričík, M.; Bačová, M.; Čonková, M.; Kršák, B. *Udržateľné Nakladanie s Komunálnym Odpadom [Sustainable Municipal Waste Management]*; Vysoká škola báňská-Technická Univerzita: Ostrava, Czech Republic, 2019; p. 304.
43. Stričík, M.; Bačová, M.; Čonková, M. Motivation of citizens of the Slovak Republic to separate municipal waste. *Waste Forum* **2019**, *4*, 399–413. Available online: http://www.wasteforum.cz/cisla/WF_4_2019.pdf (accessed on 9 December 2021).
44. Stričík, M.; Tkáčová, V. Dopad zo zavedenia žetónového systému nakladania s komunálnym odpadom vo vybraných obciach SR [Impact from introduction of a municipal waste management system in selected municipalities of the Slovak Republic]. In *Odpady: Minimalizácia, Zhodnocovanie, a Zneškodňovanie*; Wolters Kluwer SR: Bratislava, Slovakia, 2019; Volume 19, pp. 5–10.
45. Moser, S.; Kleinhückelkotten, S. Good intents, but low impacts: Diverging importance of motivational and socioeconomic determinants explaining pro-environmental behavior, energy use, and carbon footprint. *Environ. Behav.* **2018**, *50*, 626–656. Available online: https://doi.org/10.1177/0013916517710685 (accessed on 9 December 2021).
46. Obce, S.R. [Munincipalities of the Slovak Republic]. Available online: http://www.sodbtn.sk/obce/index_kraje.php/index_kraje.php (accessed on 15 May 2020).
47. Trnka, A. *Základné Štatistické Metódy Marketingového Výskumu*; Trnava Fakulta Masmediálnej Komunikácie: Trnava, Slovakia, 2016; ISBN 978-80-8105-768-7.
48. Microsoft Corporation. *Microsoft Excel*; Microsoft Corporation: Redmond, WA, USA, 2016.
49. *IBM SPSS Statistics for Windows, 20*; IBM Corp.: Armonk, NY, USA, 2011.
50. GRETL [Gnu Regression, Econometrics and Time-series Library]. Open-Source Software. 2018. Available online: http://gretl.sourceforge.net/#ack (accessed on 9 December 2021).
51. What Is Cluster Analysis? Available online: https://www.displayr.com/what-is-cluster-analysis/ (accessed on 12 May 2020).
52. KMO and Bartlett's Test. Available online: https://www.ibm.com/support/knowledgecenter/SSLVMB_26.0.0/statistics_casestudies_project_ddita/spss/tutorials/fac_telco_kmo_01.html (accessed on 12 May 2020).
53. Gavora, P.; Koldeová, L.; Dvorská, D.; Pekárová, J.; Moravčík, M. *Elektronická Učebnica Pedagogického Výskumu*; Univerzita Komenského: Bratislava, Slovakia, 2010; Available online: http://www.e-metodologia.fedu.uniba.sk/--80--223--2951--4 (accessed on 15 May 2020)ISBN 978–80–223–2951–4.
54. Croasmun, J.T.; Ostrom, L. Using Likert-Type Scales in the Social Sciences. *J. Adult Educ.* **2011**, *40*, 19–22.

55. Validation. Available online: https://www.ibm.com/support/knowledgecenter/SSLVMB_24.0.0/spss/tree/idh_idd_tree_validation.html#idh_idd_tree_validation (accessed on 15 May 2020).
56. Banga, M. Household knowledge, attitudes and practices in solid waste segregation and recycling: The case of urban Kampala. *Zamb. Soc. Sci. J.* **2011**, *2*, 4. Available online: https://www.nswai.org/docs/Household%20Knowledge,%20Attitudes%20and%20Practices%20in%20Solid%20Waste%20Segregation%20and%20Recycling%20The%20Case%20of%20Urban%20Kampala.pdf (accessed on 9 December 2021).
57. Kumara, A.S.; Pallegedara, A. Household waste disposal mechanisms in Sri Lanka: Nation-wide survey evidence for their trends and determinants. *Waste Manag.* **2020**, *114*, 62–71. Available online: https://doi.org/10.1016/j.wasman.2020.06.028 (accessed on 9 December 2021).
58. Slovak Government. Act No. 409/2006 Amending the Waste Act No. 223/2001 Col. on Waste. Order of the Government of the Slovak Republic: Bratislava, Slovakia, 2006; Available online: https://www.slov-lex.sk/pravne-predpisy/SK/ZZ/2006/409/ (accessed on 19 October 2019).
59. Slovak Government. Act No. 79/2015 Col. on Waste and on the Amendment of Certain Acts. Order of the Government of the Slovak Republic: Bratislava, Slovakia, 2015; Available online: https://www.slov-lex.sk/pravne-predpisy/SK/ZZ/2015/79/20211201 (accessed on 19 October 2019).
60. Slovak Government. Act No. 329/2018 Col. on Waste Disposal Fees and on the Amendment of the Act no. 587/2004 on the Environmental Fund and on the Amendment of certain Acts. Order of the Government of the Slovak Republic: Bratislava, Slovakia, 2018; Available online: https://www.slov-lex.sk/pravne-predpisy/SK/ZZ/2018/329/20210315 (accessed on 19 October 2019).

sustainability

MDPI

Article

Optimization and Spatiotemporal Differentiation of Carbon Emission Rights Allocation in the Power Industry in the Yangtze River Economic Belt

Dalai Ma, Yaping Xiao * and Na Zhao

School of Management, Chongqing University of Technology, Chongqing 400054, China; madalai@cqut.edu.cn (D.M.); zhaona1997@2020.cqut.edu.cn (N.Z.)
* Correspondence: ashley_air@163.com

Abstract: Reasonable allocation of carbon emission rights aids in the realization of the goal of carbon emission reduction. The purpose of this paper is to examine how carbon emission rights in the power sector in the Yangtze River Economic Belt (the YREB) are distributed. The YREB spans China's eastern, central, and western areas. The levels of development and resource endowment differ significantly across regions, resulting in great heterogeneity in the YREB provinces' carbon emission rights distribution in the power sector. The ZSG–DEA model is used in this paper to re-adjust the power sector's carbon emission quotas in each province to achieve optimal efficiency under the country's overall carbon emission reduction target. The results show that: (1) In most provinces, the power sector's initial distribution efficiency is inefficient. Only Zhejiang and Yunnan have reached the production frontier, with Jiangxi and Chongqing having the lowest distribution efficiency. In the future, we should concentrate our efforts on them for conserving energy and lowering emissions; (2) The initial distribution efficiency of the power sector in the YREB's upstream, midstream, and downstream regions is considerably different. Most upstream and downstream provinces have higher carbon emission quotas, while most midstream provinces have less, implying that the power sector in the midstream provinces faces greater emission reduction challenges; (3) The carbon emission quotas of the power industry varies greatly between provinces and shows different spatial features over time. In the early stage (2021–2027), the carbon emission quota varies substantially, while for the later stage (2027–2030), it is rather balanced. Zhejiang, Jiangsu, Sichuan, and Yunnan are more likely to turn into sellers in the market for carbon emission trading with larger carbon emission quotas. While Jiangxi and Chongqing are more likely to turn into buyers in the market for carbon emission trading with fewer carbon emission quotas. Other provinces' carbon emission quotas are more evenly distributed. To successfully achieve China's emission reduction target by 2030, the YREB should promote regional collaboration, optimize industrial structure, accelerate technical innovation, establish emission reduction regulations, and provide financial support based on local conditions.

Keywords: carbon emission rights allocation; ZSG-DEA; China's power sector; YREB; spatiotemporal differentiation

Citation: Ma, D.; Xiao, Y.; Zhao, N. Optimization and Spatiotemporal Differentiation of Carbon Emission Rights Allocation in the Power Industry in the Yangtze River Economic Belt. *Sustainability* **2022**, *14*, 5201. https://doi.org/10.3390/su14095201

Academic Editors: Nicola Raimo, Filippo Vitolla, Ornella Malandrino and Benedetta Esposito

Received: 17 March 2022
Accepted: 20 April 2022
Published: 26 April 2022

1. Introduction

Since the reform and opening up of China, its economy has been rapidly expanding and has grown to become the world's second largest (Ma and Cai) [1]. As a result, massive amounts of energy, especially fossil fuels, are consumed. The source of most carbon emissions is fossil fuels. In 2017, carbon emissions in China were responsible for 28 percent of total global emissions, far outnumbering the second-largest emitter, the United States (15%) (BP2018) [2]. China presented the dual carbon goal of "striving to peak carbon dioxide emissions by 2030 and achieve carbon neutrality by 2060" at the 75th United Nations General Assembly to promote global decarbonization, which not only brought huge pressure on achieving carbon emission reductions to China, but also made higher requirements for

China's power industry to develop in a low-carbon manner. As a pillar industry in China (Zhou et al.) [3], the power industry needs to meet the demand for electricity generated by daily economic development, industrial production, and urbanization. However, China's current power generation mode is primarily thermal (Li et al.) [4], and the amount of power generated by clean power generation methods is far from meeting society's needs, resulting in a large amount of CO_2. Nearly 40% of China's carbon emissions are attributed to the power sector (Yu et al.) [5], which has put enormous pressure on energy conservation and environmental protection. Therefore, reducing carbon emissions in the power sector is critical to achieve China's goal of reducing carbon emissions. Different regions and industries have different emission reduction targets under the overall goal of "dual carbon" because different regions have varying levels of technological development and are at various stages of development. To achieve maximum efficiency, the ZSG-DEA model can adjust each province's carbon emission allowances in accordance with the country's overall carbon reduction targets.

The YREB connects 11 Chinese provinces, with a population and economic aggregate that exceeds 40% of the country's total (Li et al.) [6], having a significant influence on China's overall social and economic development. However, the YREB has faced severe resource and environmental problems because of long-term high-intensity industrial economic development, particularly the increasingly serious carbon emission problem. The YREB is rich in hydropower and mineral resources. It is an important hydropower and pithead thermal power supply area in China, as well as a major transmission source for "West-East Power Transmission". According to statistics data [7], its power generation accounts for 37.8 percent of China's total, producing a significant amount of carbon emissions, making it a key area for reduction of carbon emissions. The YREB connects China's eastern, central, and western regions. The level of industrial development and resource endowment varies greatly across these regions. However, does the allocation of carbon emission rights in the YREB's power sector also show significant differences? This is an important issue that needs to be addressed.

The remainder of this paper is laid out as follows. Section 2 examines the relevant studies on carbon emission rights allocation. The research methods and data are described in Section 3. And the results of the study are listed in Section 4. Finally, Section 5 summarizes the research and presents implications for policy.

2. Literature Review

Low-carbon development has grown in importance as a research topic in academic circles of various countries in the recent past. The allocation of carbon emissions has also attracted increasing attention from scholars. The main point of contention in existing research is the selection of principles and methods for allocating carbon emissions rights. Most scholars recognize the principles of fairness and efficiency at the level of distribution. The United Nations Framework Convention on Climate Change (UNFCCC) [8] identified the principle of fairness and proposed "shared but distinct responsibilities" in addressing changes in the climate. Fairness, as a relatively broad concept, encompasses not only egalitarianism and historical emission responsibility, but also the ability to reduce emissions. As a result, when allocating carbon emission rights, scholars typically consider population, historical carbon emissions, and economic level. Pan et al. [9] proposed a distribution scheme based on per capita cumulative emissions to create a global carbon emission space that is fair. Zhu et al. [10] proposed that the development performance of various industries be considered to reflect the fairness of carbon emission rights allocation. In addition, scholars frequently use multiple indicators to allocate carbon emissions, because the principle of fairness necessitates the use of multiple indicators. To simulate carbon allowance allocation in the Beijing-Tianjin-Hebei region, Han et al. [11] created a comprehensive index and used a comprehensive weighting method: GDP per capita, cumulative CO_2 emissions, and energy consumption per unit of industrial added value were chosen to represent carbon emission reduction capability, potential, and responsibility. Fang et al. [12] discussed the optimal allocation of carbon emission rights based on energy equity, as well as the method for optimizing the allocation scheme under GDP constraints, population, fossil energy, and ecological production land. According to their findings, the importance of fossil energy resources and ecological production land was greater. Furthermore, as a widely used indicator for evaluating fairness, the Gini coefficient is frequently used to ensure that carbon emission allocation results are equitable (Fang et al. [12], Guo et al. [13], He et al. [14]). Through the above research, it is found that the carbon emission allocation principle is more concerned with absolute fairness and ignores the perspective of efficiency, which is not conducive to a reasonable and effective distribution of carbon emissions.

People are becoming more aware, as research into the distribution of carbon emission rights advances, that the so-called "absolutely fair" distribution of carbon emission rights does not benefit all countries and regions (He and Zhang [14], Kong et al. [15], He et al. [16]). The principle of equity considers differences in low-carbon development levels across provinces, but they were not practical. It cannot effectively motivate provinces with better low-carbon development while restricting provinces with backward low-carbon development, and it cannot improve overall efficiency. Another important principle to consider when allocating carbon emission rights is the efficiency principle of profit maximization (Du et al.) [17], which according to Zhou et al. [18], is the highest economic return for the least amount of resources. Carbon emission rights can help a country's economy grow as a valuable resource, but they are restricted and should be distributed in a scientific and rational manner. Therefore, Zhou [19], Qin et al. [20] and Liu et al. [21] studied the optimal allocation of carbon dioxide emissions by the DEA method, cooperative game model and nonlinear programming method respectively.

Scholars have studied the method of allocating carbon emission rights based on the principles of fairness and efficiency extensively. Methods for allocating carbon emission rights in the past have included the grandfather method (Schmidt and Heizig) [22], the benchmark method (Sartor et al. [23], Zhang et al. [24]), the auction method (Burtraw and McCormack) [25], the indicator method (He and Zhang [14], Zhao et al. [26]) and others. These methods, to some extent, ensure the fairness of the distribution of carbon emission rights among decision-making units, but they ignore another distribution principle—efficiency. They considered differences in low-carbon development levels across provinces, but they were not practical. It cannot effectively motivate provinces with better low-carbon development while restricting provinces with backward low-carbon development, and it cannot improve overall efficiency. In contrast, data envelopment analysis (DEA), an optimization method aimed at improving overall system efficiency, has been introduced into the study of carbon emission rights allocation. Because countries and regions often set carbon emission targets, the total amount of carbon emissions should be limited within a certain range when allocating carbon emission rights. In this situation, how can you achieve maximum efficiency? The zero sum gains DEA (ZSG–DEA) model was proposed by Lins et al. [27] as a viable solution to this problem. It has since become a widely used method for allocating carbon emission rights. Gomes and Lins et al. [28] and Chiu et al. [29] used the ZSG–DEA model to investigate the distribution of carbon emission rights. Furthermore, the ZSG–DEA model has been used by some researchers in China to examine carbon emission allowances at the provincial level (Cai and Ye [30], Yang et al. [31], Cui et al. [32]) and at the industry level (Chen et al. [33]). However, there has been little research done on the distribution of carbon emission rights across a large part of China. Zhuang et al. [34] also mentioned that future research in different geographic clusters in China could be conducted to build a more appropriate carbon dioxide emission allocation mechanism.

In response to the aforementioned issues, this paper achieves breakthroughs in two aspects. Firstly, the power sector is the leading source of carbon emissions, but few academics have been concerned about its issue of carbon emission quotas. Therefore, the allocation of carbon emission rights in the YREB's power industry is the subject of this paper. The analysis of its carbon emission rights allocation can provide more precise information for the rational allocation of carbon emission quotas in the power sector. Secondly, few studies have focused on the temporal and spatial differentiation of carbon emission rights allocation in the power industry, whereas this paper does. Paying attention to this will aid in resolving the problem of heterogeneity in the distribution of carbon emission rights in the power sector in the provinces of the YREB, as well as assisting provinces in formulating accurate carbon emission reduction targets.

This paper first forecasts the input-output variables of the power industry in the YREB from 2021 to 2030. The ZSG–DEA model is then used to calculate the carbon emission rights distribution efficiency in the YREB power industry, iterate carbon emission allowances, optimize carbon emission allowances, and establish a reasonable allocation scheme. Unlike previous research, which has focused solely on carbon emission allocation in 2030, this paper examines carbon emission allocation in each year. After that, the temporal and spatial evolution characteristics of carbon quotas are analysed to compare differences in carbon emission reduction responsibilities and emission reduction paths among provinces over the last decade. The results can then enable recommendations for achieving the YREB's low-carbon development as well as improvements to the power carbon market trading system.

3. Materials and Methods

3.1. ZSG–DEA Model

Data Envelope Analysis (DEA) was proposed in 1978 by American operations researcher Charnes et al. [35] and is a widely used method in academia to assess the relative efficiency of homogeneous decision-making units. The conventional DEA model assumes that each decision-making unit's inputs and outputs are independent of one another. When the DEA model is used in the distribution field, however, it is constrained by the requirement that a certain input indicator (or output indicator) keeps the total amount unchanged. The traditional DEA model will fail in this case. Lins and Gomes et al. [28] proposed a zero-sum gains DEA model, which we call the ZSG–DEA model, in response to this problem. The ZSG–DEA model places all decision-making units on a new aim while keeping the sum of the changed variables constant, because the inputs (or outputs) of decision-making units that were previously technically ineffective under traditional DEA have been reconfigured. The allocation of quotas among provinces in China is competitive, based on the premise of a certain amount of carbon emission quotas. In other words, an increase in emissions in some provinces results in a decrease in emissions in others, reflecting the zero-sum gains concept of constant total emissions. When the zero-sum income concept is applied to the power industry, the total carbon emissions of the power industry are limited. Adjust the distribution of power carbon emission rights in all provinces on a regular basis to achieve the optimal distribution, which will also promote more benign economic development.

Since the ZSG–DEA model was proposed, its application in carbon emission rights allocation has been continuously improved. The focus of the debate is on the treatment of input and output values. Scholars have proposed ZSG–DEA models with competitive input (Cui et al. [32], Yang et al. [31], Fang et al. [36]) or competitive output (Zhuang et al. [34]). Because there are, in reality, both competitive and non-competitive inputs and outputs, the ZSG–DEA model considering the goal of maximizing global efficiency was proposed. Based on the model setting of Wu et al. [37], this paper also considers the expansion (or reduction) of non-competitive output or input based on the distribution of output or input with competitive relationship and proposes an improved ZSG–DEA model. Assuming that there are n decision making units (DMU_j) (j = 1, . . . , n). Each decision making unit has m competitive input, s non-competitive inputs and q outputs, respectively denoted by x_{ij} (i = 1, . . . , m), y_{rj} (r = 1, . . . , s), and z_{pj} (p = 1, . . . , q). where λ_j represents the weight of

DMU_j and the specific decision-making unit is represented by j_0. For the ZSG–DEA model with competitive relationships between inputs, it can be expressed as Equation (1).

$$\min h_{j0}$$
$$s.t.\begin{cases} \sum\limits_{j=1}^{n} \lambda_i x_{ij}[1+\frac{x_{ij0}(1-h_{j0})}{\sum\limits_{j\neq j_0} x_{ij}}] \leq h_{j0}x_{ij0} \\ \sum\limits_{j=1}^{n} \lambda_j y_{rj} \leq h_{j0}x_{rj0}, \sum\limits_{j=1}^{n} \lambda_j z_{pj} \geq z_{pj0} \\ \sum\limits_{j=1}^{n} \lambda_j = 1 \\ \lambda_j \geq 0, j=1,\ldots,n \end{cases} \tag{1}$$

Among them, h_{j0} represents the efficiency value of the DMU_0. If DMU_0 is an inefficient DEA unit, in order to achieve DEA effectiveness, it must reduce the use of i-th input by $u_0 = x_{ij0}(1-h_{j0})$ and share this amount of input proportionally to other decision-making unit by $\frac{x_{ij0}(1-h_{j0})}{\sum\limits_{j\neq j_0} x_{ij}}$. The quantity obtained by the other decision-making unit is $\frac{x_{ij0}(1-h_{j0})}{\sum\limits_{j\neq j_0} x_{ij}} x_{ij}$. As all DMUs are reducing the proportion of input at the same time, the reallocation of i-th input to DMU $_j$ is:

$$x'_{ij} = \sum_{j\neq j_0} [\frac{x_{ij0}(1-h_{j0})}{\sum\limits_{j\neq j_0} x_{ij}} x_{ij}] - x_{ij}(1-h_j) \tag{2}$$

3.2. Data Source and Processing

3.2.1. Input-Output Indicators

Labour, capital, and energy consumption are all common inputs in the industrial production function. Output indicators include output value and various industrial pollutants. The power sector of the 11 YREB provinces was the basic decision-making unit in the construction of the model to measure the distribution efficiency of carbon emission rights in China's power industry. This study was based on the indicator settings of Zhou et al. [19] and Zhuang et al. [34], with power labour input, power capital, and power energy as input variables, power output value as the desired output, and power carbon emissions as the undesired output. The differences in resources and economic levels between provinces were considered in these indicators. At the same time, they strived for the highest power output value and the least amount of pollution with the least amount of labour, capital, and energy, reflecting the fairness and efficiency principle of carbon emission rights. The data were derived from the China Statistical Yearbook [7], China Energy Statistical Yearbook [38], and China Provincial Statistical Yearbook. Table 1 shows an explanation for each input-output variable.

Table 1. Input-output variables of power carbon emission rights allocation efficiency.

Variable Classification	Specific Variable	Variable Explanation
Input variable	Power labour input	Employment of power, thermal and supply sectors
	Power capital	Actual capital stock of power, thermal and supply sectors based on 2005
	Power energy	Power consumption
Output variable	Power output value	Sales value of power, thermal and supply sectors based on 2005
	Power carbon emissions	Estimated power CO_2 emissions by regions

(1) Power labour force: Employment is often used to represent labour force indicators. Because there are no special statistics on human resource investment in the power industry.

This paper replaced the number of employees in the power industry with the number of employees in the power, thermal and supply sectors, and the data in 2020 was obtained by the moving average method. The average annual population growth rate was calculated for the period 2011–2020. Assuming that the population growth rate remained unchanged from 2021 to 2030, and the proportion of provinces was consistent with 2020, the population of China's power industry from 2021 to 2030 was predicted.

(2) Power capital: This paper, like many previous studies such as Zhuang et al. [34], adopted power capital stock to measure power capital. For the initial capital stock, this paper used the method of Hall et al. [39]. The formula is $K_{i0} = I_{i0}/(\delta + g_i)$, where I_{i0} represents the total fixed capital; δ represents the depreciation rate, taking 9.6% in this paper; g_i represents the average GDP growth rate in each province. The "perpetual inventory method" was used to calculate the capital stock of each province every year. The calculation formula is $K_{i,t} = I_{i.t} + (1 - \delta)K_{i,t-1}$, where $K_{i,t}$ is the capital stock of the i-th province during the t-th period and $I_{i,t}$ is the investment of i-th province in t-th period. Then the capital stock should be adjusted to a constant price of 2005. The capital stock for 2021–2030 was predicted by the average growth rate for 2011–2022.

(3) Power energy: Energy was represented by power consumption, and we predicted power consumption from 2021 to 2030 from the average growth rate.

(4) Power output value: Power output value is the "good" output brought by the power production process. The expected output in this study is the industrial sales output value of the power industry in each province after deflator, with 2005 as the base period. In addition, the output data from 2021 to 2030 was forecast based on the power industry's average output growth rate from 2011 to 2020. Due to the lack of special statistics on the output value of the power industry, this paper replaced the output value of the power industry with the sales value of power, thermal and supply, and the data from 2017 to 2020 was obtained by using the moving average method.

(5) Power CO_2 emissions: This paper focused on the distribution of carbon emission rights, so carbon emission rights were included as an undesirable output in the distribution efficiency model. The main methods of the DEA model in dealing with undesirable output include undesirable output as an input method, hyperbolic method, reciprocal conversion method, conversion vector method, directional distance function method and SBM model method, etc. Other methods may be confronted with the problem of ineffective solutions, so this paper adopted the CCR model with undesired output as an input to deal with the issue of carbon emissions. The calculation methods of CO_2 emissions and energy consumption of each province in China over the years are as in Equation (3). This paper employed the reference method based on terminal consumption in the energy balance table of various regions, which was listed in the *2007 IPCC Guideline on National GHG Inventories* (IPCC, 2007) [40]. Each energy type was calculated based on their individual carbon dioxide emission coefficients, which eliminates the calculation error caused by rough classification.

$$CO_{2i} = \sum_j E_{ij} \times EFj \times O_j \tag{3}$$

where CO_{2i} epresents the total CO_2 emissions from the i-th province in Mt (100 million tons); E_{ij} is the physical consumption of the j-th energy in the i-th province, measured in tons (t) or cubic meters (M^3); EF_j denotes the carbon emission coefficient of the j-th energy, expressed in t CO_2/t or t CO_2/M^3. The coefficient of the j-th energy converted into standard coal is represented by O_j. Tables 2 and 3 show the carbon emission coefficients of various energy sources as well as the reference coefficients of standard coal. The data came from the China Statistical Yearbook [7].

Table 2. Carbon emission coefficients of various energy sources (t carbon/t standard coal).

Energy Type	Raw Coal	Coke	Crude Oil	Fuel Oil	Gasoline	Kerosene	Diesel Fuel	Natural Gas	Electricity
Carbon emission coefficients	0.7476	0.1128	0.5854	0.6176	0.5532	0.3416	0.5913	0.448	2.2132

Table 3. Reference coefficients of standard coal for various energy sources.

Energy Type	Raw Coal	Coke	Crude	Fuel Oil	Gasoline	Kerosene	Diesel Fuel	Natural Gas	Electricity
Standard coal coefficient	0.7143 tce/t	0.9714 tce/t	1.4286 tce/t	1.4286 tce/t	1.4714 tce/t	1.4714 tce/t	1.4571 tce/t	13.30 tce/10^4 m^3	1.229 tce/10^4 kwh

3.2.2. Calculation of Initial Carbon Emissions Allowance

Because of the large differences in economic performance, natural resources, and historical carbon emissions among provinces, focusing solely on distribution efficiency will result in an imbalance of provinces' carbon emission reduction responsibilities. As a result, we used historical cumulative carbon emissions as the initial distribution standard to ensure the fairness of carbon emission right distribution. The exact calculation procedure was as follows.

To begin, national total carbon emissions and GDP were used to calculate the carbon emission intensity per unit of GDP from 2011 to 2020. Second, from 2021 to 2030, the carbon emission intensity target value was calculated using the goal of "reducing national carbon emission intensity by 65 percent (compared to 2005) by 2030". Finally, the historical cumulative proportion of carbon emissions from 2011 to 2020 in each province was used as the basis for the allocation of carbon emission rights from 2021 to 2030 in the power industry.

4. Results and Discussion

Based on Equation (1), we used DEA to calculate the initial value of the power sector in each province in the YREB from 2021 to 2030. Table 4 shows the initial efficiency of carbon emission rights allocation. The initial allocation efficiency of carbon emission rights in each province was low, as shown in Table 4, and there were significant differences between provinces.

Table 4. Efficiency of carbon emission rights allocation in the power sector in the YREB from 2021 to 2030.

Province	2021	2022	2023	2024	2025	2026	2027	2028	2029	2030
Shanghai	0.7829	0.7675	0.7519	0.7783	0.7803	0.7794	0.6917	0.6879	0.6869	0.6841
Jiangsu	0.8605	0.8571	0.8515	0.8401	0.8497	0.8576	0.8639	0.8687	0.8718	0.8734
Zhejiang	1.0000	1.0000	1.0000	1.0000	1.0000	1.0000	1.0000	1.0000	1.0000	1.0000
Anhui	0.8317	0.7960	0.7643	0.7332	0.7028	0.6733	0.6446	0.6168	0.5898	0.5638
Jiangxi	0.4840	0.4464	0.4117	0.3796	0.3499	0.3251	0.3032	0.2834	0.2654	0.2489
Hubei	0.7435	0.7355	0.7042	0.6685	0.6346	0.6024	0.5719	0.5429	0.5154	0.4892
Hunan	0.7276	0.7242	0.7034	0.6783	0.6540	0.6307	0.6082	0.5864	0.6333	0.5453
Chongqing	0.6788	0.6674	0.6523	0.6382	0.6258	0.6150	0.6071	0.6002	0.5940	0.5886
Sichuan	0.8422	0.8097	0.7784	0.7482	0.7191	0.6707	0.6640	0.6379	0.6128	0.5887
Guizhou	0.7942	0.7819	0.7548	0.7256	0.6976	0.6976	0.6448	0.6199	0.5960	0.5730
Yunnan	1.0000	1.0000	1.0000	1.0000	1.0000	1.0000	1.0000	1.0000	1.0000	1.0000

Figure 1 depicts the agglomeration characteristics of the initial distribution efficiency of each province, showing the differences in each province's initial distribution efficiency. The initial distribution efficiency for most provinces were inefficient, and only two provinces, Zhejiang and Yunnan, had an efficiency of 1, reaching the DEA frontier. Because Zhejiang is a frontier region for efficient energy production and Yunnan is an environmentally sound region, during the first carbon emission rights allocation scheme, these two provinces achieved high levels of energy efficiency. Although the efficiency of Jiangsu province had not reached the effective frontier, it had risen above 0.8, implying that there was still room for growth. The remaining provinces had efficiencies ranging from 0.3 to 0.8, implying that the initial allocation of carbon emission rights to the power sector was inefficient in these provinces. Among them, Jiangxi had the lowest efficiency value, which was lower than 0.5, followed by Chongqing, which was lower than 0.7. We should concentrate our efforts in these two areas on conserving energy and lowering emissions in the future. In addition, the efficiency of carbon emission rights allocation showed a downward trend in most provinces over time (see Table 1), indicating that allocating carbon emission quotas using the historical method not only made the allocation efficiency low, but also decreased the allocation efficiency over time. As a result, carbon quotas must be recalculated to achieve maximum efficiency.

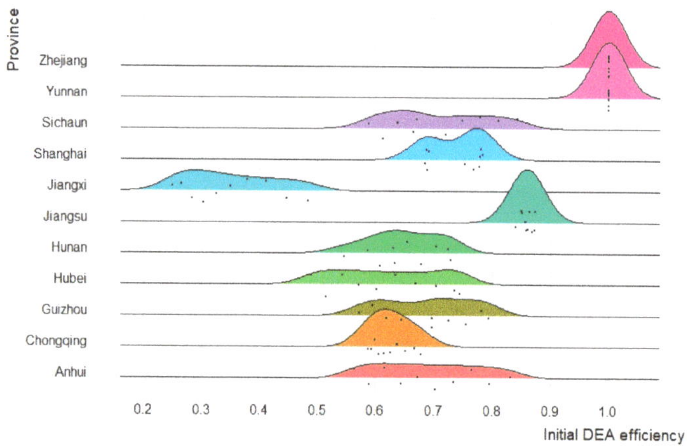

Figure 1. The initial carbon emission rights efficiency of the power sector in the provinces in the YREB from 2021 to 2030.

According to their geographical locations, the YREB can be divided into three regions: upper, middle, and lower reaches (Xing et al.) [41]. The lower reaches consist of Shanghai, Zhejiang, Jiangsu, and Anhui; the middle reaches consist of Hubei, Hunan and Jiangxi provinces; the upper reaches are made up of Chongqing, Sichuan, Guizhou, and Yunnan Figure 2 depicts the trend of initial carbon emission allocation efficiency in the YREB's upper, middle, and lower reaches from 2021 to 2030. Overall, the initial carbon emission rights distribution efficiency in these three regions showed a downward trend, and there were significant differences. The YREB's downstream had the highest power efficiency, followed by the upstream, and the efficiency in the middle reaches was the lowest. This is because that most downstream provinces have advanced economic development and the power industry's technology is relatively advanced. These provinces were early adopters of new energy power generation technology, laying a good foundation for low-carbon development. The high efficiency of the upstream provinces lies in their good ecological environment. At the same time, with the strong support of national policies, the low-carbon economy of these provinces has been well developed. The provinces in the Yangtze River's middle reaches have developed power generation technology late, and have accumulated more carbon emissions, limiting the low-carbon development of these provinces.

The results of the research into the efficiency of initial allocation rights to carbon emissions show that DEA efficiency cannot be achieved with an initial allocation based on historical emissions, so the initial allocation must be adjusted iteratively employing the ZSG–DEA model. We scaled the allocation using Equation (2) until each province had an efficiency value close to 1, and the total carbon emissions remained constant as each iteration progresses. The basic principle was to keep total carbon emissions constant, adjust initial carbon emissions correctly, keep other input-output variables constant, and iterate continuously until the carbon emission allocation efficiency approaches 1. We only show the adjustment process in 2021 due to space constraints, see Table 5. The carbon emission rights for the provincial power sector changed dramatically in the first two iterations, and the carbon emission adjustments varied widely across provinces. However, in the third iteration process, the adjustment amount of each province's carbon emission rights for the power sector were typically zero, which means that the final carbon emission quotas for the power sector in each province gradually tended to stabilise as the iteration progressed.

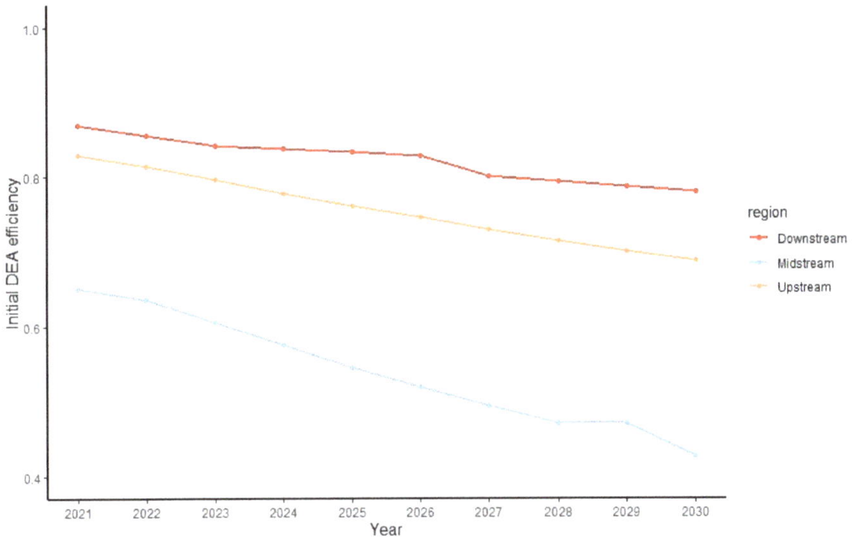

Figure 2. The initial carbon emission rights efficiency of the power sector in the upper, middle, and lower reaches of the YREB from 2021 to 2030.

Table 5. Optimization of allocation efficiency of carbon emission rights of power sector in the YREB in 2021.

Province	Initial Quota (Million Tons)	Initial Efficiency	The First Iteration Value	First Iteration DEA Score	The Second Iteration Value	Second Iteration DEA Score	Final Quota (Million Tons)	Final Efficiency
Shanghai	48.0822	0.7829	45.9293	0.8235	39.9394	0.9881	39.5591	0.9998
Jiangsu	176.7200	0.8476	174.6755	0.9314	169.1400	0.9966	168.9275	1.0000
Zhejiang	127.2981	1.0000	151.1066	1.0000	159.7891	1.0000	160.2785	1.0000
Anhui	65.4445	0.8317	65. 6307	0.9647	66.8642	0.9973	66.8731	0.9999
Jiangxi	38.3796	0.4840	24.6997	0.8726	22.8662	0.9927	22.7640	1.0000
Hubei	58.7681	0.7435	53.4204	0.9186	51.8097	0.9957	51.7297	1.0000
Hunan	50.9007	0.7276	45.5586	0.9374	45.1416	0.9966	45. 1183	1.0000
Chongqing	30.4770	0.6788	25.9780	0.9301	25.5896	0.9961	25.5648	1.0000
Sichuan	71.1966	0.8422	72.1122	0.9738	74.1675	0.9986	74.2807	1.0000
Guizhou	41.4484	0.7942	40. 1791	0.9542	40.5458	0.9975	40.5637	1.0000
Yunnan	50.3941	1.0000	59.8190	1.0000	63.2562	1.0000	63.4499	1.0000

Figure 3 shows the efficiency after each iteration. It is clear that more provinces were approaching the DEA frontier as the iterative process continued. In particular, only two provinces, Zhejiang and Yunnan reached the production frontier in the initial distribution.

In the first iteration, the distribution efficiency of most provinces exceeded 0.9, and after three iterations, almost all provinces had reached the production frontier. All the efficiency values were 1, indicating that all provinces' reallocated carbon allowances were nearly optimal after the third iteration.

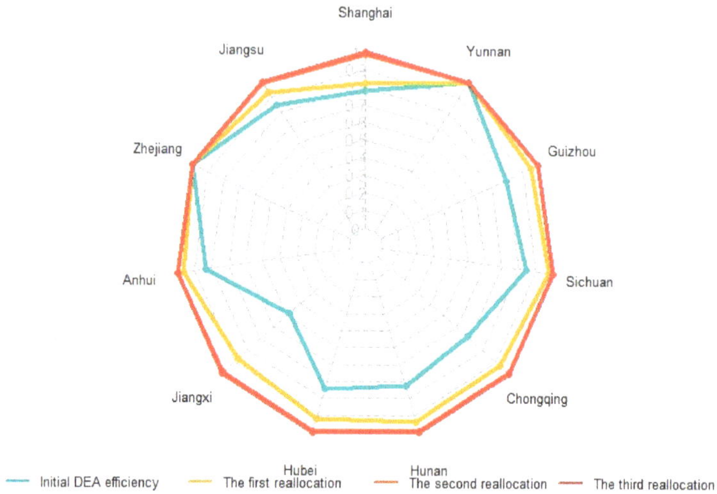

Figure 3. Distribution efficiency of the power sector in each province after each iteration in 2021.

Figure 4 depicts the adjusted amount of carbon emission allowances after three iterations. Carbon emission quotas must be increased in some provinces, while others must decrease quotas to achieve maximum efficiency. Provinces with higher initial efficiency had the largest increases in carbon emission quotas, including Zhejiang and Yunnan. Carbon emission allowances must be lowered in most provinces, with Jiangxi experiencing the greatest reduction. Because the adjustment equals zero, high-performing provinces should receive more carbon allowances from other provinces, whereas less efficient provinces should further reduce their carbon allowances, which means tightening CO_2 controls and setting higher emission reduction targets.

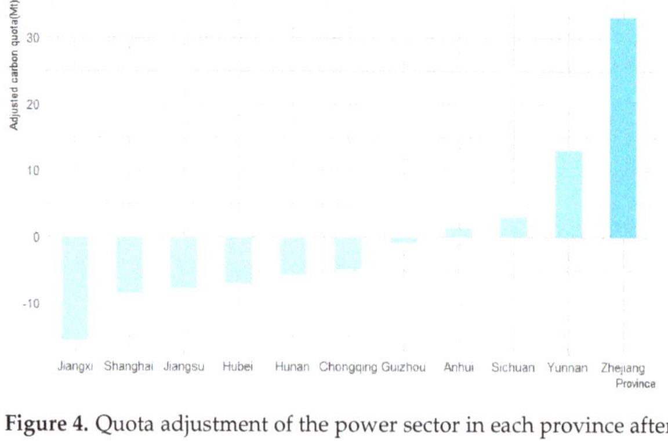

Figure 4. Quota adjustment of the power sector in each province after each iteration in 2021.

To clarify the power sector's pressure to reduce emissions in the provinces of the YREB, we used the iterative amount to achieve the optimal efficiency as the carbon emission quota that each province needs to be assigned. The amount of carbon emission quota means the carbon emission reduction pressure faced by each province. Four representative years were chosen, 2021, 2024, 2027, and 2030, and the spatial distribution of carbon emission allowances in the YREB's power industry from 2021 to 2030 was obtained using the ArcGIS 10.2, as shown in Figure 5. Power carbon emission quotas are colour-coded and divided into six levels ranging from low to high. Overall, there were significant differences in power industry's carbon emission quotas across provinces, and they showed different spatial characteristics over time. During the previous period (2021–2027), the proportion of carbon emission quotas in Yunnan, Guizhou, and Hunan increased over time; the later period (2027–2030) was relatively balanced, and each province's carbon emission quotas reached a relatively stable state.

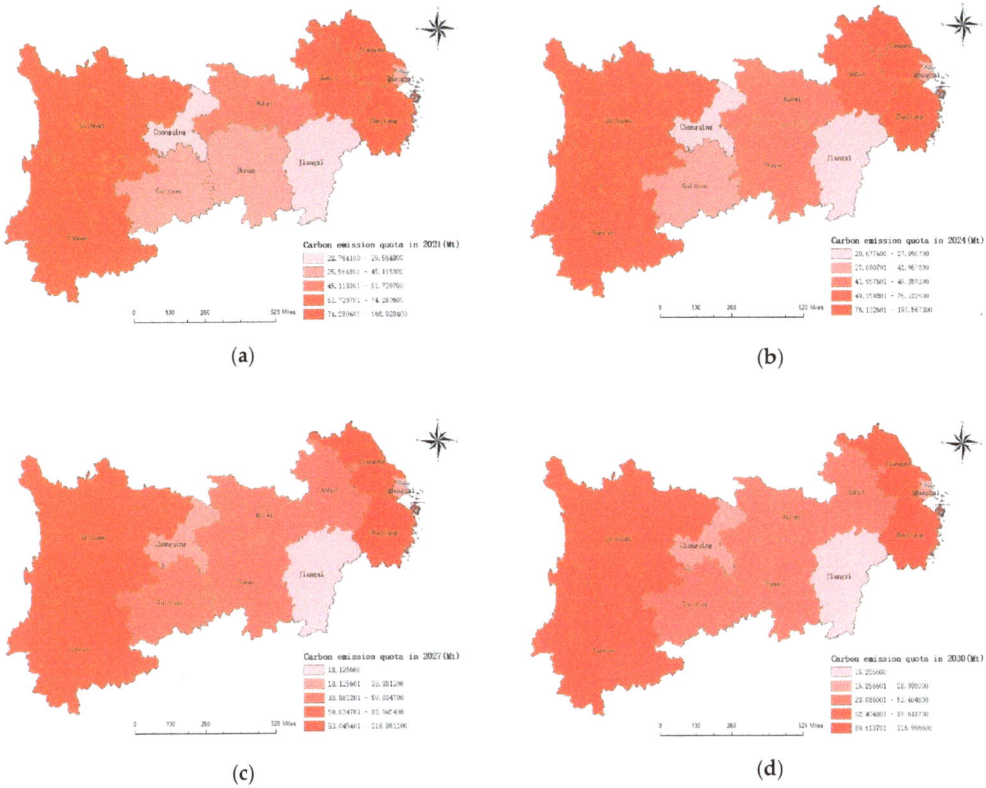

Figure 5. Spatial distribution of carbon emission quotas in the YREB's power sector. (**a**) Carbon emission quota of the power sector in each province in 2021 (Mt); (**b**) Carbon emission quota of the power sector in each province in 2024 (Mt); (**c**) Carbon emission quota of the power sector in each province in 2027 (Mt); (**d**) Carbon emission quota of the power sector in each province in 2030 (Mt).

Specifically, most provinces in the YREB's upper and lower reaches were distributed more carbon emission rights, and most provinces in the middle reaches were distributed less. This means that the power sector in the middle reaches of the province faces a greater challenge in emission reduction, because of the large number of thermal power generation tasks in the YREB's middle reaches and the excessive accumulation of carbon emissions.

Specific to each province, Zhejiang and Jiangsu were distributed the highest carbon emission quotas in these four times. On the one hand, Zhejiang and Jiangsu have strong emission reduction capabilities due to the economy's rapid growth and the use of technologies for advanced power generation. On the other hand, their economic development is not dependent on energy supply because of their relatively well-developed industrial structure and primarily high-tech industries. Following completion of their own emission-reduction tasks, the two provinces can sell excess carbon emission rights to provinces with low-carbon emission rights, as well as provide technical assistance and financial subsidies. Followed by Zhejiang and Jiangsu, Sichuan and Yunnan were also allocated high carbon emission quotas. Their economies are not as developed, but their environment is better. Furthermore the proportion of thermal power generation in Sichuan and Yunnan is small, as is the burden of carbon emissions. Jiangxi and Chongqing were allocated the fewest carbon emission quotas. These two provinces are under greater pressure to reduce emissions and should implement a variety of emission-control measures. The main feature of energy consumption structure in Jiangxi is coal-based, which is the main reason for increasing the pressure on Jiangxi to reduce emissions. At the same time, the power generation technology adopted in Jiangxi is still relatively backward, and it is strongly dependent on high-energy energy. Chongqing's pressure to reduce emissions stems primarily from its high-emission industrial structure. The industrial economy is the backbone of Chongqing's development as a city dominated by heavy industry, but it also brings high energy consumption and emissions. These two provinces should take stronger emission reduction measures, including optimization of energy structure and industrial structure, as well as technological innovation. The remaining provinces were allocated carbon emission quotas that were relatively balanced and low. This means they face stricter carbon emission restrictions. To meet their 2030 emission reduction targets, these provinces should implement a variety of emission-cutting measures. Based on the above analysis, Zhejiang, Jiangsu, Sichuan and Yunnan are more likely to be sellers in the carbon emissions trading market, while Jiangxi and Chongqing are more likely to be buyers in the carbon emissions trading mar

5. Conclusions and Policy Recommendations

In this paper, the carbon emission allocation efficiency of the power industry in 11 provinces of the YREB was calculated using the ZSG–DEA model from 2021 to 2030 Each province's carbon emissions quotas were redistributed to achieve maximum efficiency The results show that:

(1) The power sector's initial distribution efficiency was inefficient in most provinces only Zhejiang and Yunnan had reached the production frontier. Jiangxi had the lowest distribution efficiency, which was less than 0.5, and Chongqing's efficiency was less than 0.7. In these two provinces, we should concentrate on energy conservation and emission reduction. The efficiency of carbon emission rights allocation in most provinces showed a downward trend from 2021 to 2030, indicating that allocating carbon emission quotas using the historical method not only reduced allocation efficiency, but also decreased over time.

(2) The initial distribution efficiency of the power sector varied greatly between the YREB's upper, middle, and lower reaches. The downstream region had the highest power efficiency, followed by the upstream, and the middle region had the lowest. Because of the redistribution of carbon emission rights, most provinces in the upper and lower reaches had more carbon emission rights than middle reaches, implying that the power sector in the middle reaches faces greater emission reduction challenges.

(3) The carbon emission rights of the power industry varied greatly across provinces, and it exhibited different spatial characteristics over time. The early stage's carbon emission quota (2021–2027) varied greatly, while the later stage (2027–2030) was relatively balanced Carbon emission quotas were higher in Zhejiang, Jiangsu, Sichuan, and Yunnan, which are more likely to become carbon trading market sellers. Jiangxi and Chongqing had lower carbon emission quotas and are thus more likely to participate in the carbon emission trading market as buyers. Other provinces' carbon emission quotas were relatively balanced.

This paper makes the following policy recommendations based on the research findings to help the YREB's power sector meet the 2030 emission reduction target.

(1) Differentiated carbon emission reduction goals and strategies must be established by local governments. Different targets should be set based on the different emission reduction pressures in each province. Provinces with a developed economy and a reasonable energy structure can set loose emission reduction targets, whereas provinces with a

developing economy and a heavy reliance on energy should set strict emission reduction targets. Furthermore, provincial power departments should fully exploit their own resource advantages and develop their own emission reduction strategies.

(2) The government should increase its investment in science and technology, as well as speed up the development of new power generation technologies such as clean energy and a reduction in thermal power generation. Provincial power departments should be encouraged to implement appropriate advanced power generation technologies, and the government can provide technical assistance to some provinces with weak economies and slow technological development.

(3) The industrial structure should be adjusted to speed up industrial upgrading. The unreasonable industrial structure seriously restricts the low-carbon development of various provinces. Further adjustments and optimization of the industrial structure are required, as well as the avoidance of energy-intensive industries and the active encouragement of the development of strategic emerging industries.

(4) The government can create fiscal policies to lower technology-related costs and provide financial assistance to provinces that are having difficulty reducing emissions.

(5) Regional cooperation should be strengthened, and carbon emission trading market should be improved. Most provinces are eligible to join the carbon emissions trading market. Through formulating reasonable market policies and regulating carbon emissions quota trading effectively, it is possible to achieve a coordinated economic and environmental development.

Author Contributions: Conceptualization, methodology, software, investigation, writing—original draft, writing—review and editing, project administration, and funding acquisition (D.M.); conceptualization, methodology, data curation, visualization, funding acquisition (Y.X.); validation, writing—review and editing, and funding acquisition (N.Z.). All authors have read and agreed to the published version of the manuscript.

Funding: This research was funded by [the Project of Chongqing Social Science Planning Project of China] grant number (No. 2020QNGL38) and The APC was funded by [the Humanities and Social Sciences Research Program of Chongqing Education Commission of China] grant number (No. 20SKGH169).

Institutional Review Board Statement: Not applicable.

Informed Consent Statement: Not applicable.

Data Availability Statement: All the data can be found in http://www.stats.gov.cn/tjsj/ndsj/ (accessed on 15 March 2022).

Acknowledgments: We are particularly grateful for the financial support from the Project of Chongqing Social Science Planning Project of China (No. 2020QNGL38), Humanities and Social Sciences Research Program of Chongqing Education Commission of China (No. 20SKGH169), and we are grateful to anonymous reviewers and editors for their comments and suggestions on this paper.

Conflicts of Interest: The authors declare no conflict of interest.

References

1. Ma, M.; Cai, W. What drives the carbon mitigation in Chinese commercial building sector? Evidence from decomposing an extended Kaya identity. *Sci. Total Environ.* **2018**, *634*, 884–899. [CrossRef] [PubMed]
2. BP. *The BP Statistical Review of World Energy 2018*; BP: London, UK, 2018.
3. Zhou, J.G.; Guang, F.T.; Du, S.J. Decomposing the Decoupling of carbon emissions and economic growth in China's power industry. *Pol. J. Environ. Stud.* **2017**, *26*, 2407–2418. [CrossRef]
4. Li, Y.M.; Wang, Q. Exploring Carbon Emissions in China's Electric Power Industry for Low-Carbon Development: Drivers, Decoupling Analysis and Policy Implications. *Pol. J. Environ. Stud.* **2019**, *28*, 3353–3367. [CrossRef]
5. Yu, A.; You, J.; Rudkin, S.; Zhang, H. Industrial carbon abatement allocations and regional collaboration: Re-evaluating China through a modified data Envelopment analysis. *Appl. Energy* **2019**, *233*, 232–243. [CrossRef]
6. Li, D.; Zhang, J. Measurement and analysis of ecological pressure due to industrial development in the Yangtze River Economic Belt from 2010 to 2018. *J. Clean. Prod.* **2022**, *353*, 131614. [CrossRef]
7. National Bureau of Statistics of China (NBSC). *China Statistical Yearbook*; China Statistics Press: Beijing, China, 2001–2019.

8. UNFCCC. *United Nations General Assembly. United Nations Framework Convention on Climate Change*; UN: New York, NY, USA, 1992.
9. Pan, X.Z.; Teng, F.H.; Wang, G. Sharing emission space at an equitable basis: Allocation scheme based on the equal cumulative emission per capita principle. *Appl. Energy* **2014**, *113*, 1810–1818. [CrossRef]
10. Zhu, B.Z.; Jiang, M.X.; He, K.J.; Chevallier, J.; Xie, R. Allocating CO_2 allowances to emitters in China: A multi-objective decision approach. *Energy Policy* **2018**, *121*, 441–451. [CrossRef]
11. Han, R.; Tang, B.J.; Fan, J.L.; Liu, L.C.; Wei, Y.M. Integrated weighting approach to carbon emission quotas: An application case of Beijing-Tianjin-Hebei region. *J. Clean. Prod.* **2016**, *131*, 448–459. [CrossRef]
12. Fang, G.C.; Liu, M.H.; Tian, L.X.; Fu, M.; Zhang, Y. Optimization analysis of carbon emission rights allocation based on energy justice-The case of China. *J. Clean. Prod.* **2018**, *202*, 748–758. [CrossRef]
13. Guo, J.; Zhao, M.K.; Wu, X.H.; Shi, B.B.; Gonzalez, E. Study on the distribution of PM emission rights in various provinces of China based on a new efficiency and equity two-objective DEA model. *Ecol. Econ.* **2021**, *183*, 106956. [CrossRef]
14. He, W.J.; Zhang, B. A comparative analysis of Chinese provincial carbon dioxide emissions allowances allocation schemes in 2030: An egalitarian perspective. *Sci. Total Environ.* **2021**, *765*, 142705. [CrossRef] [PubMed]
15. Kong, Y.C.; Zhao, T.; Yuan, R.; Chen, C. Allocation of carbon emission quotas in Chinese provinces based on equality and efficiency principles. *J. Clean. Prod.* **2019**, *211*, 222–232. [CrossRef]
16. He, W.J.; Liu, D.; Wang, C. Are Chinese provincial carbon emissions allowances misallocated over 2000—2017? Evidence from an extended Gini-coefficient approach. *Sustain. Prod. Consum.* **2022**, *29*, 564–573. [CrossRef]
17. Du, J.; Pan, M.; Chen, Y.; Duan, Y.R. An efficiency-based allocation of carbon emissions allowance: A case study in China. *J. Clean. Prod.* **2020**, *251*, 119346. [CrossRef]
18. Zhou, H.J.; Ping, W.Y.; Wang, Y.; Wang, Y.Y.; Liu, K.L. China's initial allocation of interprovincial carbon emission rights considering historical carbon transfers: Program design and efficiency evaluation. *Ecol. Indic.* **2021**, *121*, 106918. [CrossRef]
19. Zhou, P.; Sun, Z.R.; Zhou, D.Q. Optimal path for controlling CO_2 emissions in China: A perspective of efficiency analysis. *Energy Econ.* **2014**, *45*, 99–110. [CrossRef]
20. Qin, Q.D.; Liu, Y.; Huang, J.P. A cooperative game analysis for the allocation of carbon emissions reduction responsibility in China's power industry. *Energy Econ.* **2020**, *92*, 104960. [CrossRef]
21. Liu, H.X.; Lin, B.Q. Cost-based modelling of optimal emission quota allocation. *J. Clean. Prod.* **2017**, *149*, 472–484. [CrossRef]
22. Schmidt, R.C.; Heitzig, J. Carbon leakage: Grandfathering as an incentive device to avert firm relocation. *J. Environ. Econ. Manag.* **2014**, *67*, 209–223. [CrossRef]
23. Sartor, O.; Palliere, C.; Lecourt, S. Benchmark-based allocations in EU ETS Phase 3: An early assessment. *Clim. Policy* **2014**, *14*, 507–524. [CrossRef]
24. Zhang, Y.J.; Wang, A.D.; Tan, W. The impact of China's carbon allowance Allocation rules on the product prices and emission reduction behaviors of ETS covered enterprises. *Energy Policy* **2015**, *86*, 176–185. [CrossRef]
25. Burtraw, D.; McCormack, K. Consignment auctions of free emissions allowances. *Energy Policy* **2017**, *107*, 337–344. [CrossRef]
26. Zhao, R.; Min, N.; Geng, Y.; He, Y.L. Allocation of carbon emissions among industries/sectors: An emissions intensity reduction constrained approach. *J. Clean. Prod.* **2017**, *142*, 3083–3094. [CrossRef]
27. Lins, M.; Gomes, E.G. Olympic ranking based on a zero sum gains DEA model. *Eur. J. Oper. Res.* **2003**, *148*, 312–322. [CrossRef]
28. Gomes, E.G.; Lins, M. Modelling undesirable outputs with zero sum gains data envelopment analysis models. *J. Oper. Res. Soc.* **2008**, *59*, 616–623. [CrossRef]
29. Chiu, Y.H.; Lin, J.C.; Su, W.N.; Liu, J.K. An Efficiency Evaluation of the EU's Allocation of Carbon Emission Allowances. *Energy Sources Part B Econ. Plan. Policy* **2015**, *10*, 192–200. [CrossRef]
30. Cai, W.G.; Ye, P.Y. A more scientific allocation scheme of carbon dioxide emissions allowances: The case from China. *J. Clean. Prod.* **2019**, *215*, 903–912. [CrossRef]
31. Yang, M.; Hou, Y.R.; Ji, Q.; Zhang, D.Y. Assessment and optimization of provincial CO_2 emission reduction scheme in China: An improved ZSG-DEA approach. *Energy Econ.* **2020**, *91*, 104931. [CrossRef]
32. Cui, X.Y.; Zhao, T.; Wang, J. Allocation of carbon emission quotas in China's provincial power sector based on entropy method and ZSG-DEA. *J. Clean. Prod.* **2021**, *284*, 124683. [CrossRef]
33. Chen, L.L.; He, F.; Wang, J.J. Allocative Efficiency of Carbon Emission Allowances among Sectors in China. *Pol. J. Environ. Stud.* **2018**, *27*, 557–563. [CrossRef]
34. Zhuang, M.; Yong, G.; Sheng, J.C. Efficient allocation of CO_2 emissions in China: A zero sum gains data envelopment model. *J. Clean. Prod.* **2016**, *112*, 4144–4150.
35. Charnes, A.; Cooper, W.W.; Rhodes, E. Measuring the efficiency of decision making units. *Eur. J. Oper. Res.* **1978**, *2*, 429–444. [CrossRef]
36. Fang, K.; Zhang, Q.F.; Long, Y.; Yoshida, Y.; Sun, L.; Zhang, H.R.; Dou, Y.; Li, S. How can China achieve its Intended Nationally Determined Contributions by 2030? A multi-criteria allocation of China's carbon emission allowance. *Appl. Energy* **2019**, *241*, 380–389. [CrossRef]
37. Wu, J.; An, Q.X.; Yao, X.; Wang, B. Environmental efficiency evaluation of industry in China based on a new fixed sum output data envelopment analysis. *J. Clean. Prod.* **2014**, *74*, 96104. [CrossRef]
38. National Bureau of Statistics of China (NBSC). *China Energy Statistical Yearbook*; China Statistics Press: Beijing, China, 2001–2019.

9. Hall, R.E.; Jones, C.I. Why do some countries produce so much more output per worker than others? *Q. J. Econ.* **1999**, *114*, 83–116. [CrossRef]
0. IPCC (Intergovernmental Panel on Climate Change). *2007 IPCC Guidelines for National Greenhouse Gas Inventories;* IPCC: Geneva, Switzerland, 2007.
1. Xing, Z.; Wang, J.; Zhang, J. Total-factor ecological efficiency and productivity in Yangtze River Economic Belt, China: A non-parametric distance function approach. *J. Clean. Prod.* **2018**, *200*, 844–857. [CrossRef]

sustainability

Article

Mining in Poland in Light of Energy Transition: Case Study of Changes Based on the Knowledge Economy

Ryszard Marszowski * and Sebastian Iwaszenko

Department of Acoustics, Electronics and IT Solutions, Central Mining Institute, 40-166 Katowice, Poland; siwaszenko@gig.eu
* Correspondence: rmarszowski@gig.eu; Tel.: +48-32-259-21-23

Abstract: Implementing climate policy goals, such as achieving climate neutrality by 2050, requires significant transformation of the mining sector—which for some countries and regions where mining is still the basis of the economy and part of the identity is very difficult and results in significant social and economic costs. Focusing on regional aspects and characteristics of the mining sector in Poland and its impact on energy security, the paper provides insight into the noticeable process of mining modernization in the globalized world economy through its transition from the industrial era to a knowledge-based economy and the impact of these changes on regional development. The described process is directly related to implementation of innovative and new technical concepts and technological solutions for the mining industry. The indicated changes imply the need to redefine operating principles and organizational models in the mining industry in order to build responsive solutions based on innovations—shaping modern (intelligent) mining of the future—while at the same time being part of the transformation of (post-)mining regions into a multi-industry region. All the described elements are proposed as supporting elements of the transformation process—to ensure full use of the technological and infrastructural potential during the energy transition process.

Keywords: energy transition; knowledge economy (knowledge-based economy); Industry 4.0; transformation of mining in Poland

Citation: Marszowski, R.; Iwaszenko, S. Mining in Poland in Light of Energy Transition: Case Study of Changes Based on the Knowledge Economy. *Sustainability* **2021**, *13*, 13649. https://doi.org/10.3390/su132413649

Academic Editors: Nicola Raimo, Filippo Vitolla, Ornella Malandrino and Benedetta Esposito

Received: 18 November 2021
Accepted: 8 December 2021
Published: 10 December 2021

1. Introduction

The ongoing changes in energy systems—resulting from, among other things, the move towards climate neutrality, especially noticeable in Europe—poses a huge challenge for the mining sector. One of the critical challenges for global mining is need for a skillful and innovative promotion of modern attitudes and persuading both decision makers and miners that innovation and the resulting changes are necessary. Additionally, most importantly, that departure from the mindset prevailing in the industrial era is purposeful and inevitable, because only turning to the future—by caring for self-development and the environment—can guarantee success [1]. The primary determinant in shaping this thesis is the paradigm shift in the development of the current model of the global economy as a result of its dynamic transition from the era of the Industrial Revolution to a knowledge-based economy, determined in its development by a new process called Revolution 4.0 [2]. From the content of the above thesis emerges the image of the future, in which humans, industry and digitization and automation processes developing with hitherto unknown dynamics will shape the surrounding reality. They can lead to the disappearance of the world dominated by the era of the Industrial Revolution and the transition to a new era shaped by technologies determined by data flow and their analysis. Given the indicated megatrends and transformation driven by the change in the meaning and role of knowledge, the term "knowledge society" emerged [3], in which creation and utilization of competitive and innovative solutions have become an essential capital. It seems that the automation process, together with the deepening phenomenon of demographic depression in the context of increasing competitiveness of economies, which is particularly noticeable in the depopulation of cities and regions that are post-mining or in transition, will determine the need to permanently raise the level of education, competences and to create and master new knowledge. Although modern institutions and their global reach provide people with incomparably higher chances of experiencing a safe and satisfying life than in any of the pre-modern systems, modernity has also a dark side which is the loss of the individual between the global aspirations of modernity and the local dimension of human existence, the breakdown of space–time relationships, the reorganization of social relations within the network society or the constant struggle with the processes of marginalization, differentiation and exclusion [4].

In this light, the future of mining industry is determined by a few trends, the most important being energy transition. The article presents the current state and possibilities of technological changes in the mining sector in Poland. In the context of trends in the EU climate policy, the Polish energy economy, based on the generation of electricity from solid fuels, requires a deep transformation. In order to successfully implement an energy transition system, engineering, technological and social approaches need to be considered simultaneously and on different levels—individuals, occupational groups, regions and nations, as well as political, economic and energy security aspects. On the one hand, there are the necessary changes implemented for gradual phasing out of coal mines with the application of social safeguards, ensuring that the welfare of the individual does not become a victim of technological development promoting alternative energy sources. At the same time, the social contract for energy transformation must go beyond a single interest group, encompassing not only workers of the mining sector, but also many other stakeholders from mining and post-mining regions. On the other hand, the transition should also take into account the opportunities to build on existing knowledge held in enterprises and technological infrastructure for the effective change of power generation mix [5].

The paper analyzes the history and possible future of mining sector transformation with the view of the "Green Deal" and transformations in the energy sector. The influence of driving forces caused by the pandemic situation are considered. The possible perspective and chances for the mining, and particularly coal mining, sector are targeted. To the authors' best knowledge, this is the first attempt to address this topic in the presented context.

2. Materials and Methods

In order to achieve the main goal of the article, which is to analyze the meaning of the future of mining, the phenomenon of a knowledge-based economy and the assumptions, theses and megatrends adopted in the introduction to the article, various research methods were used. The indicated research methods focus on available cognitive methods, aspects of the knowledge economy and Industry 4.0 in the context of mining corresponding to contemporary challenges.

Critical analysis of the literature covered numerous complementary works and focused mainly on processes and megatrends crucial for energy transition determined by the mentioned globalization, knowledge-based economy and Industry 4.0, shaping the state and development perspectives of the mining industry. The critical analysis covered both professional and scientific literature, including theoretical publications as well as publications discussing research results and forecasts in numerous cognitive sections. The conclusions and recommendations resulting from them are aimed at strengthening the utilitarian part of the conducted analyses.

The study also used studies devoted to the development perspectives of the studied areas in the light of analyses of strategic documents regarding their future. These documents were, in particular, the multi-annual development strategies for Europe, the country and the voivodship, in various quality sections. In addition to formal and legal issues related to the development perspectives of the studied areas—including, among others in applicable law—the studies referred to European and national documents in which medium- and long-term goals and tasks in the field of economic and social policies were outlined, especially in the context of ongoing and forecast mining transition processes. Scenario forecasts in the perspective of the knowledge-based economy focus on predictions of both future internal and external development conditions in order to achieve the desired target state, i.e., the justified need and vision of the mining industry responding to contemporary challenges. When presenting forecasts, it is assumed ex ante that the specificity of the area under study sets preferences outlining organizational and systemic solutions directly related to the future education, development and transition models. It is also assumed that the current state of the coronavirus epidemic in the territory of the Republic of Poland will accelerate the implementation—as an unfavorable circumstance—of numerous elements of the anticipated forecasts.

The implementation of climate policy is also a technological challenge. However, the term "technology" has a very broad meaning, so it is hardly possible to give a simple answer for technological readiness for upcoming changes. The problem arises where the used hardware machinery is concerned. Though the mining sector puts great effort into continuous exchange of its equipment and exchanging it with up to date, technically fluent solutions, the process is costly and requires both time and well-qualified staff. Despite the anticipated necessity of dramatic changes in the methods of raw material excavation and processing, one seems undebatable: the increasing role of digitalization and information science. Poland has more than satisfactory capabilities in preparing numerous, well-educated staff in areas such as IT, electronics and telecommunications. The same is true for mechanical engineering and geoscience. However, the problem which has to be solved is the relatively poor competitiveness of the Polish labor market. Many highly qualified specialists are able to find better opportunities for professional development abroad. Despite that, it should be noticed that there is a suitable and satisfactory potential both in technology and human resources already present. It is possible to lean on it for the upcoming necessary transformations. However, the implementation of such a scenario requires appropriate policy and incentives on the part of the state.

The unexpected outbreak of the pandemic dramatically affected many areas of life. It is difficult to find an area of life that would not experience perturbation related to the actions taken to reduce the health effects of the pathogen. The mining industry is no exception here. The pandemic has shifted the focus from working in plants and offices to working over telecommunications lines. The shift is not only technical. In fact, it changed the perception of work: it does not matter where and how the work is provided, what matters is the result delivered to the employer. It allowed focusing on the delivered value, limiting controlling the conditions, which is less important. There is natural way for such a movement in many occupations and services, but at a first glance it may seem impossible for mining industry to organize the work this way. Though it is still difficult to limit the necessity of personal engagement of workers in the mining activities, it can be noticed that development of remote control, remote sensing and autonomous machinery makes personal engagement in the excavation process less and less required. The use of machines allows people to be withdrawn from the most safety-critical areas of operation. It becomes possible to supervise the production first by the use of remote control and finally by the autonomous execution of work by the devices. However, this does not mean withdrawing people from the mining process completely. However, their role, scope of duties and required competences are changing. In this light, the COVID-19 pandemic appears as a factor that accelerated and catalyzed the ongoing process of digitization and the next level of production automation. Forcing the necessity to protect health and life made it possible to see the possibilities of development based on IT solutions of underground mining. The presented conditions allow one more regularity to be observed: even in the mining industry, dominated for many years by work consisting mainly in physical effort, there is a re-evaluation. In automated, computerized mining, it is not muscle strength, but knowledge and competences, that play and will play a fundamental role. However, these processes would take place regardless of the epidemic threat, although its role in their intensification cannot be overestimated.

3. Discussion

3.1. Polish Hard Coal Mining—An Overview of State and Perspectives in Light of Energy Security Needs

The mining industry is still one of the most significant sectors in shaping the global economic situation. In countries with low and medium economic efficiency, it ensures the creation of new jobs, reduces poverty phenomena and improves the conditions for economic development [6]. Despite ongoing changes in approach and efforts to achieve climate neutrality, in Poland both hard coal and lignite are still very important parts of development, as Poland has significant coal resources that may serve as an essential stabilizer of the country's energy security, which is of particular importance to the dependence of the Polish economy on gas imports (over 70%) and crude oil (over 95%). More precisely in terms of hard coal mining, Poland ranks 10th in the global rankings and first in the European Union. The Polish energy mix is unique in Europe because it is heavily based on domestic solid energy resources (hard coal and lignite) [7] and, for Poland, hard coal is a guarantee of energy security, still being the primary source of energy [8]. Therefore, for many years one of primary objectives of Polish energy policy, in the field of security of fuel and energy supplies, has been rational and effective management of coal deposits located in the territory of the Republic of Poland [8].

Additionally, as noted by P. Czaja (2014), it should be remembered that mining in Poland is not only about coal. Polish mining is primarily mineral resources—almost 430 million tons of mining per nation with 38 million inhabitants is more than the European average per citizen. Most of the raw materials, without which modern technology cannot function, such as critical or rare earth elements, come from mining [9]. The global comparison of the share of hard coal and lignite in global reserves of energy resources (64%), crude oil (18%) and natural gas (18%) also shows that coal is the largest carrier of primary and electric energy [10]. According to forecasts by the International Energy Agency, the demand for electricity is still increasing, especially in Asia, and coal remains the largest source of generated electricity [11]. Not only will the demand for energy resources increase but also—according to global forecasts—demand for other mineral resources [12]. Due to a forecasted increase in the global demand for electricity, it is necessary to develop energy production from all fossil fuels as well as renewable and atomic energy. Although the decline of coal-based energy production is an ongoing reality in Europe, it is highly probable that, globally, hard coal will remain the most critical fuel in the next few decades, mainly for the production of electricity with the simultaneous development of new combustion technologies and reducing the costs of mining and emissions to the atmosphere [10]. From the Polish government's point of view, "the most important in this dimension is to cover the growing demand for fuels and energy in connection with the forecasted economic growth, while ensuring uninterrupted energy supplies. It is important to maintain a high index of energy independence, increase diversification of the energy mix and diversification of directions of supplies of imported fuels" [13], guaranteeing appropriate quality parameters This is especially important in light of the paradox noticeable in the Polish domestic coal market. On the one hand, the share of coal in electricity generation in 2019 was nearly five percentage points lower than the year before. The trend of decreasing hard coal output in domestic mines, which has been visible for years, also continues—last year it decreased by 2 million tons. Other hand, the demand for steam coal remains high. A large part of it, almost 20% of On the domestic consumption, is supplemented with imported coal. Ten million tons originate from Russia, and the remaining 3 million tons come from such countries as Colombia, the USA or Kazakhstan [5].

On one hand, many experts, in their opinion about the best approach to the energy transition process and generally the best decisions for the future of Polish mining sector, agree with Lisowski (2016), who—in his thesis concerning a recovery program for Polish hard coal mining—referred to the Polish raison d'état, noting that further country development requires a base of strategic areas of the economy and energy security of the country on its own substantial reserves of hard coal and lignite, and on the labor resources offered by society and not only on imports of oil, gas, nuclear technologies and fuels or renewable energy technologies [14]. This is particularly important given the fact that, under Polish conditions, development of gas sources means the need to increase imports. Domestic natural gas production, i.e., 143.3 PJ in both 2018 and 2019, accounts for less than 20% of the total consumption of this raw material in the economy of the country (19.8% in 2018 and 18.9% in 2019, respectively), whereas the rest needs to be imported, mainly from the east. The high import prices of this raw material limit the effectiveness of its use in the power sector, and the relatively little supplier diversification poses a threat to their reliability [7].

On the other hand, due to the changes in the global, and especially European, coal market, as well as changing conditions—including those resulting from the European Green Deal [15] and Energy Roadmap 2050 [16]—transformation of the mining sector is essential. Although, with the development of carbon capture and storage (CCS) and other emerging clean technologies, coal could continue to play an important role in a sustainable and secure supply in the future, there is a need for investments to enable technological transformation. Additionally, in light of the fact that Polish hard coal reserves suitable for economically and technically efficient production are diminishing [17] and some used technologies are outdated and impacting the environment, there is a growing need for effective and sustainable management of resources and utilization of new technologies and eco-innovations necessary to eliminate or minimize the negative effects of mining processes [12,18].

In the light of the theses outlined above, it seems that the mining industry, in line with the issue of energy security, should manage its own resources and transition to cleaner forms of energy based on reliable information on its impact on the regional socio-economic environment and on the global environment and climate. As the information is treated as the third most fundamental quantity, next to energy and matter, with a decisive impact on society, its forms of intercourse and cooperation [19] and adequately shaped system of obtaining and transmitting information between the mining enterprise and its surroundings can create a kind of mechanism of coexistence and relations between these entities. Changes in the environment have a key impact on the organization, especially because they occur quickly and are hardly predictable or even unpredictable. These include, among others:

- globalization processes and, consequently, the need for enterprises to operate on the international market;
- intensifying competition, leading to a focus on customer needs and the relationship: results—outlays;
- the disappearance of many existing markets and creation of new ones;
- the emergence of new organizational forms, being a consequence of networking, numerous mergers and alliances, which result in organizing work in the form of multi-task, interdisciplinary teams;
- the fast pace of development and implementation of new technologies and operating techniques, which result in shortening of product implementation cycles and product life cycles [19].

These changes concern all aspects of business operations, taking place in the civilization, cultural and socio-economic area, leading to a turbulent environment, which shapes the decision-making processes. Therefore, the coexistence mechanism should fit into the indicated dimensions and make the relations between the mining enterprise and the environment dependent on these dimensions.

The question that arises in this context is what information should be the key content shaping the indicated mechanism? It seems that the factors characteristic for creative thinking about the relations between the mining company and its social and economic environment, especially in terms of regional development, will be an appropriate level of information aggregation consistent with the expectations of the mining company and its environment and ensuring, through information, growth of knowledge and the ability to respond effectively to the challenges facing the mining company and the region.

At the same time, for the smooth transition of the region, the information obtained should have a deeper meaning in the decision-making process, especially in the context of the accuracy of the phenomena and processes diagnosed. Acquired, collected, ordered and aggregated information should guarantee its practical application, allowing the identification and analysis of trends, forces, events and phenomena that may be of fundamental importance for building and functioning of the coexistence mechanism and relationships between the mining enterprise and its environment [20].

The general information-specific features and formulated theses and forecasts above—concerning the mining restructuring process—should create premises guaranteeing the desire to stimulate and maintain harmony and balance between the goals of the mining enterprise and its environment—in the social, economic, environmental and infrastructural dimension. Anticipating the broad spectrum of the impact of hard coal mining on its surroundings—taking into account the previously indicated areas: society, economy, environment and infrastructure—it seems that developing a model for gathering information on the relationship between a mining enterprise and its environment will require an alternative to the existing solutions' interdisciplinary approach. It seems that this function can be considered as crucial and horizontal—apart from energy security—and a challenge for the future of global and Polish mining in the knowledge-based economy.

3.2. *Polish Mining at the End of Industrial Revolution Era—Selected Aspects*

The term "Industrial Era" is derived from the worldwide process of "Industrial Revolution", which can be defined by its division into three periods called the First, Second and Third Industrial Revolutions. The first Industrial Revolution developed in the period from the late 18th to the second half of the 19th century. This period was characterized above all by the dynamic development of inventions, including that of a steam engine in 1769, which determined the industrialization process [21]. The beginning of the Second Industrial Revolution was the turn of the 19th and 20th century, when humankind became acquainted with further inventions (e.g., the method of oil refining, the internal combustion engine or the light bulb), which determined further development and modernization of copper and aluminum in metallurgy and the chemical industry—including crude oil processing—and limited the use of low-calorific coal in transport [22]. Two critical indicators of the Third Industrial Revolution that occurred in the last three decades of the 20th century are automation and computerization. This period was characterized by subsequent inventions leading to the high-technology industry and the production of atomic energy. In this phase of the Industrial Revolution, natural resources and their convenient location started losing importance as the key factors determining the allocation of industrial centers [23].

In Poland, the 19th and 20th centuries—that is, the Second and Third Industrial Revolution—were the period of dynamic development of the mining industry, which determined the flourishing of other industries, including railways and metallurgy [24]. The beginnings of the first mines were characterized by numerous unfavorable actions related to improper management, focused mainly on maximizing profits from production with irrational management of deposits. The technical condition of the mines in this period caused a very low level of occupational safety. This was the result of a lack of investments, which, with the vast supply of cheap labor, became unprofitable [25]. A radical change in Polish mining took place after the end of the Second World War. In the former political and economic system of the socialist state, the development of mining became even more decisive for the development of other strategic industrial areas. The discovery, investigation and documentation of hard coal deposits determined the formation of new mining areas and mines. Poland in the 1970s was in the fifth position in the world and the third in Europe in terms of hard coal mining. This is also the period in which the extraction of the raw material was maximized at the expense of work safety, even while attempts to modernize the mining industry were made. The time of change in the political and economic system of the country, i.e., the period after 1989, brought a new paradigm for the development of Polish mining. As a result, the more than 20-year period of permanent restructuring of the Polish hard coal mining industry led to the increase in rationalization and efficiency of managing coal deposits, so that these resources would serve the next generations of Poles. Activities related to the diversification of energy sources and the search for new opportunities to obtain energy, including from hard coal, have been undertaken at the national policy level. The vision of the development of the Polish hard coal mining industry was based on the assumption that, after 2015, mining will be a competitive sector and successfully operate in the realities of the market economy, characterized by a high degree of occupational safety, modernity and innovation of the production process and low degree of negative impact on the environment [26]. Although measures to meet environmental and safety standards are becoming increasingly important for the hard coal mining sector in Poland in the context of social and economic development of the whole country [18], the government's forecasts, assuming development rather than a decrease in output, turned out to be overly optimistic [27]. Although coal will remain the basic fuel in the power sector until 2030, it is planned to reduce its share in electricity generation to 56–60% coal. Therefore, it is necessary to ensure optimization of extraction and use of the raw material [13]. Currently (October 2020), a new plan for the mining industry is being developed, which will contain solutions acceptable to the social side and at the same time give a chance to continue extraction of coal by those mines that are profitable and necessary to provide the minimum of energy security.

At the same time, mining companies, striving to implement the policy of sustainable development—combining activities for economic success with care for the natural and social environment—more and more often exceed the minimum legal requirements, into those resulting directly from the implementation of corporate social responsibility (CSR) [28]. These, combined with the stability and flexibility of employment and work systems, can foster the scientific and technological development of mining regions, minimizing the social costs of energy transition process.

3.3. *Polish Mining in the Context of Changes Resulting from Revolution 4.0 and the Knowledge-Based Economy*

As has already been noted, along with the end of the Industrial Revolution, Polish hard coal mining entered the era of the knowledge-based economy in a state of profound organizational and technical changes. The described condition of the Polish mining industry is still evolving, determined by two processes—already mentioned several times—the Industrial Revolution 4.0 and a dynamically developing knowledge economy. The terms "knowledge economy" and "knowledge-based economy" (KBE) were introduced at the end of 20th century and a significant reorientation from a material-consumption economy to an economy based on information and knowledge has started [29–32]. According to the basic definition of this term, economic development is correlated with the appropriate use of knowledge [33], whereas its sub-dimensions cover aspects of economic and institutional regime, education and skills, information and communication infrastructure and the innovation system [34,35].

From this perspective, it is worth noting that the phenomena and processes related to automation cause transformations within companies and other institutions, related to areas of employment, human resource management and education [36]. As noted in numerous works [37–41], automation is a huge challenge for whole societies, because it forces changes in companies in terms of production processes, as well as operating principles, organizational models and information methods to build sensitive solutions based on innovation. This results in the need to adapt the entire labor market and education systems [36] in order to minimize the negative social impact of automation [42], improve skills [43] and professional flexibility [44] and build and support the competitiveness of emerging smart businesses of the future [45,46].

The described condition and development perspectives increasingly determine the disappearance of the impact on the development of economies of factors such as capital and labor resources in the areas of productivity, competitiveness and efficiency in favor of the increase in impact on the abovementioned areas of knowledge in the field of technical sciences, economics, organization and management. As indicated in the literature, relations shaped in this way enable the competitiveness of economies to increase, which additionally determines the following vital factors: innovative technologies and products as well as efficient management [47,48]. In the context of the aforementioned factors—especially focusing on the future of mining in the knowledge-based economy—it is worth anticipating a dynamically progressing process of increasing significance of the phenomenon of digitization in the economic processes, changing the way the society operates and accelerating development of the digital society [49,50]. Digital knowledge becomes the foundation of the power of the global economy and technological possibilities empower individuals and their ability to contribute to and participate in decision making processes [51].

According to the results of research on the artificial intelligence of scientists from the universities of Oxford and Yale, in 45 years machines will outperform people in all aspects of intelligence, and in 120 years all work will be automated [52]. As already mentioned, in the area of the labor market, digitization determines the disappearance and emergence of new professions. The 2016 study of the American Pew Research Center showed that 65% of Americans are convinced that in 50 years robots will do most of the work that people currently do, but at the same time 80% believe that their own profession in five decades will still exist [53]. In Poland, within the study developed by DELab University of Warsaw, 54% of the surveyed Poles acknowledged that in the future they will have to work in a few occupations in order to survive [54]. Based on the results of reports by the Organization for Economic Co-operation and Development (OECD), which analyzed 39 countries participating in the survey of adults skills (PIAAC) [55,56], it has been estimated that the average percentage of work susceptible to automation is about 48%, and 52% in Poland [57] and, importantly, even specialties which are the basis for the development of the modern business services sector may be at risk [58]. At the same time, in a PwC international analysis of the potential long-term impact of automation, the average for Poland is less than 35%, but 49% for people with a low education level [59]. In this light, it is clear that the future belongs to so-called smart enterprises (intelligent systems), whose strategic resources combine access to (and ability to use) data and information, broad knowledge and creativity [30,60]. However, enterprise intelligence is complicated to measure and even more challenging to manage [61]. One of the interrelated elements, the basis for effective knowledge management, is importing/extracting knowledge from the environment.

This condition determines the decline in the influence of such factors as capital and labor resources in the areas of productivity, competitiveness and efficiency on increasing the impact on the areas mentioned above of knowledge in the field of technical sciences, economics, organization and management. As the literature shows, relationships formed in this way enable the growth of competitiveness of economies, which is additionally determined by the following key factors: innovative technologies and products as well as efficient management. In conclusion, the dynamics of changes taking place in globalized economies are determined by the resources and accuracy of people's knowledge, the quality of work, education and training, the ability to think quickly and innovatively and to implement new solutions in production, distribution and services [62,63]. Knowledge also becomes a critical potential assigned to a better-educated unit that serves as an animator of the development of a knowledge-based society. In parallel with the process of developing a knowledge-based economy, the paradigm shaping the organizational and technical functionality of hard coal mining follows [64]. The primary determinant of the process is the change referring to the already mentioned concept of the Fourth Industrial Revolution commonly known as Industrial Revolution 4.0 or Industry 4.0 [65,66]. Although the concept is variously defined, it can briefly be defined as an advanced stage of business development triggered by digital transformation [67], in which value chains, products, services and business models change [68,69].

The Industry 4.0 focuses on combining IT models to strengthen their impact [70]. At the same time, it aims to involve people in the work of digitally controlled machines and the universal introduction of practical wireless networks and information and communication technologies. The described process is directly related to the implementation of innovative and new technical concepts and technological solutions in industry, with simultaneous and wider than before ICT instruments, and a new concept of business management. Due to the need to redefine the production management model, the company moves away from mass production, responding to the needs of a precisely defined group of recipients for personalized production and focusing on the needs of a precisely specified customer. As a result of these activities, the enterprise diversifies production and transitions to the method of production management related to so-called agile production, i.e., permanent improvement, immediate reaction, a cyclical increase in quality, social responsibility and focusing on the needs of the recipient [71]. The indicated method of management implies many new terms in the environment relating to technical and technological solutions, such as the Industrial Internet of Things (IIoT) [72,73], incremental technologies and augmented reality and simulation and information techniques using large databases and systems for their processing, which in reality redefine the previously binding processes and concepts, shaping the directions of progress and civilization's development based on the knowledge society. Along with these processes, the importance of cyber-physical systems, which correctly connect the computational area with physical processes, is growing dynamically. These are, in particular, forms of complex systems and models for monitoring and controlling physical processes operating in a feedback loop, where physical processes are the source of data for calculating the object control signal [74].

The processes mentioned above and trends based on them determine a new quality in the perception and understanding of the future of Polish mining. As there are four pillars in the knowledge economy: education and training, IT infrastructure, economic and institutional conditions and potential for innovation, therefore, the analysis of activities in the mining industry also took into account the above aspects. New IT solutions bring with them a whole new range of cost-saving possibilities [75]. Currently, more and more often, assumptions are being formulated that lead to plans to build smart mines in Poland, i.e., e-mines, in which system solutions connected with remote control of machines from the surface will be applied. Intelligent mines imply the possibility of effective control and monitoring of machines based on information and communication solutions. Remote and local control, aggregation, transmission, visualization, archiving and data analysis, as well as generating reports, are critical tasks in the area of application of information and communication technologies. The effect of these activities (in the e-mine) will primarily be an increase in the quantitative and qualitative safety of work. This will result from the fact that the staffing is far away from areas threatening safety and human life, in which only machines will be present. Therefore, the main goal of an intelligent mine is to optimize unit production costs through high technical and economic efficiency of technological processes, while striving to reduce the negative impact on the natural environment. The concept of an intelligent mine is dominated by the belief in the increase in human safety, resulting from the reduction in participation in the production process, which will be dominated by analytical skills. People will focus on controlling machines and monitoring them in subsequent phases. Thus, the highest priority in the e-mine vision was assigned to the knowledge and competences of future mining staff, whose work will focus mainly on remote control centers as well as underground and surface monitoring of machinery and equipment [76]. The process described above is becoming realistic in the real conditions of the organizational and technical functionality of the Polish hard coal mining industry. Strategic plans of KGHM Polska Miedź S.A. (hereinafter referred to as KGHM S.A.) foresee that new "intelligent" technologies and production management systems based on online communication between elements of the production process and advanced data analysis are the key factors determining success or failure in business [77]. The scope of tasks related to information management at KGHM includes the following issues: analysis and standardization of the needs reported by process owners, standardization of technical solutions (process sensing), optimization of resource use (hardware, programming), data quality management (verification, interference elimination) and supervision over the processes of generating and distributing information. Against this background, critical directions of activities at KGHM S.A. will be broadband data transmission in underground excavations, media monitoring power supply, ventilation, drainage, systems for locating and identifying machines and people underground, robotization of production and auxiliary processes and multi-dimensional analysis of data from production processes [78]. Therefore, the main objectives of the mine of the future's vision are: the improvement of operational and investment efficiency, cost effectiveness and safety, whereas the overriding goal in the mine of the future remains to ensure the highest possible level of safety. The adoption of pioneering innovations, accompanied by extensive use of data analytics with the use of the latest solutions in the area of business intelligence and big data, should enable a complete redesign of the traditional mining process [79].

Similarly, Jastrzebska Spółka Weglowa S.A. (hereinafter referred to as JSW S.A.) has as its future undertaking the goal "Towards JSW 4.0". According to the assumptions adopted by JSW Innowacje S.A., the essential tool determining the achievement of the leading business goal, which is optimization of efficiency, is the technological enhancement of the entire production process from coal to coke, including coal deposit exploration, analytics of the extraction process, online knowledge about extracted coal, optimal increase in underground work safety, inclusion of all production stages in one model subject to a continuous supply of real-time source data and the construction of analytical and data-mining models [80].

In this context, it is essential to ensure constant development of the competences and qualifications of mining sector employees by a high level and availability of continuous vocational training, also aimed at improving knowledge and skills in using the latest technologies available, in order to fully use arising opportunities [76]. One of the important elements is all kinds of cooperation between universities and research units and mining companies, to the benefit of all parties. In Poland, these are often agreements aimed at long-term cooperation in the field of education, training, joint work on innovations [81] or joint implementation of research projects. In the Silesian voivodeship these included, among others, projects implemented as part of Horizon 2020 aimed at technological development and the introduction of IT innovations (such as IlluMINEation, SIMS, RockVader) and projects implemented under the Research Fund for Coal and Steel, aimed at improving environmental results (such as MERIDA, RECOVERY, TEXMIN, STRATEGY CCUS, STAMS, MapROC), occupational safety and health using modern technologies (such as ROCD and INDIRES) and improving economic efficiency (such as INESI, COMEX, Coal2GAS).

The aforementioned approaches and solutions clearly show that, in the mining sector, globalization and the knowledge economy also caused material capital, natural resources or labor force to lose their primacy as the leading economic resources and were replaced by knowledge accumulated in person. Human capital is currently a key economic resource that is recognized as a development determinant for all other factors, most fully shaping global development [82]. Its size will determine the current and future situation of national economies and inequalities between them [83]. Therefore, the view that increasing human capital resources through a proper education system should become the imperative of modern times can be considered as fully legitimate. In this approach, education, shaping ethical values and prevailing cultural norms reduce the costs of business activity, including by minimizing wasting resources and increasing the importance of responsibility for the work performed. It implies a more dynamic and stable economic development. Secondly, well-developed social capital, social norms and trust imply more effective anticipation of human behavior and minimize the behavior of interest groups [84], limiting the unpredictability of socio-economic processes and transaction costs. It allows the society to manage the occurring phenomena more correctly and confidently, and the exchange of resources, skills and information becomes more efficient [85].

In this context—from the perspective of investing in the development of human resources—it can be said that the knowledge economy and the Industrial Revolution 4.0 are key determinants shaping the future of Polish mining, in which knowledge accumulated by humans will dominate.

3.4. Space Mining—The Way for the Future

The development of technology is constantly accompanied by an increase in the demand for mineral resources. Both the growing demand for consumer goods and the increased demand for energy put pressure on the mining industry. It is necessary to supply ever greater amounts, but also more and more diversified mineral resources. The development of mining techniques, the progressive automation and digitization of mining, continuous improvement and also the continuous increase in the requirements for staff competences, referred to in the previous paragraphs, allow for a significant increase in the efficiency of the mining process and thus meeting the growing demand. In the long run, even despite further technological development, it does not seem possible to maintain this state of affairs. Natural resources that can be obtained relatively easily are limited. The very process of obtaining them is always associated with some form of interference with the environment. Therefore, it is necessary to try to solve this problem early. It seems that the right direction of exploration is to consider the possibility of acquiring mineral resources beyond the terrestrial globe, i.e., the development of space mining [86–88]. Although the goal set in this way seems abstract and far from practical, it should be noted that for many years space has been perceived as an area of potential mining exploitation. The process to start the economically viable exploitation of celestial resources has therefore basically begun. Space mining can also be seen not only as a way to avoid the inevitably impending crisis related to the possibilities of extracting minerals on Earth, but also as a natural consequence and the culmination of the technology development and staff competences presented in the previous paragraphs. In space mining, one finds the culmination of technology developed for the needs of traditional (on Earth) forms of deposit exploitation. Here, first of all, digital technologies are directly translated. The exploitation of resources identified on both larger (the Moon, Mars) and smaller (asteroids) celestial bodies will require advanced automation of processes. In unfavorable conditions, with delays in signal transmission resulting from the fundamental laws of physics, the use of autonomous devices capable of carrying out independent work and independently solving at least typical problems that arise during it acquires special importance. This area is a special opportunity for the Polish economy. The creation of conditions and incentives conducive to the development of methods and tools of artificial intelligence seems to be one of the key actions that should be taken and, in fact, already is being taken. Although most of the research and research and development works undertaken in these areas do not refer directly to space mining, they nevertheless develop solutions that can be relatively easily adapted to this area of application.

Another aspect that will require undertaking development works is the technologies of mining the material and enriching the obtained output. These technologies must be adapted to the conditions of the exploited celestial body. It seems, however, that the presence of the atmosphere and significant gravity, as in the case of Mars, will allow the use of solutions structurally similar to those currently used on Earth. However, the challenge will be the lack of an atmosphere and the virtually no gravity found on asteroids. It seems, however, that the competences obtained during the development of devices intended for use in terrestrial conditions can be used to develop innovative solutions dedicated to space mining. It will be necessary to solve the problem of the economically efficient transport of excavated material to the vicinity of the Earth, development of a technology for enrichment of minerals in space conditions and ensuring proper service of the machinery and equipment used, so that it is possible to maintain the continuity of production.

Meeting many of the technological and organizational challenges listed here is beyond the capabilities of the vast majority of countries or enterprises. It is therefore necessary to promote far-reaching cooperation in this area. The Luxembourg Space Agency, the European Space Agency and the National Aeronautics and Space Administration are particularly active in this field. Poland is also involved in the initiatives of these organizations, especially the LSA. Regardless, the current transition from fossil fuels creates specific conditions in countries such as Poland that can and should be used as a lever to accelerate space-oriented development. The transformation that we are already witnessing should be directed with a vision of long-term actions, and not only solving problems that arise on a regular basis. Only this approach will allow for the redirection and appropriate training of staff and the creation of the potential to gain a leadership position in the newly emerging space mining market. This is an opportunity that, in particular, Poland must not miss. Although the transformation of Polish mining towards space mining, as has already been mentioned, will not be possible without extensive international cooperation it is worth defining the areas in which Poland can be at the forefront of countries providing key solutions for space mining. Achieving such a goal will allow the use of the wealth of experience and many years of mining tradition, on the one hand, and obtaining a stable and inalienable role of a key state in the newly emerging industry.

4. Conclusions

Based on the theses presented in the article, conducted literature research and described megatrends related to the knowledge-based economy and the Industrial Revolution 4.0, it can be assumed that traditional hard coal mining determined by industrialization will diminish. The process of digitization and automation of the mining industry is inevitable and necessary to maintain any efficiency in the mining sector. At the same time, it leads to the departure from a world based on work and traditional industry, as mentioned earlier, and to the transition to a time when work, progress and prosperity are built on the basis of innovation and technological development. The changes taking place focus on selected social and economic aspects and on enhancing of entrepreneurial skills and thus enabling mining regions to become multi-industry regions [89]. For example, in the Silesian voivodeship, the development strategy and an established list of priorities for the region's pro-technological development shape the convergence process between technologies for the energy and mining industry with biological environmental protection technologies, nanotechnologies, production and processing of materials and information technologies as well as cognitive sciences [90]. These also should enable the full use of the diversified potential of different companies and institutions to ensure energy security of the country and the development of innovation in the mining and energy sector, in line with the knowledge economy principles. All these aspects have to be taken into consideration when preparing regional transition plans to minimize negative social and economic costs

According to the above thesis, a considerable role should be attributed to education, not only as a tool for transferring knowledge and skills, but also a key determinant for shaping social attitudes and behavior. In this light, processes related to the knowledge-based economy, the Industrial Revolution 4.0 and the automation of the economy may determine, to an increasing extent, the unexpected and previously unknown phenomena and processes in the social, environmental and economic sphere. The vision of the future of hard coal mining fits in with the above theses. It seems that a breakthrough factor shaping the future of this industry—coherent with the knowledge-based economy and Industry 4.0 logic—is the implementation of principles of intelligent mining. As an example, the assumptions and work in progress on "e-mines" are described in this article. At the same time, intelligent mining is a broader organizational transformation, not just a "digital mine". This process also involves incorporating new solutions into all aspects of a mining company's operations, affecting the way decisions are made, the way employees are involved and the range of necessary skills of employees [91]. Achieving the goal of an intelligent mine can be comprehensively defined by the available model for the development of the mining offer in the direction of moving away from the traditional "miner shift" to permanent development based on the most modern and integrated solutions supporting the automation of extraction of raw material. In this process, knowledge and appropriately profiled qualified staff must form the most valuable resource shaping the future of hard coal mining [80]. A decisive role in creating the development of intelligent mining may be played by overcoming mental barriers to the implementation of the principles of a knowledge-based economy and the Industrial Revolution 4.0. As many sources mentioned in this article prove, the pursuit of development determined by the abovementioned processes leads directly to economic growth in the dimension of companies, regions, state and local communities, favoring social inclusion and strengthening people's sense of full life [92]. Therefore, it is worth integrating existing and innovative solutions already implemented in the mining sector to decrease its carbon emissions as part of a long-term transformation plan.

Author Contributions: Conceptualization, R.M. and S.I.; Data curation, R.M. and S.I.; Formal analysis, R.M. and S.I.; Funding acquisition, S.I.; Methodology, R.M.; Supervision, R.M. and S.I.; Validation, R.M. and S.I.; Writing—original draft, R.M. and S.I.; Writing—review and editing, R.M. and S.I. All authors have read and agreed to the published version of the manuscript.

Funding: This research was funded by the Ministry of Science and Higher Education (Poland), grant number 11126010.

Institutional Review Board Statement: Not applicable.

Informed Consent Statement: Not applicable.

Data Availability Statement: Not applicable.

Conflicts of Interest: The authors declare no conflict of interest.

References

1. Drebenstedt, C. Responsible mining—Approaches and Realization. In Proceedings of the 22nd World Mining Congress—Innovations and Challenges in Mining, Istanbul, Turkey, 2008; Available online: https://scholar.google.com.hk/scholar?hl=en&as_sdt=0%2C5&q=Responsible+mining%E2%80%94Approaches+and+Realization&btnG= (accessed on 1 December 2021).
2. Wee, D.; Cattel, J. *Industry 4.0. How to Navigate Digitization of the Manufacturing Sector*; McKinsey Digital: Washington, DC, USA, 2015.
3. Drucker, P.F. The rise of the knowledge society. *Wilson Q.* **1993**, *17*, 52.
4. Giddens, A. *Konsekwencje Nowoczesności*; Wydawnictwo Uniwersytetu Jagiellońskiego: Kraków, Poland, 2008.
5. Macuk, R. Transformacja Energetyczna w Polsce-Edycja 2020 (Energy Transition in Poland-Edition 2020). Available online: https://www.forum-energii.eu/pl/analizy/transformacja-2020 (accessed on 12 January 2021).
6. Ranosz, R. Górnictwo i jego znaczenie w gospodarce światowej. *Gospod. Surowcami Miner.* **2014**, *30*, 5–20. [CrossRef]
7. Gawlik, L. The Polish power industry in energy transformation proces. *Miner. Econ.* **2018**, *31*, 229–237. [CrossRef]
8. Ministry of Economy. *Energy Policy of Poland Until 2030*; Ministry of Economy: Warsaw, Poland, 2009.
9. Mining in Poland is Not only Coal-Interview with Prof. P. Czaja. 2014. Available online: https://www.agh.edu.pl/nauka/info/article/gornictwo-w-polsce-to-nie-tylko-wegiel-wywiad-z-prof-p-czaja/ (accessed on 21 August 2020).
10. Tajduś, A.; Czaja, P.; Kasztelewicz, Z. Rola węgla w energetyce i strategia polskiego górnictwa węgla brunatnego w I połowie XXI wieku. *Górnictwo i Geologia* **2011**, *3*, 343–363.
11. IEA. *World Energy Outlook 2019*; IEA: Paris, France, 2019. Available online: https://www.iea.org/reports/world-energy-outlook-2019 (accessed on 4 September 2020).
12. Dubiński, J. Sustainable development of mining mineral resources. *J. Sustain. Min.* **2013**, *12*, 1–6. [CrossRef]
13. Ministry of State Assets. *Poland's National Energy and Climate Plan for the Years 2021–2030 (NECP PL)*; Ministry of State Assets: Warsaw, Poland, 2019.
14. Lisowski, A. Recovery plan of hard coal mining industry and better use of coal in the Polish economy in the future. *Stow. Inżynierów i Tech. Górnictwa* **2015**, *7*, 1–7.
15. European Commission's Communication. The European Green Deal, COM/2019/640 Final. Available online: https://www.eea.europa.eu/policy-documents/com-2019-640-final (accessed on 10 October 2021).
16. European Commission's Communication. Energy Roadmap 2050. COM (2011) 885 Final. Available online: https://eur-lex.europa.eu/LexUriServ/LexUriServ.do?uri=COM:2011:0885:FIN:EN:PDF (accessed on 6 September 2021).
17. Wilczyński, M. *Zmierzch Węgla Kamiennego w Polsce*; Instytut na Rzecz Ekorozwoju: Warsaw, Poland, 2013.
18. Hałasik, K.; Kulczycka, J. CSR, Environment-Friendly Investments and Innovations—the Three Elements Necessary to Build a Modern and Strong Coal Mining Company? January 2016 E3S Web of Conferences 10. Available online: https://www.researchgate.net/publication/309226061_CSR_environment-friendly_investments_and_innovations (accessed on 1 December 2021).
19. Ansoff, H.I. *Strategic Management*; PWE: Warsaw, Poland, 1985.
20. Majerska, K. Acquiring economic information. In *Institute of Scientific Information and Bibliological Studies of the University of Warsaw*; Uniwesytet WarszawASKI, Szkoła Głowna Gospodarstwa Wiejskiegiego: Warsaw, Poland, 2015.

21. Myszczyszyn, J. Rola koksu i żelaza w industrializacji świata. In *Kultura i Historia*; Uniwersytet Marii Curie-Skłodowskiej Zakład Teorii Kultury i Metodologii Nauk o Kulturze: Lublin, Pland, 2009.
22. Briggs, A.; Clavin, P. *Europa Dwóch Stuleci: 1789–1989*; Ossolineum: Wrocław, Poland; Warszawa, Poland; Kraków, Poland, 2000.
23. Trinnaman, J.; Clarke, A. *Living in One World*; World Energy Council: London, UK, 2001.
24. Zejno, M. Istota przemian industrialnych w XIX wieku. *Studia Gdańskie Wizje i Rzeczywistość* **2015**, *XII*, 169–178.
25. Kotarba, W. Historia górnictwa jako historia wypadków na przykładzie kopalni Flora w Dąbrowie Górniczej. *Hered. Minariorum* **2017**, *4*, 261–272.
26. Marszowski, R. *Restructuring Actions to Coal Mining and Its Impact on Socio-Economic Environment. State and Perspectives*; International Social and Humanities Studies Humanum: Warszawa, Poland, 2017.
27. Derski, B. Wydobycie Węgla w Polsce Bliżej Końca Niż Mówią Politycy. 2020. Available online: https://wysokienapiecie.pl/27254-wydobycie-wegla-w-polsce-blizej-konca-niz-mowia-politycy/ (accessed on 21 October 2020).
28. Hop, N.; Kudełko, J. Społeczna odpowiedzialność biznesu jako element strategii rozwoju przedsiębiorstwa górniczego w. In *Dzieje Górnictwa—Element Europejskiego Dziedzictwa Kultury*; Zagwożdżon, P.P., Madziarz, M., Eds.; Oficyna Wydawnicza Politechniki Wrocławskiej: Wrocław, Poland, 2013; Volume 5, pp. 85–110.
29. Drucker, P.F. *The Age of Discontinuity; Guidelines to Our Changing Society*; Harper and Row: New York, NY, USA, 1969.
30. Drucker, P.F. *Post-Capitalist Society*; PWN: Warsaw, Poland, 1999. (In Polish)
31. Smith, K. *What is the 'Knowledge Economy'? Knowledge Intensity and Distributed Knowledge Bases*; UNU-INTECH Discussion Paper Series; United Nations Univeristy, Institute for New Technologies: Tokyo, Japan, 2002.
32. David, P.A.; Foray, D. Economic Fundamentals of the Knowledge Society. *Policy Futures Educ.* **2003**, *1*, 20–49. [CrossRef]
33. Mańkiewicz, P.J. Gospodarka oparta na wiedzy jako determinanta konwergencji gospodarczej Polski i wysoko rozwiniętych krajów Unii Europejskiej. In *Annales Universitatis Mariae Curie-Skłodowska Lublin–Polonia*; UMCS: Lublin, Poland, 2016; Volume 3, pp. 130–140.
34. Saisana, M.; Munda, G. *Knowledge Economy: Measures and Drivers*; European Commission Joint Research Centre Scientific and Technical Reports; Office for Official Publications of the European Communities: Luxembourg, 2008.
35. Bratianu, C.; Dinca, V.M. Knowledge Economy Dimensions. *Rev. De Manag. Comp. Int. Rev. Int. Comp. Manag.* **2010**, *11*, 210–221.
36. Marszowski, R.; Drobek, L. Automation of industrial processes and its impact on the labor market and education-selected aspects. *Sci. Pap. Sil. Univ. Technol. Organ. Manag. Ser.* **2019**, *128*, 171–184. [CrossRef]
37. Sevinc, O. Shades of Automation in the Labor Market. *Procedia Comput. Sci.* **2020**, *158*, 485–489. [CrossRef]
38. Boyd, J.; Huettinger, M. Smithian insights on automation and the future of work. *Futures* **2019**, *111*, 104–115. [CrossRef]
39. Gardberg, M.; Heyman, F.; Norbäck, P.-J.; Persson, L. Digitization-based automation and occupational dynamics. *Econ. Lett.* **2020**, *189*, 109032. [CrossRef]
40. Ghimire, R.; Skinner, J.; Carnathan, M. Who perceived automation as a threat to their jobs in metro Atlanta: Results from the 2019 Metro Atlanta Speaks survey. *Technol. Soc.* **2020**, *63*, 101368. [CrossRef] [PubMed]
41. Ivanov, S.; Kuyumdzhiev, M.; Webster, C. Automation fears: Drivers and solutions. *Technol. Soc.* **2020**, *63*, 101431. [CrossRef]
42. Mayer, F. Social impact of automation trends and issues: An human centred systems engineering perspective. *IFAC Proc. Vol.* **2008**, *41*, 5251–5256. [CrossRef]
43. Wolcott, E.L. Employment inequality: Why do the low-skilled work less now? *J. Monet. Econ.* **2020**, in press. [CrossRef]
44. Doyle, A. What Is Workplace Flexibility? Definition and Examples of Workplace Flexibility. Available online: https://www.thebalancecreers.com/workplace-flexibility-definition-with-examples-2059699 (accessed on 6 November 2020).
45. Borkowska, S.; Bohdziewicz, P. *Menedżer u Progu XXI Wieku*; Wyższa Szkoła Humanistyczno–Ekonomiczna w Łodzi: Łódź, Poland, 1998.
46. Kubik, K. Współczesne przedsiębiorstwa wobec wyzwań globalnej konkurencji. In *Zeszyty Naukowe Uniwersytetu Przyrodniczo-Humanistycznego w Siedlcach Vol. 92 Seria: Administracja i Zarządzanie*; Uniwersytet Przyrodniczo-Humanistyczny: Siedlce, Poland, 2012.
47. Madrak-Grochowska, M. Konkurencyjność gospodarek opartych na wiedzy. Propozycja pomiaru. In *Ekonomia i Prawo*; Polszakiewicz, B., Boehlke, J., Eds.; Uniwersytet Mikołaja Kopernika w Toruniu: Toruń, Poland, 2013. [CrossRef]
48. Skrzypek, A. Managing knowledge in theory and practice. Insight from Polish Quality Award laureates. *Małopolska Sch. Econ. Tarnów Res. Pap. Collect.* **2015**, *1*, 153–167.
49. EU publications. *Understanding the European Union's Policy—Digital Agenda for Europe*; European Commission Directorate-General for Social Communication Citizens' Information Department: Brussels, Belgium, 2014.
50. *Plan Działań Ministra Cyfryzacji—Załącznik do Programu Zintegrowanej Informatyzacji Państwa*; Ministerstwo Cyfryzacji RP: Warsaw, Poland, 2016.
51. Lemaire, A. Forging a Digital Society, OECD Yearbook. 2016. Available online: https://www.oecd.org/innovation/forging-a-digital-society.htm (accessed on 22 September 2020).
52. Grace, K.; Salvatier, J.; Dafoe, A.; Zhang, B.; Evans, O. When Will AI Exceed Human Performance? Evidence from AI Experts. *J. Artif. Intell. Res.* **2017**, *62*, 729–754. [CrossRef]
53. Smith, A. Public Predictions for the Future of Workforce Automation; Pew Research Center, Internet & Technology: Washington, DC, USA, 2016; Available online: https://www.pewresearch.org/internet/2016/03/10/public-predictions-for-the-future-of-workforce-automation/ (accessed on 27 November 2020).

54. *Aktywni+ Przyszłość Rynku Pracy 2017*; Raport na Podstawie Badania Przeprowadzonego Przez DELab UW na Zlecenie Gumtree Polska w Ramach Programu Gumtree.pl Start do Kariery; Uniwersytet Warszawski: Warszawa, Poland, 2017.
55. OECD. *Skills Outlook 2013: First Results from the Survey of Adult Skills*; OECD Publishing: Paris, France, 2013. [CrossRef]
56. OECD. *Skills Matter: Additional Results from the Survey of Adult Skills*; OECD Skills Studies; OECD Publishing: Paris, France, 2019. [CrossRef]
57. Nedelkoska, L.; Quintini, G. *Automation, Skills Use and Training, OECD Social, Employment and Migration Working Papers*; OECD Publishing: Paris, France, 2018. [CrossRef]
58. Burski, J.; Chłoń-Domińczak, A.; Palczyńska, M.; Rynko, M.; Śpiewanowski, P. *The Skills of Poles—Results of the International Adult Skills Survey (PIAAC)—Report 2013*; Instytut Badań Edukacyjnych: Warszawa, Poland, 2013.
59. Hawksworth, J.; Berriman, R.; Goel, S. *Will Robots Really Steal Our Jobs?: An International Analysis of the Potential Long Term Impact of Automation*; Pricewaterhouse Coopers: London, UK, 2018. Available online: https://www.pwc.co.uk/services/economics-policy/insights/the-impact-of-automation-on-jobs.html (accessed on 9 November 2020).
60. D'Monte, L. Building a Smart Enterprise with Digital Technology. 2017. Available online: https://www.livemint.com/Technology/A7dp3ij28USw0udunajTJN/Building-a-smart-enterprise-with-digital-technology.html (accessed on 4 December 2020).
61. Sopińska, A.; Wachowiak, P. Modele Zarządzania Wiedzą w Przedsiębiorstwie. 2006. Available online: https://www.e-mentor.edu.pl/artykul/index/numer/14/id/275 (accessed on 3 December 2020).
62. Kołodko, G. *Świat na Wyciągnięcie Myśli*; Prószyński i S-ka: Warszawa, Poland, 2010.
63. Marszowski, R.; Drobek, L.; Hetmańczyk, P.; Markowska, M. Education in the times of Demographic Change and Globalization. Case Study on the example of the Silesian Voivodeship. *Sustainability* **2020**, *12*, 5688. [CrossRef]
64. Marszowski, R. Młode kadry górnicze—Bariery i determinanty rozwoju i awansu zawodowego w fazie Rewolucji Przemysłowej 4.0. *Przegląd Górniczy* **2019**, *3*, 15–22.
65. Rojko, A. Industry 4.0 Concept: Background and Overview. *Int. J. Interact. Mob. Technol.* **2017**, *11*, 77–90. [CrossRef]
66. Roblek, V.; Meško, M.; Krapež, A. *A Complex View of Industry 4.0*; SAGE Open: Thousand Oaks, CA, USA, 2016; Volume 6, p. 2158244016653987.
67. Varela, L.; Araújo, A.; Ávila, P.; Castro, H.; Putnik, G. Evaluation of the Relation between Lean Manufacturing, Industry 4.0, and Sustainability. *Sustainability* **2019**, *11*, 1439. [CrossRef]
68. Dziurdzia, M.; Matuszewski, M. *Przemysł 4.0 Czyli Wyzwania Współczesnej Produkcji*; PWC: Warszawa, Poland, 2017.
69. Dobrowolska, M.; Knop, L. Fit to Work in the Business Models of the Industry 4.0 Age. *Sustainability* **2020**, *12*, 4854. [CrossRef]
70. Karabiegović, I. The role of industrial and service robots in the 4th industrial revolution—Industry 4.0. *Acta Tech. Corviniensis Bull. Eng.* **2018**, *11*, 11–16.
71. Fertsch, M.; Cyplik, P.; Hadaś, Ł. *Logistyka Produkcji. Teoria i Praktyka*; Instytut Logistyki i Magazynowania: Poznań, Poland, 2010.
72. Jeschke, S.; Brecher, C.; Song, H.; Rawat, D.B. (Eds.) *Industrial Internet of Things, Springer Series in Wireless Technology*; Springer International Publishing: Cham, Switzerland, 2017.
73. Patel, K.K.; Patel, S. Internet of Things-IOT: Definition, Characteristics, Architecture, Enabling Technologies, Application & Future Challenges. *Int. J. Eng. Sci. Comput.* **2016**, *6*.
74. Lee, E. Physical Systems: Design Challenges. In *Electrical Engineering and Computer Sciences*; University of California: Berkeley, CA, USA, 2008.
75. Humphrey, D. Mining in the Knowledge-based Economy. *Miner. Energy Raw Mater. Rep.* **2001**, *16*, 3–8. [CrossRef]
76. Stecuła, K.; Brodny, J.; Puzik, K. The state and perspectives of applying the achievements of the fourth industrial reVolution in the mining industry. *Conf. Innov. Manag. Prod. Eng. IZIP Artic.* **2018**, *1*, 334–345.
77. Biernacki, R. *KGHM. Vision of the Mine of the Future in Strategic Activities of Kghm Polska Miedź SA*; IMF: Lubin, Poland, 2017.
78. The Role of Innovation in KGHM, Wrocław. 2019. Available online: https://www.interregeurope.eu/fileadmin/user_upload/tx_tevprojects/library/file_1558960476.pdf (accessed on 30 September 2020).
79. Project illuMINEation—Bright Concepts for a Safe and Sustainable Digital Mining Future. 2020. Available online: https://cordis.europa.eu/project/id/869379 (accessed on 27 November 2020).
80. Ozon, D. *Strategia Rozwoju JSW S.A. Przez Pryzmat Innowacyjnych Działań w Procesie Węgiel–Koks*; JSW S.A.: Jastrzębie-Zdrój, Poland, 2017.
81. Heder, A.; Tkocz, M. The Functioning of Coal Mining in a Knowledge-Based Economy, Prace Komisji Geografii Przemysłu Polskiego Towarzystwa Geograficznego. *Stud. Ind. Geogr. Comm. Pol. Geogr. Soc.* **2013**, *21*, 78–93.
82. Grodzicki, J. *The Role of Human Capital in the Development of the Global Economy*; Gdańsk University Publishing House: Gdańsk, Poland, 2003.
83. Bergheim, S. Human Capital is the Key to Growth—Success Stories and Policies for 2020 (1 August 2005). Deutsche Bank Research Current Issues Working Paper. Available online: https://ssrn.com/abstract=774825 (accessed on 12 November 2020).
84. Krueger, A. The Political Economy of the Rent-Seeking Society. *Am. Econ. Rev.* **1974**, *64*, 291–303.
85. Pogonowska, B. Social Capital—An attempt to reconstruct the conceptual category. In *Theoretical Aspects of Social Capital*; Januszek, H., Ed.; Zeszyty Naukowe AE in Poznan; Wydawnictwo AE: Poznan, Poland, 2004; No. 42.
86. Hellgren, V. Asteroid Mining: A Review of Methods and Aspects. In *Student Thesis Series INES nr 384*; Lund University: Lund, Sweden, 2016.

87. Dallas, J.A.; Raval, S.; Gaitan, J.P.A.; Saydam, S.; Dempster, A.G. Mining beyond Earth for Sustainable Development: Wi Humanity Benefit from Resource Extraction in Outer Space? *Acta Astronaut.* **2020**, *167*, 181–188. [CrossRef]
88. Xu, F. The Approach to Sustainable Space Mining: Issues, Challenges, and Solutions. *IOP Conf. Ser. Mater. Sci. Eng.* **2020**, *73* 012014. [CrossRef]
89. Urząd Marszałkowski Województwa Śląskiego. *Regional Innovation Strategy of the Śląskie Voivodeship for the Years 2013–2020*; Urzą Marszałkowski Województwa Śląskiego: Katowice, Poland, 2012.
90. Urząd Marszałkowski Województwa Śląskiego. *Technology Development Program of the Silesian Voivodeship for 2019–2030*; Urzą Marszałkowski Województwa Śląskiego: Katowice, Poland, 2018.
91. Jacobs, R.; Nel, J.; Plessis, J.; Vawda, Y.; Snyman, F.; Arnd, R. Intelligent Mining Delivering Real Value, Deloitte. 2018. Availabl online: https://www2.deloitte.com/global/en/pages/energy-and-resources/articles/intelligent-mining-deloitte.html (accesse on 5 October 2020).
92. Wronka-Pospiech, M. Innowacje społecznie—Pojęcia i znaczenie w. *Studia Ekon. Zesz. Nauk. Uniw. Ekon. Katowicach* **2015**, *21* 124–136.

sustainability

Article

The Evolution of Sustainability Ideas in China from 1946 to 2015, Quantified by Culturomics

Danqing Zhang [1,2], Guowen Huang [1,3,*], Jiaen Zhang [4,*], Xiaoyu Hou [5], Tianyi Zhou [5], Xianyuan Chang [5], Ying Ge [5] and Jie Chang [5]

1 School of Foreign Studies, South China Agricultural University, Guangzhou 510642, China; zhangdq@zspt.edu.cn
2 Quality Education Center, Zhongshan Polytechnic, Zhongshan 528400, China
3 Center for Ecolinguistics, South China Agricultural University, Guangzhou 510642, China
4 Guangdong Laboratory for Lingnan Modern Agriculture, College of Natural Resources and Environment, South China Agricultural University, Guangzhou 510642, China
5 College of Life Sciences, Zhejiang University, Hangzhou 310058, China; xyhou@zju.edu.cn (X.H.); tyzhou@zju.edu.cn (T.Z.); xychang@zju.edu.cn (X.C.); geying@zju.edu.cn (Y.G.); jchang@zju.edu.cn (J.C.)
* Correspondence: flshgw@scau.edu.cn (G.H.); jeanzh@scau.edu.cn (J.Z.)

Abstract: Economy and ecology are two main aspects of human sustainable development. However, a comprehensive analysis of the status and trends of economic and ecological cognition is still lacking. Here, we defined economic and ecological concepts as cultural traits that constitute a complex system representing sustainability ideas. Adopting a linguistic ecology perspective, we analysed the frequency distribution, turnover and innovation rates of 3713 concepts appearing in China's mainstream newspaper, *People's Daily*, from 1946 to 2015. Results reveal that: (1) In the whole historical period, there were more economic concepts than ecological concepts both in amount and category. Economic concepts experienced stronger cultural drift than ecological concepts tested by the neutral model of cultural evolution; (2) popular economic concepts became more diversified, but popular ecological concepts became more uniform; (3) both economic concepts and ecological concepts attained more variation in their own disciplinary domains than in cross-disciplinary domains; and (4) as a platform of both giving information and opinion, a newspaper is subjected to cultural selection, especially reflected in the change in ecological concepts under the context of Chinese ecological civilization construction. We concluded with a discussion of promoting vibrant and resilient ecological knowledge in fostering sustainability activities and behaviours.

Keywords: cultural evolution; linguistic ecology; neutral model; selection and drift; economic concepts; ecological concepts

Citation: Zhang, D.; Huang, G.; Zhang, J.; Hou, X.; Zhou, T.; Chang, X.; Ge, Y.; Chang, J. The Evolution of Sustainability Ideas in China from 1946 to 2015, Quantified by Culturomics. *Sustainability* **2022**, *14*, 6038. https://doi.org/10.3390/su14106038

Academic Editors: Nicola Raimo, Filippo Vitolla, Benedetta Esposito and Ornella Malandrino

Received: 11 February 2022
Accepted: 13 May 2022
Published: 16 May 2022

1. Introduction

A core aim of sustainability science is to link sustainability knowledge with actions and behaviours that lead to real sustainability [1–3]. While over 30 years have passed since the Brundtland report [4], unsustainable behaviours and collective actions have resulted in ever-worsening environmental challenges, and practical actions are urgently needed. Recent studies on sustainability knowledge and concepts have found that, despite incremental progress in studies of sustainability science, the role of sustainability knowledge and concepts in affecting societal change remains uncertain [5]. Economy and ecology are two main dimensions in sustainable development. Despite the increasing understanding and improvement of the concept of sustainable development, a comprehensive analysis of the status and trends of economic and ecological cognition is still lacking [6]. In the discipline of human ecology, cultural evolutionary studies [7–12] have used concepts, tools and methods from evolutionary biology to interpret massive quantitative data by exploring the interaction between genes and cultures to explain human behaviours and have succeeded in offering insights into behaviour–culture–societal change [13,14]. Linguistic ecology is deeply rooted in human ecology and focuses on 'the interactions between any given language and its environment' [15]. Based on Darwin's Selection analogies between language and species evolution [16], examining word creation and extinction in a language system [17,18] and analysing language transmission bias [19–21] could unravel the underlying forces and changing patterns of cultural evolution.

In the last two decades, the random-copying model of the neutral theory of population genetics [22,23] has been used as a null hypothesis and model in testing value-neutral cultural traits [24–28], cultural drift and selection forces. This model has already been applied to various cultural domains in tracking social change, such as baby names, dog breeds, music, commonly used words and public media in popular culture [25,29–33], and academic terminology, technological patents, cryptocurrency and colour terms in academic culture [32,34–37]. However, this approach has not been applied to the analysis of sustainable cultural dynamics.

Concepts, cultural behaviours, or artefacts in cultural evolution are regarded as cultural replicators [38]. Cultural selection is most likely to operate in the semantic dimension of language organization [39]. Concepts are 'simple' noun phrases composed of a determiner and head noun that represent human recognition and construction of the world [40]. Different concepts may occupy a specific niche, which may influence the frequency change and dynamics of concepts related to social identity [41]. Therefore, economic concepts reflect the fundamental activities of human survival and the essence of the Anthropocene. Ecological concepts emphasize well-being and sustainable development. The conflict between economic growth and ecological preservation has long existed in both developed and developing countries. Through an evolutionary analysis of ecological and economic concepts, we attempt to answer the following research questions (RQ): RQ 1, how have ecological and economic concepts changed throughout China's different development phases? RQ 2, what changes have occurred in mainstream sustainability ideas? RQ 3, what is the underlying evolutionary force of sustainability ideas? and RQ 4, what approaches to policy and management might better transmit sustainability ideas to advance the role of sustainability?

In this paper, we tracked the dynamics of sustainable concepts for the past 70 years, covering the duration from the foundation of PR China to its establishment as the world's second-largest economy. We used culturomics analysis to detect the drift and selection in ecological and economic concept usage in China. The concepts were drawn from the *People's Daily*, the largest newspaper group (with a circulation of over 3 million) and mainstream media in China. The aim of this study is to determine the interactions among governmental policy and strategy, public opinion, and personal decisions at the population level, in dealing with the trade-off among ecological restoration, environmental protection, and economic growth.

2. Materials and Methods

2.1. Data Collection

Public media, as a transmitter [7], mirrors the shift in public attention and preferences, and accelerates the transmission of concepts. Assuming that no concept is intrinsically more valuable than another and envisioning the copying of concepts as 'replication' and the invention of new concepts as 'mutation', this process is analogous to the population-genetics mechanism of random drift [22,42,43] and transmission biases would indicate cultural selection [7,8,11]. If random copying is used as the null hypothesis, frequencies should exhibit a right-skewed distribution in which a few traits are very popular, and most traits remain rare [32,33,36]. In contrast to random copying, frequency-dependent copying generates an exponential decay distribution, or conformist bias, copying popular or rare traits. Frequency trimming or anti-conformist bias will produce irregular 'humped' distribution, avoiding the copying of popular traits or rare traits [44]. The latter becomes indistinguishable from random copying when trimming is applied to common traits [45].

As another neutral model, the turnover rate model can differentiate selection types and determine when cultural transmission is biased [32]. In addition, the diachronic dynamics of cultural variants can be analysed by coefficient of variation.

Public media has a close relationship with public opinion and governmental actions. A change in concepts might account for news attention being a 'threshold event' or 'tipping point' [43]. In this analysis, concepts were taken from the *People's Daily* of China. This newspaper provides information on the policies and viewpoints of China and is also the best representation of the Chinese ecological philosophy and sociocultural tradition. A total of 3713 concepts and their frequency data were retrieved from the BLCU Corpus Center (BCC), representing a time span of 70 years, from 1946 to 2015 [46]. The BCC corpus is a large, full-text retrieval corpus with approximately 15 billion words, including 2 billion words from the *People's Daily*. The query function in diachronic retrieval provides concepts' occurrences data from the newspaper. A Boolean query of 'Ecological + Noun' AND 'Noun + Ecology' AND 'Economic + Noun' AND 'Noun + Economy', where N has the same function as Noun, was applied to all years from 1946 to 2015.

2.2. Caculations and Analysis

In this paper, we used both the progeny distribution and the turnover rate models. The progeny distribution model focuses on frequency and probability and, in our case, is defined as the cumulative fraction. In the calculation of the concept cumulative fraction, we first chose a single-year segment and calculated the concept frequency ($\boldsymbol{P_i}$) of that year,

$$\boldsymbol{P_i} = \frac{\boldsymbol{n_i}}{\boldsymbol{N}} \quad (1)$$

where $\boldsymbol{n_i}$ is the occurrence of the concept, and $\boldsymbol{N}$ is the total occurrences of all concepts. The cumulative fraction ($\boldsymbol{P_{ci}}$) is given by

$$\boldsymbol{P_{ci}} = \sum_{i=s}^{sm} \boldsymbol{P_i} \quad (2)$$

where $\boldsymbol{s}$ is the series number, which is ordered from low to high frequency ($\boldsymbol{P_i}$). For example, $\boldsymbol{s}$1 is series one, with the lowest frequency; $\boldsymbol{s}$2 is series two, with the second lowest frequency; ... and $\boldsymbol{sm}$ is the series maximum with the highest frequency.

Turnover model: We followed the previous work by Evans and Giometto [38] in defining turnover z in the top $\boldsymbol{y}$ chart. The list of the $\boldsymbol{y}$ most-popular concepts is defined as the sum of the number of concepts existing in the top chart plus the number of new concepts entering the top chart at the same time step. The $\boldsymbol{y}$ list in this research varies from 14 to 35 (depending on the sufficiency of data).

The turnover can be described by generic function (3),

$$z = \boldsymbol{a}\boldsymbol{y}^{\boldsymbol{b}} \quad (3)$$

where $\boldsymbol{a}$ is constant value and $\boldsymbol{b}$ is turnover exponent. Exponent $\boldsymbol{b}$ = 0.86 is the judgment criteria of the neutral model [38].

Concept diachronic change: The least square method was used to carry out linear regression. Yearly concept frequency was calculated by dividing the concept occurrence with total number of phrases of the newspaper. Here, slope k serves as an indicator of growth rate in a certain time slice.

Concept mutation: Any concepts that did not appear in the previous year were considered new concepts, and the total number of phrases was retrieved from BCC [46]. The calculation began in 1975, with a time span of 5 years (data before 1975 were insufficient).

Concept coefficient of variation: Concepts with structures such as 'noun + ecology', 'ecological + noun', 'noun + economy', 'economic + noun' (semantically parallel to Chinese concepts) were retrieved from the British National corpus (BNC). To use BNC concept results as a reference for the international recognition of sustainability ideas, the most popular concepts in ecological and economic domains were selected based on their total occurrences across 70 years. By taking the overlapping concepts with BNC as a selection criterion, a hot-concept list of 5 concepts for each category (the above four conceptual structures) was formed respectively. The standard deviations σ of each concept occurrence from 2008 to 2013 was calculated by using the following equation,

$$\sigma = \sqrt{\sum_{i=1}^{n} \frac{(v - \overline{v})^2}{n - 1}} \quad (4)$$

where v is the occurrence of a concept from 2008 to 2013, $\overline{v}$ is the average occurrence of the concept from 2008 to 2013, and n is the observation year segment from 2008 to 2013. Using σ from Equation (4), the coefficient of variation (CV) of a concept was then calculated by

$$CV = \frac{\sigma}{\overline{v}} \quad (5)$$

3. Results

3.1. Diachronically Change of Economic and Ecological Concepts

The diachronic frequency dynamics of concepts in the mainstream newspaper, *People's Daily*, show that the popularity of China's economic activities increased rapidly, with an average occurrence of 1689 concepts annually in the past 70 years, but with considerable changes (Figure 1a). From 1946 to 1953, the frequency of economic concepts showed a rapid growth trend, with an average annual growth of 4211 occurrences. The peak of word frequency in 1953 (55,566 occurrences) may relate to the first Five-Year Plan of the national economy. After that, from 1954 to 1967, the frequency of word use decreased by 1217 occurrences a year and fell to the bottom of the valley in 1966. This decline may be the consequences of economic downturn and the severe damage to economic development from natural disasters, such as droughts, floods and typhoons, in China from 1958 to 1961. The Four Modernization Goals strategy and the third Five-Year Plan boosted economic development and the word frequency increased again. It reached a peak in 1992 (185,942 occurrences), when China established the reform goal of the market economy system. In addition, from 1992 to 2004, it declined rapidly with a frequency decrease of 4345 occurrences a year. From 2005, it underwent a rapid growth, of 5233 occurrences a year (from 2005 to 2015), and finally reached 163,847 occurrences in 2010 before declining again.

For the dynamics of ecological concepts, the growth rate in ecological word frequency was very low in the early stage (Figure 1b). Within the 30 years after 1946, the average word frequency increased by only three occurrences a year. Since China's reform and opening up in 1978, it increased slowly, at a rate of 120 occurrences a year from 1978 to 1992. After China's attendance at the United Nations Conference on Environment and Development in 1992, the concept frequency began to rise rapidly again and had an increase of 2834 occurrences on average a year (from 1993 to 2006), then reached a peak (28,610 occurrences) in 2007. Ecological civilization was first written into the Report of 17th National Congress of the Communist Party of China (CPC) in 2007, which might account for the large amount of concept usage. In 2009, Ecological Civilization became an integral part of the national strategy and the frequency of ecological concepts increased again, by an average rate of 858 occurrences a year from 2010 to 2015. In 2015, it reached 26,105 occurrences. Overall, the growth rate of ecological concepts was only 1/6 of that of economic concepts. China's economic activities were always hotter than its ecological activities.

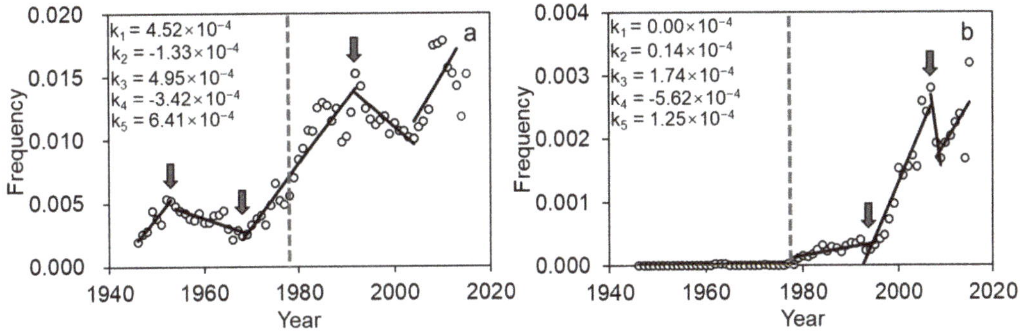

Figure 1. Changes in the frequency of economic concepts and ecological concepts from 1946 to 2015 in *People's Daily*. (**a**) Economy-related concepts, (**b**) ecology-related concepts. Dotted line: China's reform and opening up in 1978; Arrow: historical events (from left to right), the first Five-Year Plan in 1953 (1a), the third Five-Year Plan in 1966 (1a), the reform goal of the market economic system in 1992 (1a), United Nations Conference on Environment and Development in 1992 (1b), The 17th National Congress of the CPC in 2007 (1b). k-slope value.

3.2. Dynamics of Hot Economic and Ecological Concepts

The top five high-frequency concepts show the differences in China's economic and ecological concerns (Figure 2). In the economic field, the frequency of all popular concepts is higher than 10,000 occurrences. Economic policy and the economic situation (the blue line in Figure 2a) were essentially parallel in the first seven years. The occurrences of the economic structure (the black line with a white circle in Figure 2a) and economic system (the black line with a grey circle in Figure 2a) fluctuate by an order of magnitude. The concepts in Figure 2b show clear groupings: world economy, rural economy, collective economy and China economy in one group, and commodity economy (the red line in Figure 2b) in another, lower-frequency group. The *CV* value of commodity economy ranks the highest among all the hot economic concepts (Table 1). These popular concepts also reflect different foci in the economic field. Economic structure, economic policy, economic system and economic situation are elements of economic development, while world economy, rural economy, collective economy and China economy are concepts with different scales of economic observation. The fifth concepts in the two categories are cash crops and commodity economy, respectively. Cash crops (in Chinese, the concept cash crops shares the same characters with economic crop; however, the meanings of these two concepts are different. To avoid misunderstanding of the concept connotation, we use cash crops instead of economic crop in the analysis) are agricultural products with economic value and commodity economy is a sub-type of economy.

In contrast with the economic field, the ecological field shows stronger sensitivity to policies, such as ecological environment and ecological civilization, as well as to ecological problems that are people-oriented and affect human survival and development, such as agricultural ecology, marine ecology and forest ecology. In terms of historical changes, the frequency of the ecological environment (the red line in Figure 2c) is one order of magnitude higher than that of other concepts. Ecological environment first appeared in the 1960s and gradually increased since then. In 1981, it became the most popular concept and continued to rank first for the next almost 35 years (except for the years of 2012 and 2013). In contrast, ecological civilization did not exist before 1995 but increased significantly since 2005. After 2007, it experienced a four-year stable period and ranked first in 2012 and 2013 with yearly occurrences reaching 931 and 1023, respectively. The frequency of ecological civilization increased by two orders of magnitude in 10 years and resulted in high values of *CV* (Table 1). Despite such increases, no other concept attained a frequency as high as ecological environment. When ecology is the central term of the concept, the frequency fluctuation is varied. Concepts such as marine ecology (the blue line in Figure 2d) showed a stable and periodic increase in frequency and agricultural ecology (the red line in Figure 2d) declined continuously and periodically, with both resulting in high values of *CV* (Table 1).

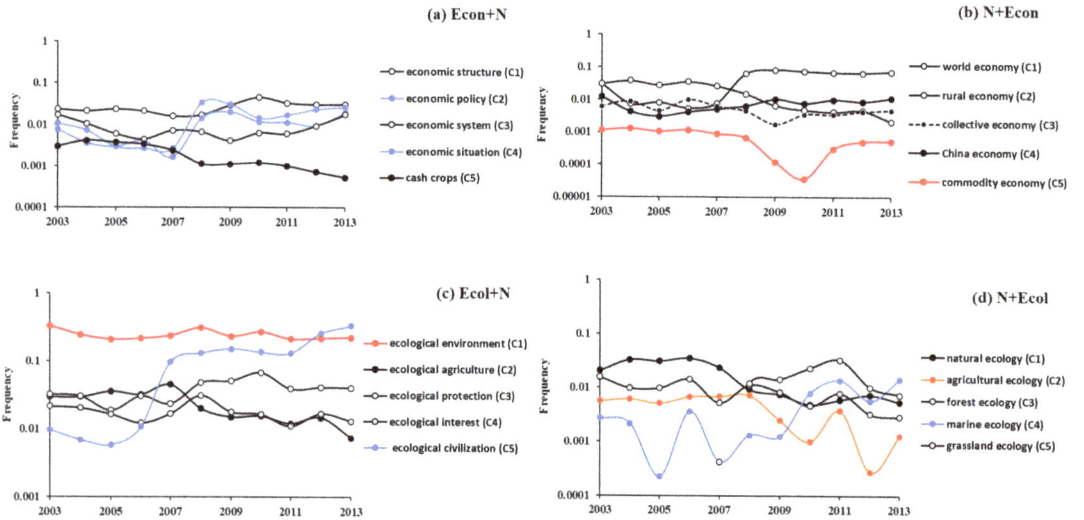

Figure 2. Frequency of the top 5 concepts from 2003 to 2013. Shown are the four structures for the concepts: (**a**). Econ (economic) + Noun, (**b**) Noun + Econ (economy), (**c**) Ecol (ecological) + Noun, and (**d**) Noun + Ecol (ecology). Y axes are logarithmic. C in parentheses stands for concept.

Table 1. Coefficient of variation of the top-five concepts during 2008–2013, tracked in Figure 2.

Category	Concept 1	Concept 2	Concept 3	Concept 4	Concept 5
Econ-N	0.45	0.32	0.43	0.26	0.40
N-Econ	0.28	0.48	0.29	0.29	0.61
Ecol-N	0.20	0.30	0.33	0.20	0.61
N-Ecol	0.25	0.72	0.36	0.81	0.68

Concepts 1 to 5 can be identified in their respective category.

3.3. Mutation Rate of Economic and Ecological Concepts

New-concept mutation or concept innovation has an impact on turnover. Before proceeding to turnover, we calculated the concept mutation rate (μ) as a measurement of innovation. The mutation rate of concepts is always higher for the economy than for ecology, which indicates that new phenomena or new elements appear faster in the economic field (Table 2).

From 1975 to 2015, the mutation rate of both economic and ecological concepts increased as a whole. The mutation rates of both concepts reached the highest value in 2015, with 54 economic concepts and 38 ecological concepts per 100,000 phrases. After 2000, the difference between economic concepts and ecological concepts decreased. In 2000, the mutation rate of economic concepts (approximately 0.4×10^{-3}) was twice that of ecological concepts (approximately 0.15×10^{-3}) but decreased to only a 1.1-fold difference between these concepts in 2005. Before 2000, the number of mutations of the economic concepts increased rapidly and then decreased afterwards until 2015. The growth rate of economic concepts was the largest (309%) from 1980 to 1985. The mutation rate of ecological concepts changed unnoticeably before 1995 and increased rapidly thereafter, but then levelled off. The significant growths in ecological concepts from 1980 to 1985 (354%) and from 1995 to 2000 (232%) finally brought the ecological concepts' usage to equal to economic concepts.

Table 2. The mutation of economic concepts and ecological concepts in *People's Daily*, from 1975 to 2015.

Year	$N\mu$ (Econ)	$N\mu$ (Ecol)	N (*All Phrases*)	μ (Econ)	μ (Ecol)
1975	1187	20	36,589,803	0.00%	0.00%
1980	3117	437	45,310,424	0.01%	0.00%
1985	12,763	1985	60,402,011	0.02%	0.00%
1990	15,717	3043	63,133,238	0.02%	0.01%
1995	22,982	3630	63,236,892	0.04%	0.01%
2000	33,410	12,038	76,579,938	0.04%	0.02%
2005	25,313	21,145	70,794,106	0.04%	0.03%
2010	19,961	16,029	43,465,357	0.05%	0.04%
2015	22,983	16,425	42,318,024	0.05%	0.04%

Econ denotes econ-N and N-econ, Ecol denotes ecol-N and N-ecol, $N\mu$ is the number of new concepts, N is the total number of concepts of the year, and μ is mutation rate.

3.4. Cumulative Frequency Distribution of Economic and Ecological Concepts

From 1980 to 2010, the cumulative frequency distribution of economic concepts and ecological concepts in China was observed by fitting a power-law distribution over time in four time segments (Figure 3). In 1980, the power β of economic concepts was −0.46, showing a long tail distribution (Figure 3a); only a few economic concepts are highly popular (in frequencies approaching 10%), whereas most concepts are presented at low- and medium-frequencies (at or below 1%). High-frequency ecological concepts account for the large proportion of the cumulative frequency, which resembles a "winner-take-all" distribution. In 1990, the β of economic concepts decreased to −0.5 and further changed to −0.6 by 2000 (Figure 3b,c) with an decrease in both the frequency and proportion distribution of high-frequency concepts. However, the β of ecological concepts changed to −0.08 in both 1990 and 2000, indicating that the word-usage process was becoming increasingly closer to power-law distribution. By 2010, the power β of economic concepts increased to −0.26 (Figure 3d). The growth in both the number and frequency of economic popular concepts (in frequencies approaching 10%) and their reduction in cumulative frequency distribution show an increasing evenness in economic concepts' usage, which may indicate a stronger neutral force of random copying. As for ecological concepts, the β value decreased to −0.09. Although there is a slight tendency toward a more even cumulative frequency distribution, more-segmented usage frequency implies that the selection of popular ecological concepts became stronger.

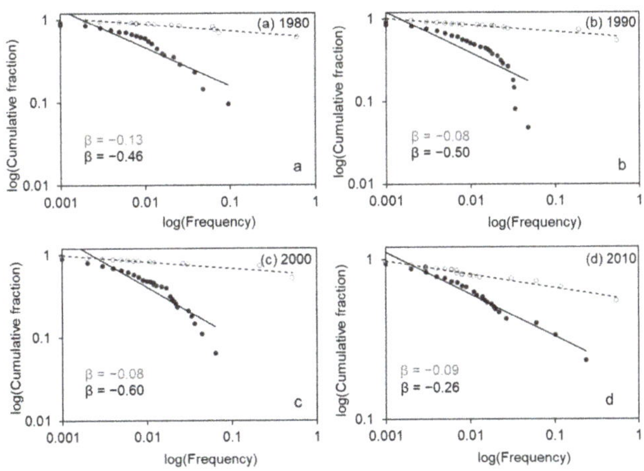

Figure 3. Cumulative frequency distribution of all economic and ecological concepts in *People's Daily*. Double logarithmic axes. The filled circle represents the distribution of economic concepts and the open circle represents the distribution of ecological concepts. (**a**) 1980, (**b**) 1990, (**c**) 2000, and (**d**) 2010. β-power exponent.

3.5. Turnover of Economic and Ecological Concepts

From 1946 to 1978, the turnover exponent of economic concepts was 0.73. The curve is below the theoretical line of the neutral model. At this stage, the change in high-frequency concepts was faster than that in medium- and low-frequency concepts (Figure 4a). In other words, the change in economic concepts was very active, and popular economic concepts were not 'fixed'. Economic concepts were subjected to anti-conformist selection. There was no significant difference between the turnover profile of economic concepts and the neutral prediction value after the reform and opening-up in 1978 (Figure 4b). At this stage, the turnover profile of economic concepts was close to neutral. Notably, compared with the period prior to the year of reform and opening-up, the turnover of economic concepts decreased as a whole following the reform. The turnover exponent value of economic concepts in *People's Daily* was lower than the neutral expectation ($b = 0.86$). The turnover exponent of ecological concepts before the reform and opening-up policy was 0.92 (Figure 4c). After the reform and opening-up, the b value of ecological concepts remained higher than the neutral value of 0.86 and increased to 1.25 (Figure 4d). In these two stages, the change in high-frequency ecological concepts was slower than that in medium- and low-frequency concepts. In other words, concepts with a high frequency, or popular ecological concepts, were 'fixed'. The transmission of ecological concepts was affected by conformist bias.

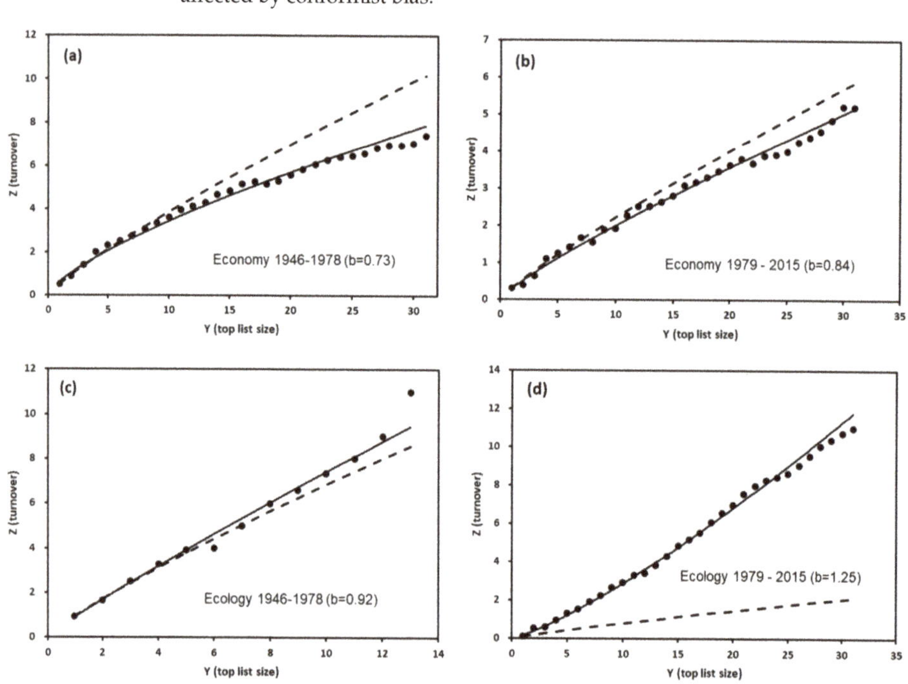

Figure 4. Turnover in the popularity of ecology-related and economy-related concepts. The dashed lines represent $b = 0.86$, and solid lines are fit lines to the data. (**a**) Economy-related concepts (including both Econ-N and N-Econ concepts) from 1946 to 1978, (**b**) economy-related concepts from 1979 to 2015, (**c**) ecology-related concepts (including both Ecol-N and N-Ecol concepts) from 1946 to 1978, and (**d**) ecology-related concepts from 1979 to 2015.

4. Discussion

The frequency of ecological and economic concepts reflects social recognition and populational behaviour about sustainability. From a linguistic ecology perspective, a connection between China's economic development and language usage can be established.

From formal governmental documents and administrative policies to daily expressions about people's livelihoods and social lives, developing the economy appears to be the top priority [47]. One phenomenon observed in this study was the 'emergence' of economic concepts within a short period, which manifests the vigour and creativity of China's economy. The cultural dynamics underlying the usage of early economic concepts may be best described as anti-conformist bias and concepts maintained a high proportion of innovation rates since the foundation of PR China. With the further development of China's economy and its gradual integration into the international market and world economic system, the innovation rate remained high but the entry and loss of concepts within a certain frequency range may imply the stabilization of concept niches. Although the overall average frequency did not appear to be maintained at a similar level, the turnover rate of popular concepts continued to move closer to the expected value of the neutral model, which indicates a stronger random copying process in economic-concept transmission. This finding does not nullify the intrinsic value of concepts but emphasizes the weakness of the selection intensity involved, which is caused by the mixed forces of national and international economic development and new understanding of economic growth. As in the neutral theory of population genetics, most evolutionary changes are caused by the random genetic drift of mutant alleles that are selectively nearly equivalent [23]. The evolutionary analysis of economic concepts demonstrated in these findings implies that these concepts have also experienced cultural drift and a neutral evolutionary process.

The blossoming of ecological ideas started four decades (around the 1980s) after the founding of PR China. Sustainable development ideas have emerged across the international community since the 1980s, and China has taken a series of actions to protect the environment and has brought forth a series of policies and regulations to strengthen its ecological civilization-construction strategy. Public media quickly responded in reporting governmental decisions and resonated with social preference and public attention. The growth rate of innovation in ecological concepts surpassed that of economic concepts in a short period and may potentially incubate a more-vigorous language-meaning system of ecological ideas than economic ideas. Despite the beneficial political environment and policies incentives, a slight downward trend appeared during 2007 to 2009 in ecological concepts (which also appeared from 1992 to 2003 in economic concepts) which may need a more in-depth observation in the future. Cumulative frequency distribution implies that the prevalent selection force underlining ecological concepts' transmission dynamics is primarily conformity bias, which indicates the selection of a few popular concepts and a lower degree of diversity compared with economic ideas. These popular ecological concepts are directly derived from national strategic slogans and ecological construction priorities, and the frequencies of popular concepts are significantly higher than those of other concepts, resulting in a 'winner-takes-all' distribution [25,34]. The turnover rate in the later period indicates that the force of the frequency-dependent copying bias was stronger than in the earlier period, which means that more focused attention was being directed to a certain specific area year after year in the process of constructing an ecological civilization.

When ecology and economy serve as the head noun (N-ecol or N-econ), the ***CV*** values are substantially higher than when they function as the modifier (ecol-N or econ-N). This result indicates that concepts rooted in ecological and economic disciplines might have experienced greater ups and downs due to a stronger selection force. In the past 70 years, the higher ***CV*** value in the ecological domain than in the economic domain implies a stronger cultural selection force underlying ecological ideas' development when taking the dynamic change into consideration (Figure 1). Combining with the result of the ecological turnover profile, a relation can be built between concept structure and context/content bias transmission. The higher value of the ***CV*** in Noun-ecology suggests content bias may account more for the transmission bias in ecological concepts than context bias, which reminds us to think of the influence that ecological science might have on public language usage and the understanding of sustainability.

The diachronic changes in popular concepts also differ in the top list of concepts. The government's strategy and strong will to improve China's ecology are expressed by concepts such as ecological environment and ecological civilization. The diversity and dynamics of economic concepts tend to experience negative frequency selection. In addition, popular concepts and their dynamics reflect that domestic culture exerts a stronger centripetal force, which leads to distinctive language-usage preferences, than the centrifugal ... force from overseas in ideas' convergence. The low level of overlapping among high-frequency popular concepts fully demonstrates the unique nationality, regionalism, and sociality features of China. Meanwhile, different changing processes in concepts also reveal the evolution direction towards protecting the environment and developing the economy. Moreover, the evolutionary process of concept usage can lead to a new perspective on the social understanding of the trade-off between economy and ecology. In our case, both domains were strongly affected by the governmental policies, but ecological concepts appeared to experience conformity bias, which means that popular concepts might be copied from the governmental sphere to society and into public daily use. In contrast, economic concepts seemed to experience anti-conformist bias and neutral processes in the recent data, which indicates an abundance of innovation and unbiased transmission from all sectors of society and in people's daily language usage.

5. Conclusions

In China, the dynamics of both economic and ecological concepts in mainstream newspapers reflect governmental and social activity hotspots for sustainability. The reform and opening-up policy profoundly changed the structure of the Chinese economy and incentivized a more diversified and vigorous economic-development system. The periodical and consistent emergence of abundant economy-related concepts after the reform and opening-up reflects an increasingly healthier and more resilient economic development pattern. The shifting of popular economic concepts reveals the mobility of the system and a greater innovation potential in economic areas. In comparison with economic concepts, the popular ecological concepts reveal a stronger conformist tendency and more force from political decisions than independent social forces in shaping their dynamics. A governmental initiative in ecological construction and protection accelerated the transmission of ecological concepts and enhanced the recognition of sustainability concepts in a much shorter period than that of economic concepts. However, the lower level of innovative and creative new concepts in ecological areas implies a shortage of vitality in developing ecological ideas. Nonetheless, we also assert that, if the data were expanded to include today, the result might be different due to the ecological restoration achievements that China has made in the intervening years. With the approaching of 2030 Agenda of Sustainable Development and China's entry into a new phase of ecological civilization, stimulating and incubating more diversified and resilient ecological knowledge and ideas, and transferring them into daily usage, will facilitate the linkage between knowledge, action and will, and foster the transition to a sustainable society and the attainment of sustainable development goals.

Author Contributions: Conceptualization, D.Z., G.H. and J.Z.; methodology, D.Z., Y.G. and J.C.; validation, J.Z.; formal analysis, D.Z. and X.H.; data curation, D.Z.; writing—original draft, D.Z.; writing—review and editing, D.Z. and J.Z.; visualization, D.Z., T.Z. and X.C.; supervision, G.H. and J.Z.; funding acquisition, J.Z. All authors have read and agreed to the published version of the manuscript.

Funding: Guangdong Laboratory for Lingnan Modern Agriculture Project (Grant No. NK2021001).

Institutional Review Board Statement: Not applicable.

Informed Consent Statement: Not applicable.

Data Availability Statement: The data presented in this study are available on request from the corresponding authors.

Acknowledgments: We thank R. Alexander Bentley for comments on an early draft. D.Z. thanks Jiang Chang for Python data-science support and Zhaoji Shi for statistics analysis advice.

Conflicts of Interest: The authors declare no conflict of interest.

References

1. Clark, W.C.; Dickson, N.M. Sustainability science: The emerging research program. *Proc. Natl. Acad. Sci. USA* **2003**, *100*, 8059. [CrossRef] [PubMed]
2. Clark, W.C. Sustainability Science: A room of its own. *Proc. Natl. Acad. Sci. USA* **2007**, *104*, 173–1738. [CrossRef] [PubMed]
3. Miller, T.R. Sustainability science. In *Companion to Environmental Studies*; Castree, N., Hulme, M., Proctor, J.D., Eds.; Routledge: London, UK, 2018; pp. 460–464.
4. Brundtland, G.H. *Our Common Future: Report of the World Commission on Environment and Development*; UN-WCED: Geneva, Switzerland, 1987.
5. Apetrei, C.I.; Caniglia, G.; von Wehrden, H.; Daniel, J. Lang Just another buzzword? A systematic literature review of knowledge-related concepts in sustainability science. *Glob. Environ. Chang.* **2021**, *68*, 102222. [CrossRef]
6. Axelsson, R.; Angelstam, P.; Degerman, E.; Teitelbaum, S.; Andersson, K.; Elbakidze, M.; Drotz, M.K. Social and cultural sustainability: Criteria, indicators, verifier variables for measurement and maps for visualization to support planning. *AMBIO* **2013**, *42*, 215–228. [CrossRef] [PubMed]
7. Cavalli-Sforza, L.L.; Feldman, M.W. *Cultural Transmission and Evolution: A Quantitative Approach*; Princeton University Press: Princeton, NJ, USA, 1981.
8. Boyd, R.; Richerson, P.J. *Culture and the Evolutionary Process*; University of Chicago Press: Chicago, IL, USA, 1985.
9. Neiman, F.D. Stylistic variation in evolutionary perspective. *Am. Antiq.* **1995**, *60*, 7–36. [CrossRef]
10. Shennan, S. *Genes, Memes and Human History: Darwinian Archaeology and Cultural Evolution*; Thames & Hudson: London, UK, 2002.
11. Richerson, P.J.; Boyd, R. *Not by Genes Alone: How Culture Transformed Human Evolution*; Chicago University Press: Chicago, IL, USA, 2005.
12. Mesoudi, A. *Cultural Evolution How Darwinian Theory Can Explain Human Culture and Synthesize the Social Sciences*; Chicago University Press: Chicago, IL, USA, 2011.
13. Kandler, A.; Creman, E.R. Analysing cultural frequency data: Neutral theory and beyond. In *Handbook of Evolutionary Research in Archaeology*; Prentiss, A.M., Ed.; Springer: Cham, Switzerland, 2019.
14. Prentiss, A.M. *Handbook of Evolutionary Research in Archaeology*; Springer: Cham, Switzerland, 2019.
15. Haugen, E. The Ecology of Language. In Proceedings of the Conference Toward the Description of the Languages of the World, Burg Wartenstein, Austria, 1–8 August 1970.
16. Darwin, C. *The Descent of Man, and Selection in Relation to Sex*; Princeton University Press: Princeton, NJ, USA, 1981.
17. Petersen, A.M.; Tenenbaum, J.N.; Havlin, S.; Stanley, H.E.; Perc, M. Languages cool as they expand: Allometric scaling and the decreasing need for new words. *Sci. Rep.* **2012**, *2*, 943. [CrossRef]
18. Ruck, D.; Bentley, R.A.; Acerbi, A.; Garnett, P.; Hruschka, D.J. Role of neutral evolution in word turnover during centuries of English word popularity. *Adv. Complex Syst.* **2017**, *20*, 1750012. [CrossRef]
19. Pagel, M.; Atkinson, Q.D.; Meade, A. Frequency of word-use predicts rates of lexical evolution throughout Indo-European history. *Nature* **2007**, *449*, 717–720. [CrossRef]
20. Newberry, M.G.; Ahern, C.; Clark, R.; Plotkin, J.B. Detecting evolutionary forces in language change. *Nature* **2017**, *551*, 223–226. [CrossRef]
21. Pagel, M.; Beaumont, M.; Meade, A.; Verkerk, A.; Calude, A. Dominant words rise to the top by positive frequency-dependent selection. *Proc. Natl. Acad. Sci. USA* **2019**, *116*, 7397–7402. [CrossRef]
22. Crow, J.F.; Kimura, M. *An Introduction to Population Genetics Theory*; Harper & Row: New York, NY, USA, 1970.
23. Kimura, M. *The Neutral Theory of Molecular Evolution*; Cambridge University Press: London, UK, 1983.
24. Bentley, R.A.; Shennan, S.J. Cultural transmission and stochastic network growth. *Am. Antiq.* **2003**, *68*, 459–485. [CrossRef]
25. Hahn, M.W.; Bentley, R.A. Drift as a mechanism for cultural change: An example from baby names. *Proc. R. Soc. Lond. Ser. B Biol. Sci.* **2003**, *270* (Suppl. S1), S120–S123. [CrossRef] [PubMed]
26. Bentley, R.A.; Lipo, C.P.; Herzog, H.A.; Hahn, M.W. Regular rates of popular culture change reflect random copying. *Evol. Hum. Behav.* **2007**, *28*, 51–158. [CrossRef]
27. Reali, F.; Griffiths, T.L. Words as alleles: Connecting language evolution with Bayesian learners to models of genetic drift. *Proc. Biol. Sci.* **2010**, *277*, 429–436. [CrossRef] [PubMed]
28. Leroi, A.M.; Lambert, B.; Rosindell, J.; Zhang, X.; Kokkoris, G.D. Neutral syndrome. *Nat. Hum. Behav.* **2020**, *4*, 780–790. [CrossRef]
29. Herzog, H.A.; Bentley, R.A.; Hahn, M.W. Random drift and large shifts in popularity of dog breeds. *Proc. R. Soc. Lond. B.* **2004**, *271*, S353–S356. [CrossRef]
30. Bentley, R.A. Random drift versus selection in academic vocabulary: An evolutionary analysis of published keywords. *PLoS ONE* **2008**, *3*, e3057. [CrossRef]

31. Dediu, D.; Cysouw, M.; Levinson, S.C.; Baronchelli, A.; Christiansen, M.H.; Croft, W.; Evans, N.; Garrod, S.; Gray, R.; Kandler, A. et al. Cultural Evolution of Language. In *Cultural Evolution: Society, Technology, Language and Religion*; Richerson, P.J., Christiansen, M.H., Eds.; The MIT Press: Cambridge, MA, USA, 2012; pp. 303–334.
32. Acerbi, A.; Bentley, R.A. Biases in cultural transmission shape the turnover of popular traits. *Evol. Hum. Behav.* **2014**, *35*, 228–236. [CrossRef]
33. Sindi, S.S.; Dale, R. Culturomics as a data playground for tests of selection: Mathematical approaches to detecting selection in word use. *J. Theor. Biol.* **2016**, *405*, 140–149. [CrossRef]
34. Bentley, R.A.; Hahn, M.W.; Shennan, S.J. Random drift and culture change. *Proc. R. Soc. Lond. B.* **2004**, *271*, 1443–1450. [CrossRef]
35. ElBahrawy, A.; Alessandretti, L.; Kandler, A.; Pastor-Satorras, R.; Baronchelli, A. Evolutionary dynamics of the cryptocurrency market. *R. Soc. Open Sci.* **2017**, *4*, 170623. [CrossRef] [PubMed]
36. Carrignon, S.; Bentley, R.A.; Ruck, D. Modelling rapid online cultural transmission: Evaluating neutral models on Twitter data with approximate Bayesian computation. *Palgrave Commun.* **2019**, *5*, 83. [CrossRef]
37. Brand, C.O.; Acerbi, A.; Mesoudi, A. Cultural evolution of emotional expression in 50 years of song lyrics. *Evol. Hum. Sci.* **2019**, *1*, e11. [CrossRef]
38. Evans, T.; Giometto, A. Turnover Rate of Popularity Charts in Neutral Models. *arXiv* **2011**, arXiv:1105.4044. Available online: https://arxiv.org/abs/1105.4044 (accessed on 10 January 2022).
39. Biber, D.; Gray, B. *Grammatical Complexity in Academic English Linguistic Change in Writing*; CUP: Cambridge, UK, 2016.
40. Altmann, E.G.; Pierrehumbert, J.B.; Motter, A.E. Niche as a determinant of word fate in online groups. *PLoS ONE* **2011**, *6*, e19009. [CrossRef]
41. Gillespie, J.H. *Population Genetics: A Concise Guide*; Johns Hopkins University Press: Baltimore, MD, USA, 1998.
42. Wright, S. Evolution in Mendelian populations. *Genetics* **1931**, *16*, 97–159. [CrossRef]
43. Hansen, A. Environment and the news media. In *Companion to Environment Studies*; Castree, N., Hulme, M., Proctor, J.D., Eds.; Routledge: London, UK, 2018.
44. Mesoudi, A.; Lycett, S.J. Random copying, frequency-dependent copying and culture change. *Evol. Hum. Behav.* **2009**, *30*, 41–48. [CrossRef]
45. Michel, J.-B.; Shen, Y.K.; Aiden, A.P.; Veres, A.; Gray, M.K.; Pickett, J.P.; Hoiberg, D.; Clancy, D.; Norvig, P.; Orwant, J.; et al. Quantitative Analysis of Culture Using Millions of Digitized Books. *Science* **2011**, *331*, 176–182. [CrossRef]
46. Xun, E.; Rao, G.Q.; Xiao, X.Y.; Zang, J.J. Construction of BCC corpus in the context of big data. *Corpus Linguist.* **2016**, *3*, 93–109.
47. Liu, J. China's Road to Sustainability. *Science* **2010**, *328*, 50. [CrossRef]

sustainability

MDPI

Article

Forest Degradation Index: A Tool for Forest Vulnerability Assessment in Indian Western Himalaya

Shashidhar Kumar Jha [1,*], Ajeet Kumar Negi [2], Juha Mikael Alatalo [3], Vignesh Prabhu [4], Mani Bhushan Jha [1] and Hemant Kumar [5]

1 Climate Program, World Resources Institute, Delhi 110016, India
2 Department of Forestry and Natural Resources, Hemvati Nandan Garhwal (A Central) University, Srinagar 246174, Uttarakhand, India
3 Environmental Science Center, Qatar University, Doha 2713, Qatar
4 Center for Study of Science, Technology, and Policy, Bengaluru 560094, Karnataka, India
5 Center for Rural Development & Innovative Sustainable Technology, IIT-Kharagpur, Kharagpur 721302, West Bengal, India
* Correspondence: shashidharkj@gmail.com or shashidhar.jha@wri.org

Abstract: The global climate is showing altered temperatures and precipitation levels. Forests can be a stabilizing force in climate change. They regulate the nutrient cycle, protect species and diversity, and support livelihoods that drive holistic growth. Presently, the forest ecosystem's capacity to withstand change is being undermined by the rate of change, along with anthropogenic pressures and the specificities of mountainous regions. Here, we attempted to design a 'forest vulnerability index' using field measurements and household surveys. A total of 71 quadrants were laid out, and 545 respondents were interviewed in 91 villages along the altitudinal gradient (altitude < 1200 m asl (Zone A), 1200–1800 m asl (Zone B), and >1800 m asl (Zone C)) of the Pauri district of Uttarakhand, India. The village-level data were normalized and combined to represent climate change impacts and the dimension of vulnerability. The IPCC (2014) protocol was used to assess forest vulnerability. The highest vulnerability was recorded in Zone 'B', and higher sensitivity, higher climate change impacts, and lower adaptive capacities were recorded in Zone 'B' and 'C'. The approach is comparable within the district and between the states. In enhancing our shared understanding of forest degradation, the results are of value to policy/decision-makers, implementers, and adaptation funding agencies, who can use them to assess the scale, cause, and actions for adaptation.

Keywords: climate change; forest vulnerability; indicators; Himalaya

Citation: Jha, S.K.; Negi, A.K.; Alatalo, J.M.; Prabhu, V.; Jha, M.B.; Kumar, H. Forest Degradation Index: A Tool for Forest Vulnerability Assessment in Indian Western Himalaya. *Sustainability* **2022**, *14*, 15655. https://doi.org/10.3390/su142315655

Academic Editors: Filippo Vitolla, Nicola Raimo, Ornella Malandrino and Benedetta Esposito

Received: 18 September 2022
Accepted: 10 October 2022
Published: 24 November 2022

1. Introduction

The Himalayas are an ecologically rich range of mountains range that serve humans with a wide range of goods and services (Rasul 2014 [1]; Badola et al., 2015 [2]). The ecosystem hosts various natural resources and diverse habitats, with a considerable altitudinal variation. The naturally rich repository nourishes more than seven countries (Government of Uttarakhand, 2014 [3]), and is associated with social and economic benefits within rural Himalayan communities (Hoy et al., 2016 [4]). Overall, the Himalayas provide a plethora of goods and services to mountain communities and millions of downstream inhabitants. The ecosystem has become critical in contributing to the continuous flow of services.

Himalaya is projected to be extremely sensitive to climate change (Indian Network for Climate Change Assessment, 2010 [5]; Chaturvedi et al., 2011 [6]; Singh and Hietala 2014 [7]), as the changes are likely to alter the regulation of terrestrial biogeochemical processes, such as litter decomposition, nitrification, denitrification, nitrogen mineralization, nutrient uptake, fine root dynamics, and soil respiration and productivity (Jha 2020 [8]). Historical patterns have had the following effects: glaciers receding (Kumar et al., 2015 [9]), deterioration in water resources (Li et al., 2016 [10]), drying out of traditional water sources

(Palazzoli et al., 2015 [11]), forest degradation (Jha 2020 [8]), biodiversity loss (Jha 2020 [8]), alteration of agricultural systems (Jethi et al., 2016 [12]), reduction in crop yields (Jethi et al., 2016 [12]), an upward shift of apple orchards, increased vulnerability of winter cropping (ADB, 2010 [13]), loss of tree species (Wani et al., 2013) [14], changes in bird types and populations, and migration (Jha 2020 [8]). Climate-driven changes in ecosystem structure, function, and species distributions negatively affect Himalayan forests (Lewis 2006 [15]; IPCC, 2007 [16]). Some studies claim that elevated atmospheric CO_2 concentration, increased temperature, and altered precipitation has, at least to some extent, had a positive impact (productivity and diversity) on the forest ecosystem. This varies, however, according to species' tolerance (Das 2004) [17] and the availability of water and nutrients.

For a long time, Himalayan forests have been exposed to severe, natural, and anthropogenic catastrophes, such as landslides, forest fires, earthquakes, and flash floods (Kala, 2014) [18]. Anthropogenic interventions such as deforestation, infrastructure development, and urbanization have further aggravated environmental degradation, which has affected the quantity and quality of ecosystem services (Ahmed et al., 1990) [19]. The degradation of ecosystem services has severe repercussions and is likely to be exacerbated by climate change (Ravindranath et al., 2006 [20]).

For survival throughout the Himalayan range, communities rely on forests and forest-based resources (fodder, fuelwood, medicinal plants, fruits, and other associated items) (Jadin et al., 2016 [21]; Jha et al., 2018 [22]). In the last few years, increased anthropogenic activities have put pressure on the Himalayan ecosystems, causing changes in forest ecosystems (Jha et al., 2018 [22]; Jha 2020 [8]). As populations continue to expand, together with their reliance on forest products, forest ecosystems are now experiencing decreased density and diversity (Sharma and Vetaas 2015 [23]; Bruggeman et al., 2016 [24]) and changes in distribution (Upgupta et al., 2015 [25]; Chakraborty et al., 2016 [26]), biodiversity (Giam 2017 [27]), productivity (Alekhya et al., 2015 [28]), and forest degradation (Jha 2020 [8]). Further, the reliance on forest resources and the requirement for a life support system in the Himalayas varies with altitude; for example, the quantum of fodder is required more at low- and mid-altitude, and fuelwood is needed at high altitudes irrespective of the administrative boundary. Similarly, agriculture and allied activities are less extensive and beneficial at lower altitudes. At the same time, constraints increase with altitude. The challenges for water, even for drinking purposes, also increase with altitude. It is comparatively easy to manage water supply even in scarce seasons at lower altitudes because of the distance to headquarters, institutional set-up, and other facilities.

Forest degradation is exacerbated by climate change, natural and manmade disasters, and resource utilization patterns, implying the need for sustainable forest management in the Himalayan area. Given the importance of forests in sustaining local communities, several studies have conducted evaluations of forest-based goods and services in the Himalayan region. However, limited research has been undertaken to evaluate forest degradation using altitude-specific data. Studies have assessed the status of the forest using primary productivity (Upgupta et al., 2015 [25]; Sharma et al., 2018 [29]; Kumar et al., 2019 [30]), forest fire index (Jha et al., 2018 [22]), biomass and carbon stock (Jha 2020 [8]), factors that are highly specific and which, to a certain extent, limit and confine the assessment. These approaches could be strengthened by integrating multiple factors, enabling a clear identification of the actual status of the forest. Adequate knowledge of the determinants of forest degradation through time and space is critical for forest management and creating policies to improve local livelihoods and strengthen community disaster preparedness, particularly in the face of climate change.

The vulnerability assessment uses a bottom-indicator-based approach from national to even local levels. The national-level indexes for vulnerability assessment, such as HDI, HWI, PVI, etc., use different parameters, such as life expectancy, sustainability, and quality of life for defining vulnerability at a larger scale (state or nation) (Table 1). The number of variables in this category is relatively few. However, it compares the relative position of a state or country. Country rankings provide helpful information on the relative level of sustainability. Contrastingly, a single value is used for comparing vulnerability on a national level. Moreover, a county has considerable variability regarding exposure, susceptibility, impact, and other developmental parameters (Cutter and Finch 2008) [31]. However, the data used may not adequately represent the processes determining vulnerability (Eriksen and Kelly 2007 [32]); generally, it results in a restricted vulnerability measure.

Table 1. Indexes used for the assessment of vulnerability and adaptive capacity.

S. No.	Indexes	Brief Description
1	Environmental Vulnerability Index (Kaly et al., 1999) [33]	This index is based on three aspects of vulnerability: risks to the environment (natural and anthropogenic), the innate ability of the environment to cope with the risks (resilience), and ecosystem integrity (the health or condition of the environment as a result of past impacts).
2	Index of Human Insecurity (Lonergan et al., 2000) [34]	IHI was developed as a classification system that distinguishes the perception of vulnerability and insecurity of different countries, indicators from environment, economy, society, and institutions were used to calculate overall vulnerability.
3	Human Well-being Index (Prescott-Allen 2001) [35]	HWI measures the community (political rights, crime, internet users, peace and order) and social equity (gender and income). The index incorporates development/sustainability goals in five categories ranging from 'good' to 'bad'.
4	Environmental Sustainability Index (World Economic Forum, 2002) [36]	This index measures progress towards environmental sustainability of 142 countries through 20 indicators.
5	Water Vulnerability Index (Sullivan 2002 [37])	An interdisciplinary approach that produces an integrated assessment of water stress and scarcity, linking physical estimates of water availability with socioeconomic variables that reflect poverty.
6	Country-level Risk (Brooks and Adger 2003) [38]	Includes several proxies for risk associated with climate variability and change, based on numbers killed and affected by climate-related disasters.
7	Predictive Indicators of Vulnerability (Adger et al., 2004) [39]	The PIV link social vulnerability with climate adaptation.
8	Social Vulnerability Index (Vincent 2004) [40]	SVI is an index of five composite sub-indices: economic well-being and stability (20%), demographic structure (20%), institutional stability and strength of public infrastructure (40%), global interconnectivity (10%) and dependence on natural resources (10%)
9	Index of Social Vulnerability to Climate Change (Adger and Vincent 2005) [41]	A social vulnerability index with emphasis on water availability.
10	Environmental Sustainability Index (Esty et al., 2005) [42]	A composite index using 76 indicators to assess 146 nations' sustainability trajectory based on five major categories: environmental system, environmental stresses, human vulnerability to environmental stresses, the human capacity to respond to environmental change, and global stewardship.
11	Prevalent Vulnerability Index (Cardona 2005) [43]	Focuses on social, economic, institutional, and infrastructural capacity to recover from natural hazards or the lack thereof.
12	Disaster Risk Index (ISDR 2004) [44]	One of the United Nations' three global initiatives to raise awareness for integrated disaster mitigation. Aims to 'improve understanding of the relationship between development and disaster risk', 'enable the measurement and comparison of relative levels of physical exposure to hazard, vulnerability and risk', identify vulnerability indicators', and map international patterns of risk.
13	Human Development Index (UNDP, 2008) [45]	Includes life expectancy, health, education, and standard of living indicators for an overall assessment of well-being and human development on a country scale.

Table 1. *Cont.*

S. No.	Indexes	Brief Description
14	Livelihood Vulnerability Index (Hahn 2009 [46])	Combines seven components or indicators for vulnerability: livelihoods, socio-demographics, social networks, health, natural disasters and climate variability, food, and water security assimilated into the dimension of vulnerability.
15	Household Social Vulnerability Index (Vincent and Cull 2010) [47]	An index of household-level social vulnerability to climate change, based on the multiple dimensions of the vulnerability identified in the sustainable livelihood framework.
16	Livelihood Vulnerability Index and the Livelihood Effect Index (Urothody and Larsen 2010 [48])	Both indexes reflect the relative differences between the two VDCs in terms of vulnerability to climate change impacts and factors contributing to it.
17	Climate Vulnerability Index (Pandey and Jha 2011 [49])	These indicators were assimilated into indices to the dimension of vulnerability and the IPCC (2007) protocol was used for vulnerability assessment.
18	Poverty and Vulnerability Assessment (Gerlitz et al., 2014) [50]	Combines general poverty predictors with relevant indicators in mountain contexts.
19	Climate Vulnerability Index (Aryal et al., 2014) [51]	An index used to assess and compare the vulnerability of transhumant communities in Nepal.
20	Multidimensional Livelihood Vulnerability Index (Gerlitz et al., 2016 [52])	A measure used to explore and describe livelihood vulnerability to climatic, environmental, and socio-economic change in the HKH region.
21	Adaptation Capability Index (Pandey et al., 2016 [53])	An approach to recognizing social and natural factors contributing to successful adaptation that addresses several household functions, such as social networking, livelihood strategies, resource availability, and accessibility.
22	Climate Vulnerability Index for Water (Pandey et al., 2015 [54])	An index to assess the climate-related water vulnerability of households in rural and urban settings, comprising three components—exposure, sensitivity, and adaptive capacity, and 14 sub-components.
23	Future Vulnerability Index (Upgupta et al., 2015 [25])	Index to assess forest ecosystem vulnerability to climate change across Himachal Pradesh under 'current climate' and 'future climate' scenarios.
24	Inherent Vulnerability (Shukla et al., 2016 [55])	Indices to assess the inherent vulnerability (internal) of agricultural communities at the village level based on sensitivity and adaptive capacity.
25	Climate Vulnerability Index, and Current Adaptive Capacity Index, (Pandey et al., 2017 [56])	These indices assess vulnerability using five forms of capital, i.e., human, natural, financial, social, and physical capital.
26	Forest Vulnerability Index (Thakur et al., 2020 [57])	An index for forest vulnerability using field-based observations with an integrated approach of multiple indicators.
27	Socio-ecological Vulnerability Index (Jha et al., 2021 [58])	An index to estimate socio-ecological vulnerability.
28	Climate Vulnerability Index (Jha 2020 [8])	An index used to calculate the vulnerability of sectors (socio-economic, agriculture, forest, and water), and the contribution of sectors to the overall vulnerability.
29	Vulnerability Index (DST, 2020 [59])	This index identifies the most vulnerable states and districts in India concerning current climate risk and the main drivers of vulnerability.
30	Vulnerability Index and Resilience Index (Jha et al., 2021) [58]	The indices assess the socio-ecological vulnerability and resilience of mountain communities residing in capital-constrained environments. Site-specific indicators were integrated into a sustainable livelihood framework in an attempt to calculate vulnerability and resilience jointly.
31	Ecosystem Services Index (Jha et al., 2022) [60]	This index quantifies ecosystem services.

Vulnerability indicators at the regional and local levels are also used in impact assessment. They are also instrumental in identifying threats or threatened systems, raising awareness, and prioritizing vulnerable sectors (Harley et al., 2008 [61]). The vulnerability indicators are further used to compare the relative vulnerability of places (Pandey and Jha, 2011 [49]), groups or communities (Sinha and Jha, 2017) [62], and sectors (Jha et al., 2018 [22]). Moreover, vulnerability indicators are widely used to prioritize vulnerable sectors and formulate adaptation options. Jha et al. (2017) [63] utilized vulnerability indicators to assess government-sponsored adaptation programs. The changes in household-level indicators of agriculture, water, and economic status due to Government-sponsored adaptation programs, i.e., MGNREGA, are aggregated and assessed.

Generally, the regional indicators-based assessment such as the Livelihood Vulnerability Index (Hahn et al., 2009 [46]); Livelihood Vulnerability Index and Livelihood Effect Index (Urothody and Larsen 2010 [48]); Climate Vulnerability Index (Pandey and Jha 2011 [49]); Poverty and Vulnerability Assessment (Gerlitz et al., 2014 [50]); Inherent Vulnerability (Shukla et al., 2016 [55]); Forest Vulnerability Index (Thakur et al., 2020 [57]); Vulnerability Index (DST, 2020 [59]); Vulnerability Index and Resilience Index (Jha et al., 2021) [58]; etc., identify and analyze the indicators with higher significance in related sectors and then normalize them into indices or components or sub-components or sectors (Table 1). Finally, these indices were assimilated into the dimension of vulnerability. The details on vulnerability indices are comprehensively discussed in Table 1, and vulnerability is discussed in Table 2. The tools for understanding forest degradation in the Himalayas are limited, and the available literature analyzed the degradation with minimal inclusion of ground surveys and issues that emphasize the needs of locals. Furthermore, several districts in the study state, Uttarakhand, have a wide range of altitudes. The study district is spread over 5400-km squared and has an altitudinal range of 300 to 3000 m above sea level. Therefore, understanding the present state of forest degradation along the altitude becomes critical. Our study explores forest degradation in the context of altitude and dependency. The study's primary aim was to examine forest vulnerability along the altitudinal gradient in the Pauri District, Uttarakhand, India. We addressed the following objectives based on broad themes: (i) to identify indicators contributing to forest degradation, (ii) to assess forest vulnerability along the altitudinal gradient, and (iii) to identify altitude-specific indicators for adaptation actions under impending forest degradation.

Table 2. Definitions of vulnerability.

S. No.	Author(s)	Definitions
1	Gabor and Griffith (1980) [64]	Threats (including chemical agents and the ecological situation of the communities, and their level of emergency preparedness) to which people are exposed.
2	Timmerman (1981) [65]	The degree to which a system acts adversely to the occurrence of a hazardous event.
3	UNDRO (1982) [66]	The degree of loss to a given element or set of elements at risk resulting from the occurrence of a natural phenomenon of a given magnitude.
4	Susman et al. (1984) [67]	The degree to which different social classes are at different risk.
5	Pijawka and Radwan (1985) [68]	The degree to which hazardous materials threaten a particular population (risk) and the capacity of the community to reduce the risk or adverse consequences of hazardous materials releases.
6	Kates et al. (1985) [69]	The capacity to be wounded.
7	Bogard (1988) [70]	The inability to take effective measures to insure against losses.
8	Mitchell (1989) [71]	Vulnerability is the potential for loss.
9	Liverman (1990) [72]	Distinguishes between vulnerability as a biophysical condition and vulnerability as defined by society's political, social, and economic conditions.

Table 2. *Cont.*

S. No.	Author(s)	Definitions
10	Downing (1991) [73]	It refers to a consequence (e.g., famine) rather than a cause (e.g., are vulnerable to hunger); it is a relative term that differentiates among socioeconomic groups or regions rather than an absolute measure of deprivation.
11	Dow (1992) [74]	The differential capacity of groups and individuals to deal with hazards based on their positions within physical and social worlds.
12	Smith (1992) [75]	The risk from a specific hazard varies through time and according to changes in physical exposure and/or human vulnerability (the breadth of social and economic tolerance available).
13	Alexander (1993) [76]	A function of the costs and benefits of inhabited areas at risk from natural disasters.
14	Cutter (1993) [77]	The likelihood that an individual or group will be exposed to and adversely affected by a hazard.
15	Watts and Bohle (1993) [78]	Exposure, capacity, and potentiality.
16	Blaikie et al. (1994) [79]	The characteristics of a person or group in terms of their capacity to anticipate, cope with, resist, and recover from the impact of a natural hazard.
17	Bohle et al. (1994) [80]	The aggregate measure of human welfare that integrates environmental, social, economic, and political exposure to a range of potentially harmful perturbations.
18	Dow and Downing (1995) [81]	Biophysical, demographic, economic, social, and technological factors such as population ages, economic dependency, racism, and age of infrastructure in association with natural hazards.
19	Eakin and Luers (2006) [82]	The lack of human capacity to avoid or reduce such adverse situations.
20	Gilard and Givone (1997) [83]	The sensitivity of land use to the hazard phenomenon.
21	Comfort et al. (1999) [84]	The circumstances that place people at risk while reducing their means of response or denying them available protection.
22	Weichselgartner and Bertens (2000) [85]	The condition of a given area concerning hazard, exposure, preparedness, prevention, and response characteristics to cope with specific natural hazards.
23	Adger (2006) [86]	The state of susceptibility to harm from exposure to stresses associated with environmental and social change and from the absence of capacity to adapt.
24	Ciurean et al. (2013) [87]	The inability to withstand the effects of a hostile environment.
25	Wolf et al. (2013) [88]	A measure of possible future harm.
26	IPCC (2014) [89]	The degree to which a system is susceptible to and unable to cope with adverse effects of climate change, including climate variability and extremes.
27	Hong et al. (2016) [90]	A weak resistance and low resilience of ecosystems in response to external interference including natural and artificial factors at a specific spatial scale.
28	Jha et al. (2022) [60]	A suite of human-ecological factors reflects exposure (climate stress on households), sensitivity (the degree to which a system is exposed), and adaptive capacity (internal or external capacity to cope).

Source: Adapted from Cutter (1996) [91], Weichselgartner (2001) [92], Hogan and Marandola (2005) [93], Adger (2006) [86], Wolf et al. (2013) [88], Paul (2013) [94], Berrouet et al. (2018) [95].

Forest Degradation Indicators: Selection and Definition

An indicator constitutes information that is either readily available or easily obtained pragmatically. It is a variable and a measure of system behavior in the context of essential and perceptible attributes (Holling et al., 1978 [96]). Indicators have been used in several ways by authors: Marcus (1983) [97] described a sign as 'defined as something which stands for something to somebody in some respect or capacity'; as a proxy—a function from an observable variable (McQueen and Noak 1988 [98]); a statistical measure (Tunstall 1992 [99]); a parameter—provides information on the state of the phenomenon and defines it as a value that is measured or observed (OECD, 1993 [100]); a piece of information (Bakkes et al., 1994 [101]); an index (Hammond et al., 1995 [102]); and a sub-index or component of an index (Ott 1978 [103]; Hahn et al., 2009 [46]; Pandey and Jha, 2011 [49]). In this study, an indicator is a perceptible local variable or piece of information chosen for its importance to livelihoods, the local economy, conservation, and resource management (availability, accessibility, and usability). Table 3 has more details on the indicators.

Table 3. Indices and indicators for climate change impacts and dimension of vulnerability (sensitivity and adaptive capacity) along the altitudinal gradient in Pauri District, Uttarakhan.

	Indices	Indicators	Explanation	Zone A	Zone B	Zone C
Climate Change Impacts	Temperature index	Increased summer temperature	Alters the phenology of forest tree species and also increases the probability of forest fire	0.760	0.100	0.900
		Increased February-March temperature	Alters phenology of forest tree species, agricultural and horticultural production	0.750	0.800	0.830
	Rainfall Index	Decreased number of rainy days	Lesser rainy days distress forest tree species and water availability	0.230	0.540	0.710
		Delayed monsoon rain	Results in water scarcity and alters the physiological character of the forest	0.110	0.290	0.340
	Extreme Event Index	Increased temperature resulting in forest fire	Increased temperature strengthens forest fire. It drives forest degradation and limits forest productivity	0.500	0.730	0.860
		Increased temperature and drought incidence	Water scarcity and restricted long-term water availability for agriculture and related activities	0.120	0.670	0.890
		Increased intensity and decreased frequency of rainfall	Increases runoff and erosion	0.990	0.920	0.770
		Uncertain extreme climatic events	These cause structural degradation and functional deterioration of forests	0.850	0.960	0.960

Table 3. *Cont.*

	Indices	Indicators	Explanation	Zone A	Zone B	Zone C
			Dimension of Vulnerability			
Sensitivity	Natural Resource Degradation Index	Frequency of forest fire (increased or decreased)	One of the major drivers of forest degradation, limits forest productivity, and restricts access to forest	0.480	0.950	0.870
		The intensity of forest fire (increased or decreased)	One of the major drivers of forest degradation, limits forest productivity, and restricts access to forest	0.430	0.900	1.000
		Lopped trees (Number/Hectare)	The higher number of lopped trees shows higher anthropogenic interference	0.651	0.609	0.477
		Grazing (%)	Reflects higher anthropogenic interferences, limits alter the ecological balance and affect the health of the forest	0.667	0.667	0.667
		Time spent on collection of forest resources	Shows that the distance between village and forest is increasing because of anthropogenic interference	0.500	0.900	0.960
		Distance travelled for collection of forest resources	A distant forest reflects more time spent on resource collection	0.490	0.910	0.910
	Resource susceptibility index	Concentration of dominance		0.373	0.633	0.545
		Evenness	Represents health, vigor and functioning of forest ecosystem	0.512	0.458	0.533
		Susceptible species	Susceptible species more prone to fire	0.620	0.560	0.170
		Regeneration rate	Higher regeneration represents better health	0.870	0.900	0.540
		Pure forest and dominant species	Pure forests are susceptible to change, and the fire-susceptible dominant species result in the loss of forest	0.420	0.870	0.590
		Age of the nearby forest (even or odd)	Even-aged forests are more susceptible to change	0.500	0.500	0.400
		Dependency on the natural water source	Natural water sources are susceptible to forest degradation and climate change	0.140	0.290	0.370

Table 3. *Cont.*

	Indices	Indicators	Explanation	Zone A	Zone B	Zone C
Adaptive Capacity	Resource collection index	Fodder and fuelwood collection	The natural support system is more susceptible to climate change and higher dependency (% respondents) on susceptible ecosystems results in vulnerability	0.430	0.400	0.450
		Collection of NTFP	NTFP (apart from fodder and fuelwood) is an additional income source and is also considered a food supplement	0.350	0.290	0.270
	Resource availability index	Sufficient fodder and fuelwood	Availability of and accessibility to sufficient fodder and fuelwood sources strengthen capacity to withstand change	0.430	0.600	0.780
	Strategy index	Individuals (Number/Hectare)	The higher the number of individuals, the higher the productivity of the forest	0.488	0.447	0.396
		Biodiversity (H)	Maintains structure and function of an ecosystem	0.593	0.389	0.350
		Species Richness	Supports species' ecological interactions and geographical ranges	0.557	0.482	0.389
		Biomass (Tons/Hectare)	Absolute reflection of forests' health; the higher the biomass, the better the health	0.505	0.447	0.412
		Alternatives to fodder during fodder scarce season (Summer and monsoon) (cultivation of fodder, use of previously stored fodder, or increased use of straw)	The availability of alternatives represents the self-sufficiency of a household	0.330	0.370	0.330
		Strategies during fuelwood scarcity (shifting to energy-efficient techniques like improved stoves, LPG, etc.)	Strategies reduce dependency and strengthen households' adaptive capacity	0.320	0.490	0.520

The term 'vulnerability' has scientific origins in geography and natural disaster studies. However, this concept has become crucial in fields such as natural hazard and disaster management, ecology, public health, poverty and development, secure livelihoods and famine, sustainability science, land change, and climate effects and adaptation. Several studies (Table 2) characterized vulnerability based on their research appropriateness. A simple, concise meaning derived from its Latin origins is frequently employed in the literature. The root of vulnerability is 'vulnerare', meaning 'to wound'; thus, Kates et al. (1985) [69] defined vulnerability as 'the capacity to be wounded'. Amartya Sen (1981) [104] emphasized the cause of the vulnerability. Defining it as a 'lack of entitlement (food security, sustainable livelihoods, social structures, etc.) due to restricted access to resources (institutional, political, and technological)'.

The IPCC defines vulnerability in its Second Assessment Report as 'the extent to which climate change may damage or disrupt a system', and states that system vulnerability 'depends not only on a system's sensitivity but also on its ability to adapt to changing climatic settings' (Watson et al., 1996 [105]). According to the report, socioeconomic systems are 'usually more vulnerable in developing nations with less favorable economic and institutional circumstances'. The existence of an inter-relationship between wealth (strengthens adaptive capacity) and poverty (begets vulnerability) in a region discussed deliberately in the Regional Impacts of Climate Change: An Assessment of Vulnerability (IPCC, 1997 [106]). The location and context specificity of vulnerability with uncertainty was highlighted by Handmer (1999) [107]. He underlined that uncertainty might be addressed by bold policy initiatives that address social and economic concerns while taking into account human behavior, culture, and the institutional capabilities of affected communities. At COP-6, Watson defined vulnerability as 'the amount to which a natural or social system is vulnerable to suffering damage from climate change', and it is a result of climate change magnitude, sensitivity, and adaptability (IPCC, 2000) [108].

Several definitions of vulnerability have been offered since 1980. It is generally defined as the ability to anticipate, resist, cope with, and respond to a hazard (Blaikie et al., 1994) [79]. It occurs due to various factors including disaster management, public health, poverty, food security, ecology, and climate change, including climate variability, change, and extremes (Fussel 2007 [109]). Goulden (2010) [110] defines the natural perspective of vulnerability as 'a function of the character, magnitude, and rate of climate variation to which a system is exposed, its sensitivity and adaptive capacity'. The potential implications of climatic fluctuation and change are referred to as exposure; sensitivity is the extent to which the exposure affects rural communities; and adaptive capacity is the system's ability to adapt, tolerate, or recover from the effects of the exposure (Ebi et al., 2006 [111]). However, the definition does not precisely define the relationship among dimensions of vulnerability (Schauser et al., 2010 [112]). Sometimes researchers describe vulnerability as an exposure susceptibility–resilience complex when they view sensitivity as susceptibility and adaptive capability as resilience (Balica et al., 2012 [113]). Vulnerability measures a system's (a community's) defenselessness or propensity to be injured or hurt by climate change. The IPCC defines vulnerability as 'the propensity or predisposition to be negatively affected'. Vulnerability includes several factors, such as sensitivity or susceptibility to injury and a lack of ability to cope and adapt (IPCC, 2014 [89]). The IPCC 2014 report defines vulnerability as a characteristic, intrinsic system attribute. The exterior property, i.e., exposure, is inherent. Thus, it is not taken into consideration for the assessment. Exposure is an external element that increases vulnerability and is regarded as an inseparable component of vulnerability assessment, driving sensitivity and adaptive capacity.

There have been few vulnerability assessment studies based on the IPCC 2014 framework, with most using the IPCC 2007 framework and measuring exposure as an inseparable component of vulnerability (Simane et al., 2016 [114], Kumar et al., 2016 [115], Jha et al., 2017 [63], Jha et al., 2018 [22], Pandey et al., 2018 [116], Jha 2020 [8]). The indicators used for vulnerability assessment are shown in Table 1. These investigations are based on the queries 'who is vulnerable?', 'what is vulnerability?', and 'vulnerability to what?' (Malone and Engle 2011 [117]), whereas new vulnerability assessment paradigms strive to answer the question 'vulnerability to what?' Updated assessments are not related to exposure and are a system's distinguishing feature, demonstrating its current internal condition (Sharma et al., 2013 [118]). Vulnerability, in this construct, is regarded as an inherent attribute of the system, consisting of sensitivity and adaptive capability.

The current study evaluates vulnerability as a pre-existing, system-specific attribute. We measured vulnerability using hazard-specific indicators for sensitivity and adaptive capability after identifying likely and prospective dangers. Climate change is currently regarded as a hazard, and factors that directly or indirectly affect and/or influence sensitivity and adaptation ability were considered for the evaluation. The indications were aggregated into indices, which were then categorized as 'possible repercussions of climate change'.

2. Methodology

2.1. Study Area

The research was carried out in the Pauri district of Uttarakhand, India. The district (29°20′–30°15′ N; 78°10′–79°20′ E) has an average elevation of 1800 m above sea level (m asl) (range 210–7817 m asl) and occupies a land area of 5230 km^2 (FSI, 2019 [119]). The district's population is 686,527, with a density of 129 persons per km^2, and the sex ratio is 1103 females per 1000 males. The district has reported a negative population growth rate (−1.51%). The district has an 82.59 percent literacy rate (males 93.18 percent, females 73.26 percent), compared to 74.04 percent in India (males 82.14 percent, females 65.46 percent) (Census of India, 2011 [120]). With a mean annual temperature of 25–30 °C (45 °C in June and 1.3 °C in January) and a mean annual rainfall of 2180 mm, the area has a sub-temperate to temperate climate, with over 90% of precipitation falling during the monsoon season (July–September).

The district has a hilly topography. The fluvial valleys have a convex shape, with steep valleys, interlocking spurs descending the main river, and terraced agricultural fields on the valley sides' moderate slopes. The people, who stay in the area, are primarily involved in agriculture, rely on forest resources for their survival, and are locally known as Garhwali. Tourism and animal husbandry are supplementary income sources in the region. The Garhwal region is predominantly rural, with scattered housing patterns and a mixed economy. The landholding is scattered and small, and per capita land availability is 0.68 ha in the hilly districts and 1.77 ha in the plain districts. Out of the total reported land, only 14.02% is under cultivation, and more than 55% is rainfed. The existence of different topographical and orographic elements has resulted in exceptional biodiversity in the region. The state is well-forested and consists of about 34,652 square kilometers (64.78%) of forest cover; the study district has 3269 square kilometers (61.34 %) of forested area. Out of the total, 519 km square (15.87%) is very dense forest; 1954 km square (59.77%) is moderately dense forest; and 672 km square (24.35%) is open forest. The district contributes 13.48% of forest area to the state, and its share in national forest cover is 03.45%. The per capita forest area in the district is 0.0047 km square per person, much higher than the state's per capita forest area, i.e., 0.0024 km^2 per person (ISFR, 2021) [121]. The forest type (Figure 1b) in the district shows a considerable higher difference. The forest of the study district is differentiated into moist Shivalik sal forest, deciduous forest, and ban-oak forest.

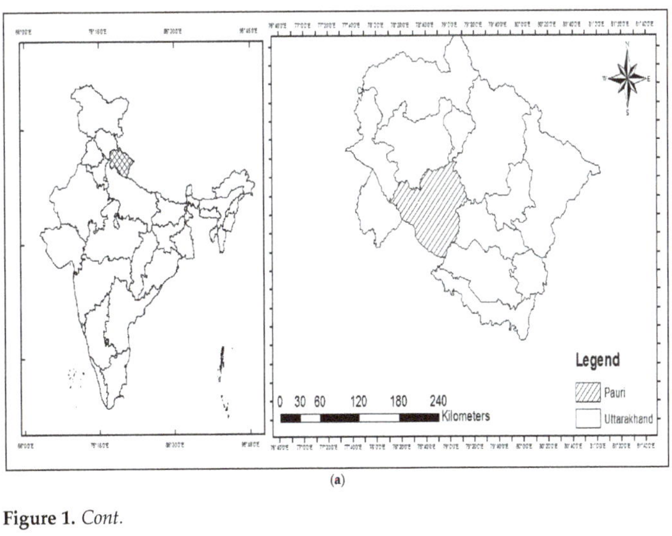

(a)

Figure 1. *Cont.*

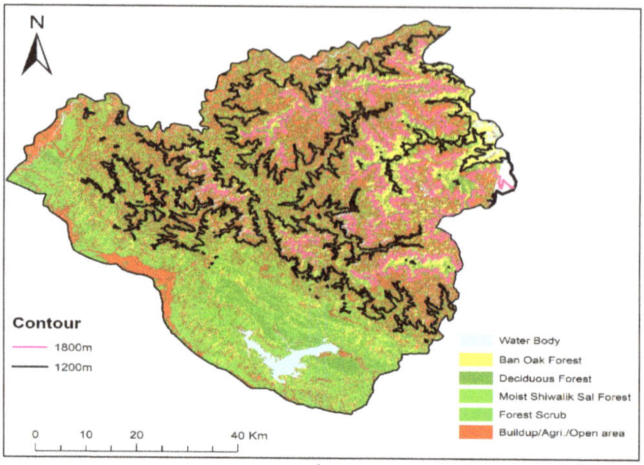

(**b**)

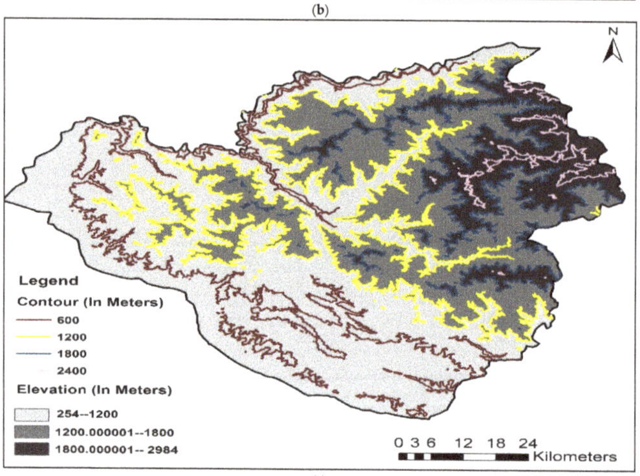

(**c**)

Figure 1. (**a**) Study site location district and state, (**b**) land use and land cover map depicting forest type along the contour map of the study district, (**c**) contour and elevation map of the study district.

The moist Shivalik sal forest, i.e., Zone A, comprises the Tarai–Bhabar area of the adjacent Gangetic plain, and outer and sub-Himalayan tract up to 1200 m asl. The forest types of these zones consist of (a) *Shorea robusta* forest, (b) *Acacia-Dalbergia* forest, (c) *Acacia-Shorea robusta* forest, (d) *Holoptelea* forest, (e) Bamboo (*Dendroclamus*) brakes, and (f) *Anogeissus latifolia-Spondias pinnata* forest and scrubs, associated with various evergreen and deciduous tree species. Zone B, 1200–1800 m asl has mountainous topography and the forest type is a deciduous forest comprising of (a) Pine forest, (b) Oak forest, (c) Pine-Oak forest, (d) Oak-*Lyonia* forest, (e) Deodar Forest, (f) Deodar-Pine forest, and (g) mixed forest. Zone C, i.e., more than 1800 m asl usually known as the sub-Himalayan peaks, has a ban-oak forest type consisting of (a) oak forest, (b) deodar forest, (c) rhododendron forest, (d) spruce forest, and (e) mixed forest.

2.2. Sampling Strategies

Inter- and intra-structural indicators of forest degradation were examined over an altitudinal gradient for vulnerability assessment. The patterns of land use and resource utilization were explored in detail. As basic requirements differ with altitude in the mountains, the present investigation divided the district into three zones (A–C) based on altitude and forest distribution. The zones were defined as altitude < 1200 m asl (zone A), altitude 1200–1800 m asl (zone B), and altitude > 1800 m asl (zone C) (Figure 1c).

Questionnaires and face-to-face interviews, ideally with the head of the household, were used to gather information on households in randomly selected villages in each zone. Multiple villages in each zone were chosen to retain the villages' variability. Large samples are difficult to gather in mountainous areas, since households are sparsely scattered and typically engage in their livelihood throughout the day. In the study district, 91 villages (30 in Zone A, 32 in Zone B, and 29 in Zone C) were surveyed. The villages were chosen because of the inhabitants' reliance on the forest. A minimum of five and a maximum of ten respondents from each participating village were chosen for the interview to avoid similarities in responses. For data collection, 545 respondents were questioned from the three designated altitudinal zones (182 in Zone A, 187 in Zone B, and 176 in Zone C).

With the help of a local person, the interviews were conducted in Hindi and the local language. The questionnaire addressed topics and concerns regarding the forest and its ability to provide a livelihood. A phytosociological study was also conducted to determine the forest's condition in density, regeneration, species richness, variety, lopped branches/trees, and biomass. Along the altitudinal gradient, the investigation was conducted in the open (OF), moderately dense (MDF), and dense forests (DF). To prevent uniformity and cover more area in each zone, a total of 71 quadrants (19 in Zone A (OF 5, MDF 9, DF 5), 22 in Zone B (OF 8, MDF 10, DF 4), and 30 in Zone C (OF 19, MDF 5, DF 6)) of 500 m^2 each (20 m × 25 m) were set out at intervals of 500 m elevation. The primary data and field measurements were transformed into indicators and incorporated into the forest vulnerability indices.

2.3. Analytical Framework

A bottom-up, indicator-based approach was used in the study (Hahn et al., 2009 [46]). The underlying assumption was that local knowledge and advice are essential to identifying drivers of vulnerability and indicators that may help the forest ecosystem become less vulnerable. These indicators are sympatric, indicating susceptibility at a local level (Jha et al., 2017 [63]). Multidimensional Vulnerability (Sullivan and Meigh 2005 [122]), Livelihood Vulnerability Index (Hahn et al., 2009 [46]), Livelihood Effect Index (Urothody et al., 2010 [48]), Climate Vulnerability Index (Pandey and Jha 2011 [49]), Capacity Assessment Index (Jha et al., 2017 [63]), and Vulnerability and Capacity Assessment (Sinha and Jha, 2017 [62]) are two vulnerability assessment methodologies that include indications from several components to show both sectoral and overall vulnerability. The current study integrated field measurements with primary data (household questionnaires) to develop a method for assessing forest vulnerability. Most indicator-based techniques employ balance-weighted averages, assuming that each indicator contributes equally to the aggregate index (Sullivan 2002 [37]; Pandey and Jha 2011 [49]; Jha et al., 2017 [63]) and are thus comparable and relevant.

Our approach was designed to assess the vulnerability of forests that provide abundant goods and services to communities. Household dependency on the forest was studied through resource degradation, susceptibility in terms of sensitivity, forest resource availability, collection, and strategies as adaptive capacities. Previous studies on forest degradation in the Himalayas are limited. The vulnerability of natural systems can be attributed to multidimensional factors (Jha 2020 [8]), with each dimension associated either positively or negatively (Jha et al., 2017 [63]). Change or variation in a single dimension may or may not influence others and a change in the dimension of vulnerability can impact households differently depending on their existing endowments (Pandey and Jha, 2011 [49]).

The present study employed a participatory evaluation technique that included participatory assessment (household surveys, participatory rural appraisals, focus group discussions) and field measurements (phytosociological analysis) (Table 3). Diversity was studied along with richness, regeneration, individuals, evenness, the concentration of dominance (CD), and biomass to represent the forest ecosystem's long-term health, vigor, and functioning (Solbrig 1991 [123]; Lee and Chun 2016 [124]). These variables are also important ecological characteristics, strongly correlated with environmental and anthropogenic variables (Gairola et al., 2008 [125]; Rawat and Chandra 2012 [126]). Indicators were created through primary qualitative and quantitative data. These indicators were identified through a combination of literature specific to the area or similar regions (e.g., Sharma et al., 2009 [127]; Urothody et al., 2010 [48]; Pandey and Jha 2011 [49]; Tse-ring et al., 2012 [128]; Sandhu and Sandhu 2014 [129]; Pandey et al., 2015 [54]; Gerlitz et al., 2016 [52]; Pandey et al., 2016, 2017 [53,56]; Jha et al., 2017 [63]) and through expert consultation. The indicators were originally in different units or scales, and they were standardized depending on their functional relationship with vulnerability, e.g., whether vulnerability increased with an increase in the value of the indicator (positive relationship; Equation (1)) or decreased with an increase in the value of the indicator (negative relationship, Equation (2)). The current state of the forest ecosystem was then described in terms of vulnerability using the IPCC vulnerability assessment technique (IPCC, 2014 [89]), which consists of eight indices and 30 site-specific indicators. The higher the indicator's value, the bigger the vulnerability, it was thought.

$$\text{Index}_{\text{sv}} = \frac{S_v - S_{min}}{S_{max} - S_{min}} \tag{1}$$

$$\text{Index}_{\text{sv}} = \frac{S_{max} - S_v}{S_{max} - S_{min}} \tag{2}$$

where S_V is the average value of the indicator at the village level, and S_{min} and S_{max} are the minimum and maximum values of the indicator.

The indicators were averaged after standardization using Equation (3) to calculate the score for the indexes:

$$M_V = \frac{\sum_{i=1}^{n} Index}{n} \tag{3}$$

where, M_V is one of the indices for the dimension, Index is the value of the ith indicator, and n is the number of indicators for the dimension.

Dimensions of vulnerability (sensitivity (Se) and adaptive capacity (Ac)) are calculated separately by taking a simple average of indexes using Equation (4):

$$N_v = \frac{\sum_{i=1}^{n} Dimension\ of\ Vulnerability}{n} \tag{4}$$

where N_V is one of the dimensions of vulnerability, the dimension of vulnerability is the value of the ith indexes, and n is the number of indexes for the dimension.

The overall vulnerability is calculated with sensitivity and adaptive capacity, consisting of site-specific indicators in the context of climate change using Equation (5). Here the data for indicators were gathered through household surveys which capture opinions or perceptions of the communities and field measurements in the nearby forest.

$$Vulnerability = Sensitivity - Adaptive\ Capacity \tag{5}$$

3. Results and Discussion

A detailed analysis of the historical climate trend (1951–2018) using INRM multi-model ensembles, along with communities' perception of climate, and forest vulnerability based on a bottom-up indicator-based approach using IPCC assessment protocol was employed in this investigation. The result is presented in two sections, the first section discusses the historical climate trend and communities' perceptions while forest vulnerability is discussed in the second section.

3.1. Historical Climate Trends

According to INRM multi-model ensembles (RCP 4.5), from 1951 to 2018, the average rainfall of the district decreased with low confidence for pre-monsoon, post-monsoon, and winter months, while the decrease was more significant in monsoon months. In contrast, very heavy precipitation and simple rainfall intensity index have been increased. The minimum temperature for winter, pre-monsoon, and monsoon seasons indicated an increasing trend with low confidence. At the same time, the post-monsoon minimum temperature is rising with high confidence. The average maximum temperature for winter and monsoon decreased with low confidence, while the pre- and post-monsoon maximum temperature increased. The district's maximum and minimum daytime temperatures indicated increasing trends (Figure 2).

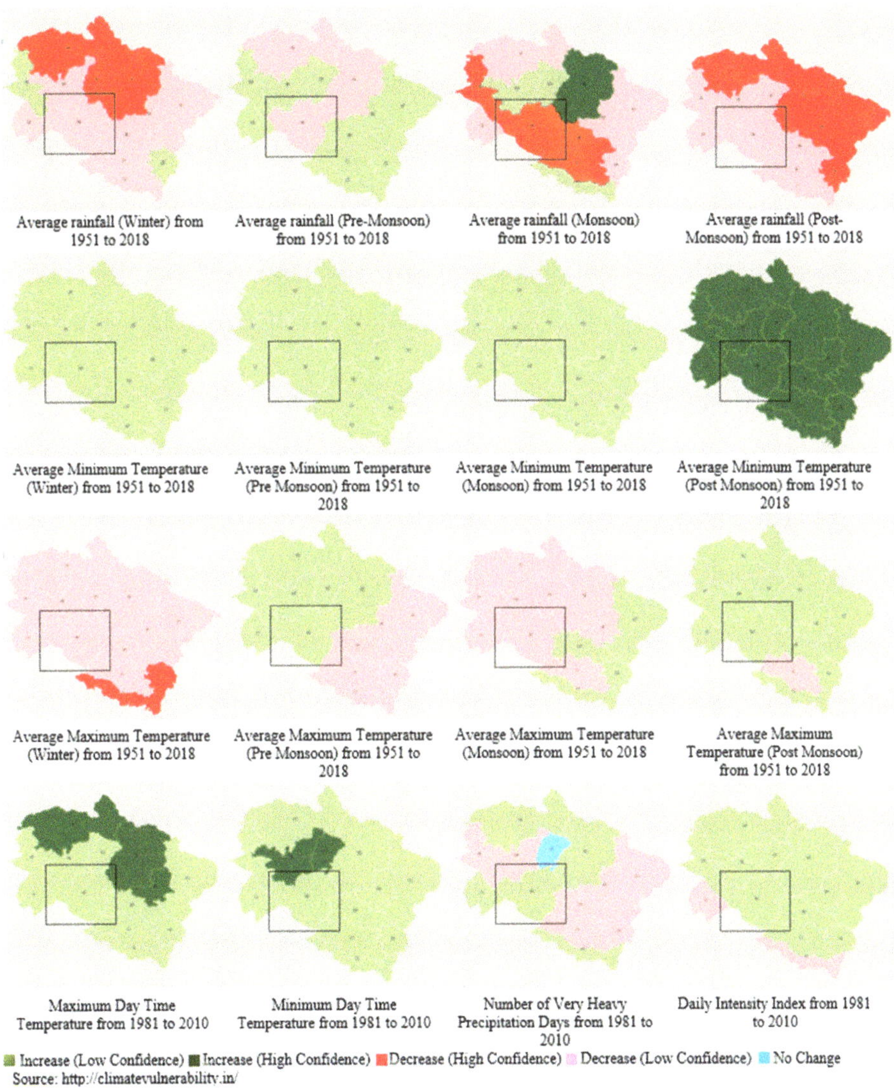

Figure 2. Climate trend in Uttarakhand highlighting study district in a box (district number 06).

3.1.1. Climate Perception

Further, communities' perception was also studied with the help of indicators. The indicators or set of information were normalized and converted into indices and finally assimilated into components of climate. Climate perception was recorded with the help of changes in the indicators of maximum and minimum temperature, rainfall, and extreme events over the previous 15–20 years.

3.1.2. Temperature Perception (Maximum and Minimum)

The local communities confirmed the variation in the district's climate. They noted that the climate was changing, with temperature increases and variations in rainfall (frequency and intensity). These were analyzed using a temperature index (A—0.76; B—0.45; C—0.87), rainfall index (A—0.17; B—0.42; C—0.53), and extreme event index (A—0.62; B—0.82; C—0.87). Each of the indices consisted of a set of indicators, with the temperature index consisting of 'increased summer temperature' and 'increased February-March temperature'. The most significant increases in summer temperature were reported in Zone C (0.90), followed by Zone A (0.76) and Zone B (0.10). Official data for India (MoEF, 2012 [130]) have also reported substantial increases in temperature in the Himalayan region. The study district's mean, minimum and maximum temperatures, both annually and during the summer, indicate increasing trends (Jha et al., 2020 [131]). Furthermore, this increasing trend has been highlighted above 4000 m asl. in eastern Nepal and Tibet (Shrestha and Devkota 2010 [132]). The overall temperature in the Himalayan area is expected to rise at a pace of 0.06 degrees Celsius every year (Shrestha et al., 2012 [133]).

The most significant increases in February-March temperature were reported in Zone C (0.83), followed by Zone B (0.80) and Zone A (0.75). Indeed, a previous study has similarly found an increase in January-March temperatures in the Eastern Himalayan area (Sharma 2012 [134]). Jha et al. (2020) [131] reported that winter temperatures in the Pauri district are also increasing aside from the annual average. Increased winter temperatures in the Himalayas directly influence the permafrost cover and snow melting rate. Accelerated melting will likely increase river discharge, indirectly affecting water availability (Nepal and Shrestha 2015 [135]). Based on the current warming and GHG emission rates (ICIMOD 2019 [136]), the Himalayas could lose two-thirds of its glaciers by 2100.

3.1.3. Rainfall Perception

The rainfall index, comprising a decreased number of rainy days and delayed monsoon rain, reflected the most significant overall variation in rainfall in Zone C (0.53), followed by Zone B (0.42), and Zone A (0.17), indicating an increasing trend from Zone A to C. The scores for decreased number of rainy days and delayed monsoon rain were 0.23 and 0.11 (Zone A), 0.54 and 0.29 (Zone B), and 0.71 and 0.34 (Zone C).

The World Bank (2012) [137] has reported a decline in monsoon rainfall and an increase in the frequency of severe rainfall events since the 1950s. Several studies in the Indian Himalayan region suggest temperature increases, more seasonal variations, milder winters and variations in the magnitude, intensity, and duration of precipitation (Goswami et al. 2006 [138]; Sharma et al., 2009 [127]; Xu et al., 2009 [139]; Pandey and Jha, 2011 [49]; Pandey et al., 2017 [56]; Jha 2020 [8], Jha et al., 2020 [131]).

3.1.4. Extreme Events Perception

Meanwhile, the extreme event index was composed of increasing temperatures that triggered forest fires, increased temperature and drought incidence, increased rainfall intensity and frequency, and unexpected, extreme climatic events. Previous research reporting climatic data (Jha et al., 2020 [131]) and residents' perceptions of climate variance in the district have verified this (Rao et al., 2018 [140]; Jha et al., 2020 [131]). Furthermore, official statistics from India (MoEF, 2012 [130]) have revealed visible alterations in the Himalayan climate pattern.

Scores were reported in direct proportion to the altitude (Zone A—0.62; B—0.82; C—0.87). The first two indicators, higher incidence of forest fire and drought owing to increasing temperature, followed a similar pattern, with higher scores in Zone C (0.86 and 0.89) than in Zone B (0.73 and 0.67) and Zone A (0.50 and 0.12). Zone A (0.99) had higher scores for increased rainfall intensity and decreased rainfall frequency than Zone B (0.92) and Zone C (0.92). (0.77). Finally, extreme climatic event uncertainty was found to be almost equivalent in Zone B (0.96) and C (0.96) but greater in Zone A. (0.85) (Table 3).

With increasing altitude, the overall impact increased (Zone A—0.51; B—0.56; C—0.75). The increased effect in Zone C might likely be attributed to its closeness to and reliance on the climate-sensitive natural support system. IPCC (2014) [89] clearly states that natural support systems are relatively more vulnerable to climate change, making those dependent on vulnerable ecosystems more susceptible. Moreover, variation in climate parameters and its uncertainty also alters composition and structure (Gaire et al., 2017 [141]), phenology (Bajpai et al., 2016 [142]), budburst and flowering (Amano et al., 2010 [143]), and forest productivity (Alekhya et al., 2015 [28]).

3.2. Forest Vulnerability

The forest vulnerability index, which includes characteristics or dimensions of vulnerability such as sensitivity and adaptive capacity, was employed in this investigation.

3.2.1. Sensitivity

Sensitivity along the altitudinal gradient in the Pauri district was assessed using the natural resource degradation index and resource susceptibility index (Figure 3). Natural resource degradation was reported to be very similar in Zone B (0.82) and C (0.81) and comparatively less in Zone A (0.54). This index consists of grazing (%), time spent collecting fodder and fuelwood, distance travelled for resource collection, especially fodder and fuelwood, forest fire frequency and intensity, and lopped trees (number/hectare). Grazing was a significant cause of degradation and was found in similar volumes (0.667) in all three zones (A to C). Furthermore, forest degradation reduced services and impacted hydrological functioning by contributing to rainfall interception, infiltration, purification, evapotranspiration, and groundwater recharge (Locatelli, 2016 [144]).

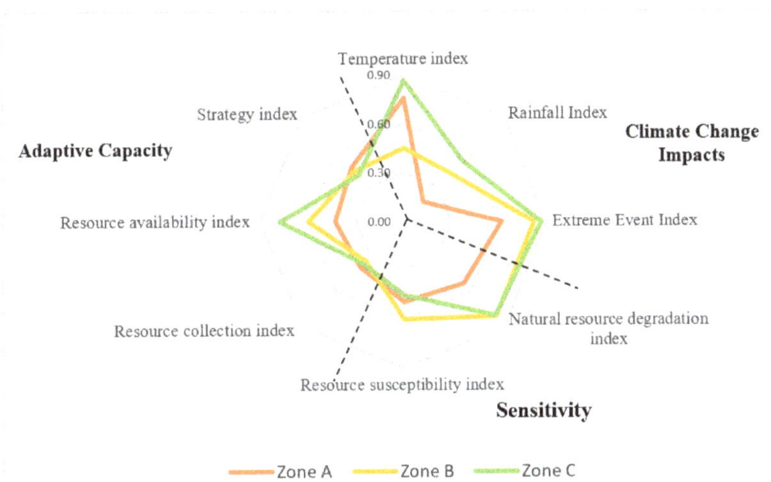

Figure 3. Indices of climate change impacts, sensitivity, and adaptive capacity along the altitudinal gradient in the Pauri district.

The increasing demand for resources and their extensive harvest undermines the forest in the designated zones. Zone C (5 to 6 h/day, up from 2 to 3 h/day) experienced the largest increase in average time spent collecting fodder and fuelwood compared to Zone B (3 to 4 h/day, up from 1.5 to 3 h/day) and A (1 to 1.5 h/day, up from 0.5 to 1.2 h/day). Indeed, it is probable that the time spent on collection (fodder and fuelwood) has increased due to the increased travel distance required. In the past, respondents traveled around 3–5 km/day to collect resources, but the distance has grown to 0–4 km/day. Zone C (7–10 km/day, up from 3–5 km/day) had the largest increase, and overall, the normalized values for time spent and distance traveled were highest in Zone C (0.96 and 0.91), followed by Zone B (0.90 and 0.91) and Zone A (0.50 and 0.49) (Table 3).

Zone B (0.95) had the most increased frequency of forest fires, followed by Zone C (0.87) and Zone A (0.48), whereas Zone C (1.00) had the highest increased intensity, followed by Zone B (0.90) and Zone A (0.48). Pauri is among the districts most affected by these fires, with approximately 38% of the state's forest fires reported to have occurred here. The Kotdwar block of the district had the highest number of forest fires (712) (Hussain, 2018) [145], the bulk of which were human-caused for reasons such as encouraging fodder gathering (Jha et al., 2018 [22]).

The number of lopped trees was determined along an altitudinal gradient, with Zone A (0.65) being the most, followed by Zone B (0.61) and Zone C (0.48). Forest degradation is said to be exacerbated by the livelihood concerns of millions of impoverished people living in and around the forest (Maikhuri et al., 2001 [146]; Davidar et al., 2010 [147]), which is believed to be highest in Zone C due to greater dependency.

Zone B (0.60) has the highest score for the resource susceptibility index, followed by Zone A (0.49) and Zone C (0.45). This index was composed of seven indicators: concentration of dominance (CD), evenness, fire-susceptible species, regeneration, dominant species, forest age, and reliance on natural water resources. CD was found to be greater in Zone B (0.63) than in Zone C (0.54) and A (0.37), while evenness was found to be quite comparable in Zone A (0.51) and C (0.53), and lower in Zone B (0.45). Fire-susceptible species decreased with altitude; the maximum number was recorded in Zone A (0.620). The dominant forest tree species in Zone A and B were susceptible to fire, as demonstrated by Jha et al. (2018, 2020) [22,131]. Indeed, the significance of forest composition to fire occurrences has been studied previously and was initially highlighted by Shank and Noorie (1950) [148]. The Himalayan area has seen a 90 percent rise in forest fires and increased numbers of fires started intentionally (Levine et al., 1999) [149].

The area under forest was recorded highest in Zone B (0.87), followed by Zone C (0.59) and Zone A (0.42). Regeneration was limited in Zone A (0.87) and B (0.90) due to the dominance of *Pinus roxburghii,* while more effective regeneration was reported in the oak forests of Zone C (0.54). The forests of Zone A (0.50) and B (0.50) were evenly aged, and slight variation was recorded in Zone C (0.40). Other significant determinants of vulnerability can be found in water availability, accessibility (Rajesh et al., 2014 [150]), and storage (Connor 2015) [151]. It may be anticipated that households who rely on a natural water supply (e.g., river, spring, etc.) are more sensitive to rising climate catastrophe and related repercussions. Dependence on a natural water source was found to be highest in Zone C (0.37), followed by Zone B (0.29), with Zone A (0.14) being nearly half that of Zone B (Table 3). Deforestation, rising global temperature, intensified precipitation, and seasonal droughts are other site-specific variables related to water scarcity (Tambe et al., 2011) [152], all of which contribute to natural springs drying up and stream base flow dropping (Rawat et al., 2011) [153].

3.2.2. Adaptive Capacity

Forests provide a broad set of goods and services for sustenance in the mountains (Jha et al., 2018 [22]) and strengthen their capacity to withstand changes. Adaptive capacity in the Pauri district was assessed using 03 indices and 09 indicators. Greater adaptive capacity was reported in Zone A, followed by Zone B and C. The resource collection index consists of fodder and fuelwood collection and non-timber forest products (NTFPs) (Figure 3). The majority of the households in the study area collect fodder and fuelwood from the forest; the collection was reported to be slightly higher in Zone C than in Zone A and Zone B, and fuelwood was found to be preferable for cooking, water heating, space heating, etc., for economic reasons (Kanagawa and Nakata, 2007) [154]. Extraction of fuelwood was reported to be higher in winter (November to February) due to higher consumption levels and limited labor requirements in agriculture and allied sectors, while minimum extraction was reported in the monsoon season due to the availability of green fodder in close proximity (Bhatt and Sachan 2004) [155]. Dependence on the collection of NTFPs was significantly less throughout the district, although a few households in each zone reported collecting NTFPs from a nearby forest. NTFP accessibility and collection were lowest in Zone A, producing a higher normalized score for the zone. At the same time, the harvest of NTFPs was comparatively superior in Zone B and C.

The availability of resources, i.e., fodder and fuelwood, was found to be highest in Zone A, followed by Zone B and C, with more dependents in Zone B and C than in Zone A. A higher number of dependents indicates greater competition and a lower proportion of households acquiring sufficient fodder and fuelwood. Indeed, Jha et al. (2020) [131] have reported similar findings about resource accessibility, availability, and sufficiency. The strategy index, which comprised six indications, was determined to be greatest in Zone C, followed by Zone B and Zone A. Zone C had the most individual trees (number/hectare), followed by Zone B and Zone A; Zone C also had the highest density of trees. The dense forest in Zone A had a stem density of 716 trees/ha, which is within the range described by Sharma et al. (2011) [156] for Indian forests and Saxena and Singh (1982) [157] for a forest in the Kumaun Himalaya. However, this result was lower than the range reported by Gairola et al. (2011) [158] for a forest in the Western Himalayas. The aggregate tree density in Zone B varied from 286 to 907 trees/ha, and the tree density of Quercus leucotrichophora (273 trees/ha) in Zone B (1200–1800 m asl) was lower than reported by Gairola et al. (2011) [158] and Pandey (2001) [159]. The dominating species in Zone C was found to be *Quercus leucotrichophora*, a dominance at high altitudes that have also been documented by Sharma et al. (2009) [127] (Garhwal Himalaya). Detailed analysis on individuals (trees/hectare), above-ground biomass density (AGBD), species richness, and biodiversity in dense, moderately dense, and open forests in the selected zones (A, B, C) of the study district is presented in Figure 4.

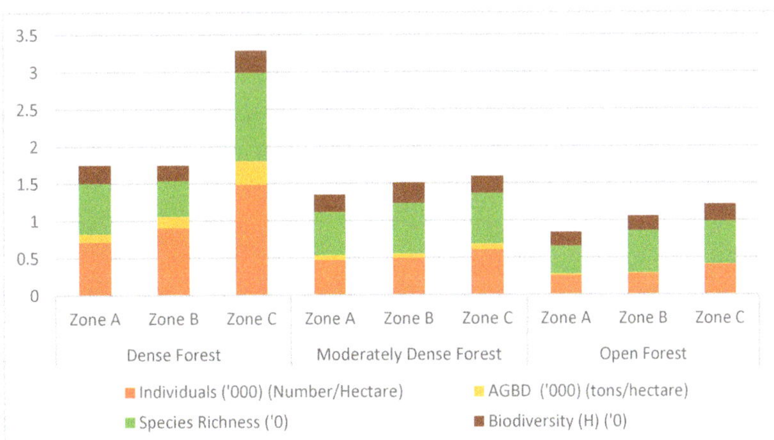

Figure 4. Individuals, above-ground biomass density (AGBD), species richness, and biodiversity in dense, moderately dense, and open forests along the altitudinal gradient of the Pauri district.

The biomass production increased with altitude, from 43 tons (t)/ha in Zone A, to 47.50 t/ha in Zone B, to 88 t/ha in Zone C. Biomass production by *Pinus roxburghii* in Zone A (112 t/ha) was shown to be more than that indicated by Rana (1985) [160], and similar to that reported by Chaturvedi and Singh (1986) [161] and Chaturvedi (1983) [162] for the Kumaun region of India. *Pinus roxburghii* contributed more than 57 percent of the biomass in Zone B. The above-ground biomass density for *Pinus roxburghii* (153.44 t/ha) was lower in the present study than that reported by Gairola et al. (2011) [158] (183.05 t/ha) and Kumar et al. (2019) [163] (213 t/ha) for Pauri Garhwal. Haripriya (2000) [164], on the other hand, observed an even lower above-ground biomass density for *Pinus roxburghi* (69.50 t/ha). The highest above-ground biomass density in Zone C was recorded for *Cedrus deodara*. In a previous study, Sundriyal et al. (1994) [165] reported an above-ground biomass density range of 368–682 t/ha in the higher altitude forests of Eastern Himalaya. Moreover, the above-ground biomass density estimated for *Cedrus deodara* forest was lower than that reported by Sharma et al. (2010) [166] for the Garhwal forest.

Species richness and biodiversity were reported to be higher in Zone C than Zone B and Zone A. Species richness increased with the density of the forest, suggesting that dense forest has the highest degree of species richness. According to Singh et al. (1994) [167], the maximum species richness was found in *Pinus roxburghii* mixed broadleaf forests, whereas the lowest was found in high-elevation forests. The species richness value observed in this study did not differ significantly from the values reported by Semwal et al. (2010) [168] and Raturi (2012) [169].

The highest biodiversity was found in the dense forest of Zone C, followed by the moderately dense forest of Zone B. Indeed, similar findings on biodiversity and species richness have been presented by Kumar and Ram (2005) [170] and Pandey (2003) [171] in Kumaun Himalaya. However, the values recorded in the present study were higher than those of Sanjeev and Sankhayan (2006) [172] in Mussoorie Dehradun, Dhar et, al. (1997) [173], and Kumar and Ram (2005) [170] in Kumaun Himalaya, and lower than those of Gairola et al. (2011) [158] in Garhwal Himalaya. The lowest diversity value was found in Zone A. This lower degree of biodiversity might be attributed to high anthropogenic disturbances. Semwal et al. (2010) [168] reported a relatively low species diversity in the Himalayan forest, while a higher range was reported by Singh and Singh (1987) [174] in the Chir pine mixed forest of Central Himalaya, by Rawat and Chandhok (2009) [175], and by Raturi (2012) [169] for temperate mixed forests and sub-tropical forests. Biodiversity can also vary based on biogeography, habitat, and disturbance (Sagar et al., 2003 [176]). According to Srivastava et al. (2008) [177], community characteristics can differ in aspect, slope, and altitude, even within the same vegetation type.

Availability of fodder during the scarce season was reported to be about similar in Zone A and C and lower in Zone B, but access to fuelwood during the scarce season was found to decrease with altitude. As previously discussed, most of the households in this region possess traditional livestock, which is directly or indirectly dependent on the forest for fodder. The storage of fodder was common traditional practice for the villagers, who were generally found to collect more fodder after the monsoon and then store it for the fodder-scarce season. The fodder storage practice was generally more prevalent in Zone C, which was believed to be one of the best solutions to combat fodder scarcity. Furthermore, several strategies have been adopted to reduce forest dependency in Zone A, including adopting efficient technologies. However, relatively few alternatives and strategies for fodder scarcity were adopted in Zone B and C. Almost all villages surveyed had fuelwood alternatives, i.e., LPG, while fuelwood remained the most economical choice for cooking, water, and space heating. Fuelwood extraction increased in winter (November to February) due to higher consumption and limited labor requirements in agriculture and allied sectors. There was also a tendency to store fuelwood for the upcoming months. On the other hand, minimum fuelwood extraction was reported in the monsoon season. Overall, there were fewer strategies to combat fuelwood scarcity in Zone B and C.

3.2.3. Vulnerability

The highest vulnerability was recorded in Zone B due to higher sensitivity and lower adaptive capacity (Figure 5). The households in Zone B disproportionately relied on the forest to sustain their livelihood. Climate change threatens the Himalayan Forest, and the dominance of *Pinus roxburghii* exacerbates sensitivity and increases vulnerability. Limited understory vegetation in pine forests restricts the availability of fodder and fuelwood and constitutes a potential cause of lopping. Moreover, pine needles are highly susceptible to forest fires, and the incidence of forest fires is comparatively higher in this zone. The possession of livestock is a traditional practice, and most households will have 2–3 livestock and subsequently depend on the nearby forest for fodder and grazing. The higher anthropogenic pressure for resource collection is a prime cause of forest degradation, indicated by the increased distances between the forest and the villages and the increased time spent collecting forest resources. The anthropogenic pressure then alters the zone's biodiversity, richness, and biomass and is reflected by a lower adaptive capacity.

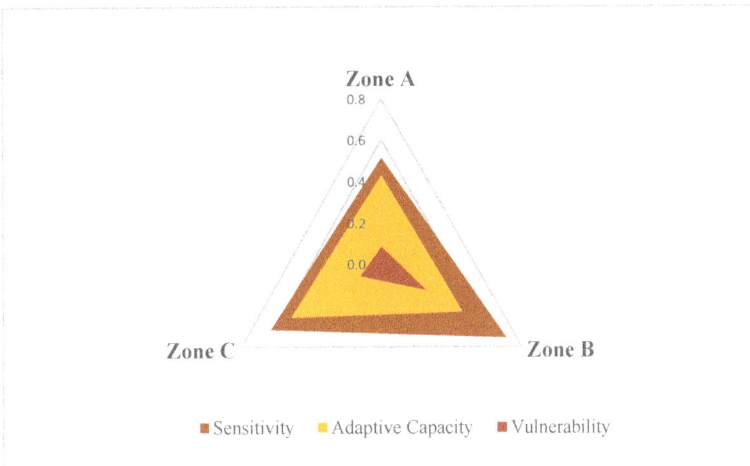

Figure 5. Sensitivity, adaptive capacity, and vulnerability of the Himalayan forest along the altitudinal gradient in the Pauri district.

The vulnerability in Zone C is lower than in Zone B and higher than in Zone A (Figure 5). The vulnerability of Zone C is attributed to the availability, collection, and degradation of natural resources, while the zone has a comparatively better adaptive capacity to withstand climate change. Although this zone reports sufficient availability of natural resources, they are not easily accessible, and those that are accessible are currently degrading due to higher anthropogenic pressures. The lowest vulnerability was reported in Zone A, owing to lower dependency on natural resources and availability of alternatives representing a better adaptive capacity.

4. Conclusions

It is evident that the temperature (maximum and minimum) of the district has increased, and rainfall has decreased in the last few decades. The communities along the altitudinal gradient also accepted and identified the changes in the climate. The greater ability of Zone C communities to identify changes is thought to be due to their geophysical position; they live at a higher altitude, mostly in cold weather condition, and thus can quickly recognize variations in climate such as prolonged summer and variation in temperature range.

Lower natural resource degradation in Zone A is attributed to less reliance on natural resources as well as combatively active law enforcement. The higher altitude and some inaccessible pockets of the mountain limit the reach of forest officials, while at the same time, natural resource alternatives are limited, and communities are also dependent on available resources for sustaining livelihood, so degradation is greater. The forest is a renewable resource, and the degradation could be reduced, and services restored, but biodiversity loss is irreversible. Thus, conservation-centric forest management, with maximized involvement of local communities in collaboration with research or academic institutions and forest departments, is likely to be highly effective. Forests in Uttarakhand are under great pressure (natural and anthropogenic), demanding afforestation programs with site-specific and climate-adapted multi-species initiatives and protective measures for species-rich ecosystems. Community-owned biodiversity hotspots should be more emphasized, such as the *Nagdev*, *Ekeshwer*, and *Tarkeshwer* sacred groves in the study district, and *Taxus baccata* forest of Uttarkashi district; the local community should be made aware of the need for ecological conservation. This approach can promote the sustainable extraction (time, part, quantity, etc.) of forest resources, which is currently one of the major drivers of forest degradation. The larger area covered by a conservation-centric plan in an altitudinal plan may limit the supply of forest produce; additionally, the inclusion of a community-centric forest management plan may contribute to meeting demand. As a result, it is critical to adopt and implement a forest management plan that incorporates conservation and a community-centric approach.

Establishing fodder banks and up-gradating energy consumption patterns (e.g., using improved smokeless stove (chulha), solar cookers, pine bricks, etc.) in Zone B and C could, to a certain extent, reduce dependency on the forest. Agroforestry practices with particular emphasis on fast-growing, demand-oriented tree species in Zone B possess significant potential for reducing the gap between supply and demand and could be an important mitigation strategy. Furthermore, all the zones require upgraded fodder storage techniques (hay and silage), particularly Zone C. The forests of Uttarakhand are also highly vulnerable to forest fires, which are major sources of GHG and, therefore, contributors to reductions in the terrestrial carbon sink. Consequently, there is a need to introduce fire management techniques and innovative monitoring and warning approaches in as little time as possible, which should be included in the policy of the Government of Uttarakhand.

The results revealed that collecting fuelwood and fodder was common in the district. Maximum degradation was recorded in the moderately dense forest of Zone B due to anthropogenic disturbances such as lopping, grazing, fodder, and fuelwood collection. Furthermore, the forest in Zone A and B was dominated by *Pinus roxburghii*, which hinders understory regeneration. The maximum above-ground biomass density in the forest of Zone C was recorded in *Cedrus deodara*. Overall, maximum above-ground biomass density was recorded in the dense forest of Zone C. The variation in biomass at different altitudinal ranges was attributed to forest type, species, age, other environmental factors, and edaphic factors. Therefore, vulnerability reduction strategies should be centered on developing natural capital and tailored to a specific sector and site (i.e., altitude). Development of the natural resource base should be targeted throughout the study district, strengthening alternatives to natural resources prioritized in Zone B.

Author Contributions: Conceptualization, S.K.J.; Data curation, S.K.J.; Formal analysis, S.K.J. and V.P.; Investigation, S.K.J., A.K.N., J.M.A. and V.P.; Methodology, S.K.J.; Supervision, A.K.N. and J.M.A.; Validation, M.B.J.; Writing—original draft, S.K.J., A.K.N. and J.M.A.; Writing—review & editing, S.K.J., A.K.N., J.M.A., V.P., M.B.J. and H.K. All authors have read and agreed to the published version of the manuscript.

Funding: This research received no external funding.

Institutional Review Board Statement: The authors are thankful to HNB (A Central) Garhwal University, Uttarakhand, India for research support.

Informed Consent Statement: Written informed consent has been obtained from the respondent(s) to conduct household survey and publish this paper.

Data Availability Statement: The datasets generated during and/or analyzed during the current study are part of Ph.D. research, collected by authors.

Conflicts of Interest: The authors declare no conflict of interest.

References

1. Rasul, G. Food, water, and energy security in South Asia: A nexus perspective from the Hindu Kush Himalayan region. *Environ. Sci. Policy* **2014**, *39*, 35–48. [CrossRef]
2. Badola, R.; Hussain, S.A.; Dobriyal, P.; Barthwal, S. Assessing the effectiveness of policies in sustaining and promoting ecosystem services in the Indian Himalayas. *Int. J. Biodivers. Sci. Ecosyst. Serv. Manag.* **2015**, *11*, 216–224. [CrossRef]
3. Government of Uttarakhand (GoU). *Uttarakhand Action Plan on Climate Change (UAPCC)*; GoU: Dehradun, India, 2014.
4. Hoy, A.; Katel, O.; Thapa, P.; Dendup, N.; Matschullat, J. Climatic changes and their impact on socio-economic sectors in the Bhutan Himalayas: An implementation strategy. *Reg. Environ. Chang.* **2016**, *16*, 1401–1415. [CrossRef]
5. Indian Network for Climate Change Assessment. *Climate Change and India: A 4X4 Assessment a Sectoral and Regional Analysis for 2030s*; Government of India: New Delhi, India, 2010.
6. Chaturvedi, R.K.; Ranjith, G.; Jayaraman, M. Impact of climate change on Indian forests: A dynamic vegetation modeling approach. *Mitig. Adapt. Strateg. Glob. Chang.* **2011**, *16*, 119–142. [CrossRef]
7. Singh, R.B.; Hietala, R. *Livelihood Security in Northwestern Himalaya—Case Studies from Changing Socio-Economic Environments in Himachal Pradesh, India. Advances in Geographical and Environmental Sciences (AGES)*; Springer: Tokyo, Japan, 2014.
8. Jha, S.K. Assessment of Climate Change Vulnerability and Adaptation of Forest-Dependent Communities in Western Himalaya. Ph.D. Thesis, DoF&NR, HNB-GU, Srinagar, India, 2020.
9. Kumar, P.; Kotlarski, S.; Moseley, C.; Sieck, K.; Frey, H.; Stoffel, M.; Jacob, D. Response of Karakoram-Himalayan glaciers to climate variability and climatic change: A regional climate model assessment. *Geophys. Res. Lett.* **2015**, *42*, 1818–1825. [CrossRef]
10. Li, H.; Xu, C.Y.; Beldring, S.; Tallaksen, L.M.; Jain, S.K. Water resources under climate change in Himalayan Basins. *Water Resour. Manag.* **2016**, *30*, 843–859. [CrossRef]
11. Palazzoli, I.; Maskey, S.; Uhlenbrook, S.; Nana, E.; Bocchiola, D. Impact of prospective climate change on water resources and crop yields in the Indrawati basin, Nepal. *Agric. Syst.* **2015**, *133*, 143–157. [CrossRef]
12. Jethi, R.; Joshi, K.; Chandra, N. Toward climate change and community-based adaptation-mitigation strategies in hill agriculture. In *Conservation Agriculture: An Approach to Combat Climate Change in Indian Himalaya*; Bisht, J.K., Meena, V.S., Mishra, P.K., Pattanayak, A., Eds.; Springer: Singapore, 2016; pp. 185–202.
13. ADB. *Climate Change Adaptation in Himachal Pradesh: Sustainable Strategies for Water Resources*; ADB: Mandaluyong, Philippines, 2010; ISBN 978-92-9092-060-1.
14. Wani, A.M.; Raj, A.J.; Kanwar, M. Impact of climate change on forests of Eastern Himalayas and adaptation strategies for combating it. *Int. J. Agric. For.* **2013**, *3*, 98–104.
15. Lewis, S.L. Tropical forests and the changing earth system. *Philos. Trans. R. Soc. B* **2006**, *361*, 195–210. [CrossRef]
16. IPCC. *Climate Change 2007: Synthesis Report*; Cambridge University Press: Cambridge, UK, 2007. Available online: https://www.ipcc.ch/publications_and_data/ar4/syr/en/mains1.html (accessed on 20 December 2021).
17. Das, H.P. Adaptation strategies required to reduce vulnerability in agriculture and forestry to climate change, climate variability and climate extremes. In *World Meteorological Organization (WMO). Management Strategies in Agriculture and Forestry for Mitigation of Greenhouse Gas Emissions and Adaptation to Climate Variability and Climate Change*; Report of CAgM Working Group. Technical Note No. 202, WMO No. 969; WMO: Geneva, Switzerland, 2004; pp. 41–92.
18. Kala, C.P. Deluge, disaster and development in Uttarakhand Himalayan region of India: Challenges and lessons for disaster management. *Int. J. Disaster Risk Reduct.* **2014**, *8*, 143–152. [CrossRef]
19. Ahmad, A.; Rawat, J.S.; Rai, S.C. An analysis of the himalayan environment and guidelines for its management and ecologically sustainable development. *Environmentalist* **1990**, *10*, 281–298. [CrossRef]
20. Ravindranath, N.H.; Joshi, N.V.; Sukumar, R. Impact of climate change on forests in India. *Curr. Sci.* **2006**, *90*, 354–361.
21. Jadin, I.; Meyfroidt, P.; Lambin, E.F. Forest protection and economic development by offshoring wood extraction: Bhutan's clean development path. *Reg. Environ. Chang.* **2016**, *16*, 401–415. [CrossRef]
22. Jha, S.K.; Jana, P.; Negi, A.K.; Negi, R.S. Livelihood Vulnerability Associated with Forest Fire in Pauri-Garhwal, Western Himalaya. *Open Ecol. J.* **2018**, *11*, 62–74. [CrossRef]
23. Sharma, L.N.; Vetaas, O.R. Does agroforestry conserve trees? A comparison of tree species diversity between farmland and forest in mid-hills of central Himalaya. *Biodivers. Conserv.* **2015**, *24*, 2047–2061. [CrossRef]
24. Bruggeman, D.; Meyfroidt, P.; Lambin, E.F. Forest cover changes in Bhutan: Revisiting the forest transition. *Appl. Geogr.* **2016**, *67*, 49–66. [CrossRef]
25. Upgupta, S.; Sharma, J.; Jayaraman, M.; Kumar, V.; Ravindranath, N.H. Climate change impact and vulnerability assessment of forests in the Indian Western Himalayan region: A case study of Himachal Pradesh, India. *Clim. Risk Manag.* **2015**, *10*, 63–76. [CrossRef]
26. Chakraborty, A.; Joshi, P.K.; Sachdeva, K. Predicting distribution of major forest tree species to potential impacts of climate change in the central Himalayan region. *Ecol. Eng.* **2016**, *97*, 593–609. [CrossRef]

27. Giam, X. Global biodiversity loss from tropical deforestation. *Proc. Natl. Acad. Sci. USA* **2017**, *114*, 5775–5777. [CrossRef]
28. Alekhya, V.V.L.; Pujar, G.S.; Jha, C.S.; Dadhwal, V.K. Simulation of vegetation dynamics in Himalaya using dynamic global vegetation model. *Trop. Ecol.* **2015**, *56*, 219–231.
29. Sharma, J.; Ravindranath, N.H. Vulnerability and Risk Assessment: Framework, Methods and Guideline. Indian Institute of Science. 2018. Available online: http://ihcap.in/?media_dl=1519 (accessed on 20 December 2021).
30. Kumar, M.; Kumar, R.; Konsam, B.; Sheikh, M.A.; Pandey, R. Above-And Below-Ground Biomass Production in *Pinus roxburghi* Forests along Altitudes in Garhwal Himalaya, India. *Curr. Sci.* **2019**, *116*. [CrossRef]
31. Cutter, S.L.; Finch, C. Temporal and spatial changes in social vulnerability to natural hazards. *Proc. Natl. Acad. Sci. USA* **2008**, *105*, 2301–2306. [CrossRef] [PubMed]
32. Eriksen, S.H.; Kelly, P.M. Developing credible vulnerability indicators for climate adaptation policy assessment. *Mitig. Adapt. Strateg. Glob. Chang.* **2007**, *12*, 495–524. [CrossRef]
33. Kaly, U.L.; Briguglio, L.; McLeod, H.; Schmall, S.; Pratt, C.; Pal, R. SOPAC Technical Report. In Proceedings of the Environmental Vulnerability Index (EVI) Think Tank, Pacific Harbour, Fiji, 7–10 September 1999.
34. Lonergan, S.; Gustavson, K.; Carter, B. The index of human insecurity. *AVISO* **2000**, *6*, 1–7.
35. Prescott-Allen, R. *The Well-Being of Nations*; Island Press and International Development Research Centre: Washington, DC, USA, 2001.
36. World Economic Forum. *Environmental Sustainability Index: An Initiative of the Global Leaders for Tomorrow Environment Task Force*; World Economic Forum: Cologny, Switzerland, 2002.
37. Sullivan, C. Calculating a Water Poverty Index. *World Dev.* **2002**, *30*, 1195–1210. [CrossRef]
38. Brooks, N.; Adger, W.N. *Country Level Risk Measures of Climate-Related Natural Disasters and Implications for Adaptation to Climate Change*; Tyndall Centre for Climate Change Research: Norwich, UK, 2003.
39. Adger, W.N.; Brooks, N.; Kelly, M.; Bentham, S.; Eriksen, S. *New Indicators of Vulnerability and Adaptive Capacity*; Technical Report 7; Tyndall Centre for Climate Change Research, University of East Anglia: Norwich, UK, 2004.
40. Vincent, K. *Creating an Index of Social Vulnerability to Climate Change for Africa Tyndall*; Working Paper 56; Centre for Climate Change Research: Norwich, UK, 2004.
41. Adger, W.N.; Vincent, K. Uncertainty in adaptive capacity. *Comptes Rendus Geosci.* **2005**, 399–410. [CrossRef]
42. Esty, D.C.; Levy, M.; Srebotnjak, T.; de-Sherbinin, A. *Environmental Sustainability Index: Benchmarking National Environmental Stewardship*; Yale Center for Environmental Law & Policy: New Haven, CT, USA, 2005.
43. Cardona, O.D. Indicators of Disaster Risk and Risk Management-Program for Latin America and the Caribbean: Summary Report, Inter-American Development Bank, Department of Sustainable Development, Washington. 2005. Available online: http://idea.unalmzl.edu.co (accessed on 17 September 2022).
44. UNISDR. *Living with Risk*; United Nations International Strategy for Disaster Reduction: Geneva, Switzerland, 2004.
45. UNDP. The Human Development Concept. 2008. Available online: http://hdr.undp.org/en/humandev/ (accessed on 7 December 2017).
46. Hahn, M.B.; Riederer, A.M.; Foster, S.O. The Livelihood Vulnerability Index: A pragmatic approach to assessing risks from climate variability and change—A case study in Mozambique. *Glob. Environ. Chang.* **2009**, *19*, 74–88. [CrossRef]
47. Vincent, K.; Cull, T. A Household Social Vulnerability Index (HSVI) for Evaluating Adaptation Projects in Developing Countries. In Proceedings of the PEGNet Conference 2010: Policies to Foster and Sustain Equitable Development in Times of Crises, Midrand, South Africa, 2–3 September 2010.
48. Urothody, A.A.; Larsen, H.O. Measuring climate change vulnerability: A comparison of two indexes. *Banko Jankari* **2010**, *20*, 9–16. [CrossRef]
49. Pandey, R.; Jha, S.K. Climate vulnerability index—Measure of climate change vulnerability to communities: A case of rural Lower Himalaya, India. *Mitig. Adapt. Strateg. Glob. Chang.* **2011**, *17*, 487–506. [CrossRef]
50. Gerlitz, J.Y.; Banerjee, S.; Hoermann, B.; Hunzai, K.; Macchi, M.; Tuladhar, S. *Poverty and Vulnerability Assessment—A Survey Instrument for the Hindu Kush Himalayas*; Report; ICIMOD: Lalitpur, Nepal, 2014.
51. Aryal, S.; Cockfield, G.; Maraseni, T.N. Vulnerability of Himalayan transhumant communities to climate change. *Clim. Chang.* **2014**, *125*, 193–208. [CrossRef]
52. Gerlitz, J.Y.; Macchi, M.; Brooks, N.; Pandey, R.; Banerjee, S.; Jha, S.K. The Multidimensional Livelihood Vulnerability Index—An instrument to measure livelihood vulnerability to change in the Hindu Kush Himalayas. *Clim. Dev.* **2016**, *9*, 124–140. [CrossRef]
53. Pandey, R.; Maithani, N.; Aretano, R.; Zurlini, G.; Archie, K.M.; Gupta, A.K.; Pandey, V.P. Empirical assessment of adaptation to climate change impact of mountainous households: Development and application of Adaptation Capability Index. *J. Mt. Sci.* **2016**, *13*, 1503–1514. [CrossRef]
54. Pandey, R.; Meena, D.; Aretano, R.; Satpathy, S.; Semeraro, T.; Gupta, A.K.; Rawat, S.; Zurlini, G. Socio-ecological vulnerability of smallholders due to climate change in mountains: Agroforestry as an adaptation measure. *Chang. Adapt. Socio-Ecol. Syst.* **2015**, *2*, 26–41. [CrossRef]
55. Shukla, R.; Sachdeva, K.; Joshi, P.K. Inherent vulnerability of agricultural communities in Himalaya: A village-level hotspot analysis in the Uttarakhand state of India. *Appl. Geogr.* **2016**, *74*, 182–198. [CrossRef]
56. Pandey, R.; Jha, S.K.; Alatalo, J.M.; Archie, K.M.; Gupta, A.K. Sustainable livelihood framework-based indicators for assessing climate change vulnerability and adaptation for Himalayan communities. *Ecol. Indic.* **2017**, *79*, 338–346. [CrossRef]

57. Thakur, S.; Negi, V.S.; Pathak, R.; Dhyani, R.; Durgapal, K. Indicator based integrated vulnerability assessment of community forests in Indian west Himalaya. *For. Ecol. Manag.* **2020**, *497*, 117674. [CrossRef]
58. Jha, S.K.; Negi, A.K.; Alatalo, J.M.; Negi, R.S. Socio-ecological Vulnerability and Resilience of Mountain Communities residing in Capital-constrained Environments. *Mitig. Adapt. Strat. Glob. Chang.* **2021**, *26*, 38. [CrossRef]
59. DST. *Climate Vulnerability Assessment for Adaptation Planning in India Using a Common Framework*; Government of India: New Delhi, India, 2020.
60. Jha, S.K.; Negi, A.K.; Alatalo, J.M. Quantification of ecosystem services providing socio-economic benefits to customary owners of natural resources in Pauri, western Himalaya. *Curr. Res. Environ. Sustain.* **2022**, *4*, 100121. [CrossRef]
61. Harley, M.; Horrocks, L.; Hodgson, N. *Climate Change Vulnerability and Adaptation Indicators. European Topic Centre on Air and Climate Change (ETC/ACC) Technical Paper 2008/9*; European Environment Agency: Copenhagen, Denmark, 2008.
62. Sinha, B.; Jha, S.K. *Vulnerability and Capacity Assessment to Climate Change of Malwa-Madhya Pradesh*; Bishen Singh and Mahinder Pal Singh, Publisher and Distributor of Scientific Books: Dehradun, India, 2017.
63. Jha, S.K.; Mishra, S.; Sinha, B.; Alatalo, J.M.; Pandey, R. Rural development program in tribal region: A protocol for adaptation and addressing climate change vulnerability. *J. Rural Stud.* **2017**, *51*, 151–157. [CrossRef]
64. Gabor, T.; Griffith, T.K. The Assessment of Community Vulnerability to Acute Hazardous Materials Incidents. *J. Hazard. Mater.* **1980**, *3*, 323–339. [CrossRef]
65. Timmerman, P. *Vulnerability, Resilience and the Collapse of Society: A Review of Models and Possible Climatic Applications*; Institute for Environmental Studies, University of Toronto: Toronto, Canada, 1981.
66. UNDRO. *Natural Disasters and Vulnerability Analysis*; Office of the United Nations Disaster Relief Co-Ordinator: Geneva, Switzerland, 1982.
67. Susman, P.; O'Keefe, P.; Wisner, B. Global disasters: A radical interpretation. In *Interpretations of Calamity from the Viewpoint of Human Ecology*; Hewitt, K., Ed.; Allen and Unwin: Boston, MA, USA, 1984; pp. 264–283.
68. Pijawaka, K.D.; Radwan, A.E. Transportation of hazardous materials. Risk assessment and hazard management. *Dang. Prop. Int. Mater. Rep.* **1985**, *5*, 2–11.
69. Kates, R.W.; Ausubel, J.H.; Berberian, M. (Eds.) *SCOPE Report on Impact Assessments*; Wiley: Hoboken, NJ, USA, 1985.
70. Bogard, W.C. Bringing social theory to hazards research: Conditions and consequences of the mitigation of environmental hazards. *Sociol. Perspect.* **1988**, *31*, 147–168. [CrossRef]
71. Mitchell, J.K. Hazards research. In *Geography in America*; Gaile, G.L., Willmott, C.J., Eds.; Merill: Colombus, OH, USA, 1989; pp. 410–424.
72. Liverman, D.M. Drought impacts in Mexico: Climate, agriculture, technology, and land tenure in Sonora and Puebla. *Ann. Assoc. Am. Geogr.* **1990**, *80*, 49–72. [CrossRef]
73. Downing, T.E. Vulnerability to hunger in Africa: A climate change perspective. *Glob. Environ. Change* **1991**, *1*, 365–380. [CrossRef]
74. Dow, K. Exploring differences in our common future (s): The meaning of vulnerability to global environmental change. *Geoforum* **1992**, *23*, 417–436. [CrossRef]
75. Smith, K. *Environmental Hazards: Assessing Risk and Reducing Disaster*; Routledge: London, UK, 1992.
76. Alexander, D. *Natural Disasters*; Chapman and Hall: New York, NY, USA, 1993.
77. Cutter, S.L. *Living with Risk*; Edward Arnold: London, UK, 1993.
78. Watts, M.J.; Bohle, H.G. Hunger, famine and the space of vulnerability. *GeoJournal* **1993**, *30*, 117–125. [CrossRef]
79. Blaikie, P.; Cannon, T.; Davis, I.; Wisner, B. *At Risk: Natural Hazards, People's Vulnerability, and Disasters*; Routledge: London, UK, 1994; ISBN 9780415084772.
80. Bohle, H.G.; Downing, T.E.; Watts, M.J. Climate change and social vulnerability. Toward a sociology and geography of food insecurity. *Glob. Environ. Chang.* **1994**, *4*, 37–48. [CrossRef]
81. Dow, K.; Downing, T.E. *Vulnerability Research: Where Things Stand*; National Emergency Training Centre: Emmitsburg, MD, USA, 1995.
82. Eakin, H.; Luers, A.L. Assessing the vulnerability of social-environmental systems. *Annu. Rev. Environ. Resour.* **2006**, *31*, 365. [CrossRef]
83. Gilard, O.; Givone, P. Flood risk management: New concepts and methods for objective negotiations. In *Destructive Water: Water-Caused Natural Disasters, Their Abatement and Control*; Leavesley, G.H., Lins, H.F., Nobilis, F., Parker, R.S., Schneider, V.R., van de Ven, F.H.M., Eds.; IAHS Press: Oxfordshire, UK, 1997; pp. 145–155.
84. Comfort, L.; Wisner, B.; Cutter, S.; Pulwarty, R.; Hewitt, K.; OliverSmith, A.; Krimgold, F. Reframing disaster policy: The global evolution of vulnerable communities. *Environ. Hazards* **1999**, *1*, 39–44.
85. Weichselgartner, J.; Bertens, J. Natural disasters: Acts of God, nature or society? On the social relation to natural hazards. In *Risk Analysis II*; Andretta, M.A., Ed.; WIT Press: Southampton, UK, 2000; pp. 3–12.
86. Adger, W.N.; Winkels, A. Vulnerability, poverty, and sustaining well-being. In *Handbook of Sustainable Development*; Atkinson, G., Dietz, S., Neumayer, E., Eds.; Elgar: Cheltenham, UK, 2006.
87. Ciurean, R.L.; Schröter, D.; Glade, T. Conceptual Frameworks of Vulnerability Assessments for Natural Disasters Reduction. In *Approaches to Disaster Management-Examining the Implications of Hazards, Emergencies and Disasters*; Tiefenbacher, J., Ed.; IntechOpen: London, UK, 2013; Chapter 1; pp. 1–32.

88. Wolf, S.; Hinkel, J.; Hallier, M.; Bisaro, A.; Lincke, D.; Ionescu, C.; Klein, R.J. Clarifying vulnerability definitions and assessments using formalisation. *Int. J. Clim. Chang. Strateg. Manag.* **2013**, *5*, 54–70. [CrossRef]
89. IPCC. *Climate Change 2014: Synthesis Report*; Contribution of Working Groups I, II and III to the Fifth Assessment Report of the Intergovernmental Panel on Climate Change; Core Writing Team, Pachauri, R.K., Meyer, L.A., Eds.; IPCC: Geneva, Switzerland, 2014.
90. Hong, W.; Jiang, R.; Yang, C.; Zhang, F.; Su, M.; Liao, Q. Establishing an ecological vulnerability assessment indicator system for spatial recognition and management of ecologically vulnerable areas in highly urbanized regions: A case study of Shenzhen, China. *Ecol. Indic.* **2016**, *69*, 540–547. [CrossRef]
91. Cutter, S.L. Vulnerability to environmental hazards. *Prog. Hum Geogr.* **1996**, *20*, 529–539. [CrossRef]
92. Weichselgartner, J. Disaster mitigation: The concept of vulnerability revisited. *Disaster Prev. Manag.* **2001**, *10*, 85–95. [CrossRef]
93. Hogan, D.J.; Marandola, E., Jr. Vulnerability and risk: Between Geography and Demography. In Proceedings of the Seminar Making Sense of Vulnerability, Dundee, UK, 10–11 March 2005.
94. Paul, S.K. Vulnerability Concepts and its Application in Various Fields: A Review on Geographical Perspective. *J. Life Earth Sci.* **2013**, *8*, 63–81. [CrossRef]
95. Berrouet, L.M.; Machado, J.; Villegas-Palacio, C. Vulnerability of socio—Ecological systems: A conceptual Framework. *Ecol. Indic.* **2018**, *84*, 632–647. [CrossRef]
96. Holling, C.S. *Adaptive Environmental Assessment and Management*; John Wiley Sons: Hoboken, NJ, USA, 1978.
97. Marcus, S. Towards a Semiotic Approach to Social Indicators. In Proceedings of the International Congress of Logic, Methodology and Philosophy of Science, Salzburg, Austria, 11–16 July 1983.
98. McQueen, D.; Noak, H. Health Promotion Indicators: Current status, issues and problems. *Health Promot.* **1988**, *3*, 117–125, (MWCD). [CrossRef]
99. Tunstall, D. Developing Environmental Indicators: Definitions, framework, and issues. (Draft paper). Background Materials for the World Resources Institute. In *Workshop on Global Environmental Indicators, Washington, D.C. 7–8 December 1992*; World Resources Institute: Washington, DC, USA, 1992.
100. OECD. *Organization for Economic Cooperation and Development Core Set of Indicators for Environmental Performance Reviews*; A Synthesis Report by the Group on the State of the Environment; OECD: Paris, France, 1993.
101. Bakkes, J.A.; Van-den-Born, G.J.; Helder, J.C.; Swart, R.J.; Hope, C.W.; Parker, J.D.E. *An Overview of Environmental Indicators: State of the Art and Perspectives*; Environmental Assessment Sub-Programme; UNEP: Nairobi, Kenya, 1994.
102. Hammond, A.; Adriaanse, A.; Rodenburg, E.; Bryant, D.; Woodward, R. *Environmental Indicators: A Systematic Approach to Measuring and Reporting on Environmental Policy Performance in the Context of Sustainable Development*; World Resources Institute: Washington, DC, USA, 1995.
103. Ott, W.R. *Environmental Indices: Theory and Practice*; Ann Arbor Science: Ann Arbor, MI, USA, 1978.
104. Sen, A.K. *Poverty and Famines: An Essay on Entitlement and Deprivation*; Clarendon: Oxford, UK, 1981.
105. Watson, R.T.; Zinyoera, M.C.; Moss, R.H. *Climate Change 1995: Impacts, Adaptations and Mitigation of Climate Change: Scientific-Technical Analysis*; Contribution of Working Group II to the Second Assessment Report of the Intergovernmental Panel on Climate Change; Cambridge University Press: Cambridge, UK, 1996.
106. IPCC. *The Regional Impact of Climate Change: An Assessment of Vulnerability*; Cambridge University Press: Cambridge, UK, 1997.
107. Handmer, J.W.; Dovers, S.; Downing, T.E. Societal Vulnerability to Climate Change and Variability. *Mitig. Adapt. Strateg. Glob. Chang.* **1999**, *4*, 267–281. [CrossRef]
108. IPCC. Report by Robert Watson, Chair, Intergovernmental Panel on Climate Change, to the Sixth Conference of the Parties to the United Nations Framework Convention on Climate Change, The Hague, 20 November 2000. Available online: https://archive.ipcc.ch/graphics/speeches/robert-watson-november-13-2000.pdf (accessed on 10 October 2022).
109. Fussel, H.M. Vulnerability: A generally applicable conceptual framework for climate change research. *Glob. Environ. Chang.* **2007**, *17*, 155–167. [CrossRef]
110. Goulden, M.; Porter, K. *CLICO Conceptual Framework. CLICO Project Deliverable D.1.1 Report*; University of East Anglia: Norwich, UK, 2010.
111. Ebi, K.; Kovats, R.S.; Menne, B. An approach for assessing human health vulnerability and public health interventions to adapt to climate change. *Environ. Health Perspect.* **2006**, *114*, 1930–1934. [CrossRef]
112. Schauser, I.; Otto, S.; Schneiderbauer, S.; Harvey, A.; Hodgson, N.; Robrecht, H.; Morchain, D.; Schrander, J.; Khovanskaia, M.; Celikyilmaz-Aydemir, G.; et al. *Urban Regions: Vulnerabilities, Vulnerability Assessments by Indicators and Adaptation Options for Climate Change Impacts—Scoping Study*; ETC/ACC Technical Paper 2010/12; European Topic Centre on Air and Climate Change: Bilthoven, The Netherlands, 2010.
113. Balica, S. Approaches of Understanding Developments of Vulnerability Indices for Natural Disasters. *Environ. Eng. Manag. J.* **2012**, *11*, 963–974. [CrossRef]
114. Simane, B.; Zaitchik, B.F.; Foltz, J.D. Agroecosystem specific climate vulnerability analysis: Application of the livelihood vulnerability index to a tropical highland region. *Mitig. Adapt. Strateg. Glob. Chang.* **2016**, *21*, 39–65. [CrossRef] [PubMed]
115. Kumar, P.; Geneletti, D.; Nagendra, H. Spatial assessment of climate change vulnerability at city scale: A study in Ban-galore. *India Land Use Pol.* **2016**, *58*, 514–532. [CrossRef]

116. Pandey, R.; Alatalo, J.M.; Thapliyal, K.; Chauhan, S.; Archie, K.M.; Gupta, A.K.; Jha, S.K.; Kumar, M. Climate change vulnerability in urban slum communities: Investigating household adaptation and decision-making capacity in the Indian Himalaya. *Ecol. Indic.* **2018**, *90*, 379–391. [CrossRef]
117. Malone, E.L.; Engle, N.L. Evaluating regional vulnerability to climate change: Purposes and methods. *WIREs Clim. Chang.* **2011**, *2*, 462–474. [CrossRef]
118. Sharma, J.; Chaturvedi, R.K.; Bala, G.; Ravindranath, N.H. Challenges in vulnerability assessment of forests under climate change. *Carbon Manag.* **2013**, *4*, 403–411. [CrossRef]
119. FSI. *State of Forest Report 2019*; Forest Survey of India: Dehradun, India, 2019.
120. Census of India. *Population Census 2011*; Ministry of Home Affairs, Government of India: New Delhi, India, 2011.
121. ISFR. 2021. Available online: https://fsi.nic.in/forest-report-2021-details (accessed on 17 September 2022).
122. Sullivan, C.; Meigh, J. Targeting attention on local vulnerabilities using an integrated index approach: The example of the climate vulnerability index. *Water Sci. Technol.* **2005**, *51*, 69–78. [CrossRef]
123. Solbrig, O.T. *From Genes to Ecosystems: A Research Agenda for Biodiversity*; International Union of Biological Sciences: Paris, France, 1991.
124. Lee, C.-B.; Chun, J.-H. Environmental Drivers of Patterns of Plant Diversity Along a Wide Environmental Gradient in Korean Temperate Forests. *Forests* **2016**, *7*, 19. [CrossRef]
125. Gairola, S.; Rawal, R.S.; Todaria, N.P. Forest vegetation patterns along an altitudinal gradient in sub-alpine zone of West Himalaya, India. *Afr. J. Plant Sci.* **2008**, *2*, 42–48.
126. Rawat, V.S.; Chandra, J. Tree layer vegetational analysis in temperate forest of Uttarakhand. *Nat. Sci.* **2012**, *10*, 167–171.
127. Sharma, E.; Chettri, N.; Tse-ring, K.; Shrestha, A.B.; Jing, F.; Mool, P.; Eriksson, M. *Climate Change Impacts and Vulnerability in the Eastern Himalayas*; International Centre for Integrated Mountain Development: Kathmandu, Nepal, 2009; ISBN1 978-92-9115-134-9 (printed). ISBN2 978-92-9115-136-3 (electronic).
128. Tse-ring, K.; Sharma, E.; Chettri, N.; Shrestha, A. *Climate Change Vulnerability of Mountain Ecosystems in the Eastern Himalayas*; Working Paper; ICIMOD: Lalitpur, Nepal, 2012.
129. Sandhu, H.; Sandhu, S. Poverty, development, and Himalayan ecosystems. *AMBIO* **2014**, *44*, 297–307. [CrossRef]
130. MoEF. *India Second National Communication to the United Nations Framework Convention on Climate Change*; Ministry of Environment and Forests, Government of India: New Delhi, India, 2012.
131. Jha, S.K.; Negi, A.K.; Alatalo, J.M.; Negi, R.S.; Patasaraiya, M.K. Assessment of climate change pattern in the Pauri Garhwal of the Western Himalayan Region: Based on climate parameters and perceptions of forest-dependent communities. *Environ. Monit. Assess.* **2020**, *192*, 632. [CrossRef] [PubMed]
132. Shrestha, A.B.; Devkota, L.P. Climate change in the Eastern Himalayas: Observed trends and model projections. In *Climate Change Impact and Vulnerability in the Eastern Himalayas—Technical Report 1*; ICIMOD: Kathmandu, India, 2010.
133. Shrestha, U.B.; Gautam, S.; Bawa, K.S. Widespread climate change in the Himalayas and associated changes in local ecosystems. *PLoS ONE* **2012**, *7*, e36741. [CrossRef] [PubMed]
134. Sharma, E. Chapter 2 Climate Change and its Impacts in the Hindu Kush-Himalayas: An Introduction. In *Climate Change Modeling for Local Adaptation in The Hindu Kush-Himalayan Region (Community, Environment and Disaster Risk Management, Vol. 11)*; Lamadrid, A., Kelman, I., Eds.; Emerald Group Publishing Limited: Bingley, UK, 2012; pp. 17–32. [CrossRef]
135. Nepal, S.; Shrestha, A.B. Impact of climate change on the hydrological regime of the Indus, Ganges and Brahmaputra river basins: A review of the literature. *Int. J. Water Resour. Dev.* **2015**, *31*, 201–218. [CrossRef]
136. ICIMOD. *The Hindu Kush Himalaya Assessment*; ICIMOD: Kathmandu, Nepal, 2019. [CrossRef]
137. World Bank. *Turn Down the Heat: Why a 4 °C Warmer World Must Be Avoided*; World Bank: Washington, DC, USA, 2012.
138. Goswami, B.N.; Venugopal, V.; Sengupta, D.; Madhusoodanan, M.S.; Xavier, P.K. Increasing trend of extreme rain events over India in a warming environment. *Science* **2006**, *314*, 1442–1445. [CrossRef]
139. Xu, J.; Grumbine, R.E.; Shrestha, A.B.; Eriksson, M.; Yang, X.; Wang, Y.; Wilkes, A. The melting Himalayas: Cascading effects of climate change on water, biodiversity, and livelihoods. *Conserv. Biol.* **2009**, *23*, 520–530. [CrossRef]
140. Rao, C.A.R.; Raju, B.M.K.; Rao, A.V.M.S.; Rao, K.V.; Rao, V.U.M.; Ramachandran, K.; Nagasri, K.; Dupdal, R.; Samuel, J.; Shankar, K.R.; et al. Climate change in Uttarakhand: Projections, Vulnerability and Farmers' Perceptions. *J. Agrometeorol.* **2018**, *20*, 37–41.
141. Gaire, N.P.; Koirala, M.; Bhuju, D.R.; Carrer, M. Site- and species specific treeline responses to climatic variability in eastern Nepal Himalaya. *Dendrochronologia* **2017**, *41*, 44–56. [CrossRef]
142. Bajpai, O.; Pandey, J.; Chaudhary, L.B. Periodicity of different phenophases in selected trees from Himalayan Terai of India. *Agrofor. Syst.* **2016**, *91*, 1–12. [CrossRef]
143. Amano, T.; Smithers, R.J.; Sparks, T.H.; Sutherland, W.J.A. 250-year index of first flowering dates and its response to temperature changes. *Proc. R. Soc.* **2010**, *277*, 2451–2457. [CrossRef]
144. Locatelli, B. Ecosystem Services and Climate Change. In *Routledge Handbook of Ecosystem Services*; Potschin, M., Haines-Young, R., Fish, R., Turner, R.K., Eds.; Routledge: London, UK; New York, NY, USA, 2016; pp. 481–490. ISBN 978-1-138-02508-0. Available online: https://www.routledge.com/products/9781138025080 (accessed on 20 December 2021).
145. Hussain, A. *Impact of Forest Fire on Vegetation Composition and Nutrient Status in Oak Dominated Forest in Garhwal Himalaya*; HNB-Garhwal (A Central) University: Srinagar, India, 2018.

146. Maikhuri, R.; Nautiyal, S.; Rao, K.; Saxena, K. Conservation policy—People conflicts: A case study from Nanda Devi Biosphere Reserve (a World Heritage Site), India. *For. Policy Econ.* **2001**, *2*, 355–365. [CrossRef]
147. Davidar, P.; Sahoo, S.; Mammen, P.C.; Acharya, P.; Puyravaud, J.P.; Arjunan, M.; Garrigues, J.P.; Roessingh, K. Assessing the extent and causes of forest degradation in india: Where do we stand? *Biol. Conserv.* **2010**, *43*, 37–44. [CrossRef]
148. Shank, R.E.; Noorie, E.N. Microclimate vegetation in a small valley in eastern Tennessee. *Ecology* **1950**, *11*, 531–539.
149. Levine, J.S.; Bobbe, T.; Ray, N.; Singh, A. *Wildland Fires and the Environment: A Global Synthesis*; UN Environment Program UNEP/DEIAEW/TR.99-1, Geneva 46; UNEP: Nairobi, Kenya, 1999.
150. Rajesh, S.; Jain, S.; Sharma, P.; Bahuguna, R. Assessment of inherent vulnerability of rural communities to environmental hazards in Kimsar region of Uttarakhand, India. *Environ. Dev.* **2014**, *12*, 16–36. [CrossRef]
151. Connor, R. *The United Nations World Water Development Report 2015: Water for a Sustainable World*; UNESCO Publishing: Paris, France, 2015; Volume 1.
152. Tambe, S.; Arrawatia, M.L.; Bhutia, N.T.; Swaroop, B. Rapid, cost effective and high-resolution assessment of the climate-related vulnerability of rural communities of Sikkim Himalaya, India. *Curr. Sci.* **2011**, *101*, 165–173.
153. Rawat, P.K.; Tiwari, P.C.; Pant, C.C. Climate change accelerating hydrological hazards and risks in Himalaya: A case study through remote sensing and GIS modelling. *Int. J. Geomat. Geosci.* **2011**, *1*, 687–699.
154. Kanagawa, M.; Nakata, T. Analysis of the energy access improvement and its socio-economic impacts in rural areas of developing countries. *Ecol. Econ.* **2007**, *62*, 319–332. [CrossRef]
155. Bhatt, B.P.; Sachan, M.S. Firewood consumption along an altitudinal gradient in mountain villages of India. *Biomass Bioenergy* **2004**, *27*, 69–75. [CrossRef]
156. Sharma, C.M.; Gairola, S.; Baduni, N.P.; Ghildiyal, S.K.; Suyal, S. Variation in carbon stocks on different slope aspects in sever major forest types of temperate region of Garhwal Himalaya, India. *J. Biosci.* **2011**, *36*, 701–708. [CrossRef]
157. Saxena, A.K.; Singh, J.S. A phytosociological analysis of woody species in forest communities of a part of Kumaun Himalaya. *Vegetation* **1982**, *50*, 3–22. [CrossRef]
158. Gairola, S.; Sharma, C.M.; Ghildiyal, S.K.; Suyal, S. Live tree biomass and carbon variation along an altitudinal gradient in moist temperate valley slopes of the Garhwal Himalaya (India). *Curr. Sci.* **2011**, *100*, 1862–1870.
159. Pandey, P.K. Quantitative vegetation analysis as per aspect and altitude, and regeneration behaviour of tree species in Garhwal Himalayan forest. *Annu. For.* **2001**, *9*, 39–52.
160. Rana, B.S. Biomass and net Primary Productivity in Different Brest Ecosystems Alony ah Attitudinal lradient in Kumaun Himataya. Ph.D. Thesis, Kumaun University, Naini Tal, India, 1985.
161. Chaturvedi, O.P.; Singh, J.S. The structure and function of pine forest in central Himalayan I: Dry matter dynamics. *Ann. Bot.* **1986**, *60*, 237–252. [CrossRef]
162. Chaturvedi, O.P. Biomass Structure, Productivity and Nutrient Cycling in *Pinus roxburghii* Forest. Ph.D. Thesis, Kumaun University, Naini Tal, India, 1983.
163. Kumar, M.; Savita; Singh, H.; Pandey, R.; Singh, M.P.; Ravindranath, N.H.; Kalra, N. Assessing vulnerability of forest ecosystem in the Indian Western Himalayan region using trends of net primary productivity. *Biodivers. Conserv.* **2019**, *28*, 2163–2182. [CrossRef]
164. Haripriya, G.S. Estimates of biomass in Indian forester. *Biome. Bioenerg.* **2000**, *19*, 245–258. [CrossRef]
165. Sundriyal, R.C.; Sharma, E.; Rai, L.K.; Rai, S.C. Tree structure, regeneration and woody biomass removal in a sub-tropical forest of Mamlay watershed in the Sikkim Himalaya. *Plant Ecol.* **1994**, *113*, 53–63. [CrossRef]
166. Sharma, C.M.; Baduni, N.P.; Gairola, S.; Ghildiyal, S.K.; Suyal, S. Tree diversity and carbon stocks of some major forest types of Garhwal Himalaya, India. *For Ecol Manag.* **2010**, *260*, 2170–2179. [CrossRef]
167. Singh, S.P.; Adhikari, B.S.; Zobel, D.B. Biomass productivity, leaf longevity and forest structure in the Central Himalaya. *Ecol Monogr.* **1994**, *64*, 401–421. [CrossRef]
168. Semwal, D.P.; Uniyal, P.L.; Bhatt, A.B. Structure Composition and Dominance—Diversity Relations in Three Forest Types of a Part of Kedarnath Wildlife Sanctuary, Central Himalaya. *India Not. Sci. Biol.* **2010**, *2*, 128–132.
169. Raturi, G.P. Forest Community Structure along an Altitudinal Gradient of District Rudraprayag of Garhwal Himalaya, India. *Ecol* **2012**, *2*, 76–84. [CrossRef]
170. Kumar, A.; Ram, J. Anthropogenic disturbances and plant biodiversity in forest of Uttaranchal, Central Himalaya. *Biodivers Conserv.* **2005**, *14*, 309–331. [CrossRef]
171. Pandey, J.C. Vegetation analysis in a mixed Oak-Conifer forest of Central Himalaya. *Ind J. For.* **2003**, *26*, 66–74.
172. Sanjeev, G.M.; Sankhayan, P.L. Phytosociological analysis of Arnigad micro-watershed in Mussorie hills of Garhwal Himalayas *Indian For.* **2006**, *132*, 19–29.
173. Dhar, U.; Rawal, R.S.; Samant, S.S. Structural diversity and representativeness of forest vegetation in a protected area of Kumaun Himalaya, India: Implications for conservation. *Biodivers. Conserv.* **1997**, *6*, 1045–1062. [CrossRef]
174. Singh, J.S.; Singh, S.P. Forest vegetation of the Himalaya. *Bot. Rev.* **1987**, *53*, 80–192. [CrossRef]
175. Rawat, V.S.; Chandhok, A. Phytosociological Analysis and Distribution Patterns of Tree Species: A case study from Govind Pashu Vihar, National Park, Uttarakhand. *N. Y. Sci. J.* **2009**, *2*, 58–63.

176. Sagar, R.; Raghubanshi, A.S.; Singh, J.S. Tree species composition, dispersion and diversity along a disturbance gradient in a dry tropical forest region of India. *For. Ecol. Manag.* **2003**, *186*, 61–71. [CrossRef]
177. Srivastava, K.; Tewari, A.; Shah, S.; Tewari, B. Species composition and regeneration pattern along a transect perpendicular to a river course in foot hill Deciduous tropical forest of Kumaun. *Indian J. For.* **2008**, *1*, 7–12. [CrossRef]

MDPI
St. Alban-Anlage 66
4052 Basel
Switzerland
Tel. +41 61 683 77 34
Fax +41 61 302 89 18
www.mdpi.com

Sustainability Editorial Office
E-mail: sustainability@mdpi.com
www.mdpi.com/journal/sustainability

www.ingramcontent.com/pod-product-compliance
Lightning Source LLC
Chambersburg PA
CBHW061613220926
51850CB00005B/494